IET ENERGY ENGINEERING SERIES 67

Power Grids with Renewable Energy

Other volumes in this series

Volume 1 **Power Circuit Breaker Theory and Design** C.H. Flurscheim (Editor)
Volume 4 **Industrial Microwave Heating** A.C. Metaxas and R.J. Meredith
Volume 7 **Insulators for High Voltages** J.S.T. Looms
Volume 8 **Variable Frequency AC Motor Drive Systems** D. Finney
Volume 10 **SF_6 Switchgear** H.M. Ryan and G.R. Jones
Volume 11 **Conduction and Induction Heating** E.J. Davies
Volume 13 **Statistical Techniques for High Voltage Engineering** W. Hauschild and W. Mosch
Volume 14 **Uninterruptible Power Supplies** J. Platts and J.D. St Aubyn (Editors)
Volume 15 **Digital Protection for Power Systems** A.T. Johns and S.K. Salman
Volume 16 **Electricity Economics and Planning** T.W. Berrie
Volume 18 **Vacuum Switchgear** A. Greenwood
Volume 19 **Electrical Safety: A Guide to Causes and Prevention of Hazards** J. Maxwell Adams
Volume 21 **Electricity Distribution Network Design, 2nd Edition** E. Lakervi and E.J. Holmes
Volume 22 **Artificial Intelligence Techniques in Power Systems** K. Warwick, A.O. Ekwue and R. Aggarwal (Editors)
Volume 24 **Power System Commissioning and Maintenance Practice** K. Harker
Volume 25 **Engineers' Handbook of Industrial Microwave Heating** R.J. Meredith
Volume 26 **Small Electric Motors** H. Moczala *et al.*
Volume 27 **AC–DC Power System Analysis** J. Arrillaga and B.C. Smith
Volume 29 **High Voltage Direct Current Transmission, 2nd Edition** J. Arrillaga
Volume 30 **Flexible AC Transmission Systems (FACTS)** Y.-H. Song (Editor)
Volume 31 **Embedded Generation** N. Jenkins *et al.*
Volume 32 **High Voltage Engineering and Testing, 2nd Edition** H.M. Ryan (Editor)
Volume 33 **Overvoltage Protection of Low-Voltage Systems, Revised Edition** P. Hasse
Volume 36 **Voltage Quality in Electrical Power Systems** J. Schlabbach *et al.*
Volume 37 **Electrical Steels for Rotating Machines** P. Beckley
Volume 38 **The Electric Car: Development and Future of Battery, Hybrid and Fuel-cell Cars** M. Westbrook
Volume 39 **Power Systems Electromagnetic Transients Simulation** J. Arrillaga and N. Watson
Volume 40 **Advances in High Voltage Engineering** M. Haddad and D. Warne
Volume 41 **Electrical Operation of Electrostatic Precipitators** K. Parker
Volume 43 **Thermal Power Plant Simulation and Control** D. Flynn
Volume 44 **Economic Evaluation of Projects in the Electricity Supply Industry** H. Khatib
Volume 45 **Propulsion Systems for Hybrid Vehicles** J. Miller
Volume 46 **Distribution Switchgear** S. Stewart
Volume 47 **Protection of Electricity Distribution Networks, 2nd Edition** J. Gers and E. Holmes
Volume 48 **Wood Pole Overhead Lines** B. Wareing
Volume 49 **Electric Fuses, 3rd Edition** A. Wright and G. Newbery
Volume 50 **Wind Power Integration: Connection and System Operational Aspects** B. Fox *et al.*
Volume 51 **Short Circuit Currents** J. Schlabbach
Volume 52 **Nuclear Power** J. Wood
Volume 53 **Condition Assessment of High Voltage Insulation in Power System Equipment** R.E. James and Q. Su
Volume 55 **Local Energy: Distributed generation of heat and power** J. Wood
Volume 56 **Condition Monitoring of Rotating Electrical Machines** P. Tavner, L. Ran, J. Penman and H. Sedding
Volume 57 **The Control Techniques Drives and Controls Handbook, 2nd Edition** B. Drury
Volume 58 **Lightning Protection** V. Cooray (Editor)
Volume 59 **Ultracapacitor Applications** J.M. Miller

Volume 62 **Lightning Electromagnetics** V. Cooray
Volume 63 **Energy Storage for Power Systems, 2nd Edition** A. Ter-Gazarian
Volume 65 **Protection of Electricity Distribution Networks, 3rd Edition** J. Gers
Volume 66 **High Voltage Engineering Testing, 3rd Edition** H. Ryan (Editor)
Volume 67 **Multicore Simulation of Power System Transients** F.M. Uriate
Volume 68 **Distribution System Analysis and Automation** J. Gers
Volume 69 **The Lightening Flash, 2nd Edition** V. Cooray (Editor)
Volume 70 **Economic Evaluation of Projects in the Electricity Supply Industry, 3rd Edition** H. Khatib
Volume 72 **Control Circuits in Power Electronics: Practical issues in design and implementation** M. Castilla (Editor)
Volume 73 **Wide Area Monitoring, Protection and Control Systems: The Enabler for Smarter Grids** A. Vaccaro and A. Zobaa (Editors)
Volume 74 **Power Electronic Converters and Systems: Frontiers and Applications** A.M. Trzynadlowski (Editor)
Volume 75 **Power Distribution Automation** B. Das (Editor)
Volume 76 **Power System Stability: Modelling, Analysis and Control** A.A. Sallam and Om P. Malik
Volume 78 **Numerical Analysis of Power System Transients and Dynamics** A. Ametani (Editor)
Volume 79 **Vehicle-to-Grid: Linking electric vehicles to the smart grid** J. Lu and J. Hossain (Editors)
Volume 81 **Cyber-Physical-Social Systems and Constructs in Electric Power Engineering** S. Suryanarayanan, R. Roche and T.M. Hansen (Editors)
Volume 82 **Periodic Control of Power Electronic Converters** F. Blaabjerg, K. Zhou, D. Wang and Y. Yang
Volume 86 **Advances in Power System Modelling, Control and Stability Analysis** F. Milano (Editor)
Volume 87 **Cogeneration: Technologies, Optimisation and Implementation** C.A. Frangopoulos (Editor)
Volume 88 **Smarter Energy: From Smart Metering to the Smart Grid** H. Sun, N. Hatziargyriou, H.V. Poor, L. Carpanini and M.A. Sánchez Fornié (Editors)
Volume 89 **Hydrogen Production, Separation and Purification for Energy** A. Basile, F. Dalena, J. Tong and T.N. Veziroğlu (Editors)
Volume 90 **Clean Energy Microgrids** S. Obara and J. Morel (Editors)
Volume 91 **Fuzzy Logic Control in Energy Systems with Design Applications in MATLAB®/Simulink®** I.H. Altaş
Volume 92 **Power Quality in Future Electrical Power Systems** A.F. Zobaa and S.H.E.A. Aleem (Editors)
Volume 93 **Cogeneration and District Energy Systems: Modelling, analysis and optimization** M.A. Rosen and S. Koohi-Fayegh
Volume 94 **Introduction to the Smart Grid: Concepts, Technologies and Evolution** S.K. Salman
Volume 95 **Communication, Control and Security Challenges for the Smart Grid** S.M. Muyeen and S. Rahman (Editors)
Volume 96 **Industrial Power Systems with Distributed and Embedded Generation** R. Belu
Volume 97 **Synchronized Phasor Measurements for Smart Grids** M.J.B. Reddy and D.K. Mohanta (Editors)
Volume 98 **Large Scale Grid Integration of Renewable Energy Sources** A. Moreno-Munoz (Editor)
Volume 100 **Modeling and Dynamic Behaviour of Hydropower Plants** N. Kishor and J. Fraile-Ardanuy (Editors)
Volume 101 **Methane and Hydrogen for Energy Storage** R. Carriveau and D.S.-K. Ting
Volume 104 **Power Transformer Condition Monitoring and Diagnosis** A. Abu-Siada (Editor)
Volume 106 **Surface Passivation of Industrial Crystalline Silicon Solar Cells** J. John (Editor)

Volume 107 **Bifacial Photovoltaics: Technology, Applications and Economics** J. Libal and R. Kopecek (Editors)
Volume 108 **Fault Diagnosis of Induction Motors** J. Faiz, V. Ghorbanian and G. Joksimović
Volume 110 **High Voltage Power Network Construction** K. Harker
Volume 111 **Energy Storage at Different Voltage Levels: Technology, Integration, and Market Aspects** A.F. Zobaa, P.F. Ribeiro, S.H.A. Aleem and S.N. Afifi (Editors)
Volume 112 **Wireless Power Transfer: Theory, Technology and Application** N. Shinohara
Volume 115 **DC Distribution Systems and Microgrids** T. Dragičević, F. Blaabjerg and P. Wheeler
Volume 117 **Structural Control and Fault Detection of Wind Turbine Systems** H.R. Karimi
Volume 119 **Thermal Power Plant Control and Instrumentation: The Control of Boilers and HRSGs, 2nd Edition** D. Lindsley, J. Grist and D. Parker
Volume 120 **Fault Diagnosis for Robust Inverter Power Drives** A. Ginart (Editor)
Volume 121 **Monitoring and Control using Synchrophasors in Power Systems with Renewables** I. Kamwa and C. Lu (Editors)
Volume 123 **Power Systems Electromagnetic Transients Simulation, 2nd Edition** N. Watson and J. Arrillaga
Volume 124 **Power Market Transformation** B. Murray
Volume 125 **Wind Energy Modeling and Simulation Volume 1: Atmosphere and Plant** P. Veers (Editor)
Volume 126 **Diagnosis and Fault Tolerance of Electrical Machines, Power Electronics and Drives** A.J.M. Cardoso
Volume 128 **Characterization of Wide Bandgap Power Semiconductor Devices** F. Wang, Z. Zhang and E.A. Jones
Volume 129 **Renewable Energy from the Oceans: From Wave, Tidal and Gradient Systems to Offshore Wind and Solar** D. Coiro and T. Sant (Editors)
Volume 130 **Wind and Solar Based Energy Systems for Communities** R. Carriveau and D.S.-K. Ting (Editors)
Volume 131 **Metaheuristic Optimization in Power Engineering** J. Radosavljević
Volume 132 **Power Line Communication Systems for Smart Grids** I.R.S. Casella and A. Anpalagan
Volume 139 **Variability, Scalability and Stability of Microgrids** S.M. Muyeen, S.M. Islam and F. Blaabjerg (Editors)
Volume 145 **Condition Monitoring of Rotating Electrical Machines** P. Tavner, L. Ran, C. Crabtree
Volume 146 **Energy Storage for Power Systems, 3rd Edition** A.G. Ter-Gazarian
Volume 147 **Distribution Systems Analysis and Automation, 2nd Edition** J. Gers
Volume 152 **Power Electronic Devices: Applications, Failure Mechanisms and Reliability** F Iannuzzo (Editor)
Volume 155 **Energy Generation and Efficiency Technologies for Green Residential Buildings** D. Ting and R. Carriveau (Editors)
Volume 157 **Electrical Steels, 2 Volumes** A. Moses, K. Jenkins, Philip Anderson and H. Stanbury
Volume 158 **Advanced Dielectric Materials for Electrostatic Capacitors** Q Li (Editor)
Volume 159 **Transforming the Grid Towards Fully Renewable Energy** O. Probst, S. Castellanos and R. Palacios (Editors)
Volume 160 **Microgrids for Rural Areas: Research and Case Studies** R.K. Chauhan, K. Chauhan and S.N. Singh (Editors)
Volume 166 **Advanced Characterization of Thin Film Solar Cells** N. Haegel and M. Al-Jassim (Editors)
Volume 172 **Lighting Interaction with Power Systems, 2 Volumes** A. Piantini (Editor)
Volume 905 **Power System Protection, 4 Volumes**

Power Grids with Renewable Energy

Storage, integration and digitalization

Edited by
Abdelhay A. Sallam and Om P. Malik

The Institution of Engineering and Technology

Published by The Institution of Engineering and Technology, London, United Kingdom

The Institution of Engineering and Technology is registered as a Charity in England & Wales (no. 211014) and Scotland (no. SC038698).

First published 2020

The Institution of Engineering and Technology
Michael Faraday House
Six Hills Way, Stevenage
Herts, SG1 2AY, United Kingdom

www.theiet.org

British Library Cataloguing in Publication Data
A catalogue record for this product is available from the British Library

ISBN 978-1-83953-027-2 (hardback)
ISBN 978-1-83953-028-9 (PDF)

Typeset in India by MPS Limited

No significant work can be accomplished without the moral support of the families and we wish to acknowledge that with our appreciation and dedication

To our wives,

Hanzada Sallam and Margareta Malik

Contents

About the authors

Abdelhay A. Sallam is Professor Emeritus in Port Said University, Egypt. He has published over 100 papers in international journals and conferences and 2 books, has supervised thesis of over 60 students and has had 17 years of experience in industry.

Om P. Malik is Professor Emeritus in the University of Calgary, Canada. His publication record comprises over 375 papers in international journals such as *IEEE Transactions* and *IET* (formerly *IEE*) *Proceedings*, and he has presented over 360 papers in international conferences. He is the author of two books. Under his supervision, 52 PhD and 58 Masters students have successfully completed their degrees.

Preface

Traditionally, growth and development of the electric power system has been based on the idea of establishing large size centralized generating power plants to provide electricity and meet the continuous growing demand in an optimal manner, economical and technical. The common types of these plants are that fuelled by fossil fuels, hydro-turbine power plants, or nuclear power plants. A power grid is dedicated to serve both large and small consumers with electrical energy. In developing the power grid, attention of power system planners and operators is primarily aimed at providing the consumers with electrical energy as economically as possible and with a high degree of reliability and quality of supply.

With the rising environmental concerns about global warming trends, there has been a big change in thinking about the source of energy for the generation of electricity. This has resulted in the generation of electricity from renewable sources, having attracted a tremendous amount of attention worldwide. Renewable energy, derived from imperishable resources of nature being inexhaustible, has the major advantage of being replenished or replaced naturally. Examples of potential sources of renewable energy include biomass, solar, wind, hydro, geothermal and ocean waves, among others. Geothermal and hydro sources have been exploited for a long time.

Over the past few decades, there has been a tremendous amount of progress in the development of electricity generation from renewable energy sources such as wind and solar. Solar and wind-based plants ranging in capacity from hundreds of megawatts to small distributed plants serving individual residences are in existence now. Now, these may be considered as having reached the stage of mature technologies. At the same time, additional research is going on in parallel on the development of electricity generation from other sources such as oceans and biomass.

Considered for long-term use, renewable energy is sustainable and less expensive as it does not require external feeding of fuels that cannot be replenished. There is, however, the concern of the non-availability of renewable sources under certain conditions and lack of control on the resource (fuel) input can significantly affect the production of sustainable power. Despite this, the benefits of sustainable energy outweigh its drawbacks.

Nature of the various renewable sources, such as variability and inability to accurately predict and control, unlike the fossil fuel sources of energy used in conventional generation, creates its own problems when integrating with the conventional power systems at both the grid and the distribution levels. A tremendous amount of research and development is taking place in ensuring utilization and integration of the electricity generated from renewable sources.

Lack of control on the basic input source of energy, unlike that of the conventional fossil or even hydro-based generation, and intermittence of generated electricity requires special consideration while integrating renewable energy sources in conventional power grids. Energy storage and proper interfacing play an indispensable role in the optimal application and utilization of natural resources, and efficient overall system operation of the conventional and renewable energy sources integrated in power systems.

With the increasing generation of electricity from renewable sources at all levels of the power system, operation of the power grid is becoming more complex as significant power will flow in both directions, i.e. not only from the power system to the consumer but also from the consumer to the power system. Problems like excessive voltage variations and bottlenecks may occur more frequently, and then, additional investments in the network reinforcement, in more sophisticated protective relaying and control systems, are required. Thus, grid design modification is needed to benefit from the potential of greater flexibility offered by distributed generation in the form of microgrids, flexible loads, storage systems, computer monitoring and control technologies, and communication networks. Such flexibility will require that the power grid become smarter.

The technology of renewable power generation is well covered in many existing books. Material relating to both the mature, wind and solar, and other emerging technologies for the generation of electricity from renewable sources is limited in this book to that necessary. The primary thrust of this book is to cover the requirements for proper interfacing and integration of the renewable generation technologies in the existing power grids considering the special operational characteristics of these variable generation sources with no control on the basic input source of energy unlike that of the conventional fossil or even hydro-based generation.

An outline of the contents of this book is given as follows:

- A general background to the topic of this book, new trends developing in power grids, i.e. generation of electricity from renewable sources, is provided in the introductory (Chapter 1).
- The state of development of various renewable energy sources being pursued at present is presented in brief in Part I 'Renewable Energy Sources' consisting of Chapters 2–6.
- Energy storage is one of the important aspects in the integration of renewable generation for efficient operation of the conventional and renewable generation integrated power grid. The essential role played by energy storage systems, various types of energy storage systems, such as electrical, mechanical, chemical, thermal, electromechanical, electrostatic and magnetic including superconducting, and their applications, especially, the electric vehicles is included in Chapters 7–12, Part II 'Energy Storage Systems'.

- Despite the generation and use of energy from sustainable sources having increased significantly recently, there is still a need for optimizing the application of natural resources. The renewable energy sources need to be integrated properly with the existing power network for efficient system operation. The most significant part of the book, Part III 'Renewables Integration', Chapters 13–19, covers interfacing, integration and operational aspects of renewable energy and includes energy storage, power electronics interfacing and other operational considerations.
- Communications and cybersecurity, essential for the safe and secure operation of the integrated system, are covered in Chapters 20 and 21, in the concluding Part IV 'Associated Systems'.
- All chapters are supported by a limited number of relevant references so that the interested reader can pursue more in-depth coverage of any specific topic.

The authors have drawn on their own research and many other tools such as MATLAB/Simulink and toolboxes.

Abdelhay A. Sallam
Om P. Malik

Acknowledgements

In the present day, no individual is unto self. It is the cooperative effort of many, and this book is no exception in that regard. The authors, therefore, wish to record their appreciation of the help received from various sources in preparing this book. A part of the material included is based on their research and the work of a number of graduate students, though not individually named here, who have worked under their supervision over the years and are making their own contribution to the profession. Their work and contribution are well appreciated.

The authors, however, feel it appropriate to express their gratitude to certain individuals by name for their specific contributions and help in putting this book together. These are

- Dr Azza A. El-Desouky,
- Dr Mai M. Eladany,
- Ahmed H. Elmetwaly,
- Mohamed A. Kapeel,
- Mohamed G. Anany, and
- Mahmoud M. Mashaal.

Chapter 1

Towards the new trend of power grids

Historic growth and development of the electric power system has been based on the idea of economies of scale and building bigger and bigger electricity generating plants primarily fuelled by fossil fuels with some hydro generating plants, where such possibility exists, and a few nuclear generating plants. Global warming trends and the environmental concerns have brought in a big change in thinking with the result that over the past couple of decades a tremendous amount of attention has been devoted to the generation of electricity from renewable sources such as wind, solar, oceans and biomass.

Electricity generation from wind and solar as primary sources of energy has advanced tremendously with the result of solar and wind-based plants in all sizes, ranging in capacity from hundreds of megawatts to small distributed plants serving individual residences, that are in operation today. Very significant cost reductions have also resulted in parallel with these developments.

While the state of technology in the development of the wind-based and solar plants may be considered to have reached the state of maturity, development in the generation of electricity from other renewable sources has been progressing in parallel. These are the energy in oceans, utilization of biomass and so on.

The nature of the various renewable sources such as its variability and our inability to accurately predict and control, unlike the fossil fuel-based sources of energy used in the conventional generation, creates its own problems when integrating them with the conventional sources at both the grid and the distribution levels. A tremendous amount of research and development is taking place in ensuring proper utilization and integration of the electricity generated from renewable sources. The aspects of both generation and integration are covered in the subsequent chapters in the book.

A power grid is dedicated to serve both large and small consumers with electrical energy. In developing the power grid, the focus of power system planners and operators is primarily aimed at providing electrical energy to the customers as economically as possible and with a high degree of reliability and supply quality. The term 'grid' denotes the entire electric system infrastructure, which is also commonly known as 'electric power system'. Thus, the two terms 'grid' and 'power system' are often used interchangeably [1].

A brief description of conventional power systems, recent developments and future perspectives based on new enabling technologies, advanced controllers and communication facilities is presented in this chapter.

1.1 Conventional power systems

A complete power system is a chain of three consecutive sectors. These sectors are denoted by their functions as generation, transmission and distribution, respectively. The conventional power system infrastructure comprises centralized bulk generating stations (e.g. fossil-fuel thermal, nuclear and hydro stations) supplying electricity, through extra-high-voltage (EHV) and high-voltage (HV) transmission networks (also called transmission and sub-transmission networks, respectively) to local distribution systems. Local distribution systems serve the consumers' loads are classified into three major types, industrial, commercial and residential, as each type has its own characteristics and requirements. In the conventional power systems, electricity flows in one direction starting at generating stations and terminating at the individual load points in distribution network as shown in Figure 1.1.

With the progress in the level of technology in power systems, many programmes using available new technology (e.g. flexible AC transmission system (FACTS), fast controllers and communication systems) have been applied to improve power system characteristics, especially, reliability, security and stability [2].

Power system reliability: In general, system reliability is specified by the overall ability of the system to perform its function. In a power system, the basic aspect of identifying its reliability is the power system adequacy. It relates to the presence of enough facilities within the system to satisfy the consumer load demand, i.e. it is necessary to generate sufficient energy and to use both transmission and distribution networks with sufficient power transfer capacity to enable the system to transport the energy to the load points. Therefore, adequacy is associated with static conditions excluding contingencies and is evaluated by identifying suitable indices such as the following:

- Expected frequency and duration of failure: It identifies the expected frequency of encountering a deficiency and the expected duration of the deficiencies.
- Loss of energy expectation: It identifies the expected energy that will not be supplied by the generating station when the load demand exceeds the available generating capacity.
- Outage time of consumer load points: It is the annual time of unavailability of feeding load points with energy.

Power system security: The term 'security' refers to the degree of risk in the ability of the electric power systems to withstand without serious consequences to any of a preselected set of sudden disturbances such as short circuits or unanticipated loss of system elements. From a control perspective, security is the degree of risk in the ability of a power system to reach a steady-state operating point without violating system operating constraints, mainly, thermal limits of the transmission lines and bus voltage limits. These constraints are referred to as

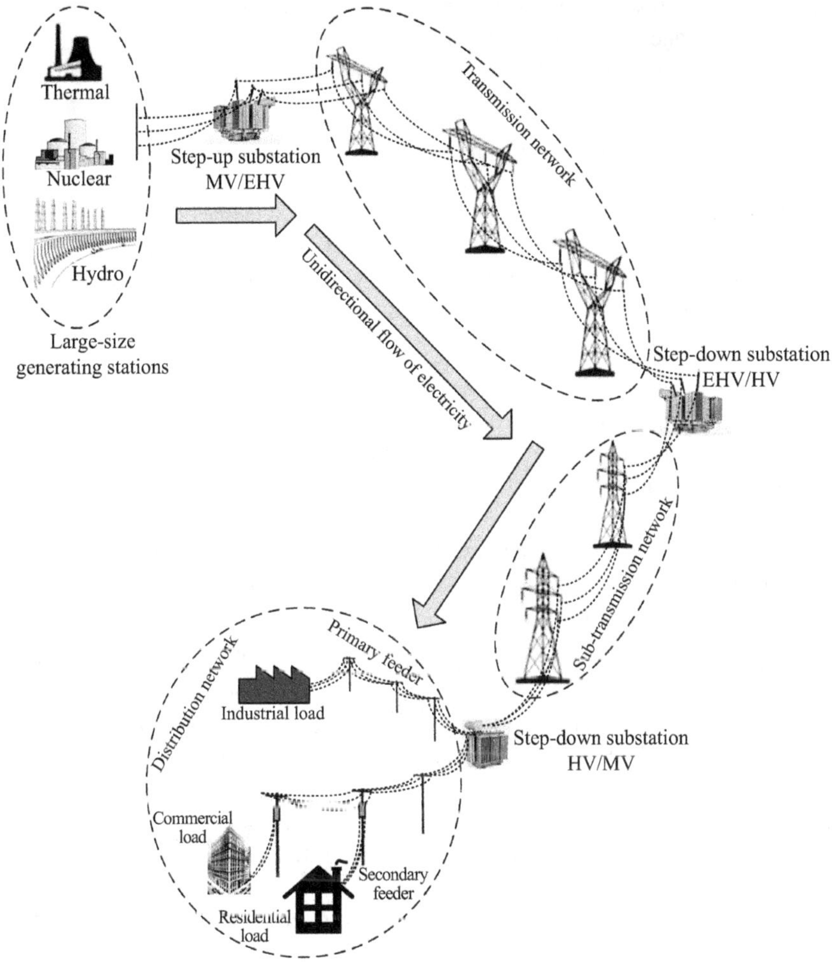

Figure 1.1 A typical conventional power system

security constraints when referred to the post contingency scenarios and can be written as

$$\sum_{i=1}^{N_g} P_{Gi} = P_D + P_{\text{loss}} \quad \text{and} \quad P_{Gi}^{\min} \le P_{Gi} \le P_{Gi}^{\max} \tag{1.1}$$

$$\left|V_j^{\min}\right| \le \left|V_j\right| \le \left|V_j^{\max}\right| \quad \text{and} \quad S_{jk} \le S_{jk}^{\max} \quad \forall \text{ branch } j-k \tag{1.2}$$

where

$P_{Gi} \triangleq$ generated real power at bus # i;
$P_D \triangleq$ total system demand;

$P_{\mathrm{loss}} \triangleq$ total real power loss in the transmission network;
$|V_j| \triangleq$ voltage magnitude at bus # j;
$S_{jk} \triangleq$ apparent power (MVA) flow in branch $j-k$ and $N_g \triangleq$ number of generators.

Power system stability: It commonly pertains to rotor angle stability. Stability phenomenon is characterized by generator rotor speed oscillations under a severe perturbation and can be analysed by solving non-linear dynamic equations describing the transient behaviour of the system under a set of credible contingencies.

Nowadays, electricity has become all pervasive in the world resulting in tremendous growth in load demand. Consequently, the conventional power system infrastructure has become

- large scale;
- multi-input/multi-output;
- distributed over large geographical areas requiring the energy needs to be routed through long distances, resulting in increased transmission losses;
- complex with many interconnections;
- costly due to the investments needed to expand generation, transmission and distribution systems while still maintaining a desired level of reliability, security and stability.

In addition, a large part of generation is based on fossil fuel (coal, oil and natural gas) as source. Inflation and astronomical increase in fuel prices has resulted in rapid increase in consumer tariffs as well as adverse effects on the environment. Their combined effects introduce considerable uncertainty in predicting the future. To satisfy this continued growth in load demand necessitates further development of the power system that is compatible with environmental considerations.

In a new energy economy, it will be beneficial to empower every consumer to conserve energy and become an energy producer by generating electricity from renewable energy sources (RESs). This will become a major issue in satisfying system adequacy as it creates a reduction in forecast demand, improves system security and stability as well as reduces greenhouse gas emissions.

Distributed generation (DG) including RESs (e.g. wind, solar, photovoltaic (PV), tidal and micro-hydro) and non-RESs (e.g. fuel cells, diesel generators and micro-gas turbines) can be interconnected together to form what is called a ‘microgrid’ (MG) that can operate in isolation, islanded or, in turn, can be connected to a utility distribution system. Therefore, power system planning and configuration must be modified to recognize this new scenario [3,4].

1.2 Microgrids

MGs are small distribution systems encompassing a group of electricity consumers (industrial, commercial and residential) connected to several distributed generators and storage units that can be interfaced by power electronics.

MG technologies help electric system evolve into one that is more efficient, is less polluting, has lower chances of transmission congestion, has reduced losses and is more flexible to provide energy consumers' wants and needs. MGs can help to make better use of energy generated, stored and used at a local level, e.g. local generation and islanding operation. In addition, in future each client can become an autonomous entity with the capability of not only demanding but also injecting power into the grid when the generated power exceeds the local needs [5,6].

Lately, utility companies have been motivated to encourage connection of local RESs at the distribution level. This can modernize the conventional power systems into multiple, interconnected distribution systems, i.e. MGs. However, it is not a simple proposition to move from the centralized system that has developed over the past more than 135 years to one that is distributed. Implementing DG comes with challenges as the present electric power infrastructure is not designed to meet problems pertaining to operational capability and quality of supply of electricity generation using RESs such as wind and solar, which are intermittent and unpredictable.

An example of MG architecture connected to an utility distribution network is illustrated in Figure 1.2. It typically comprises non-renewable (e.g. AC diesel generators) and renewable generation units (e.g. wind, solar, micro-hydro, ocean wave and tidal) in addition to energy storage systems (ESSs). These energy sources (variable frequency AC, DC and energy storage) are connected to DC bus through AC/DC converter, DC/DC converter and interface power electronic circuits. The local DC loads are fed from the DC bus. The excess of DC energy can be converted into AC energy by DC/AC inverter to be connected to AC bus at the point of common coupling (PCC). The AC bus is fed from both MG and utility grid to supply the AC loads [7]. More details are given in Chapter 14, Section 14.5.

1.3 Energy storage systems

RESs (e.g. wind speed and solar radiation) can fluctuate quickly, and at random, it is difficult to ensure a constant stable supply. So, ESSs can be used to compensate the fluctuation from RESs and maintain power quality. Installation of energy storage devices can help alleviate limits on the size of the RESs.

Services provided by storage depend on the type of storage technology and are also based on their capacity. Two types of energy storage technologies can be used: (i) bulk storage (such as compressed air energy storage, pumped hydro and large batteries) is able to provide a wide range of grid services, e.g. day-night arbitrage, peak shaving regulation and spinning and non-spinning reserves. This type of storage can be connected to utility transmission network. (ii) Short-term storage (including batteries, flywheel, superconducting magnetic energy storage (SMES) [8], plug-in electric vehicles (PEVs), etc.) is generally used for providing shorter-term regulation services to accommodate load and variation of renewable generation on a limited time scale. Therefore, the development of DG can be supported by deploying small-scale, low-voltage (LV) or medium-voltage (MV) grid-connected

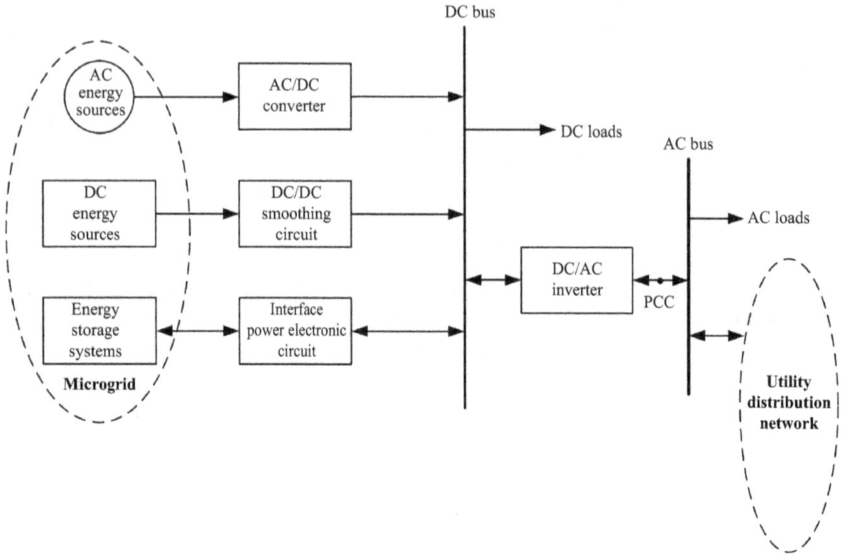

Figure 1.2 Schematic diagram of an MG connected to utility distribution network

electricity storage systems at different locations in the MGs (decentralized storage). Such storage systems could affect the management of distribution grid in different ways: (a) energy management by delivering capacity to satisfy energy arbitrage (generation and consumption), (b) optimizing the operation of distribution grid and (c) keeping the quality of service and security of supply. Under certain circumstances, non-storage-type solutions, e.g. micro-gas turbine and micro-hydro turbine, may be employed to level load.

Accordingly, a typical overall power grid perspective can be visualized as in Figure 1.3. It is to be noted that the MG is connected to the utility distribution system at the PCC. The MG encompasses RESs (wind, solar and PV array), non-RESs (fuel cells and PEVs) and ESSs such as SMES, flywheel and battery storage. PEVs demand power from the MG (as a load) in the charging state and inject power into the MG (as a source) in discharging state. For the transmission system, bulk power delivered by the compressed-air storage system can be injected to it as and when needed. More details are given in the forthcoming chapters (Part II).

1.4 Smarter grid

It is to be noted that the traditional concepts and practices of power systems have changed as different energy sources and storage systems are added. More complicated controllers, information systems and new technologies encompassing power electronics and two-way communication systems are of paramount importance for

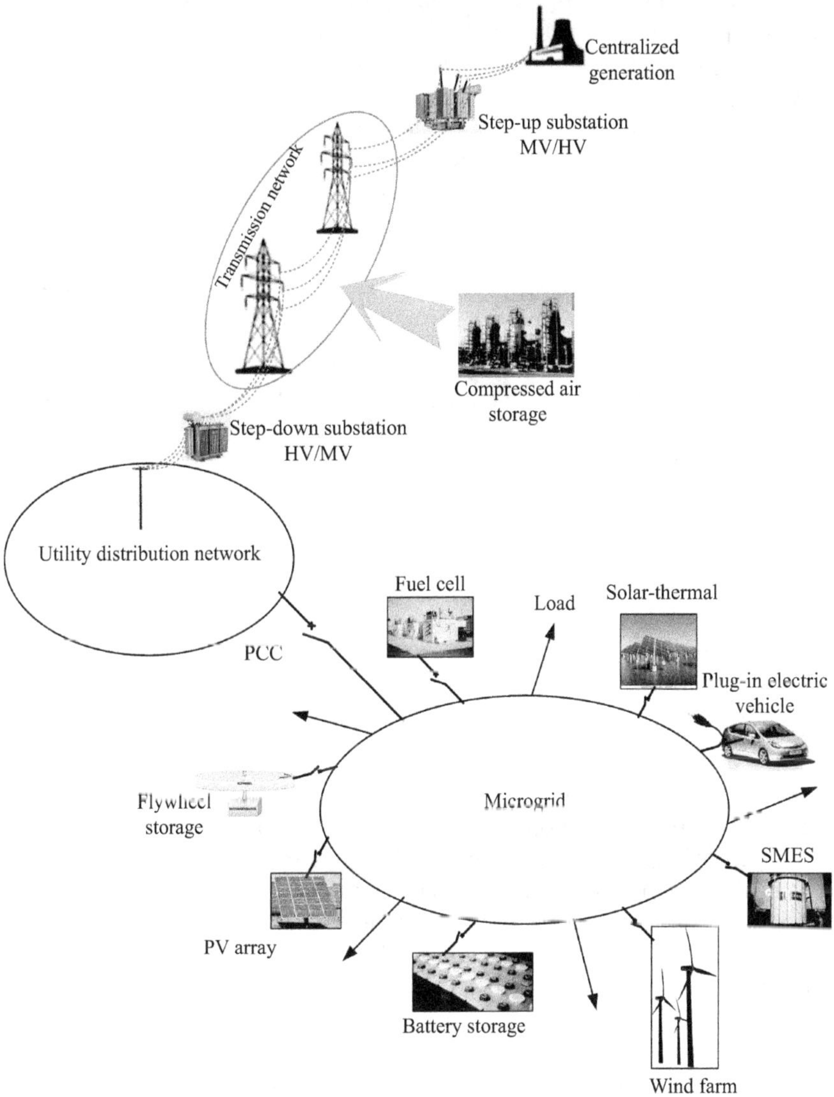

Figure 1.3 Power grid supported by MG and compressed air storage

system components' integration, interface and better utilization of the existing power system. In consequence, efforts are continuously being made to restructure the electric power industry into a more competitive industry focusing on the wholesale electricity market. In this market, generating companies produce electricity and sell it to energy service providers, which, in turn, resell the electricity to residential, industry and commercial consumers in a retail market.

Operation of the power grid shown in Figure 1.3 will become more complex as significant power will flow in both directions, i.e. not only from the power system to the consumer but also from the consumer to the power system. Problems such as excessive voltage variations and bottlenecks may occur more frequently, and then, additional investments in the network reinforcement in more sophisticated protective relaying and control systems are required. Thus, modification in the grid design is needed essentially to benefit from the potential of greater flexibility offered by DG, flexible loads, storage systems, computer monitoring and control technologies, and communication networks [9].

Such flexibility would help reinforce the conventional network, make it smarter and provide enhancement to ensure the following:

- High level of security, quality, reliability and availability of electric power.
- Optimization of the use of the existing network, thereby minimizing distribution grid extensions.
- Improved economic productivity and quality of life.
- Minimization of environmental impact while simultaneously maximizing safety and sustainability.
- To deal with this paradigm shift, active involvement of both consumers and distribution system operators (DSOs) is a must [10].

1.4.1 Consumers' role

Design of the distribution systems is based on the load analysis performed at the time. As the load levels increase and load patterns change, some sort of upgrading and more investment are required to make the design accommodate the new conditions and obviate system overloads. On the other hand, proper management of distribution system overloads can help reduce the required investment. It may be like congestion management in the transmission system where it needs distribution capacity reservation, use of retail pricing, curtailable schedules and distribution loading relief to manage the loading of the distribution facilities. Traditional load management tools have already been used by direct load control, peak shaving, peak shifting and various load management programmes. Now, demand response (DR) is taking on new realities with the push for energy conservation and demand-side management as a key strategy of distributed resource management.

The role of the consumers in the electricity market will increase, paving the way for DR in addition to the traditional tools of load management. Consumers will be able to manage and adjust their electricity consumption in response to real-time information and changing price signals. Information can be provided by dynamic feedback programs designed by suppliers to let the consumers actively manage their consumption. This alone is necessary but not enough to change electricity consumption. It only becomes enough when consumers become more aware of the value of shifting their electricity consumption. Meanwhile, grid tariffs should reflect the actual costs in a manner that motivates electric markets to introduce attractive services.

Consumers' awareness of their energy consumption can be increased by installing smart meters, which will be the responsibility of DSOs, as well as intelligent appliances and smart chargers. Smart meters as a kind of intelligent metering are a part of the physical grid infrastructure, fall under the grid operator's domain and are managed by DSOs. They offer new advantages to both the utility and the consumer by efficient use of energy, operational efficiencies, improved daily service, performing majority of operations remotely, efficient use of resources and the possibility for consumers to monitor and reduce their consumption. Therefore, the grid will be ready to move from the traditional load-following operating strategy to a load-shaping strategy in which demand-side resources are managed to meet the available generation and the grid's power delivery capabilities at any time [9].

Moreover, smart metering greatly contributes to optimizing network management as it offers the following:

- remote connect/disconnect function to manage, for instance, peak shaving programmes;
- two-way wire or wireless communication;
- advanced measurement capabilities;
- detailed monitoring of power quality;
- fault identification, fraud tempering and theft detection function; and
- automatic and self-contained interval metering capability.

1.4.2 DSOs' role

DSOs need to play a more active and prominent role in applying the right tools and to be following their basic tasks of maintaining supply reliability and service quality. DSOs are responsible for the following:

- preventing bottlenecks in the most cost-effective way through coordination with transmission system operators based on long-term network planning;
- integrating DG resources into the network in such a way that they maintain voltage quality level;
- ability to obtain flexibility from DG resources and consumers to solve grid constraints;
- application of desired technical tools to become real system operators – this necessitates supervision and control of power flows using advanced protection systems, monitoring and control strategies;
- dealing with new technologies in the grid to improve service quality and reduce costs;
- monitoring the integration of the consumers in the most effective and economical way to maintain adequacy and service level while satisfying overall energy efficiency of the grid; and
- adapting information technology (IT) systems to probable change in the grid structure.

1.4.3 Smart grid definition

New technologies are in progress on various fronts. At present, a lot of research is devoted to incorporating renewable and non-renewable distributed generation

(MGs) as well as ESSs into the power grid. So, the power grid is a system that encompasses multi-energy sources, transmission networks and distribution networks incorporating several MGs. This composition needs different components of the power grid to be integrated together to constitute the infrastructure that enables the power grid to function properly. New technologies such as digital communication systems, sensor technology, FACTS devices, supervisory control and data acquisition (SCADA), phasor measuring units (PMUs), DR programs, IT and intelligent metering technology can support this integration with a goal of satisfying the following [11]:

- procuring accurate demand forecast and renewable energy availability;
- real-time monitoring, operation and control of power system components (remotely and local);
- real-time management of power demand and economic dispatch;
- desired power quality and security level;
- energy sustainability;
- achieving cleanliness and environmental friendliness as much as possible; and
- self-healing that needs the grid to rapidly detect, analyse, respond and restore.

With the implication of advanced technologies satisfying all or some of these aspects, the power grid is being referred to as ‘smart grid’ (SG). Improvement of the conventional power grid, to behave as an SG, is based on selected aspects that each utility concerns itself with. In addition, as the new enabling technologies (information, communication and automation) are exploited intensely in the power grids, the power grid gets smarter. There is no clear or standard definition of an SG. Everyone seems to have own vision and a framework for an SG.

According to an Alberta, Canada, government report, SG is a broad concept that describes the integration of hardware, computer monitoring and control technologies, and modern communications networks into an electricity grid [1].

European Commission Task Force on Smart Grid defined SG as ‘an electricity network that can intelligently integrate the action of all users connected to it – generators, consumers and those that do both – in order to ensure economically efficient and sustainable power system with low losses, high level of power quality, security of supply and safety’ [12].

In the US Energy Independence and Security Act 2007, ‘smart grid’ refers to a modernization of the electricity delivery system so that it monitors, protects and automatically optimizes the operation of its interconnected elements – from the central and distributed generator through appliances and other household devices.

In the absence of a formal definition of ‘smart grid’, a generic description, as good as any, is as follows: ‘a smart grid is a digitally enabled electric grid that gathers, distributes and acts on information about the behavior of all components in order to improve the efficiency, reliability and sustainability of electricity devices’ [1].

An example of SG perspective of the future is depicted in Figure 1.4, where the communication, information and computer technologies provide predictive information, recommendations, and real-time monitoring and control to utilities. Each of

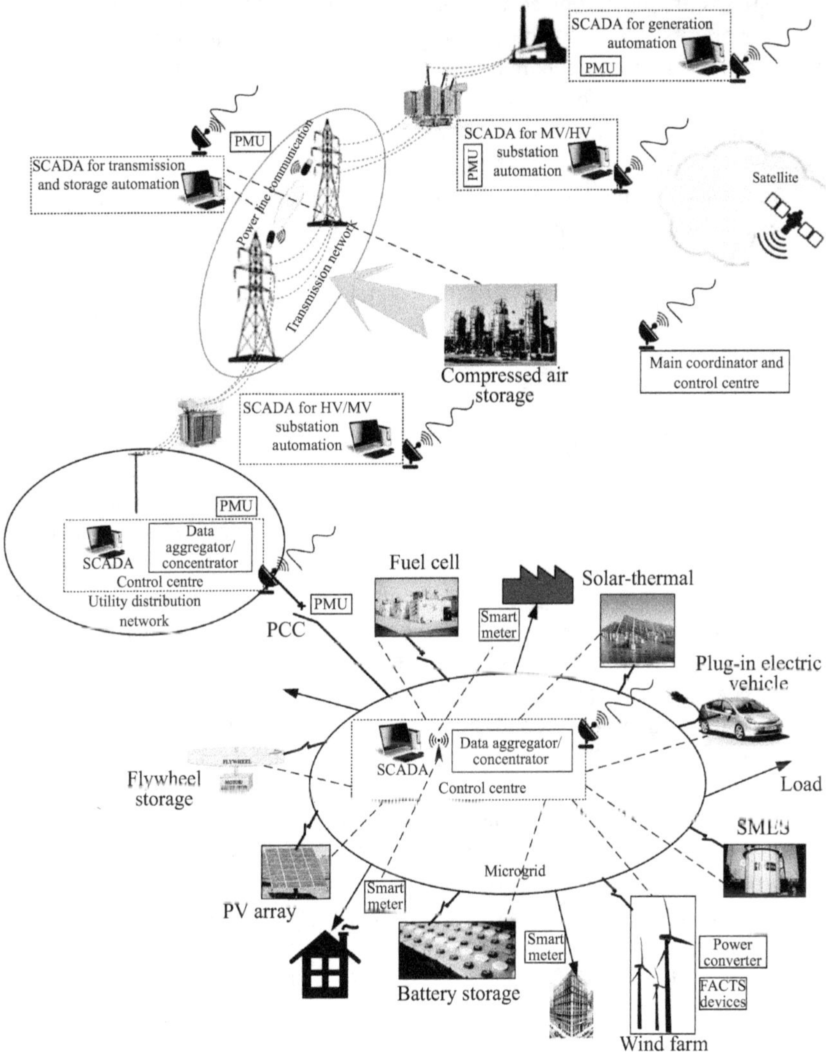

Figure 1.4 SG perspective of the future

MG, distribution system, transmission system and central generation is locally controlled by SCADA through measurements taken from data aggregator/ concentrator. Then, a coordinator and control centre unit receives data from the local control centres through satellite for decision-making necessary to integrate the function of overall system components and satisfy the aspects of SG. Some of the major elements used are phasor measurement systems, PMUs and smart meters.

PMUs are used to measure in a complex form (magnitude and direction) the electrical parameters at widely dispersed locations in the power system. The measurements are synchronized by using the global positioning system (GPS) to help the operators to specify the state of the power system and respond to system conditions in a rapid and dynamic manner. Smart meters with the aid of wireless networks are used for information exchange between users.

In addition, power line communication, by using wireless modems, is used to communicate between the substations at the terminals of transmission lines. Moreover, sensor technology is used for monitoring and measurement in order to detect components' failure and system faults to be isolated before they cause cascading effects and lead to catastrophic breakdowns. However, from what has been discussed and as reported in [13], it can be concluded that a smarter grid comprises three systems: smart infrastructure system, smart protection system and smart management system.

1.5 Power grid examination

Lack of control on the basic input source of energy, unlike that of the conventional fossil or even hydro-based generation, and intermittence of generated electricity requires special consideration while integrating RESs in conventional power grids. System frequency and voltage control are important issues and must be studied. This study needs the system to be modelled in steady state, particularly, to ensure the rated voltage at each bus in the power grid. The integration of DG (renewable and non-RESs) as a part of an MG that is connected to the local utility power grid in addition to its asynchronous operation must be examined. Consequently, the load flow analysis is required to compute the power line flows and bus voltages, and hence, the bus impedance and admittance matrices should be determined. Also, economic operation based on economic load dispatch of power grid incorporating both DG and ESSs as a part of an MG should be examined.

On the other hand, power grid security during abnormal operating conditions must be analysed as this grid accommodates different energy sources, ESSs, complicated networks and several additional assets. Thus, many security risks pertaining to this complexity as well as communication requirements, system automation, new technologies and data collection may be introduced.

All these issues in addition to more detailed description of different energy sources and energy storage systems are presented in the forthcoming chapters.

References

[1] Malik O. P. 'Evolution of power systems into smarter networks'. *Journal of Control, Automation and Power Systems*. 2013;24(1):139–47.

[2] Arefifar S. A. and Mohamed Y. A. 'DG mix, reactive sources and energy storage units for optimizing microgrid reliability and supply security'. *IEEE Trans. on Smart Grid*. 2014;5(4):1835–44.

[3] Smith C. 'Wind and solar energy – Variable generation across the land'. *IEEE Power & Energy Magazine*. 2015;13(6):16–20.

[4] Krishnan V., Ho J., Hobbs B. H., *et al.* 'Co-optimization of transmission and other supply resources – concept, review and modeling approaches'. *Energy Systems*. 2015;7(2):1–36.

[5] Basu A. K., Chowdhury S. P., Chowdhury S., Ray D. and Crossley P. A. 'Reliability study of a micro-grid power system'. *Universities Power Engineering Conference 2008, UPEC, 43rd International*. Padova, Sep 2008, pp. 1–4.

[6] Ahshan R., Iqbal M. T., Mann G. K. I. and Quaicoe J. E. 'Micro-grid system based on renewable power generation units'. *Electrical and Computer Engineering (CCECE), 2010 23rd Canadian Conference on*. Calgary, AB, Canada, May 2010, pp. 1–4.

[7] Nehrir M. H., Wang C., Strunz K., *et al.* 'A review of hybrid renewable/alternative energy systems for electric power generation: Configurations, control, and applications'. *IEEE Trans. on Sustainable Energy*. 2011; 2(4):392–401.

[8] Nakayama T., Yagai T., Tsuda M. and Hamajimam T. 'Micro power grid with SMES and superconducting cable modules cooled by liquid hydrogen'. *IEEE Trans. on Applied Superconductivity*. 2009;19(3):2062–5.

[9] Albuyeh I. F. 'Grid of the future'. *IEEE Power & Energy Magazine*. 2009; 7(2):52–62.

[10] Mallet P., Granstrom P. O., Hallberg P., Lorenz G. and Mandatova P. 'European perspectives on the future of electric distribution'. *IEEE Power & Energy Magazine*. 2014;12(4):51–64.

[11] Petinrin J. O. and Shaaban M. 'Smart power grid: technologies and applications'. *2012 IEEE Int. Conf. on Power and Energy (PECon)*. Kota Kinabalu Sabah, Malaysia, 2012, pp. 892–7.

[12] European Commission. 'European smart grid technology platform'. Luxemburg, Dec 2006.

[13] Fang X., Misra S., Xue G. and Yang D. 'Smart grid – The new and improved power grid: a survey'. *IEEE Communications Surveys & Tutorials*. 2012; 14(4):944–80.

Part I

Renewable energy sources

Chapter 2
Wind energy

One of the most mature technologies for renewable resources is the wind energy. The wind is a motion of air masses caused by the different thermal conditions of these masses. The energy produced by this motion can be converted into another form of energy, such as electric energy. This approach, known as wind power generation, has grown rapidly during the past few decades in many countries around the world.

Major catalysts for this rapid growth have arisen in the recent past: continuous technological advances in power electronics, controls and physical attributes (e.g. tower heights and blades) as well as cost reduction and significant advancements in understanding the access of wind power generation to the grid. Consequently, more attention is paid to high penetration of wind power sources in power systems as these sources are not only a means to reduce CO_2 gas emissions but also an economic alternative.

This chapter is focused on the description of wind system components and their functions being combined to convert the wind power to electrical output power. The basic relations necessary to design the wind turbine (WT) and evaluate its performance are introduced. The recommended controllers are also explained as the wind speed varies with time and yields a fluctuating character to the system.

Different terms may be used when talking about the wind system such as wind generator, wind turbine generator, wind-driven generator and wind energy conversion system. In this chapter and in most literature, the term 'wind turbine' is used. It is to be noted that the power is the energy per unit time.

2.1 Wind turbines

The main components of a WT are blades, rotor shaft, gearbox and generator (Figure 2.1). A simplified block diagram representing the WT system is shown in Figure 2.2. It is to be noted that the components gearbox, yaw drive, coupling, main shaft, etc. are represented as gear system model, whereas the generator model includes the interface system and a transformer to step up the generated voltage to the desired voltage. In addition, other parts, e.g. control panel, are all placed in a compartment known as 'nacelle'. It is to be noted that the nacelle may also contain a transformer. Inflow of wind activates the blades and rotor which in turn spins the

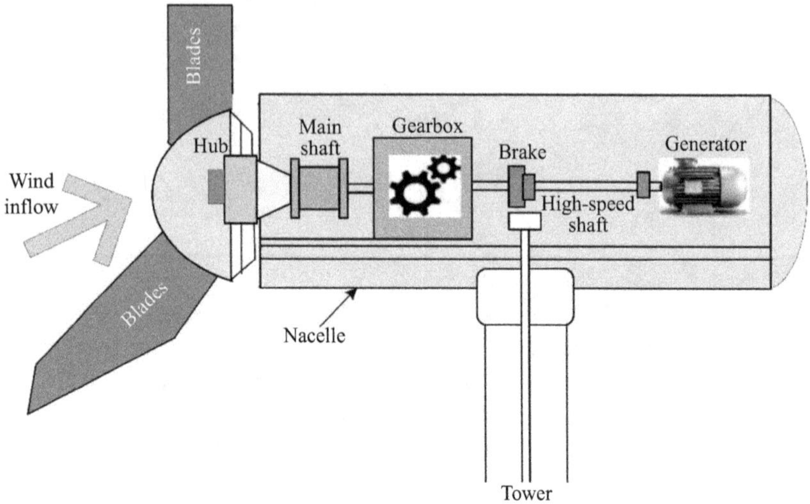

Figure 2.1 WT outline configuration

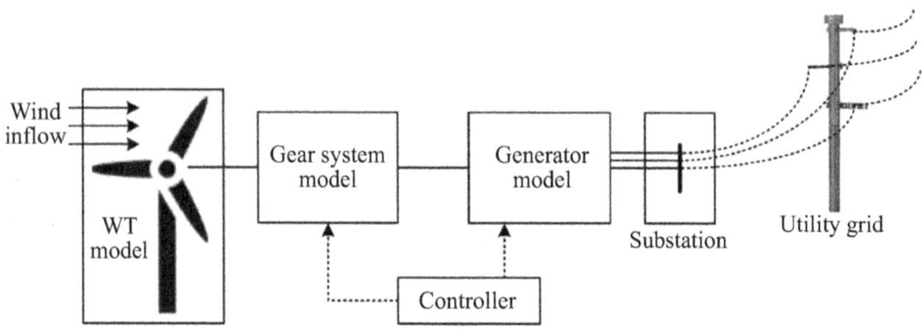

Figure 2.2 A simplified block diagram of WT system

main shaft and generator resulting in electrical output power. WTs can be installed close to the ground (onshore) or close to sea level (offshore) as shown in Figure 2.3.

Onshore WT installation is much cheaper than the offshore installation. The rotor shaft can rotate about either a horizontal axis or a vertical axis. Based on the axis of rotation, WT is classified into two types: horizontal axis wind turbine (HAWT) and vertical axis wind turbine (VAWT).

2.1.1 Horizontal axis wind turbines

HAWT is the most common design of WTs. The axis of rotation is parallel to the ground. The nacelle (with main rotor shaft and generator) is placed on the top of the tower (Figure 2.4(a) and (b)). WT must be pointed into the wind so that the blades facing into the wind provide full swept area. Small turbines are pointed by a simple vane, whereas large turbines generally use a wind sensor coupled with a servo

Figure 2.3 Onshore HAWT (on the left side) and offshore HAWT (on the right side)

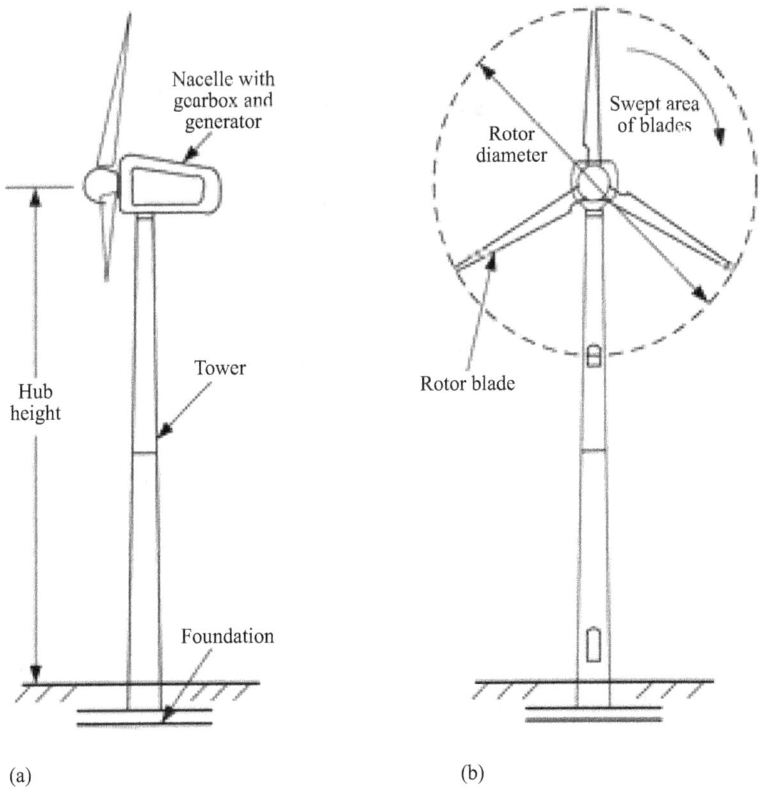

Figure 2.4 HAWT: a schematic diagram. (a) Side view and (b) front view

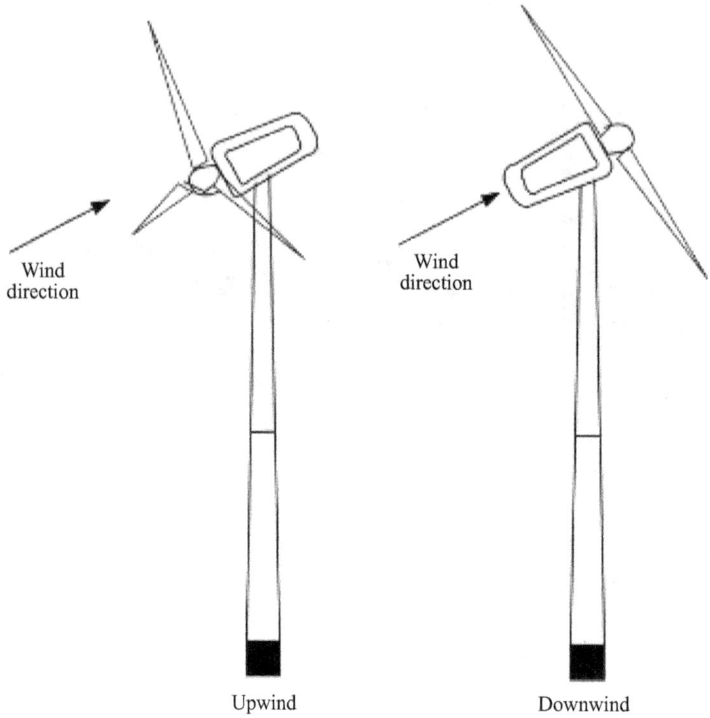

Figure 2.5 Upwind and downwind machines

motor. Most have a gearbox which turns the slow rotation of the blades into a faster rotation more suitable to drive an electrical generator. Depending on the rotor orientation, two types should be discussed: upwind machines and downwind machines (Figure 2.5).

2.1.1.1 Upwind machines

Upwind machines have the rotor facing the wind. Upwind design has a basic advantage of avoiding the wind shade behind the tower. On the other hand, the wind starts bending away from the tower before it reaches the tower itself; that is, some wind shade in front of the tower exists. Therefore, the power from the WT drops slightly each time the rotor passes the tower. Due to this, the rotor needs to be made stiff and placed a considerable distance in front of the tower to prevent the blades from being pushed into the tower by high winds. Upwind machines need a yaw mechanism to keep the rotor facing the wind.

2.1.1.2 Downwind machines

Downwind machines do not need an additional yaw mechanism for keeping the blades in line with the wind. The wind itself controls the yaw, and so, it naturally orients itself correctly with respect to wind direction. In addition, the rotor may be made more flexible; that is, the blades can be allowed to bend at high wind speeds, thus taking part of the load off the tower. Therefore, the basic advantage of the downwind machine is that it may be built somewhat lighter than an upwind machine. Downwind configuration has another advantage; it is easier to take advantage of centrifugal forces to reduce the blade root flap bending moments. This is because the blades are normally coned downwind, so centrifugal moments tend to counteract moments due to thrust. On the contrary, the basic disadvantage of downwind machine is the fluctuation in the wind power due to shadowing effects of the tower. This may give more fatigue loads on the turbine than with an upwind design.

2.1.2 Vertical axis wind turbines

The main rotor shaft of a VAWT is arranged vertically as well as the generator and gearbox can be placed down on the ground. The blades take the form as those results from holding a rope at both ends and spinning it around a vertical axis (Figure 2.6(a) and (b)). Principal advantages of this arrangement are the following: (i) there is no need for any kind of yaw control to keep blades facing into the wind to be effective; (ii) the tower itself need not be structurally as strong as that of a HAWT; and (iii) WT components can easily be serviced and maintained. On the other hand, several disadvantages of VAWT configuration also exist such as the following: (i) the blades are relatively close to the ground, where wind speeds are lower, so less wind energy is available for a given size of turbine; (ii) wind near the surface of the ground is more turbulent, so the shear stresses are more severe and the rotor experiences higher loads; (iii) because of

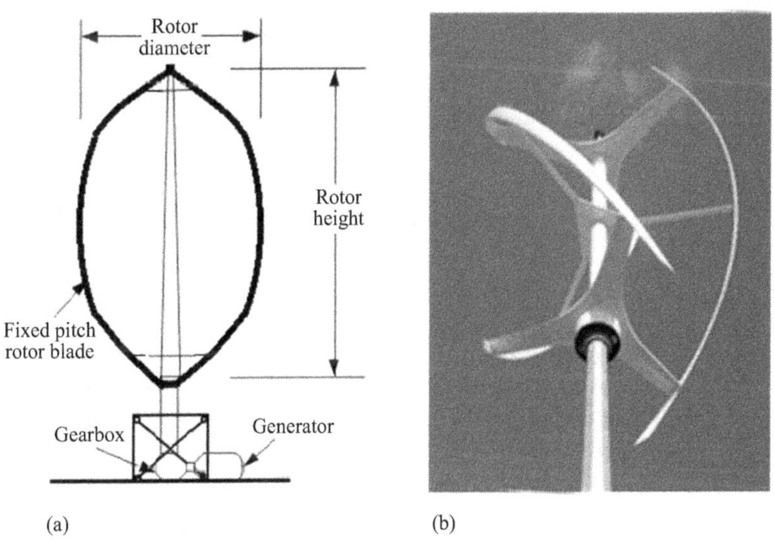

(a) (b)

Figure 2.6 HAWT: (a) schematic diagram and (b) physical picture

turbulent flow, some problems can be introduced associated with vibrations such as noise and bearing wear, so more frequent maintenance is required and otherwise the service life may be shortened; and (iv) the blades on a VAWT cannot be made to spill the high wind to protect the generator as easily as pitch-controlled blades on a HAWT.

It is seen that the WT can be designed as a HAWT or a VAWT. For a HAWT which is used in common, it will be either upwind or downwind. In all designs, the rotor with blades is the WT component that is facing into the wind to convert the wind power into rotational motion. So, the design decision relating to the number of blades should be taken. Basically, multi-bladed design presents a large swept area, and accordingly, it enables both low speed and high torque operation. For instance, in case of a rotor with one blade the turbulence negatively affects the blade efficiency but with fewer blades the turbine can spin faster before the turbulence becomes excessive and consequently the generator can be physically smaller in size. In the USA, WTs mostly have two blades, whereas most European turbines have three blades. Three blades WT presents smoother operation and lower noise, but on the other hand, the weight and cost are considerably increased.

2.2 Basic relations

2.2.1 Power from wind

With a mass of air, m, moving at speed, v, its kinetic energy, K_e, is given by

$$K_e = \frac{1}{2} m v^2 \tag{2.1}$$

Thus, the power, P, produced by this mass through an area, A, can be considered as the energy per unit time, that is,

$$P = \frac{1}{2} \frac{m}{T} v^2 = \frac{1}{2} \dot{m} v^2 \tag{2.2}$$

where T is the time.

The flow rate of the mass, m, through the area, A, is defined as the product of air density, ρ, speed, v, and cross-sectional area, A, that is,

$$\frac{dm}{dt} \triangleq \dot{m} = \rho A v \tag{2.3}$$

Incorporating (2.2) and (2.3) and considering the wind as a packet of air with mass, m, the wind power in watts, P_w, can be given by

$$P_w = \frac{1}{2} \rho A v^3 \tag{2.4}$$

where ρ is the air density (kg/m^3), v is the wind speed normal to A (m/s) and A is the cross-sectional area (m^2) through which the wind passes. It is to be noted that $\rho = 1.225$ kg/m^3 at 15 °C and 1 atm.

Hence, the wind power per unit area which is called 'power intensity' or 'specific power' P_{spec}, can be obtained in W/m^2 as

$$P_{spec} = \frac{P_w}{A} = ½\rho v^3 \tag{2.5}$$

which can be plotted as shown in Figure 2.7.

It is found from (2.4) that the wind speed is proportional to both the cube of wind speed and the swept area of turbine rotor. On the other hand, wind energy, E_w, obtained in time, T, can be expressed as

$$E_w = P_w T = ½\rho A v^3 T \tag{2.6}$$

Hence, decreasing the wind speed requires an increase in time taken to obtain the same wind energy. The swept area of turbine rotor, $A = \pi D^2/4$, where D is the blade diameter (m), and so (2.4) can be rewritten as

$$P_w = \pi \rho D^2 v^3/8 \tag{2.7}$$

Thus, the wind power is proportional to the square of blade diameter as well. This means that depending on increasing the blade diameter to increase the wind power is more cost-effective as the turbine cost may be considered as a quantity proportional to the blade diameter. More discussion of wind power dependency on

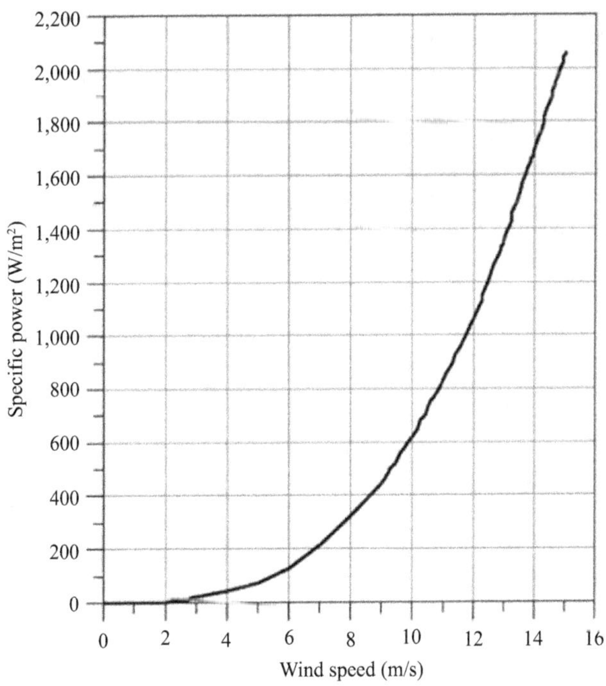

Figure 2.7 Variation of specific power (or power intensity) versus wind speed

those different variables is given in Example 2.1, in addition to the calculation of average wind energy.

Example 2.1 From (2.4), it is seen that doubling the wind speed increases the power by a factor of 8, whereas from (2.7), doubling the blade diameter increases the power to four times its value.

With respect to the wind energy, it is the product of wind power and the time as given by (2.6). So, for a turbine with constant swept area and assuming $v = 10$ m/s, the wind energy $E = ½\rho A(10)^3 T$. If v is increased to 15 m/s, the corresponding energy will be $E = ½\rho A(15)^3 T$; that is, the energy E is increased by a factor of $(1.5)^3$. So, the time T can be decreased by a factor of $(1/1.5)^3$ to obtain the same wind energy. Similarly, when v decreases and becomes 5 m/s (i.e. half its value), time T is increased by a factor of 8 which equals $(1/0.5)^3$.

Considering a period of 50 h, wind velocity is 10 m/s during the first 20 h, and then increases to 15 m/s for the rest of the time. Obviously, the wind energy is the sum of that obtained in 20 and 30 h. Assuming $A = 1$ m^2, then

$$
\begin{aligned}
E &= ½(1.225)(1)(10)^3(20) + ½(1.225)(1)(15)^3(30) \\
&= 12{,}250.0 + 62{,}015.62 \\
&= 74{,}265.62 \text{ Wh}
\end{aligned}
$$

If the average wind speed is proposed to be used to calculate the average wind energy for the total time of 50 h, wind energy is $E_{av} = ½(1.225)(1)(12.5)^3(50) = 59{,}814.45$ Wh, which is a wrong answer. Another proposed approach is to use the average cubic value of speeds, that is, $(v^3)_{av} = ½[(10)^3 + (15)^3] = 2{,}187.5$ and the corresponding energy is $E_{av} = ½(1.225)(1)(2{,}187.5)(50) = 66{,}992.18$ Wh which is again a wrong answer.

The best way to solve this problem is to calculate the average cubic speed as a summation of each cubic speed multiplied by its fraction of time, that is,

$$
\begin{aligned}
(v^3)_{av} &= \sum_i \left[v_i^3(\text{fraction of hours at } v_i)\right] \\
&= \sum_i \left[v_i^3(\text{probability } (v = v_i)\right]
\end{aligned}
\tag{2.8}
$$

Thus,

$$
(v^3)_{av} = \left(\frac{20}{50}\right) \times (10)^3 + \left(\frac{30}{50}\right) \times (15)^3 = 400 + 2{,}025 = 2{,}425
$$

and the corresponding $E_{av} = ½(1.225)(1)(2{,}425)(50) = 74{,}265.625$ Wh, which is an accurate result. Another form of (2.8) can be written as

$$
(v^3)_{av} = \frac{\sum_i \left[v_i^3 \cdot (hours\ at\ v_i)\right]}{\sum hours} \tag{2.9}
$$

It is to be noted that the term '*fraction of hours at* v_i' in (2.8) represents the probability of the wind having $v = v_i$. However, more details about statistics and probability application are given in Section 2.2.3.

For a VAWT, the swept area may be approximated as two-thirds the area of a rectangle with height equal to the rotor height (H) and width equal to the rotor diameter (D). Thus,

$$A = \frac{2}{3}DH \tag{2.10}$$

2.2.2 *Factors affecting wind power*

As seen in (2.4), the wind power is half the product of air density, swept area and cube of wind speed. Therefore, it is of interest to discuss some physical factors affecting the determination of air density as well as those having an impact on wind speed [1].

2.2.2.1 Factors affecting air density

Factors affecting air density can be specified by examining the ideal gas law that is

$$PV = nRT \tag{2.11}$$

where

- $P \triangleq$ the absolute pressure (atm), 1 atm = 101.325 kPa;
- $V \triangleq$ the volume (m^3);
- $n \triangleq$ the mass (mol);
- $R \triangleq$ the ideal gas constant = 8.2056×10^{-5} (m^3atm)/(kmol);
- $T \triangleq$ the absolute temperature (K) and K = °C + 273.15.

Assuming M_w stands for molecular weight of the gas (g/mol), it may equal 28.97 g/mol and air density can be expressed as

$$\rho = \frac{nM_w}{V} \times 10^{-3}\ \text{kg/m}^3 \tag{2.12}$$

Substitute V from (2.11) into (2.12) to obtain

$$\rho = \frac{PM_w}{RT} \times 10^{-3}\ \text{kg/m}^3 \tag{2.13}$$

Thus, both the pressure, P, and temperature, T, have an impact on determining the air density, ρ.

Pressure impact

As the air pressure is a function of altitude and the WTs may be located at sites above sea level, it is useful to determine the pressure at different heights in terms of air density.

Considering a static column of air with cross section A and taking a slice of air with thickness ds and density ρ, the difference of pressure between the top and bottom of the slice is given by

$$dp = -g\rho ds$$

Hence,

$$\frac{dp}{ds} = -g\rho \tag{2.14}$$

where g is the gravitational acceleration and equals 9.806 m/s^2.

Substituting ρ as in (2.13) into (2.14) gives

$$\frac{dp}{ds} = -\frac{gM_w}{RT} \times p \times 10^{-3} \tag{2.15}$$

Assuming T is constant throughout the air column, an acceptable assumption as typically the rate of temperature drop is about 6.5 °C/km, then, substituting the constants and unit's conversion factors at $T = 15$ °C, (2.15) can be rewritten as

$$\frac{dp}{ds} = -1.185 \times p \times 10^{-4} \tag{2.16}$$

and the solution can take the form as follows:

$$p = p_o e^{-1.185 \times 10^{-4} H} \tag{2.17}$$

where P_o is the pressure of one atm, and H is the height in meter (m).

Therefore, using (2.13) air density will be

$$\rho = \frac{p_o M_w}{RT} e^{-1.185 \times 10^{-4} H} \times 10^{-3}\ \mathrm{kg/m^3} \tag{2.18}$$

Temperature impact

As shown in (2.13) or (2.18), air density is inversely proportional to the temperature; that is, when the temperature increases, the air density decreases. The standard value of air density is 1.225 kg/m^3 at 15 °C.

2.2.2.2 Factors affecting wind speed

WTs exploit the wind energy close to the ground. So, the wind stream is turbulent and air movement experiences friction because of ground roughness. Not only due to that but also due to irregularities such as forests and buildings, wind speed is considerably slowed. On the other hand, it is very useful to increase wind speed as the wind power is proportional to the cube of wind speed; that is, it is significantly increased with an incremental increase in wind speed. Therefore, the way to get the WT facing into the wind with high speed is only to mount the turbine on a taller tower.

Impact of earth's surface roughness on wind speed is commonly characterized by an approximated expression to get wind speed variation with elevation as follows:

$$\text{In Europe}: \left(\frac{v}{v_o}\right) = \frac{\ln(H/z)}{\ln(H_o/z)} \tag{2.19}$$

where

$v \triangleq$ wind speed;

$v_o \triangleq$ wind speed at height H_o;

$H \triangleq$ rotor height;

$H_o \triangleq$ reference height (it is often a height of 10 m);

$z \triangleq$ roughness length (it equals 2×10^{-4} for water surface, 0.03 for open areas, 0.1 for farm land, 0.4 for urban districts and 1.6 for dense urban or forest) or in the United States:

$$\left(\frac{v}{v_o}\right) = \left(\frac{H}{H_o}\right)^{\alpha} \tag{2.20}$$

where $\alpha \triangleq$ the friction coefficient and roughly equals 1/7 for open terrain (it ranges from 0.1 for smooth ground or calm water to 0.4 for big city with tall buildings).

From (2.4), the wind power, P_w, at height H, referred to reference wind power, P_{wo}, at height H_o can be obtained as

$$\begin{aligned}\left(\frac{P_w}{P_{wo}}\right) &= \left(\frac{v}{v_o}\right)^3 \\ &= \left(\frac{\ln(H/z)}{\ln(H_o/z)}\right)^3 \quad \text{European form} \\ &= \left(\frac{H}{H_o}\right)^{3\alpha} \quad \text{US form}\end{aligned} \tag{2.21}$$

Example 2.2

(a) Find the air density at 1 atm and 35 °C.

(b) Find the air density at 20 °C and an elevation of a height of 2,500 m. What is its value at temperature 7 °C? Also find the power density.

(c) A wind speed of 7 m/s is indicated by an anemometer mounted at height of 10 m. Assuming a temperature of 15 °C, pressure of 1 atm and friction coefficient of 0.2, find the wind speed and specific power at a height of 40 m.

Solution

(a) Using (2.13), air density, ρ, is

$$\rho = \frac{pM_W \times 10^{-3}}{RT} = \frac{1 \times 28.97 \times 10^{-3}}{8.2056 \times 10^{-5}(273.15 + 35)} = 1.146\ \mathrm{kg/m^3}$$

It shows that air density decreases with increase in the temperature.

(b) Using (2.17), the pressure at 2,500 m height is

$$p = p_o e^{-1.185 \times 10^{-4} H} = 1 \times e^{-1.185 \times 10^{-4} \times 2{,}500} = 0.744\ \mathrm{atm}$$

It shows that the pressure decreases with increase in height.

$$\rho = \frac{pM_W \times 10^{-3}}{RT} = \frac{0.744 \times 28.97 \times 10^{-3}}{8.2056 \times 10^{-5} \times 293.15} = 0.896\ \mathrm{kg/m^3}$$

It shows the combined effect of increasing both temperature and height. If the temperature is decreased to 7 °C, air density will be

$$\rho = \frac{pM_W \times 10^{-3}}{RT} = \frac{0.744 \times 28.97 \times 10^{-3}}{8.2056 \times 10^{-5} \times 280.15} = 0.937\ \mathrm{kg/m^3}$$

The corresponding power density is:

$$\frac{P}{A} = \frac{1}{2}\rho v^3 = \frac{1}{2} \times 0.937 \times 7^3 = 170.47\ \mathrm{W/m^2}$$

(c) Using (2.20) as it is common in the United States, the velocity at 40 m height is

$$v = v_o \left(\frac{H}{H_o}\right)^{\alpha} = 7\left(\frac{40}{10}\right)^{0.2} = 9.1\ \mathrm{m/s}$$

and

$$\text{Specific power} = \tfrac{1}{2} \times 1.225 \times 9.1^3 = 461.56\ \mathrm{W/m^2}$$

The specific power at 7 m/s is ½ × 1.225 × 7^3 = 210.09 W/m^2.

So, increasing the height from 10 to 40 m yields an increase of more than the double in the specific power.

If (2.19) is used as it is common in Europe and assuming the roughness length, z, equals 0.4, the velocity and specific power are as follows:

$$v = v_o \frac{\ln(H/z)}{\ln(H_o/z)} = 7\frac{\ln(40/0.4)}{\ln(10/0.4)} = 10\ \mathrm{m/s}$$

$$\text{Specific power} = \tfrac{1}{2} \times 1.225 \times 10^3 = 612.5\ \mathrm{W/m^2}$$

2.2.2.3 Rotor efficiency

It is very important to evaluate the wind power intensity (W/m^2) to access the economic feasibility when selecting a site location. The effective power extracted from wind can be derived in terms of the airflow speed just as it arrives at the turbine, v_1, and the velocity just departing it, v_2. Assuming the wind is passing through swept area, A, of rotor blades at average speed, $v_{av} = ½(v_1 + v_2)$, then using (2.3) gives

$$\dot{m} = \rho A v_{av} = ½\rho A(v_1 + v_2) \tag{2.22}$$

The net wind mechanical power of the turbine is imposed by the difference in the kinetic energy in the wind speed just reaching and just leaving the turbine. Incorporating (2.2) and (2.22) obtains

$$P_w = \frac{1}{4}\rho A(v_1 + v_2)\left(v_1^2 - v_2^2\right) = \frac{1}{4}\rho A v_1^3\left(1 + \frac{v_2}{v_1}\right)\left(1 - \frac{v_2^2}{v_1^2}\right) \tag{2.23}$$

or

$$P_w = ½C_P\rho A v_1^3 \tag{2.24}$$

where

$$C_P \triangleq \text{power coefficient} = ½\left(1 + \frac{v_2}{v_1}\right)\left(1 - \frac{v_2^2}{v_1^2}\right) \tag{2.25}$$

Energy conversion from one form to another, e.g. from kinetic energy in the wind to mechanical power, mostly has some constraints. The analysis necessitates understanding what happens to the wind as it passes through a WT. A portion of wind kinetic energy approaching the WT at speed, v_1, is extracted by the turbine, and consequently, the wind leaving the turbine has a lower velocity, v_2, and reduced pressure. This causes the air to expand downwind of the turbine. One may think that the maximum power extracted from the wind can be achieved when the total kinetic energy of the wind is already extracted and converted into mechanical power, that is, v_2 is zero. This is not true in reality as the air would come to a complete stop behind the turbine and prevent any more of the wind to pass through the rotor. In addition, it does not make sense for v_1 equals v_2; that is, downwind velocity and upwind speed are the same and no energy is extracted by the turbine. The question now is, what is the downwind velocity for maximum power extracted by the turbine? The answer can be obtained by determining the value of C_P, in (2.25), that satisfies the condition of maximum power as in the following.

Assuming that $\lambda = v_2/v_1$, then (2. 23) and (2. 25) can be rewritten as

$$P_w = \frac{1}{4}\rho A v_1^3(1 + \lambda)\left(1 - \lambda^2\right) \tag{2.26}$$

$$C_P = ½(1 + \lambda)\left(1 - \lambda^2\right) \tag{2.27}$$

Comparison of (2.26) with (2.24) shows that the fraction of the wind's power extracted by the blades (in other words, rotor efficiency) is C_P as given in (2.27). C_P versus λ is depicted in Figure 2.8. It shows that the rotor efficiency reaches a maximum when downwind speed is one-third of its wind speed.

Differentiating (2.27) with respect to λ and equating to zero results in $\lambda = 1/3$ and the corresponding rotor efficiency (it is called 'Betz limit') is 59.26%. In practice, rotor efficiency is not as high as 59%; a typical efficiency is between 35% and 45%.

2.2.2.4 Tip-speed ratio

Tip-speed ratio (TSR) is defined as the ratio of rotor tip speed, v_{rotor}, to the wind speed, v:

$$\text{TSR} = \frac{v_{\text{rotor}}}{v} = \frac{\text{rpm} \times \pi D}{60v} \tag{2.28}$$

where

- rpm is the rotor speed (revolutions per minute);
- D is the rotor diameter (m);
- v is the wind speed (m/s).

It is usual to determine rotor efficiency as a function of TSR that depends on the rotor tip speed. If the rotor turns too slowly, its efficiency drops off as much

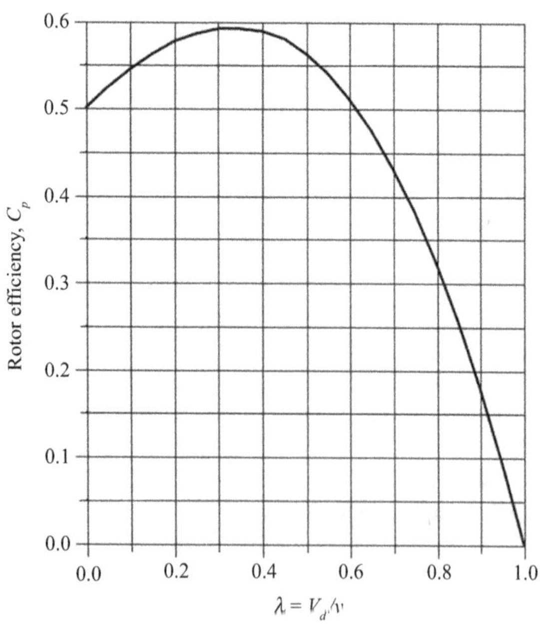

Figure 2.8 Rotor efficiency versus ratio of downwind speed to upwind speed

upwind does not affect the blades. If the rotor turns too fast, its efficiency is reduced as well because of the turbulence. Variation of rotor efficiency versus TSR for turbines with different number of blades is shown in Figure 2.9 [2].

Example 2.3 Two-blade WT, 40-m rotor diameter, produces 400 kW at a wind speed of 12 m/s. Using the standard air density value of 1.225 kg/m^3 at 15 °C and 1 atm, find the following:

(a) the rotor rpm assuming that it operates with a TSR = 5.5;
(b) the rotor tip speed;
(c) the gear ratio needed to rotate the generator at 2,000 rpm;
(d) the overall efficiency of the WT system.

Solution

(a) Using (2.28),

$$\text{rpm} = \frac{TSR \times 600v}{\pi D} = \frac{5.5 \times 60 \times 12}{\pi \times 40} = 31.5 \text{ rpm}$$

(it means about 2 s per revolution).

(b) Tip speed = (31.5 × π × 40)/60 = 65.97 m/s.

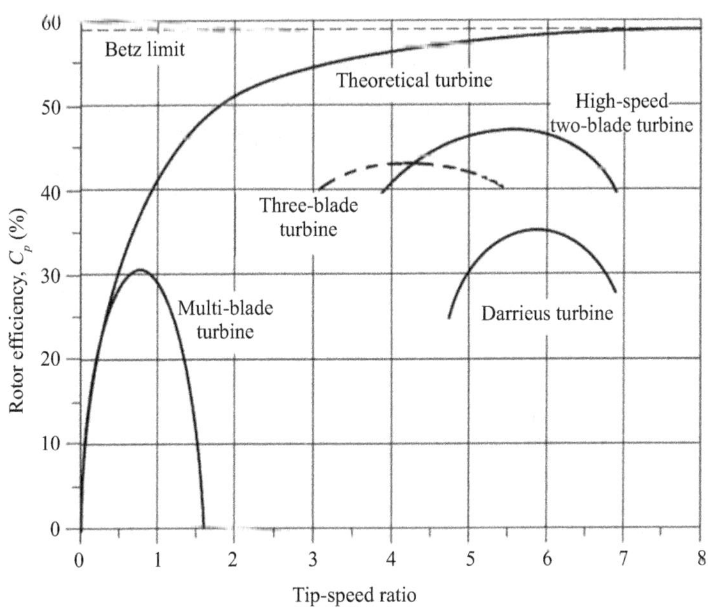

Figure 2.9 Rotor efficiency versus TSR

(c) Gear ratio $= \frac{\text{Generator speed}}{\text{Rotor speed}} = \frac{2{,}000}{31.5} = 63.5.$
(d) Using (2.4), the wind power is

$$P_w = ½\rho A v^3 = P_w = ½ \times 1.225 \times \left(\frac{\pi \times 40^2}{4}\right)(12)^3 = 1{,}330\ \text{kW}$$

Thus, the overall efficiency, η_{ovall}, is

$$\eta_{\text{ovall}} = 400/1{,}330 = 30\%$$

2.2.3 *Average wind power*

As seen in (2.8), the average of cube of wind speed can be calculated by summing the product of each wind speed and its probability. In next two subsections, additional probability characteristics for the wind may be considered to compute the average value of the cube of wind speed.

2.2.3.1 Probability density functions

Probability density function (*pdf*) is a continuous function that presents the discrete wind histogram. Its variation against wind speed variation is shown in Figure 2.10. It is to be noted that the area under the curve is equal to unity, whereas the area under the curve between two wind speeds, v_1 and v_2, equals the probability of the wind that has a speed within those two speeds. Relations next are used to, mathematically, express these features:

$$\left.\begin{aligned}\text{Probability of wind speed} &= \int_0^{\infty} f(v)dv = 1 \quad \text{for } 0 \le v \le \infty \\ &= \int_{v_1}^{v_2} f(v)dv \quad \text{for } v_1 \le v \le v_2\end{aligned}\right\} \tag{2.29}$$

where $f(v)$ is the *pdf* of wind speed.

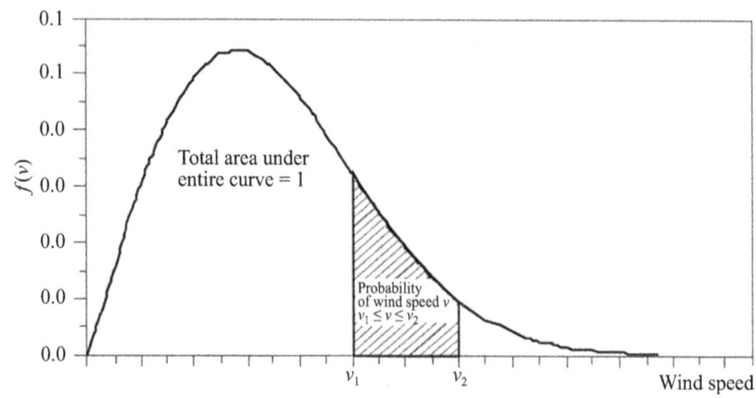

Figure 2.10 Probability density function versus wind speed

The number of hours in a period, T, at which the wind speed is within v_1 and v_2 can be computed by the product of its probability and the time, T, in hours, because $f(v)$ expresses the fraction of time the wind speed is between v_1 and v_2. Thus,

$$\text{Number of hours}/T\ (v_1 \le v \le v_2) = T\int_{v_1}^{v_2} f(v)dv \tag{2.30}$$

For instance, if $T = 8{,}760$ h, (2.30) gives the number of hours per year, the wind speed is between v_1 and v_2.

The average value of the cube of the wind speed using *pdf* can be found in the same manner as that applied for the discrete approach in (2.8) and is given by

$$\left(v^3\right)_{\text{av}} = \int_0^{\infty} v^3 f(v)dv \tag{2.31}$$

2.2.3.2 Probability and statistics

To increase confidence in the data available, a large sample of wind should be taken over many years as the wind movement is a random phenomenon. This is always difficult and shorter periods are often used. Therefore, the data are averaged over calendar months and can be described by what is called 'Weibull *pdf*' for characterizing the statistics of wind speed. This function is given as

$$f(v) = \frac{k}{c}\left(\frac{v}{c}\right)^{k-1} \mathrm{e}^{-(v/c)^k} \quad \text{for } 0 \le v \le \infty\ \%\ \text{of hours/year/(m/s)} \tag{2.32}$$

where

- k is defined as the shape parameter and is practically presented for most sites within the range of 1.5 to 2.5;
- c is known as the scaling factor and pertains to the number of days with high wind speed.

The shape of the curve representing (2.32) varies with the value of the parameter k as plotted in Figure 2.11 assuming a constant scaling factor, c.

At most sites around the world, $k = 2$, and substituting this value into (2.32) gives what is called 'Rayleigh *pdf*' as

$$f(v) = \frac{2}{c}\left(\frac{v}{c}\right)\mathrm{e}^{-(v/c)^2} \tag{2.33}$$

Rayleigh *pdf* can be plotted at various values of the scaling parameter, c, against wind speed variation as shown in Figure 2.12. It is noted that as c increases, the curve shifts towards higher wind speed.

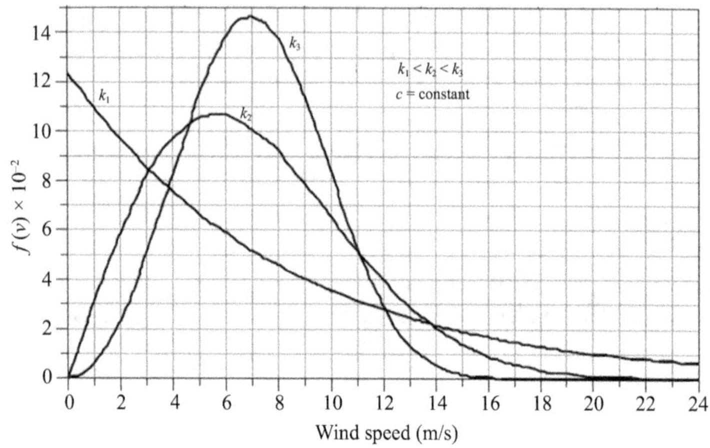

Figure 2.11 Weibull pdf at various values of k and constant c

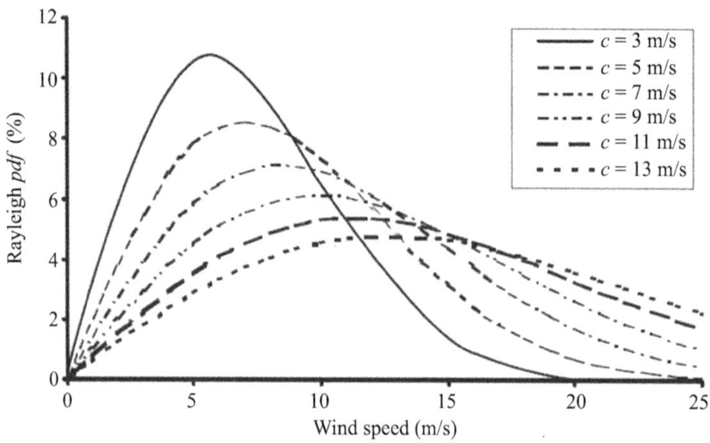

Figure 2.12 Rayleigh pdf at various values of scaling factor, c

Substituting $f(v)$ from (2.33) into (2.31), the average value of the cube of the wind speed is

$$\left(v^3\right)_{\mathrm{av}} = \int_0^\infty v^3 \frac{2v}{c^2} \mathrm{e}^{-(v/c)^2} dv = \frac{3\sqrt{\pi}}{4} c^3 \tag{2.34}$$

Consequently, the average wind power can be calculated using

$$P_{\mathrm{av}} = \frac{1}{2} \rho A \left(v^3\right)_{\mathrm{av}} = \frac{3\sqrt{\pi}}{8} \rho A c^3 \tag{2.35}$$

Example 2.4 Deduce the average wind speed using Rayleigh *pdf* and compare the result with that obtained for the value of the cube of the wind speed. Examine the impact on the average wind power.

Solution Like the equation of $(v^3)_{av}$ in (2.31), the average speed can be written as

$$v_{av} = \int_0^\infty v f(v) dv \tag{2.36}$$

Substituting Rayleigh *pdf*, $f(v)$ as in (2.33) into (2.36) gives

$$v_{av} = \int_0^\infty 2\left(\frac{v}{c}\right)^2 e^{-(v/c)^2} dv = \frac{\sqrt{\pi}}{2} c \tag{2.37}$$

Hence,

$$c = \frac{2}{\sqrt{\pi}} v_{av} \tag{2.38}$$

Combining (2.34) and (2.38) gives

$$\left(v^3\right)_{av} = \frac{6}{\pi}\left(v_{av}\right)^3 \tag{2.39}$$

Therefore, the average wind power in terms of the average of the cube of the wind speed, $(v^3)_{av}$, is $(6/\pi)$ times that in terms of average speed to power 3, that is, $(v_{av})^3$.

For further explanation, it was found in Example 2.1 that using either the arithmetic average speed $((v_1 + \ldots + v_n)/n)$ or the arithmetic average of the cube of the speed $((v_1^3 + \ldots + v_n^3)/n)$ yields a wrong result and must be corrected by multiplying each wind speed to the power of 3 by its probability.

This example emphasizes that even in case of using *pdf*, the cube of wind speed must take the form of $(v^3)_{av}$, not $(v_{av})^3$, where the ratio between them is $(6/\pi)$.

2.3 Wind characteristics

Basic relations of wind power estimation have been explained in Section 2.2. It can be, simply, seen that the rotor blades extract energy, in the form of kinetic energy, from the wind to rotate the rotor, then a gearbox is used to adapt the rotor speed at the nominal rotating speed of the generator shaft and finally the generator delivers the electrical power (Figure 2.13). In addition, the characteristics of a WT are introduced in brief in the next subsections to specify the performance of the WT.

2.3.1 Airfoils

Airfoils are aerodynamic surfaces, for example, air plane wings, propeller blades and wind rotor blades. Based on the shape of their cross section, they must generate

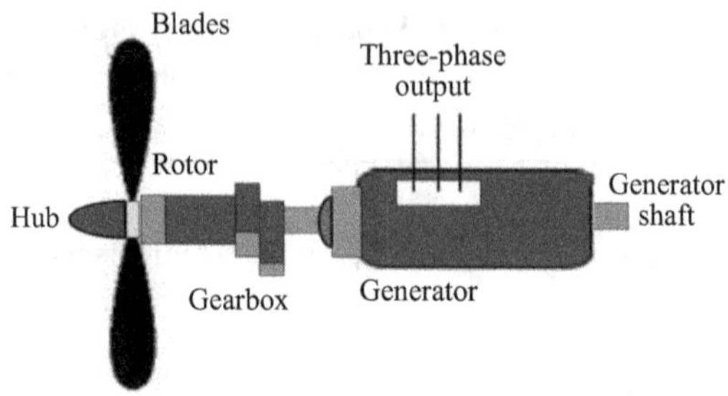

Figure 2.13 Blades–rotor–gearbox–generator set

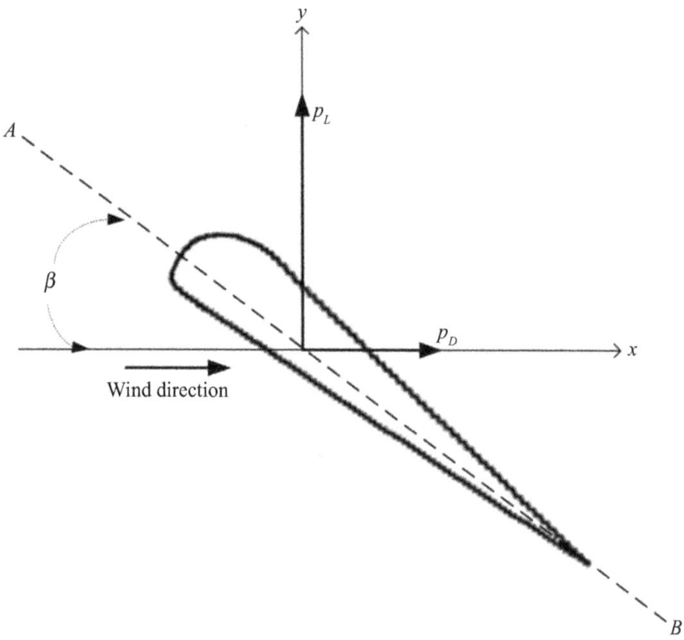

Figure 2.14 Drag and lift pressures on airfoil

a big lift with a corresponding small drag. The lifting force causes the turbine blades to rotate. It is created as the pressure on the top of airfoil is less than that under it. This difference of pressure is due to the fact that the air moving on the top travels a longer distance before rejoining again the air passing immediately under the airfoil [3].

Figure 2.14 depicts a cross section through an airfoil describing the forces on a rotating turbine blade. An arbitrary reference plane is chosen and represented in x–y frame by asymmetric axis A–B, where the region above the plane differs from that under it. The air flows relative to the airfoil along x-axis, and so, a force is acting on the foil. This force has two components: (i) drag component parallel to the wind velocity, v, along the x-axis produced by pressure p_D and (ii) lift component perpendicular to the velocity produced by pressure p_L. The angle between wind direction and the reference line is defined as the 'angle of attack' β.

Pressures p_L and p_D are determined experimentally under specified conditions for each shape of the airfoil. It is found that the pressures are related to the dynamic pressure, $\frac{1}{2}\rho v^2$, by proportionality constants: lift coefficient C_L and drag coefficient C_D, respectively.

$$\left.\begin{aligned} p_L &= \tfrac{1}{2}\rho v^2 C_L \\ p_D &= \tfrac{1}{2}\rho v^2 C_D \end{aligned}\right\} \tag{2.40}$$

Here, C_L and C_D are mainly functions of the angle of attack. A typical plot of both C_L and C_D versus the angle of attack is shown in Figure 2.15. Increasing the angle of attack improves lift component associated with an increase in drag component as well. On the other hand, if the lift increase is too much, a phenomenon called 'stall' takes place. When a blade stalls, the airflow over the top does not stick to the surface and the resulting turbulence destroys the lift.

Based on the effect of lift and drag forces, two designs of turbines are mostly used to generate electricity from wind: drag turbines and lifting turbines. The operation of drag turbine is based on the friction caused by wind on turbine blades. The wind creates a force in the direction of its blowing. The turbine blade on which the wind impinges cannot move faster than the wind itself.

Lifting turbines operate through the lifting effect produced by wind where the wind generates a force normal to the direction of its blowing. The turbine blade can move faster than the wind itself.

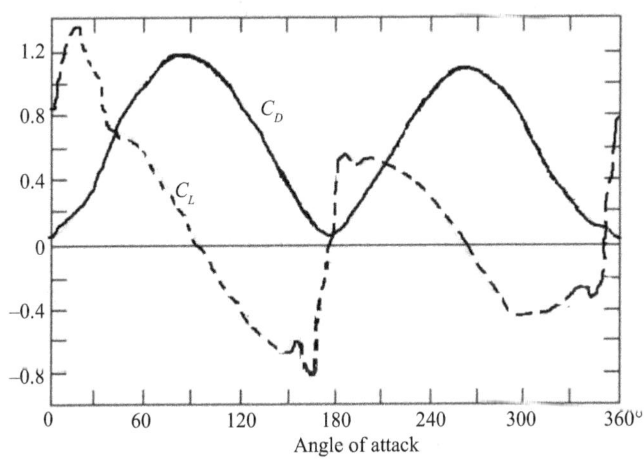

Figure 2.15 Typical lift and drag coefficient versus angle of attack

Power curve describes the relationship between wind speed and generator electrical power output. It is of crucial importance to technically evaluate the performance of a specific WT [4].

Power curve, in general form, encompasses four bands of wind speed: (i) cut-in, (ii) normal operation, (iii) high and (iv) cut-out wind speeds (Figure 2.16). The first band represents the cut-in wind speed band. It ranges from zero to what is called 'cut-in speed', which is the minimum speed of generation, $V_{C(in)}$. Below this speed, wind may not have enough power to supply both friction losses and generator field windings. At most, the power generated just supplies the friction losses. Therefore, no power is generated at wind speed below $V_{C(in)}$.

The second band ranges from $V_{C(in)}$ to the rated speed, V_r. Starting from $V_{C(in)}$ and increasing the wind speed, the power delivered by the generator tends to rise as it is proportional to the cube of wind speed. Reaching the rated wind speed V_r, the generator delivers its rated power in correspondence to its design. So, this band is the normal operation maintained by controlling the blade position with respect to the direction of wind attack.

The third band is the high wind speed. When the speed exceeds the rated value, V_r, high-speed controllers can be actuated to maintain a maximum constant output power limited by generator capacity. Pitch-control system to reduce the blade's angle of attack to shed some of wind or using stall control to increase the angle of attack to create stall can be applied until a certain speed limit is reached, V_{limit}. The fourth band, above the speed limit, corresponds to the wind having excessive speed

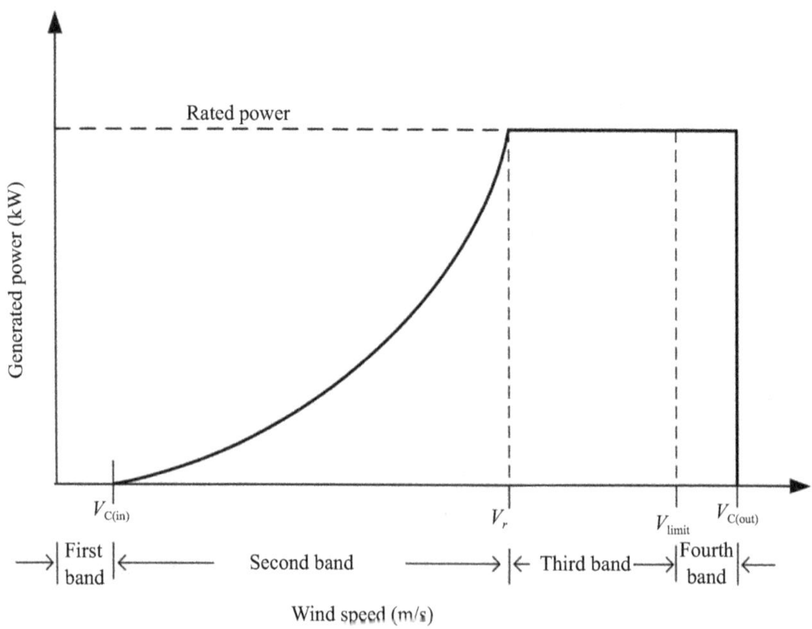

Figure 2.16 WT power curve

that is too strong and may cause mechanical damage to the generator, as well as there is a real danger to the WT. At this speed, called 'cut-out wind speed', $V_{C(out)}$ the machine must be shut down and no power is generated.

2.3.2 Capacity factor

Capacity factor, CF, is used to estimate the actual energy delivered by the WT as a percentage of the rated energy. It can be calculated as

$$CF = \frac{\text{Actual energy delivered}}{P_r \times 8{,}760} = \frac{\text{Average power}}{\text{Rated power}} \tag{2.41}$$

where P_r is the rated power of the WT generator in kW.

It is noted that CF is a convenient measure of WT capability for producing electrical power over a year or certain number of hours.

Example 2.5 A WT has 1,500 kW rated power, 64-m rotor diameter, cut-in wind speed of 5 m/s, rated wind speed of 16 m/s and cut-out wind speed of 27 m/s. Using Rayliegh statistics ($k = 2$) and considering the scaling factor c equals 12, find the following:

(a) The number of hours per year at which (i) wind speed is less than or equal to cut-in speed and (ii) the WT should be shut down when the wind speed is excessive.
(b) The energy generated per year at the rated power.

Solution

(a) As given by (2.37), the average wind speed is

$$v_{av} = \frac{\sqrt{\pi}}{2} c = 10.6 \text{ m/s}$$

From (2.29) and (2.32), the cumulative distribution function for Weibull statistics of wind speed equals to or is less than a specific value, V, and is the integration of the function from zero to that speed, that is,

$$f(0 \le v \le V) = \int_0^V \frac{k}{c}\left(\frac{v}{c}\right)^{k-1} e^{-(v/c)^k} dv \text{ for } 0 \le v \le V \tag{2.42}$$

Thus, it can be rewritten as

$$f(v \le V) = \int_0^x e^{-x} dx \tag{2.43}$$

where

$$x = \left(\frac{v}{c}\right)^k, \text{then}, \frac{dx}{dv} = \frac{k}{c}\left(\frac{v}{c}\right)^{k-1}$$

Hence,

$$f(v \le V) = 1 - e^{-(v/c)^k} \tag{2.44}$$

In case of Rayleigh statistics, (2.44) is of the form

$$f(v \le V) = 1 - e^{-(v/c)^2} \tag{2.45}$$

Substitute c from (2.37) into (2.45) to get

$$f(v \le V) = 1 - e^{-(\pi/4)(v/v_{av})^2} \tag{2.46}$$

Therefore, probability function of wind speed greater than V is the integration of the function from V to infinity, or in other words, it equals one minus the probability of wind speed less than V as the summation of the two probabilities represented by the total area under the curve of the function against wind speed is unity. This means that

$$f(v \ge V) = 1 - [1 - e^{-(\pi/4)(v/v_{av})^2}] = e^{-(\pi/4)(v/v_{av})^2} \tag{2.47}$$

Apply (2.46) to get the probability that the wind speed is $\le V_{C(in)}$ as

$$f\left(v \le V_{C(in)}\right) = 1 - e^{-(\pi/4)(v/v_{av})^2} = 1 - e^{-(\pi/4)(5/10.6)^2} = 0.16$$

The number of hours the wind$\left(v \le V_{C(in)}\right) = 8{,}760 \times 0.16 = 1{,}401.6\ \text{h/year}$.

Similarly, the hours when wind speed is greater than $V_{C(out)}$ can be obtained using (2.47).

$$\text{Hours}\left(v \ge V_{C(out)}\right) = 8{,}760e^{-(\pi/4)(v/v_{av})^2} = 8,\ 760e^{-(\pi/4)(27/10.6)^2}$$
$$= 53.6\ \text{h/year}$$

(b) As is clear from the WT power curve, the rated power is delivered when the wind speed is between rated speed, V_r, and cut-out speed, $V_{C(out)}$. Probability ($V_r \le v \le V_{C(out)}$) is the difference between probability ($v \ge V_r$) and probability ($v \ge V_{C(out)}$), that is,

$$\begin{aligned} f\left(V_r \le v \le V_{C(out)}\right) &= f(v \ge V_r) - f\left(v \ge V_{C(out)}\right) \\ &= e^{-(\pi/4)(16/10.6)^2} - e^{-(\pi/4)(27/10.6)^2} \\ &= 0.167 - 0.006 = 0.161 \end{aligned}$$

Thus, hours $\left(V_r \le v \le V_{C(out)}\right) = 8{,}760\ \times 0.161 = 1{,}410.36\ \text{h/year}$

Consequently,

The rated energy delivered by the wind turbine

$$= \text{rated power} \times \text{no. of hours}$$
$$= 1{,}500 \times 1{,}410.36 = 2.11554 \times 10^6 \text{ kWh/year}$$

2.4 Electrical generators in wind systems

WTs can be equipped with any type of rotating generators such as DC or AC generators. DC generators require maintenance and are relatively bulky and inefficient. AC generators are mostly used in wind systems and can be either asynchronous (induction) or synchronous type. Asynchronous generator may be squirrel cage induction generator (SCIG) or doubly fed induction generator (DFIG), whereas permanent magnet synchronous generator (PMSG) is commonly used as a synchronous generator.

Based on classifying the WTs as constant speed and variable speed turbines, Figure 2.17 shows the schematic diagrams of the general configuration of the three most commonly used generator–WT systems [4,5]. SCIG can be used for constant speed WTs as SCIG speed varies only little with wind speed because of the generator slip. The generator and WT rotor are coupled through a gearbox as the optimal rotor and generator speed ranges are different. Then, a transformer may be used to directly connect the generator terminals with the utility grid and a shunt capacitor bank may be used to feed the generator with the required reactive power, particularly, in stand-alone operation (Figure 2.17(a)). If the wind speed increases, automatically the angle of attack increases. At a certain wind speed, above the rated value, the angle of attack will reach the value where stall occurs. Therefore, stall control is used to keep the power produced by the turbine near the rated power.

For variable speed WT, either DFIG for limited variable speed or direct-drive PMSG for full variable speed WT may be used. In case of using DFIG, a gearbox is used. The stator is directly connected to utility grid and the rotor is connected to a converter (Figure 2.17(b)). Speed variation ranges from about 60% to 110% of the rated speed using variable speed capability to keep the TSR at the value resulting in optimal energy. If the wind speed exceeds the rated value, a pitch control is applied to reduce the power to be around its rated value.

When using PMSG, no gearbox is necessary as the generator rotates at very low speed. The total power generated flows to the utility grid through a converter to fix the varying generator frequency at the grid frequency (Figure 2.17(c)). At high wind speed, above the rated speed, the power increase over the rated value is reduced by pitching the blades. Accordingly, stall control and pitch control are mainly used with constant speed and variable speed turbines, respectively.

2.4.1 Asynchronous induction generators

Induction generators (IGs) have outstanding operation as they have very robust construction, mechanical simplicity and lowest cost, among generators. In addition,

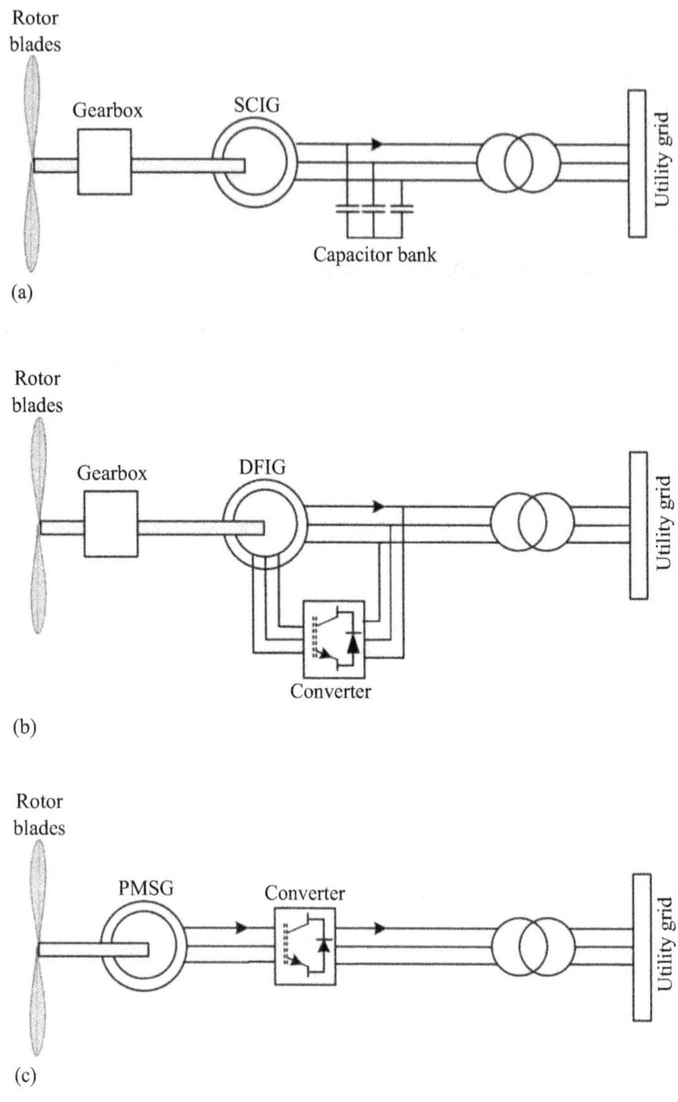

Figure 2.17 Classic WT–generator systems. (a) Constant wind speed turbine with SCIG. (b) Variable wind speed turbine with DFIG. (c) Variable wind speed turbine with PMSG

IGs provide natural protection against short circuits and damping any current surge by the magnetization path of the iron core without fear of demagnetization. On the contrary, a major disadvantage is that the IG needs a reasonable amount of reactive power as it does not contain permanent magnets (PMs) or other means to provide magnetic excitation. Therefore, reactive power must be fed externally to establish the magnetic field necessary to convert the input mechanical power into output electrical power. External sources may be the grid in interconnected applications. In stand-alone applications, reactive power may be supplied by the load itself, by a capacitor bank connected across generator terminals, or by power electronic systems.

2.4.1.1 Squirrel cage induction generators

The stator of this type of generator has a three-phase stator winding and the rotor consists of a number of copper or aluminium bars shorted together at their ends forming a cage. Applying a three-phase voltage to the stator winding, three-phase currents will flow in the stator windings and produce a rotating magnetic field. Its speed, called synchronous speed, is determined by the number of poles in the windings and the frequency of the current. When the rotor rotates at a speed other than the synchronous speed, voltage is induced in the rotor cage winding caused by the relative speed (called slip) between the rotor and the rotating stator magnetic field. This causes a current to flow in the rotor winding. Interaction of the associated magnetic field of the rotor with the stator field results in a torque acting on the rotor [2].

Relations and equivalent electric circuit at steady state

For the induction machine to operate in generator mode, the rotor must spin faster than the synchronous speed; that is, the operational speed is advanced with respect to the stator magnetic field rotation. In quantitative terms, let n_s and f_s be the synchronous mechanical speed (rpm) and synchronous frequency (Hz), respectively. The induced voltage, resulting in output power, is proportional to the relative speed difference between the electrical synchronous rotation and the mechanical rotation with a speed-slip factor, s, defined as

$$s = \frac{n_s - n_r}{n_s} \tag{2.48}$$

where n_r is the rotor speed (rpm).

Denoting f_s and f_r as the electrical stator frequency and rotor frequency, respectively, the interrelation between them can be expressed as

$$f_r = \frac{p}{120}(n_s - n_r) = sf_s \tag{2.49}$$

where $f_s = (p/120)n_s$ and p is the number of poles.

It is to be noted that in case of no relative speed difference, that is, $n_s = n_r$, the rotor frequency is DC as $s = 0$. For other speeds, s is a negative value if $n_s < n_r$, and the voltage induced on the rotor, E_r, is directly proportional to s and transferred to the stator windings. The stator and rotor are analogous to the primary and secondary of a transformer with the difference that the secondary is a rotating element. So, in the following relations, the symbols of stator and rotor parameters referred to the stator are subscripted by #1 and #2, respectively.

The voltage induced on the rotor, E_r, is established for any speed with respect to the blocked rotor voltage, E_{ro}, as

$$E_2 = E_r = sE_{ro} \tag{2.50}$$

and the stator voltage, E_1, is

$$E_1 = \gamma \frac{E_r}{s} \tag{2.51}$$

where γ is the turns ratio.

Similarly, the rotor current, I_2, is obtained by

$$I_2 = \frac{1}{\gamma} I_r \tag{2.52}$$

and

$$Z_r = \frac{E_r}{I_r} = \frac{sE_{ro}}{I_r} = R_r + jsX_{ro} \tag{2.53}$$

where X_{ro} is the 'blocked rotor' reactance and R_r is the rotor winding resistance.

From (2.51), the rotor impedance can be calculated at any rotor speed from the blocked rotor test as

$$Z_{ro} = \frac{E_{ro}}{I_r} = \frac{R_r}{s} + jX_{ro} = \frac{Z_r}{s} \tag{2.54}$$

Then, using (2.51) through (2.54), the rotor impedance referred to the primary can be obtained by

$$Z_2 = \frac{E_1}{I_2} = \gamma^2 \left(\frac{R_r}{s} + jX_{ro} \right) \tag{2.55}$$

and in terms of stator values, it can be rewritten as

$$Z_2 = \frac{R_2}{s} + jX_2 \tag{2.56}$$

Consequently, the rotor current, I_2, is

$$I_2 = \frac{E_1}{Z_2} = \frac{E_1 s}{R_2 + jX_2 s} \tag{2.57}$$

Accordingly, the equivalent circuit of the IG can be drawn as in Figure 2.18. It is valid only for perfect sinusoidal excitation and balanced conditions. The parameters R_1 and X_1 are the resistance and leakage reactance of the stator; R_2 and X_2 are the resistance and leakage reactance of the rotor; and R_m and X_m are the

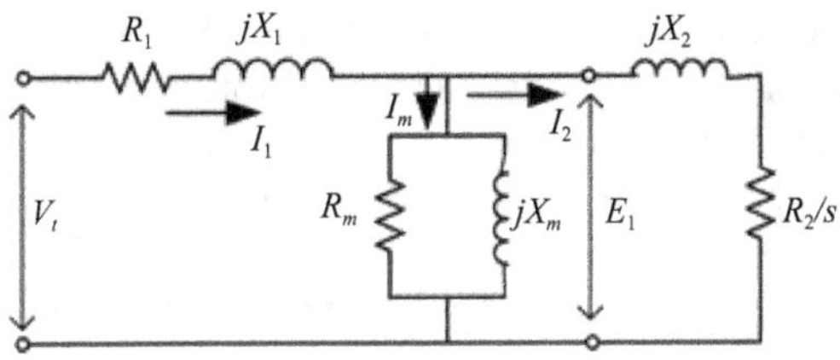

Figure 2.18 Equivalent circuit of the IG

equivalent core loss resistance and magnetizing reactance, respectively. The equivalent per-phase impedance, Z, seen at the terminals is

$$Z = Z_1 + \frac{Z_2 Z_m}{Z_2 + Z_m} \tag{2.58}$$

where

$$Z_1 = R_1 + jX_1$$

$$Z_2 = \frac{R_2}{s} + jX_2$$

$$Z_m = \frac{1}{\frac{1}{R_m} + \frac{1}{jX_m}}$$

The rotor equivalent resistance, R_2/s, depends on the rotor speed, which is a major difference with respect to a transformer equivalent circuit as the rotor voltage is subjected to a variable frequency causing E_r, R_r and X_r to be variable. However, the equivalent circuit shown in Figure 2.18 is usually considered for analysis.

Using (2.57), the apparent rotor power can be formulated by

$$S = E_1 I_2 = \frac{E_1^2 s}{R_2 + jX_2 s} \tag{2.59}$$

Therefore, it is seen that the rotor power factor depends on the slip, s, and other parameters. In respect of voltage, the quadrature leading current component is almost constant for all terminal voltages. So, the reactive power absorbed by the equivalent circuit must be supplied by an external source such as utility grid, capacitor banks or an electronic compensator.

Equivalent circuit in transient state

Machine model is necessary for examining the dynamic and transient behaviour of the machine, to develop machine control and define the machine variables in transient state as well as in steady state. Therefore, the dynamic model is used to provide information to the planners to design enabling controllers for transition from one state to another. Consequently, the dynamic model is generally represented by differential equations.

A reasonable representation of an induction machine can be obtained by considering the effects of mechanical transients and rotor electrical transients with a single time constant. The effect of stator electrical transients on system response usually can be neglected. The transient behaviour of an induction machine is represented by the equivalent circuit shown in Figure 2.19. The rate of change of the voltage behind transient reactance, X', can be described by the following differential equation:

$$\frac{dE'}{dt} = -j2\pi sE' - \frac{1}{T_o}[E' - j(X - X')I_t] \tag{2.60}$$

and the terminal current is given by

$$I_t = (V_t - E')\frac{1}{R_1 + jX'} \tag{2.61}$$

where

$T_o = \frac{X_2 + X_m}{2\pi f R_r}$ and is defined as the rotor open-circuit time constant in seconds.

X and X' are calculated from the steady-state equivalent circuit by neglecting rotor resistance. The values are

$$X = X_1 + X_m \text{ and } X' = X_1 + \frac{X_m X_2}{X_m + X_2}$$

It is to be noted that the fixed speed turbine with SCIG directly connected to the grid has a number of drawbacks. The reactive power and, consequently, the grid voltage level are uncontrollable. The power varies due to blade rotation causing voltage variations in the grid. Sensitivity to flicker is high at the resonance frequency that the generator may have. Most of these drawbacks can be avoided by using variable speed WTs with a doubly fed induction machine (DFIM) that can operate in both generating and motoring mode. Obviously, in the normal operation of a WT, it operates as a generator to deliver electric energy from the wind to the grid.

2.4.1.2 Doubly fed induction machines

DFIM is a wound rotor induction machine (WRIM). The rotor has three-phase windings connected to an external variable voltage and frequency source through

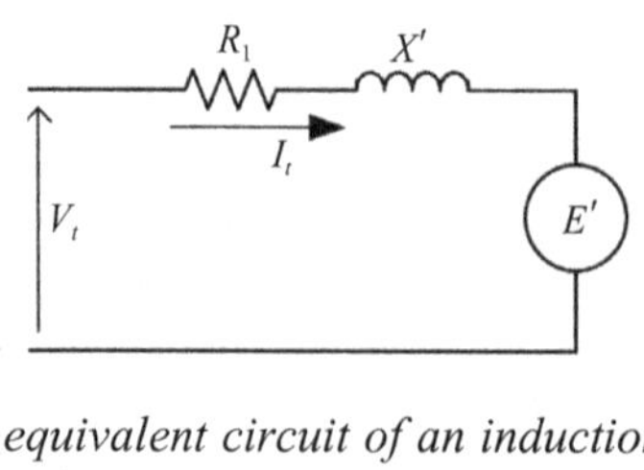

Figure 2.19 Simplified equivalent circuit of an induction machine for transient analysis

slip rings using a power electronic converter. It is also possible to alter the rotor reactance by modulating series inductors. The stator is directly connected to the utility grid or an external AC voltage source to be supplied at constant frequency and constant three-phase voltage amplitude [6].

Magnetic fields created in the air gap by both stator and rotor windings will rotate at the same speed but with a phase shift, as a function of the torque. The machine torque, active power, reactive power through the stator and rotor can be controlled by adjusting the voltage (magnitude, phase and frequency) introduced in the rotor.

Commonly, two level converters (rotor side converter 'RSC' and grid side converter 'GSC') are used. RSC generates a three-phase voltage with variable amplitude and frequency to control the machine torque as well as the reactive power exchanged between the stator and the grid. GSC exchanges with the grid the active power extracted or injected by RSC from the rotor. A schematic diagram showing this configuration is depicted in Figure 2.20. It is noted that a transformer is used to adapt the generated voltage to the point of common coupling (PCC).

Steady-state model

To ideally represent the DFIM by a simplified per-phase equivalent circuit, some assumptions are considered: the stator windings are supplied by an external voltage source or the grid at constant and balanced three-phase voltage amplitude and frequency. Similarly, the rotor is supplied independently, by a source at constant and balanced three-phase voltage such as power electronic converter. It is to be noted that the frequency of the voltage and current of the stator differs from those of the rotor. The stator frequency is fixed as the stator is directly connected to the external source, whereas the frequency of rotor voltages and currents depends on the machine speed. So, the rotor has controllable parameters such as current, voltage, frequency and phase shift with respect to the stator voltage angle. Three modes of operation for the DFIM, depending on the sign of the slip, are possible; sub-synchronous operation at positive slip, synchronous operation at zero slip and hyper-synchronous operation at negative slip.

The per-phase equivalent circuit representing the DFIM is drawn in Figure 2.21, where the rotor parameters, current and voltages are necessarily to be

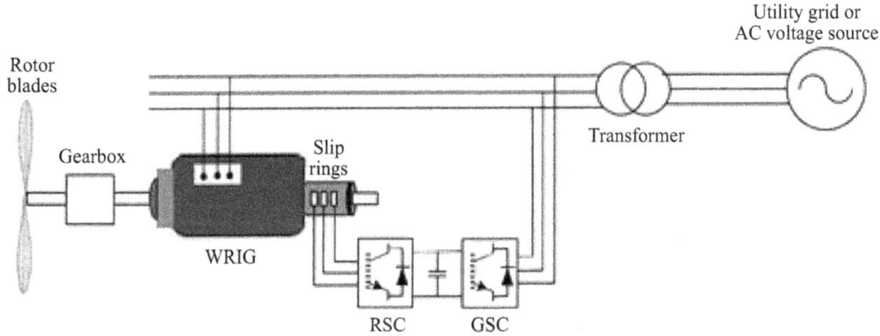

Figure 2.20 Configuration of DFIM connected to grid

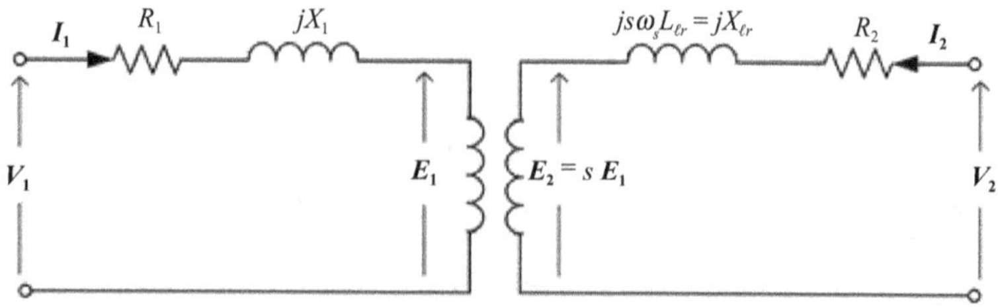

Figure 2.21 DFIM representation with different stator and rotor frequencies (rotor current and voltage are referred to the stator)

referred to the stator. The parameters used in this representation are defined as in the following. Vectors are written in bold fonts.

- $\boldsymbol{V_s} \triangleq$ the applied voltage to the stator $= V_s \angle 0°$;
- $\boldsymbol{V_r} \triangleq$ the applied voltage to the rotor $= V_r \angle \varphi°$;
- $\boldsymbol{I_s}, \boldsymbol{I_r} \triangleq$ stator and rotor current, respectively;
- $\boldsymbol{E_s}, \boldsymbol{E_r} \triangleq$ induced emf in the stator and rotor, respectively;
- $R_s, R_r \triangleq$ stator and rotor resistance, respectively;
- $N_s, N_r \triangleq$ number of turns per phase of stator and rotor windings, respectively;
- $\gamma \triangleq$ turns ratio $\triangleq N_s/N_r$;
- $L_\ell s, L_{\ell r} \triangleq$ stator and rotor leakage inductance, respectively;
- $X_\ell s, X_{\ell r} \triangleq$ stator and rotor leakage impedance, respectively.

From this definition, the stator parameters are renamed as

$$\boldsymbol{V}_1 = \boldsymbol{V}_s = V_1 \angle 0°, \boldsymbol{E}_1 = \boldsymbol{E}_s, I_1 = I_s,\ R_1 = R_s, X_1 = X_{\ell s} = \omega_s L_{\ell s}$$

and consequently, the rotor parameters, current and voltages, referred to the stator are

$$\boldsymbol{V}_2 \triangleq \text{applied rotor voltage} = \gamma \boldsymbol{V}_r = V_2 \angle \varphi°$$

$$\boldsymbol{I}_2 \triangleq \text{rotor current} = \frac{1}{\gamma}\boldsymbol{I}_r, \boldsymbol{E}_2 \equiv \text{induced voltage in the rotor} = \gamma \boldsymbol{E}_r = s\boldsymbol{E}_1$$

$$R_2 \triangleq \text{rotor resistance} = \gamma^2 R_r$$

$$X_{\ell r} \triangleq \text{rotor reactance at rotor frequency} = s\omega_s L_{\ell r}$$

As shown in Figure 2.21, the following relations can be derived.

From the stator side:

$$V_1 - E_1 = I_1(R_1 + jX_1) \text{ at stator frequency, } f_s \tag{2.62}$$

From the rotor side:

$$V_2 - sE_1 = I_2(R_2 + jX_{\ell r}) \text{ at rotor frequency, } f_r \tag{2.63}$$

$$\frac{V_2}{s} - E_1 = I_2\left(\frac{R_2}{s} + jX_2\right) \text{ at stator frequency, } f_s \tag{2.64}$$

where

$$X_2 = \frac{X_{\ell r}}{s} = j\omega_s L_{\ell r}$$

Incorporating (2.62) and (2.64) gives

$$V_1 - \frac{V_2}{s} - I_1(R_1 + jX_1) + I_2\left(\frac{R_2}{s} + jX_2\right) = 0 \tag{2.65}$$

According to (2.65), the per-phase equivalent circuit of the DFIM can be derived as shown in Figure 2.22. The induced emf in the stator can be expressed as

$$E_1 = Z_m(I_1 + I_2) = Z_m I_m \tag{2.66}$$

where Z_m is the equivalent impedance of core loss resistance, R_m, and magnetizing reactance of the machine, that is,

$$Z_m = \frac{1}{1/R_m + 1/jX_m} \tag{2.67}$$

Steady-state phasor diagram

The phasor diagram can be deduced from the equivalent circuit shown in Figure 2.22 by neglecting the mechanical power losses and using machine equations for specific operating conditions. The variables are expressed by their rms values. Currents and fluxes of both stator and rotor are calculated by

$$\left.\begin{aligned} \Psi_1 &= L_m(I_1 + I_2) + L_1 I_1 = L_s I_1 + L_m I_2 \\ \Psi_2 &= L_m(I_1 + I_2) + L_2 I_2 = L_m I_1 + L_r I_2 \end{aligned}\right\} \tag{2.68}$$

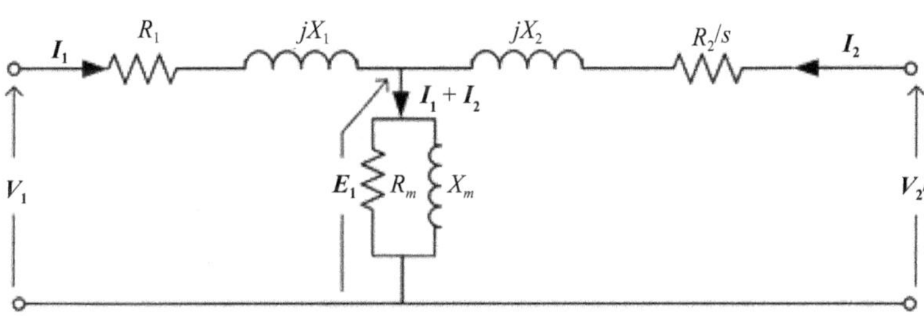

Figure 2.22 Per-phase equivalent electric circuit of the DFIM

where L_s and L_m are the stator and rotor inductances given by

$$L_s = L_1 + L_m \quad \text{and} \quad L_r = L_2 + L_m \tag{2.69}$$

Rewrite V_1 and V_2 in (2.62) and (2.64) in terms of their corresponding flux as

$$\begin{aligned} V_1 - R_1 I_1 &= j\omega_s \Psi_1 \\ V_2 - R_2 I_2 &= js\omega_s \Psi_2 \end{aligned} \tag{2.70}$$

It is easy to start the deduction of the phasor diagram for a DFIM operating in motor mode at sub-synchronous speed and, from there, derive the phasor diagram for the machine acting as a generator. In addition, the stator reactive power, Q_1, is assumed to be zero (the voltage, $\boldsymbol{V_1}$, and the current, $\boldsymbol{I_1}$, of the stator are in-phase). Therefore, from (2.68) to (2.70) the phasor diagram can be drawn and is shown in Figure 2.23 for negative slip. The stator variables are drawn in solid lines, whereas for rotor are in dotted lines.

As shown in Figure 2.23(a), it is to be noted that $\boldsymbol{\Psi_1}$ is determined by (2.70), $\boldsymbol{I_2}$ and $\boldsymbol{\Psi_2}$ can be derived from (2.68), and then the rotor voltage, $\boldsymbol{V_2}$, is obtained by (2.70). Finally, the phasor diagram indicating the phase shift angles between the parameters of both stator and rotor is shown in Figure 2.23(b). The angles obtained in this case show that (i) $\boldsymbol{\Psi_1}$ leads $\boldsymbol{\Psi_2}$; (ii) $\alpha_v \approx 90°$, $\alpha_i = \alpha_v$ as $Q_1 = 0$ and consequently $P_1 > 0$, i.e. providing power to the stator; (iii) β_v is very close to 90° as the difference is just due to $R_2 I_2$; (iv) $\beta_i + \beta_v > 90°$ and $\boldsymbol{V_2}$ is leading to $\boldsymbol{I_2}$, i.e. $P_2 < 0$ (the rotor is absorbing power) and $Q_2 > 0$ (the machine is magnetized through the rotor); and (v) $\boldsymbol{V_2}$ is leading to $\boldsymbol{\Psi_2}$ at sub-synchronous speed (alternatively, lagging at hyper-synchronous speed).

Therefore, if the machine works in generating mode as DFIG, the phasor diagram may be one of two cases: (i) generating power, $P_1 < 0$, with lagging power factor ($Q_1 > 0$), Figure 2.24(a); (ii) generating power, $P_1 < 0$, with leading power factor ($Q_1 < 0$), Figure 2.24(b). It is noted that there is no change in the relative position between $\boldsymbol{\Psi_1}$ and $\boldsymbol{\Psi_2}$, whereas the change of magnitude exists.

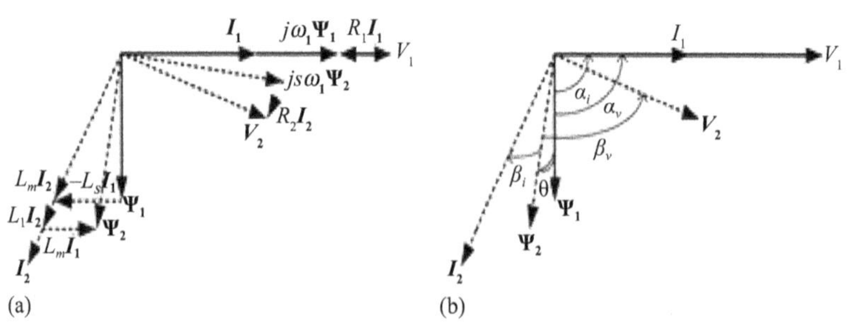

Figure 2.23 Phasor diagram of a DFIM at negative slip and zero stator reactive power (a) deduced from model equations and (b) indicating phase shifts between flux and both voltage and current for stator and rotor

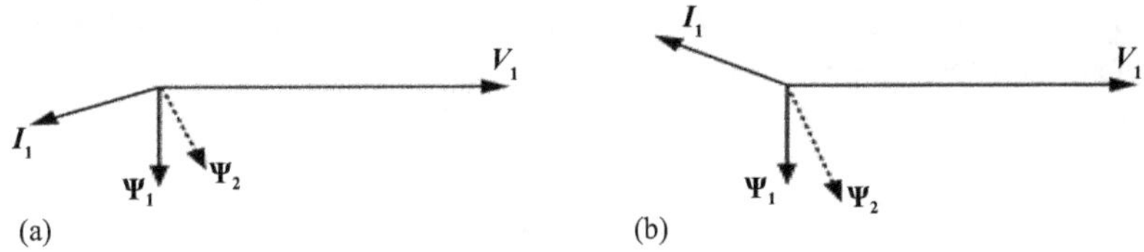

Figure 2.24 DFIG phaser diagrams: (a) $Q_1 > 0$; (b) $Q_1 < 0$

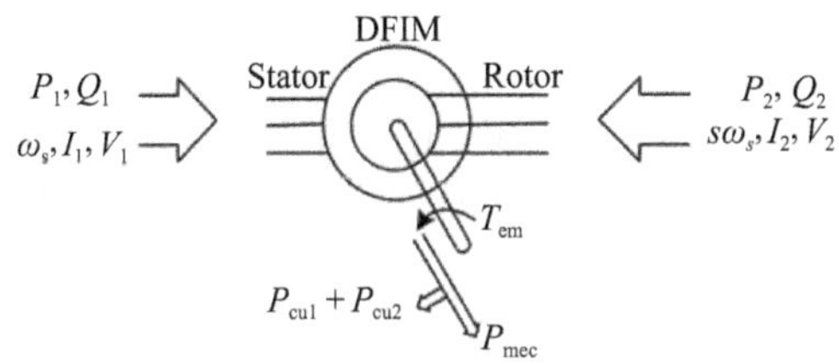

Figure 2.25 Inputs/outputs of the DFIM with motor convention

Power and torque relations

The positive sign of power flow into the machine corresponds to what is called 'motor convention' as shown in Figure 2.25.

Active power calculations. It can be seen that

$$P_1 + P_2 = P_{\text{mec}} + P_{\text{cu1}} + P_{\text{cu2}} \tag{2.71}$$

where

$P_1 \triangleq$ the three-phase active power through the stator; positive when receiving and negative for delivering. It is given by

$$P_1 = 3\mathcal{R}e\left(\boldsymbol{V}_1\boldsymbol{I}_1^*\right) = 3|\boldsymbol{V}_1||\boldsymbol{I}_1|\cos\left(\alpha_v - \alpha_i\right) \tag{2.72}$$

$P_2 \triangleq$ the three-phase rotor active power; positive for receiving and negative for delivering. It is calculated by the relation

$$P_2 = 3\mathcal{R}e\left(\boldsymbol{V}_2\boldsymbol{I}_2^*\right) = 3|\boldsymbol{V}_2||\boldsymbol{I}_2|\cos\left(\beta_v - \beta_i\right) \tag{2.73}$$

$P_{\text{cu1}} \triangleq$ the three-phase stator copper losses

$$P_{\text{cu1}} = 3R_1|\boldsymbol{I}_1|^2 \tag{2.74}$$

$P_{\text{cu2}} \triangleq$ the three-phase rotor copper losses

$$P_{\text{cu2}} = 3R_2|\boldsymbol{I}_2|^2 \tag{2.75}$$

Substituting (2.72) through (2.75) into (2.71) to obtain P_{mec}, which includes core loss corresponding to R_m in Figure 2.22 as

$$\begin{aligned} P_{\mathrm{mec}} &= 3|\boldsymbol{V}_1||\boldsymbol{I}_1|\cos(\alpha_v - \alpha_i) + 3|\boldsymbol{V}_2||\boldsymbol{I}_2|\cos(\beta_v - \beta_i) \\ &\quad - 3R_1|\boldsymbol{I}_1|^2 - 3R_2|\boldsymbol{I}_2|^2 \end{aligned} \tag{2.76}$$

As in (2.75), it is to be noted that R_2, independent of the slip, is the electric element which is responsible for rotor power losses. Referring to Figure 2.22, the equivalent rotor resistance is equal to R_2/s and, then, the difference $R_2(1 - s)/s$ is dependent on the slip and responsible for generating mechanical power. Similarly, the rotor voltage, V_2/s, can be represented by two sources; V_2 and $V_2(1 - s)/s$ as shown in Figure 2.26. As the inductances are only pertained to the reactive power, the elements dependent on the slip participate in the mechanical power. Therefore, an alternative expression for calculating the three-phase mechanical power can be written as

$$P_{\mathrm{mec}} = 3\frac{(1-s)}{s}\{R_2|\boldsymbol{I}_2|^2 - \mathcal{R}e(\boldsymbol{V}_2\boldsymbol{I}_2^*)\} \tag{2.77}$$

Based on (2.72), (2.73) and (2.76), the machine efficiency, η, can be obtained by

$$\begin{aligned} \eta &= \frac{P_{\mathrm{mec}}}{P_1 + P_2} \text{ for motoring operation, } P_{\mathrm{mec}} > 0 \\ &= \frac{P_1 + P_2}{P_{\mathrm{mec}}} \text{ for generating operation, } P_{\mathrm{mec}} < 0 \end{aligned} \tag{2.78}$$

Consequently, the electro-magnetic torque in the shaft of the machine, T_{em}, can be obtained in terms of the mechanical power, neglecting mechanical power losses, as

$$T_{\mathrm{em}} = \frac{p}{2\omega_r}P_{\mathrm{mec}} \tag{2.79}$$

where ω_r is the angular frequency of the rotor (rad/s) and p is the number of poles. P_{mec} can be substituted from either (2.76) or (2.79).

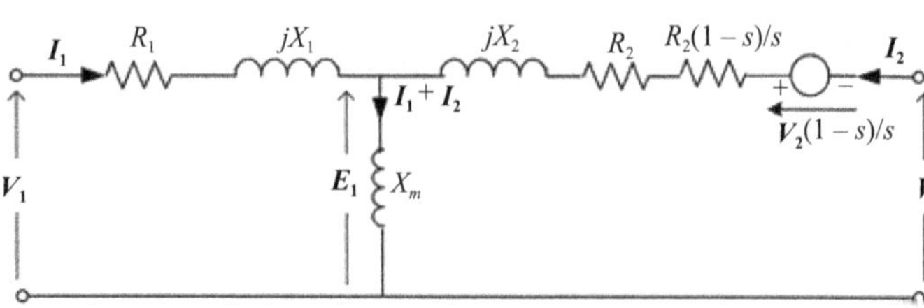

Figure 2.26 The electric equivalent circuit indicating the dependent elements on the slip

Reactive power calculations. As shown in Figure 2.22 and incorporating (2.68) and (2.70), the stator and rotor voltage can be calculated as

$$
\begin{aligned}
\boldsymbol{V}_1 &= R_1\boldsymbol{I}_1 + j\omega_s L_s \boldsymbol{I}_1 + j\omega_s L_m \boldsymbol{I}_2 \\
\boldsymbol{V}_2 &= R_2\boldsymbol{I}_2 + js\omega_s L_m \boldsymbol{I}_1 + js\omega_s L_r \boldsymbol{I}_2
\end{aligned}
\tag{2.80}
$$

Hence, the stator reactive power, Q_1, and rotor reactive power, Q_2, can be derived as follows:

$$
\begin{aligned}
Q_1 = 3\mathrm{Im}\{\boldsymbol{V}_1 \cdot \boldsymbol{I}_1^*\} &= 3\mathrm{Im}\{R_1|\boldsymbol{I}_1|^2 + j\omega_s L_s|\boldsymbol{I}_1|^2 + j\omega_s L_m \boldsymbol{I}_2\boldsymbol{I}_1^*\} \\
&= 3\omega_s L_s|\boldsymbol{I}_1|^2 + 3\omega_s L_m \mathcal{R}\mathrm{e}\{\boldsymbol{L}_2\boldsymbol{I}_1^*\} \\
Q_2 = 3\mathrm{Im}\{\boldsymbol{V}_2 \cdot \boldsymbol{I}_2^*\} &= 3\mathrm{Im}\{R_2|\boldsymbol{I}_2|^2 + js\omega_s L_m \boldsymbol{I}_1\boldsymbol{I}_2^* + js\omega_s L_r|\boldsymbol{I}_2|^2\} \\
&= 3s\omega_s L_r|\boldsymbol{I}_2|^2 + 3s\omega_s L_m \mathcal{R}\mathrm{e}\{\boldsymbol{I}_1\boldsymbol{I}_2^*\}
\end{aligned}
\tag{2.81}
$$

According to (2.72), (2.73), (2.77) and (2.81), the DFIM can operate at different modes (two generating and two motoring modes) depending on the power and speed. Of course, the DFIM in wind energy application is used as a DFIG. The power flow and the corresponding phasor diagram of each mode are depicted in Figure 2.27 with $Q_1 > 0$. Modes (a) and (b) show that the DFIG is operating at hyper-synchronous speed and sub-synchronous speed, respectively.

Commonly, the magnitude of machine parameters (voltage, current, speed, power, etc.) are normalized to be expressed as per unit values that are used for

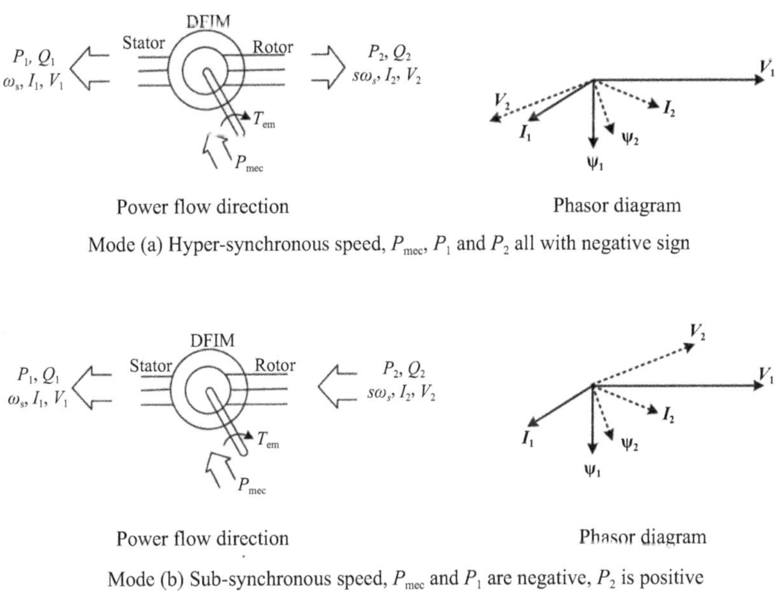

Figure 2.27 Generating modes of the DFIM with $Q_1 > 0$

numeric calculations. To do that, base quantities are chosen. For instance, if the stator rms values of voltages, currents and angular speed are chosen as basic base values, then

$$\begin{aligned}&\text{Base voltage}: V_{\text{base}} = |\boldsymbol{V}_1|_{\text{rated}}\\&\text{Base current}: I_{\text{base}} = |\boldsymbol{I}_1|_{\text{rated}}\\&\text{Base angular speed}: \omega_{\text{base}} = \omega_s\end{aligned} \tag{2.82}$$

where

- $|V_1|_{\text{rated}} \triangleq$ line to neutral rated rms stator voltage (V);
- $|I_1|_{\text{rated}} \triangleq$ rated rms stator current (A);
- $\omega_s \triangleq$ rated stator angular speed (rad/s).

From (2.82), the other base values can be deduced such as

$$\left.\begin{aligned}&\text{Base impedance}: Z_{\text{base}} = \frac{V_{\text{base}}}{I_{\text{base}}}\\&\text{Base flux}: \Psi_{\text{base}} = \frac{V_{\text{base}}}{\omega_{\text{base}}}\\&\text{Base inductance}: L_{\text{base}} = \frac{\Psi_{\text{base}}}{I_{\text{base}}}\\&\text{Base power}: S_{\text{base}} = 3V_{\text{base}}I_{\text{base}}\\&\text{Base torque}: T_{\text{base}} = \frac{p}{2}\frac{S_{\text{base}}}{\omega_{\text{base}}}\end{aligned}\right\} \tag{2.83}$$

From (2.82) and (2.83) and dividing the magnitude of each of stator parameters and rotor parameters referred to the stator by the corresponding base value, the relations (2.68), (2.70), (2.72), (2.73), (2.79) and (2.81) can be derived and are rewritten in per unit as follows:

$$\left.\begin{aligned}\boldsymbol{V}_{1\text{pu}} &= R_{1\text{pu}}\boldsymbol{I}_{1\text{pu}} + j\boldsymbol{\Psi}_{1\text{pu}}\\\boldsymbol{V}_{2\text{pu}} &= R_{2\text{pu}}\boldsymbol{I}_{2\text{pu}} + j\boldsymbol{\Psi}_{2\text{pu}}\\\Psi_{1\text{pu}} &= L_{1\text{pu}}\boldsymbol{I}_{1\text{pu}} + L_{m\text{pu}}\boldsymbol{I}_{2\text{pu}}\\\Psi_{2\text{pu}} &= L_{2\text{pu}}\boldsymbol{I}_{2\text{pu}} + L_{m\text{pu}}\boldsymbol{I}_{1\text{pu}}\\P_{1\text{pu}} &= |\boldsymbol{V}_{1\text{pu}}||\boldsymbol{I}_{1\text{pu}}|\cos(\alpha_v - \alpha_i)\\Q_{1\text{pu}} &= |\boldsymbol{V}_{1\text{pu}}||\boldsymbol{I}_{1\text{pu}}|\sin(\alpha_v - \alpha_i)\\P_{2\text{pu}} &= |\boldsymbol{V}_{2\text{pu}}||\boldsymbol{I}_{2\text{pu}}|\cos(\beta_v - \beta_i)\\Q_{2\text{pu}} &= |\boldsymbol{V}_{2\text{pu}}||\boldsymbol{I}_{2\text{pu}}|\sin(\beta_v - \beta_i)\\T_{\text{empu}} &= \text{Im}\{\Psi_{2\text{pu}}.\boldsymbol{I}^*_{2\text{pu}}\}\end{aligned}\right\} \tag{2.84}$$

Dynamic model in d-q frame

Self-inductances of the stator and rotor windings are constant because the air gap is smooth, whereas the mutual inductance between them varies with the rotor displacement relative to the stator. This variation of the stator to rotor mutual inductance makes the induction machine analysis complicated in terms of real variables as the voltage equations become non-linear. In order to eliminate this difficulty, and thus, facilitate analysis, a change of variables can be devised for stator and rotor variables. This gives a fictitious magnetically coupled two-phase machine in which the rotor circuits are not only made stationary but are also aligned with the respective stator windings. In this way, all the inductances become constant. These orthogonally placed balanced windings are known as the *d-q* windings that may be considered fixed or moving with respect to the stator [7–9].

A *d-q* frame of reference is chosen to model the DFIM. The model of the IM is based on the two-axis representation commonly known as the 'Park model'. A synchronously rotating *d-q* frame of reference is used with the direct axis *d*-axis oriented along the stator flux position. In this way, decoupled control between the electrical torque and the rotor excitation current is obtained. The frame of reference is rotating with the same speed as the stator voltage.

Referring to Figure 2.28, the transformation of stator state space variables '*abc*' to fictitious '*dqo*' variables can be obtained with the invariance of power as the necessary criterion. The equivalent two-phase windings are assumed to have $\sqrt{(3/2)}$ times as many turns per phase as the three-phase winding. The fictitious stator *dqo* variables are obtained from the stator variables *abc* through a transform defined as

$$\begin{bmatrix} f_{d1} \\ f_{q1} \\ f_{o1} \end{bmatrix} = \sqrt{\frac{2}{3}} \begin{bmatrix} \cos\theta_s & \cos\left(\theta_s - \frac{2\pi}{3}\right) & \cos\left(\theta_s + \frac{2\pi}{3}\right) \\ -\sin\theta_s & -\sin\left(\theta_s - \frac{2\pi}{3}\right) & -\sin\left(\theta_s + \frac{2\pi}{3}\right) \\ \frac{1}{\sqrt{2}} & \frac{1}{\sqrt{2}} & \frac{1}{\sqrt{2}} \end{bmatrix} \begin{bmatrix} f_{a1} \\ f_{b1} \\ f_{c1} \end{bmatrix} \tag{2.85}$$

where f is any of stator variables such as voltage, current and flux, and the components along the axis denoted by 'o' in *dqo* frame of reference can be ignored in balanced conditions (v_o and i_o will be zero). On the other hand, the same transformation is applied to the rotor quantities by replacing θ_s by $(\theta_s - \theta_r)$. Therefore, a set of equations, (2.86), results from applying this transformation to obtain the

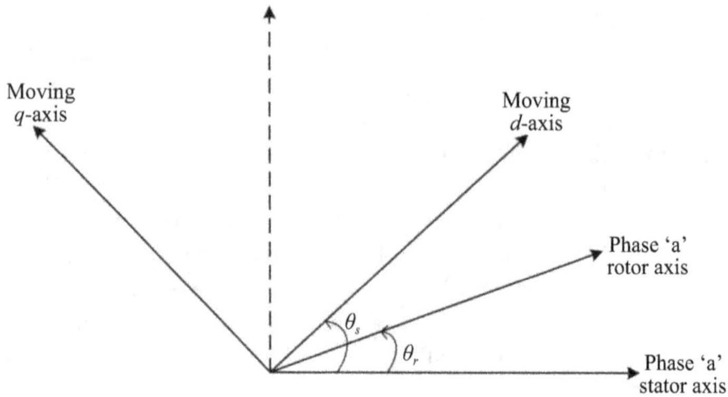

Figure 2.28 Angles of reference axes

stator and rotor voltages in the *d-q* axes with using motor convention. In addition, the corresponding electric equivalent circuit can be drawn as shown in Figure 2.29.

$$\left.\begin{aligned}
v_{d1} &= R_1 i_{d1} + \omega\varphi_{sq1} - \frac{d\varphi_{d1}}{dt} \\
v_{q1} &= R_1 i_{q1} - \omega\varphi_{sd1} - \frac{d\varphi_{q1}}{dt} \\
v_{d2} &= R_2 i_{d2} + s\omega\varphi_{sq2} - \frac{d\varphi_{d2}}{dt} \\
v_{q2} &= R_2 i_{q2} - s\omega\varphi_{sd2} - \frac{d\varphi_{q2}}{dt} \\
T_{em} &= \frac{3}{2}\left[\frac{L_m p}{2}\left(i_{q1} i_{d2} - i_{d1} i_{q2}\right)\right] \\
P_1 &= \frac{3}{2} Re\{\mathbf{v_1} \mathbf{i_1^*}\} = \frac{3}{2}\left(v_{d1} i_{d1} + v_{q1} i_{q1}\right) \\
Q_1 &= \frac{3}{2} \mathrm{Im}\{\mathbf{v_1} \mathbf{i_1^*}\} = \frac{3}{2}\left(v_{q1} i_{d1} - v_{d1} i_{q1}\right) \\
P_2 &= \frac{3}{2} Re\{\mathbf{v_2} \mathbf{i_2^*}\} = \frac{3}{2}\left(v_{d2} i_{d2} + v_{q2} i_{q2}\right) \\
Q_2 &= \frac{3}{2} \mathrm{Im}\{\mathbf{v_2} \mathbf{i_2^*}\} = \frac{3}{2}\left(v_{q2} i_{d2} - v_{d2} i_{q2}\right)
\end{aligned}\right\} \tag{2.86}$$

2.4.1.3 WT with DFIG control system

The basic configuration of a WT with DFIG connected to a three-phase grid as explained in Section 2.4.1.2 is shown in Figure 2.30. The system incorporates a power electronic interface (converter) between the generator rotor and the grid (or the load). So, the electrical power delivered by the generator to the grid can be dynamically controlled. The WT mechanical power, P_w, is converted into electric power, P_g, by the DFIG. The rotor active power, P_r, can be imparted to the grid

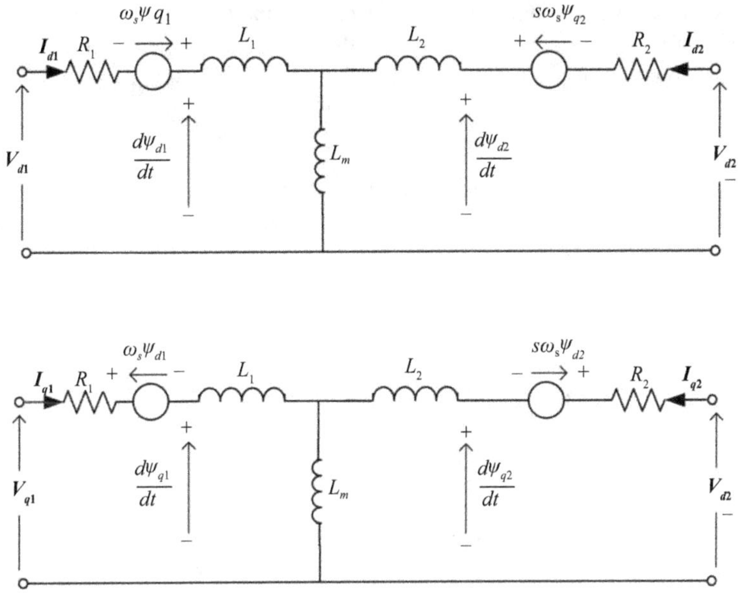

Figure 2.29 d-q equivalent circuit of the DFIM

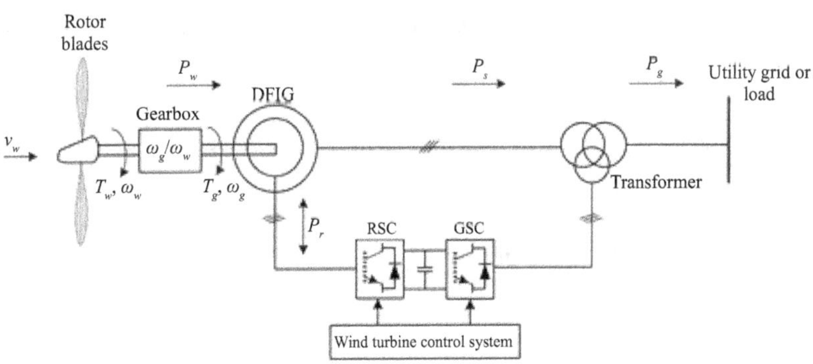

Figure 2.30 Basic configuration of WT with DFIG

(P_r < 0 at hyper-synchronous speed) or taken from the grid (P_r > 0 at sub-synchronous speed) or equals zero at stationary operation modes. Therefore, a WT control system is needed to regulate the rotor power at different wind speeds, which in turn affects the power P_g. It is to be noted that the converter can directly be connected between the stator circuit of the generator, and the grid has to be designed for the rated power of the WT, which makes it more expensive than that connected between the rotor and the grid where the converter rating is typically 25% of the total system power as well as the speed range of the generator is ±33% around the synchronous speed.

To obtain a proper controlled WT system, the mathematical model of the WT system must be formed. It usually contains the models of elements representing the basic functional components: wind and WT aerodynamic, WT drive train, DFIG, electric grid and WT control system. These elements are described briefly as follows:

- *Wind and WT:* The wind facing the swept area of WT blades is represented by its velocity, v_w, which is expressed as the summation of four components defined by

$$v_w = v_{w(\mathrm{av})} + v_{w(\mathrm{gst})} + v_{w(\mathrm{rmp})} + v_{w(\mathrm{nos})} \tag{2.87}$$

  where all components are in m/s and
  $v_{w(\mathrm{av})} \triangleq$ average wind speed;
  $v_{w(\mathrm{gst})} \triangleq$ gust wind component;
  $v_{w(\mathrm{rmp})} \triangleq$ ramp wind component;
  $v_{w(\mathrm{nos})} \triangleq$ noise wind component.

  Gust, ramp and noise components are the turbulent wind velocity components. In addition, the first two components are deterministic turbulences, whereas the noise wind component is a stochastic part to predict turbulence occurrence and correlation. The WT model can be mathematically expressed by (2.26) and (2.27) where the electric power is proportional to the cube of wind velocity.
- *WT drive train:* It can be considered as the main part of the WT dynamic structure that contributes to the interaction with the grid. WT-drive train includes the inertia of the WT, generator and gearbox connecting the two rotating shafts. When modelling the drive train, it is common practice to neglect the dynamics of mechanical parts as their responses are considerably slow in comparison to the fast-electrical dynamics on especially for machines with large inertia. So, the two-mass rotational system may be expressed by a single equation of motion given by

$$J_{WG}\frac{d\omega_g}{dt} = \frac{P_w - P_g}{\omega_g} \tag{2.88}$$

  where
  $J_{WG} \triangleq$ WT mechanical inertia + generator mechanical inertia;
  $P_w \triangleq$ mechanical power produced by turbine;
  $P_g \triangleq$ electric power delivered to the grid (or the load);
  $\omega_g \triangleq$ angular speed of the rotor.

  Integrating (2.88) yields

$$J_{WG}\int_{\omega_1}^{\omega_2} \omega d\omega = \int_{t_1}^{t_2} (P_w - P_g)dt$$

Hence,

$$\tfrac{1}{2} J_{WG}\left(\omega_2^2 - \omega_1^2\right) = \int_{t_1}^{t_2} \left(P_w - P_g\right) dt \tag{2.89}$$

It is seen from (2.89) that the instantaneous difference between the mechanical wind power and electrical power changes the rotor speed. This method of speed control has an advantage of not involving any mechanical action and is smooth in operation. On the other hand, it implies a disadvantage of continuous fluctuation of the power output to the grid due to the continuous control of the rotor speed, which is undesirable for the power system.

- *DFIG:* Its *d-q* dynamic model has been explained in Section 2.4.1.2 and described by (2.86). The active and reactive power of the DFIG can be controlled by the *q*-axis and *d*-axis components of the rotor current, respectively.
- *Electric grid:* At the PCC with the grid, the parameters such as current, voltage and power can be considered as a mirror that reflects the response of the grid to the operation of WT at normal conditions and in contingencies as well.
- *Control system:* The WTs require control systems. It is desirable to reduce the drive train loads and protect the generator and associated equipment from overloading by limiting the turbine power to the rated value up to the furling speed. At gust speeds, the machine must be stalled. At low and moderate wind speeds, the aim should be to capture power as efficiently as possible. WTs can have two types of control: one for controlling mechanical dynamics such as pitch angle control, stall control and yaw control, and the other for controlling electrical dynamics using power electronics to control generator speed and power. Electrical dynamics are much faster than mechanical dynamics, and therefore, it is possible to control the machine in a cascade structure.

However, the WT has four different speed bands for wind speed as described in Section 2.3.2. So, it requires different speed strategies. A typical example of using power electronics to control the active and reactive power of WTs is known as machine vector control as schematically depicted in Figure 2.30. It shows the power generation scheme using two bidirectional converters (RSC and GSC) and a DC link capacitor. The capacitor is charged to a small value initially and the control is initiated by injecting a small value of current in both the *d*- and *q*-axes of the rotor current. While the *d*-axis current establishes the stator flux, the *q*-axis current has a direct control of the DC link capacitor voltage. When the capacitor voltage reaches a predetermined reference value, both converters are controlled in such a way that the net power flowing to the capacitor is zero. At this point, the load switch along with a filter is turned ON. At steady state, the system operates exactly as a conventional slip power recovery drive except that the combined reactive power demand of the machine and load is supplied by the two converters instead of the grid.

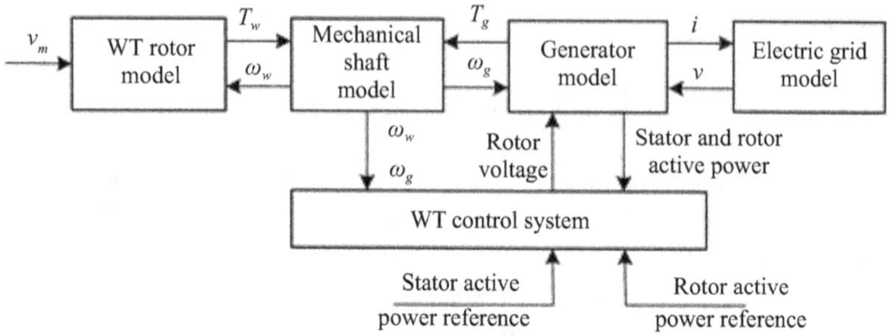

Figure 2.31 Dynamic model of WT-DFIG system connected to the grid

According to what has been explained in Section 2.3.1, the dynamic model of a WT system with DFIG can be established in compliance with the block diagram shown in Figure 2.31.

2.4.2 Permanent magnet synchronous machines

The use of PM machines has become an attractive option for a wide variety of industrial applications because of the benefits that can be gained such as follows: (i) the excitation is provided by PMs, and so, brushes and slip rings are eliminated resulting in simplification of construction and maintenance; (ii) because of the use of PMs, the excitation is current-free and lossless leading to a substantial increase in efficiency; (iii) these are higher in power density (power generated to weight ratio) as compared to other electric machines; (iv) these have better dynamic performance than machines with electromagnetic excitation (higher magnetic flux density in the air gap); (v) it's enabled machine speed control because of its unique operation mode; (vi) there is adequacy to carry out precise speed and position control on a PM machine as the armature field and rotor are always synchronized, in addition to the facility of controlling the armature field precisely [10].

However, the synchronous nature of PM machines may cause problems during start-up, synchronization and voltage regulation. The synchronous operation causes a stiff performance in case of short-circuit occurrence and if the wind speed is unsteady. In addition, the magnetic materials are sensitive to temperature and may lose the magnetic qualities at high temperature, for instance, during a fault. Also, application of PM machines is limited by the fixed excitation from PMs; that is, the magnet flux is fixed. In wind energy application, it is required to operate the PM machine over a wide range of speeds at constant power, which necessitates the use of control laws for flux weakening [11,12]. There are different topologies of PM machines presented in the literature [13]. The most common types are the radial flux machine, the axial flux machine and the transversal flux machine.

2.4.2.1 Structure, equivalent circuit and phasor diagram

PM machines have a wound stator of a typical three-phase AC machine and a rotor. The rotor is provided with a PM pole system and may have salient poles or may be cylindrical depending on the magnet mounting style. Salient poles are more common in low-speed machines and may be the most useful version for wind generator application. It is to be noted that the magnetic permanence along the direct axis (d-axis) of the magnets in a PM machine is extremely low as compared to the other types of AC machines due to the existence of PMs.

Equivalent circuit of a PM machine is shown in Figure 2.32. It is similar to that of a conventional synchronous machine except that the internal induced voltage, E_o, is due to magnets rotation rather than the rotation of the field winding; that is, the source of the flux is different, as well as the direct-axis reactance, X_d, is much smaller than the quadrature-axis reactance, X_q, because of the PMs. If the machine is without damping winding and the zero-sequence current in the stator does not exist ($v_o = 0$), the PM machine can mathematically be represented by a set of equations as follows:

$$\left.\begin{aligned} v_d &= r_a i_d + \frac{d\varphi_d}{dt} - \omega\varphi_q \\ v_q &= r_a i_q + \frac{d\varphi_q}{dt} + \omega\varphi_d + E \end{aligned}\right\} \tag{2.90}$$

where $\Psi_q = L_q i_q, \Psi_d = L_d i_d$ and E is the internal induced voltage due to the magnet generated flux linking the stator windings. The power equation is

$$P = \frac{EV}{X_d}\sin\delta + V^2\frac{X_d - X_q}{X_d X_q}\sin(2\delta) \tag{2.91}$$

where δ is the angle between E and V.

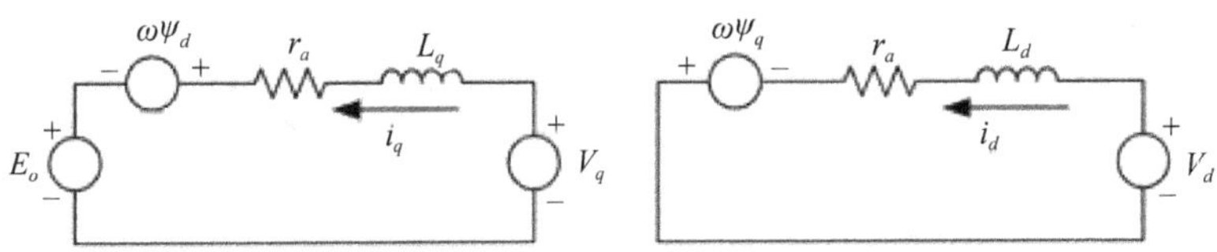

Figure 2.32 Equivalent circuit of a PM machine

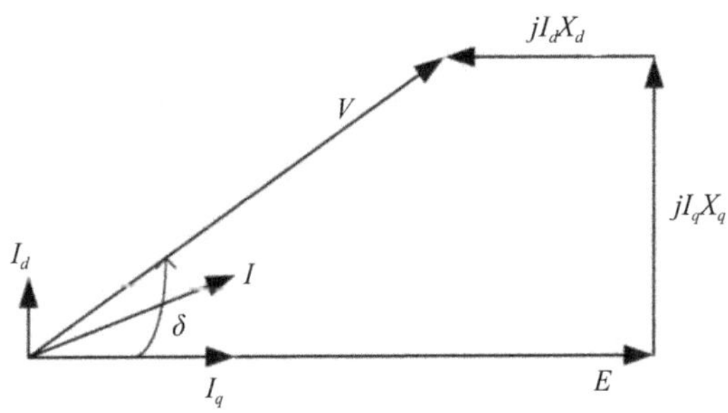

Figure 2.33 Steady-state phasor diagram of a PM machine

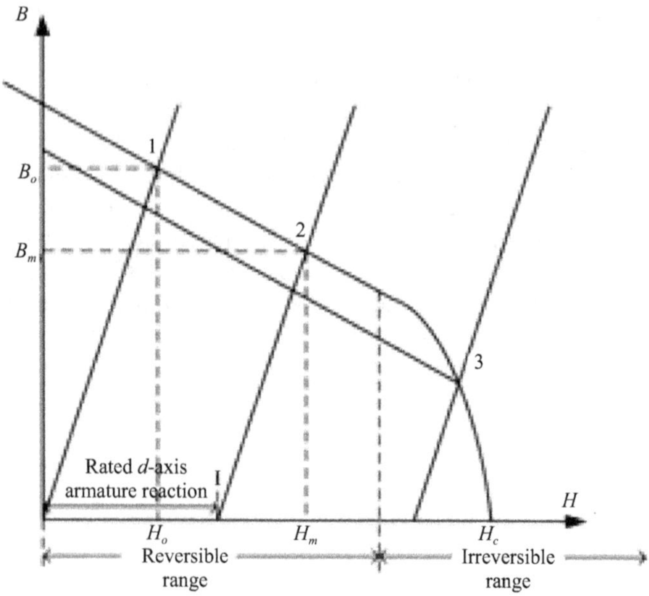

Figure 2.34 Demagnetization curve of a PM

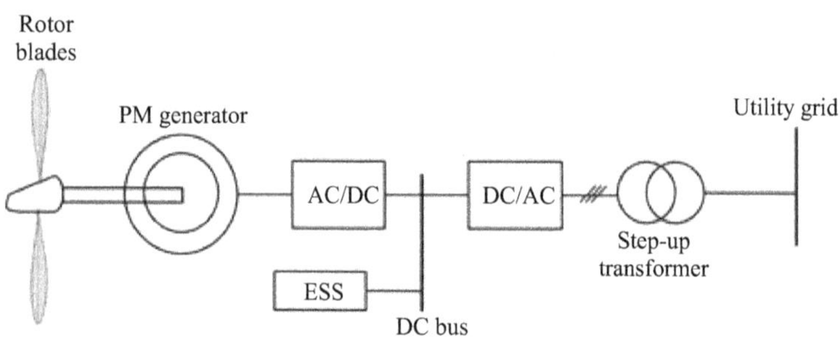

Figure 2.35 PM generator connected to utility power grid

The steady-state operation of a PM machine can be characterized by utilizing the phasor diagram shown in Figure 2.33. In addition, the characteristics of the PMs, given in Figure 2.34, help to describe the PM machine. The demagnetizing curve of the magnet can be divided into three ranges: no-load, rated-load and excessive load bounded by lines #1, #2 and #3, respectively. So, if the demagnetizing current is near or beyond line #3, the PM is in danger and may be damaged.

In transient studies, it is common to neglect the effect of speed variations, and then, the terms $\omega\Psi_q$ and $\omega\Psi_d$ in (2.90) can be eliminated.

2.4.2.2 PM machines connected to power grid

PM machines can be used as variable speed generators in variable speed wind energy systems. This means that the drive, supplying mechanical power to the generator, has a variable speed. So, the connection of PM generator to a local power grid can be implemented using full AC/DC and DC/AC power converters as shown in Figure 2.35. In this case, the variable speed PM generator produces power at variable voltage with variable frequency. Therefore, the generated power cannot directly be injected into the power grid. It needs, first, to match its frequency and voltage to those of the grid. For this reason, the variable AC frequency voltage is rectified by the AC/DC rectifier. The DC power may be used to charge an energy storage system through a DC bus, and meanwhile, it is converted to AC power at grid frequency and voltage using DC/AC inverter to be imparted to the grid through a step-up transformer.

References

[1] Masters G. M. *Renewable and efficient electric power systems*. New York: Wiley-Interscience, John Wiley & Sons, Inc.; 2004.

[2] Farret F. A. and Simoes M. G. *Integration of alternative sources of energy*. New York: IEEE Press, Wiley-Interscience, John Wiley & Sons, Inc.; 2006.

[3] Da Rosa A. V. *Fundamental of renewable energy processes*. 2nd edn. Burlington, USA: Elsevier Academic Press; 2009.

[4] Ackermann T. *Wind power in power systems*. NY: John Wiley & Sons. Inc.; 2005.

[5] Polinder H., de Haan S. W. H., Dubois M. R. and Slootweg J. G. 'Basic operation principles and electrical conversion systems of wind turbines'. *EPE Journal: European Power Electronics and Drives*. 2005;15(4):43–50.

[6] Abad G., Rodriguez J. M. A., Marroyo L. and Iwanski G. *Doubly fed induction machine*. New York: IEEE Press, John Wiley & Sons, Inc.; 2011.

[7] Huan-ping L. I. and Jin-ming Y. (eds.). 'The performance of large scale wind farm connected to external power grid'. *Power Electronics Systems and Applications, PESA 2009, 3rd Int. Conf. on*. Hong Kong, May 2009, pp. 1–5.

[8] Smajo J. 'Wind turbine system with doubly-fed induction generator and rotor power feedback control'. *WSEAS Trans. on Systems*. 2006;12 (5):2860–7.

[9] Morren J. and de Haan S. W. H. 'Ridethrough of wind turbines with doubly-fed induction generator during a voltage dip'. *IEEE Trans. on Energy Conversion*. 2005;20(2):435–41.

[10] Gieras J. F. and Wing M. *Permanent magnet motor technology – Design and applications*. 2nd edn. New York: Marcel Dekker, Inc.; 2002.

[11] Xu L., Ye L., Zhen L. and El-Antably A. 'A new design concept of permanent magnet machine for flux weakening operation'. *IEEE Trans. on Industry Applications*. 1995;31(2):373–8.

[12] Amara Y., Lucidarme J., Gabsi M., Lecrivain M., Ben-Ahmed A. H. and Akemakou A. D. 'A new topology of hybrid synchronous machine'. *IEEE Trans. on Industry Applications.* 2001;37(5):1273–81.
[13] Luo X. and Lipo T. A. 'A synchronous/permanent magnet hybrid AC machine'. *IEEE Trans. on Energy Conversion.* 2000;15(2):203–10.

Chapter 3
Solar energy

3.1 Introduction

Solar is the Latin word of the sun and solar energy is radiant light and heat from the sun that is a powerful source of energy. Many applications may use solar energy such as heating, cooling, ventilation, illumination, transport, cooking, water heating, water treatment, fuel production, electricity production, energy storage systems (ESSs) and buildings [1,2]. It is found, statistically, that the amount of solar energy from the sun falling on the earth in one hour is more than that used by everyone in the world in one year [3]. So, it is an important source of renewable energy, and solar technologies are broadly characterized as either active solar or passive solar depending on how they capture and distribute solar energy or convert it into solar power [4].

Active solar technologies are the most commonly used. The sun delivers its energy in two main forms: light and heat. Correspondingly, there are two main types of solar power systems, namely, solar photovoltaic (PV) and solar thermal systems (Figure 3.1). In case of PV, when the sun shines onto a solar panel, photons from the sunlight are absorbed by the cells in the panel. It creates an electric field across the layers and causes electricity to flow. Solar thermal technology uses the sun's energy, rather than fossil fuels, to generate low-cost, environmentally friendly thermal energy. This energy is used to heat water or other fluids and can also power solar cooling systems. It comes in many shapes and sizes and is used in a wide variety of commercial applications. Commonly seen types of solar thermal systems with limited size include domestic hot water systems, swimming pool-/hot tub-heating systems and radiant space heating systems. Large-scale solar thermal systems, concentrating solar power (CSP) technology, can be used for electricity production.

CSP is used primarily in very large plants and its technology uses mirrors to reflect and concentrate sunlight onto receivers that collect solar energy and convert it into heat to produce steam. The steam drives a turbine and generates power in the same way as conventional power plants. Other concepts are being explored and not all future CSP plants will necessarily use a steam cycle. Therefore, businesses and industry use these technologies to diversify their energy sources, improve efficiency and save money. In addition, developers and utilities are also using these technologies to produce electricity on a massive scale to power cities and small towns [3].

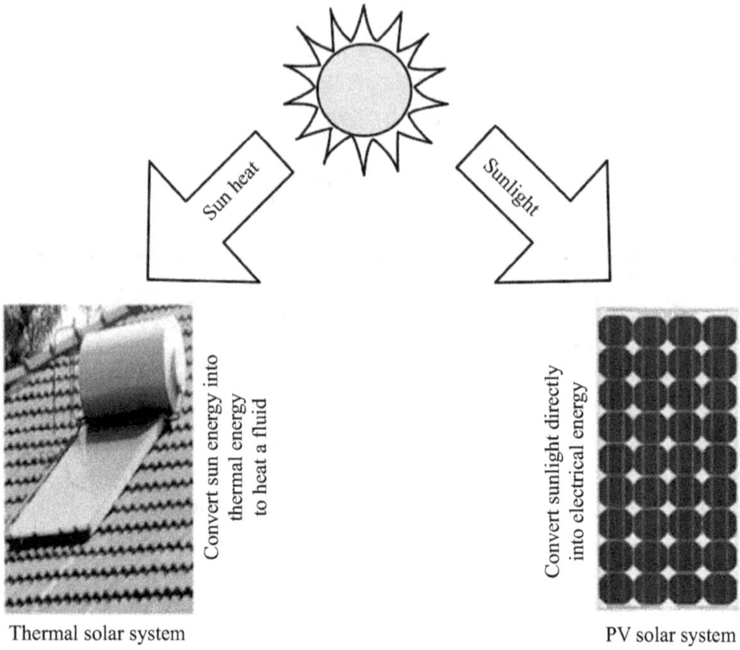

Figure 3.1 Thermal and PV solar systems

However, solar thermal systems differ from solar PV systems which generate electricity directly rather than heat.

Passive solar technologies include orienting a building to the sun, selecting materials with favourable thermal mass or light-dispersing properties and designing spaces that naturally circulate air.

Consequently, active solar technologies increase the supply of energy and are considered as supply-side technologies, whereas passive solar technologies reduce the need for alternate sources and are generally considered as demand-side technologies.

To study and analyse a utility grid supported by such active solar energy, it entails understanding how the solar energy is converted into electricity especially when using PV or CSP technologies. Characteristics of both solar PV and CSP technologies are explained in the following sections.

3.2 Solar PV technology

3.2.1 Basics

PV materials and devices (also known as solar PV) convert sunlight directly into electrical energy. A single PV device is known as a PV cell or solar cell. PV gets its name from the process of converting light (photons) to electricity (voltage), which

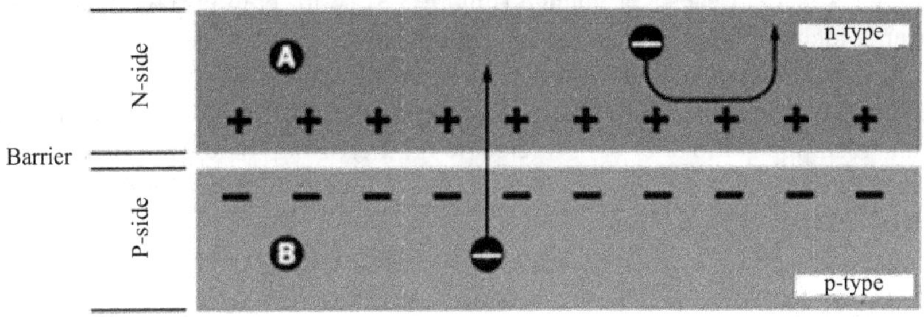

Figure 3.2 Effect of electric field in a PV cell

is called PV effect [5,6]. It is a non-mechanical device usually made from silicon alloys. Sunlight is composed of photons, or particles of solar energy. These photons contain various amounts of energy corresponding to the different wavelengths of the solar spectrum. When photons strike a PV cell, they may be reflected, pass right through or be absorbed. Only the absorbed photons provide energy to generate electricity.

Silicon (Si) crystals are all electrically neutral. In n-type Si, extra electrons are balanced out by the extra protons in the phosphorous. In p-type Si, missing electrons (holes) are balanced out by the missing protons in the boron. However, when the holes and electrons mix at the junction between n-type and p-type silicon, that neutrality is disrupted. Do all the free electrons fill all the free holes? No. If they did, then the whole arrangement would not be very useful. Right at the junction, however, they do mix and form a barrier, making it harder and harder for electrons on the N side to cross to the P side. Eventually, equilibrium is reached, and an electric field separating the two sides (A and B) exists as schematically shown in Figure 3.2 [7].

This electric field acts as a diode, allowing (and even pushing) electrons to flow from the P side to the N side, but not the other way around. It's like a hill – electrons can easily go down the hill (to the N side) but cannot climb it (to the P side). So, the electric field is acting as a diode in which electrons can only move in one direction.

When light, in the form of photons, hits the solar cell, its energy frees electron–hole pairs. Each photon with enough energy will normally free exactly one electron and result in a free hole as well. If this happens close enough to the electric field, or if free electron and free hole happen to wander into its range of influence, the field will send the electron to the N side and the hole to the P side. The P side accumulates holes and the N side accumulates electrons, which creates a voltage that can be used to deliver current to a load. This causes further disruption of electrical neutrality. If electrical contacts are attached to the top and bottom of the cell, an external current path is provided and electrons will flow out of the N side into the path and back to their original side (the P side) through the load to unite with holes that the electric field sent there. The electron flow provides the current, and the cell's electric field causes a voltage (Figure 3.3). With both current and voltage, power, which is the product of the two, is obtained.

So, solar cells, like batteries, generate direct current (DC), which is generally used for small loads (electronic equipment). When DC from PV cells is used for commercial applications or sold to electric utilities using the electric grid, it must be converted to alternating current (AC) using inverters, solid-state devices that convert DC power to AC [8].

An individual PV cell is usually small, typically producing about 1–2 W of power [9]. Modules or panels can be formed as larger units by connecting PV cells together in chains to boost the power output of PV cells. They can be used individually, or several can be connected to form arrays. One or more arrays are then connected to the electrical grid as part of a complete PV system. Because of this modular structure, PV systems can be built to meet almost any electric power need, small or large. So, PV has evolved from small-scale applications towards becoming a mainstream electricity source with efficiency varying from 3% to 31% depending on the technology, the light spectrum, temperature, design and material of the solar cell.

With a goal of improving the efficiency and achieving cost-effective alternative, there are variety of technologies using different semiconductor materials in solar cells such as [9] those given below:

1. *Crystalline silicon.* It is the most common material used in solar cells, representing approximately 90% of the modules sold today. Crystalline cells are made from ultra-pure silicon raw material such as those used in semiconductor chips. They use silicon wafers that are typically 150–200 μm (one fifth of a millimetre) thick. Typical mono-crystalline silicon and poly-crystalline silicon modules are shown in Figure 3.4(a) and (b), respectively.
2. *Thin-film PVs.* The solar cell is made by depositing one or more thin layers of semiconductor material on a supporting material such as glass, plastic or metal.

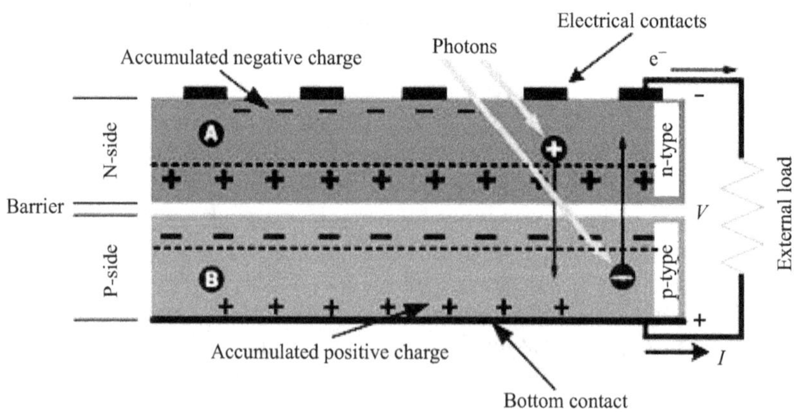

Figure 3.3 Operation of a PV cell

The two main types of thin-film PV semiconductors commonly used, so far, are cadmium telluride (CdTe) and copper indium gallium diselenide (CIGS) (Figure 3.4(c) and (d)). Both materials can be deposited directly onto either the front or the back of the module surface. This technology, after silicon, enables low-cost manufacturing processes.

3. *Organic PVs.* The solar cell is composed of carbon-rich polymers and can be tailored to enhance a specific function of the cell, e.g. sensitivity to a certain type of light. This technology has the theoretical potential to provide electricity at a lower cost than silicon or thin-film technologies, but it is less efficient and has shorter lifetimes.
4. *Concentration PV.* It focuses sunlight onto a solar cell using a mirror or lens. Thus, less PV material is required because of focusing sunlight onto a small area as well as PV materials become more efficient at energy conversion as the light becomes more concentrated. So, the highest overall efficiencies are obtained with concentration PV cells and modules. However, more expensive materials, manufacturing techniques and tracking are required. A round figure about conversion efficiency for different PV technologies is given in Table 3.1 [5].

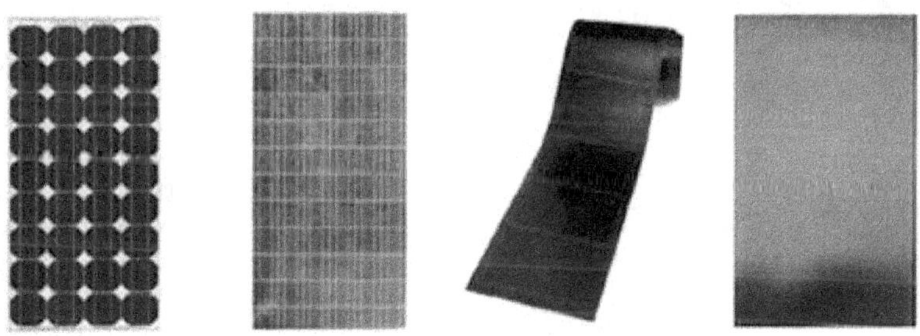

(a) (b) (c) (d)

Figure 3.4 Common PV module technologies [6]: (a) mono-crystalline silicon, (b) poly-crystalline silicon, (c) CdTe thin film and (d) CIGS thin film

Table 3.1 Conversion efficiencies of different module technologies

Module technology	**Module efficiency (%)**
Mono-crystalline silicon	12.5–15
Poly-crystalline silicon	11–14
CdTe	9–12
CIGS	10–13
Amorphous silicon (a-si)	5–7

In addition, as crystalline silicon is the most common used in PV cells, different attempts can be applied in order to reduce the cost of manufacturing.

Single-crystal silicon is not the only material used in PV cells. Poly-crystalline silicon is also used to cut manufacturing costs, although resulting cells are not as efficient as the cells resulting from single-crystal silicon. Amorphous silicon, which has no crystalline structure, is also used, again to reduce production costs. Other materials used include gallium arsenide, copper indium diselenide and CdTe. As different materials have different band gaps, they seem to be 'tuned' to different wavelengths, or photons of different energies. One-way to improve the efficiency is to use two or more layers of different materials with different band gaps. The higher band-gap material is on the surface, absorbing high-energy photons while allowing lower-energy photons to be absorbed by the lower band-gap material beneath. This technique can result in much higher efficiencies. Such cells, called multi-junction cells, can have more than one electric field.

There are a few more steps before using the PV cell. Silicon happens to be a very shiny material, which means that it is very reflective. Photons that are reflected cannot be used by the cell. For that reason, an antireflective coating is applied to the top of the cell to reduce reflection losses to less than 5%. The final step is the glass cover plate that protects the cell from the elements (Figure 3.5).

PV modules are made by connecting several cells (usually 36) in series and parallel to achieve useful levels of voltage and current and putting them in a sturdy frame complete with a glass cover and positive and negative terminals on the back. The series-parallel connection of cells allows the design of solar panels with high

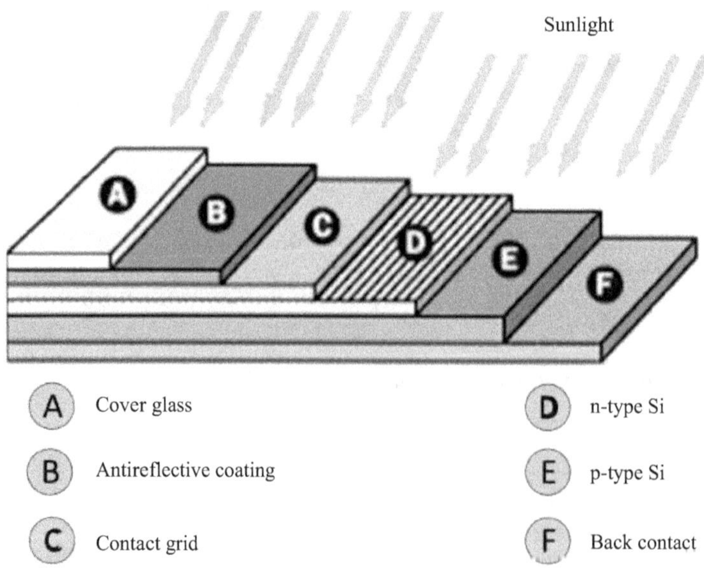

Figure 3.5 Basic structure of a generic silicon PV cell

currents and voltages in the range of kilovolts. To connect a PV system to the grid, it is necessary to include power electronic equipment, energy storage and monitoring systems.

Power electronic circuits may be used for smoothing the DC output of PV system or directly supplying a DC bus to feed an ESS and DC loads (if any). In addition, inverters are used to convert the DC current generated by the PV system into AC electricity, which may be connected to the utility grid through a step-up transformer or feeding any local AC loads (Figure 3.6). PV systems have either one inverter that converts the electricity generated by all the modules or micro-inverters that are attached to each individual module. A single inverter is generally less expensive and can be more easily cooled and serviced when needed. The micro-inverter allows for independent operation of each module, which is useful if some modules might be shaded. For instance, it is expected that inverters will need to be advanced inverters, or 'smart inverters', and allow for two-way communication between the inverter and the electrical utility. This can help balance supply and demand either automatically or via remote communication with utility operators. Allowing utilities to have this insight into (and possible control of) supply and demand allows them to reduce costs, ensure grid stability and reduce the likelihood of power outages.

The ESS, for example, batteries, allows storing the solar PV energy to be used at night or when weather elements keep sunlight from reaching PV panels. Such storage system can be used not only in homes but also for playing a crucial important role for utilities as it will help modernize and stabilize the electric grid.

A typical example of application of PV systems to buildings is shown in Figure 3.7. PV modules can be mounted on the buildings or on the ground if the land is not a constraint. For buildings, commonly, PV modules are mounted either on the roof or displacing another building component, e.g. window glass or roof/wall cladding. The building has two parallel power supplies, one from the PV system and the other from the power grid. The combined power supply feeds all the loads connected to the main AC distribution board. The ratio of PV system supply to power grid supply varies, depending on the size of the PV modules. Whenever the PV supply exceeds the building's demand, excess electricity will be exported

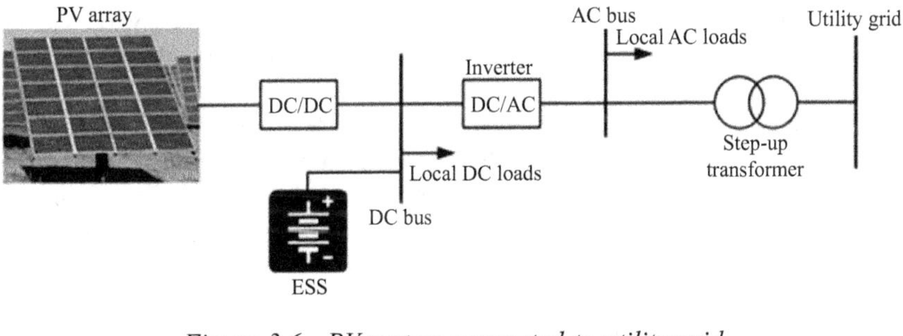

Figure 3.6 PV system connected to utility grid

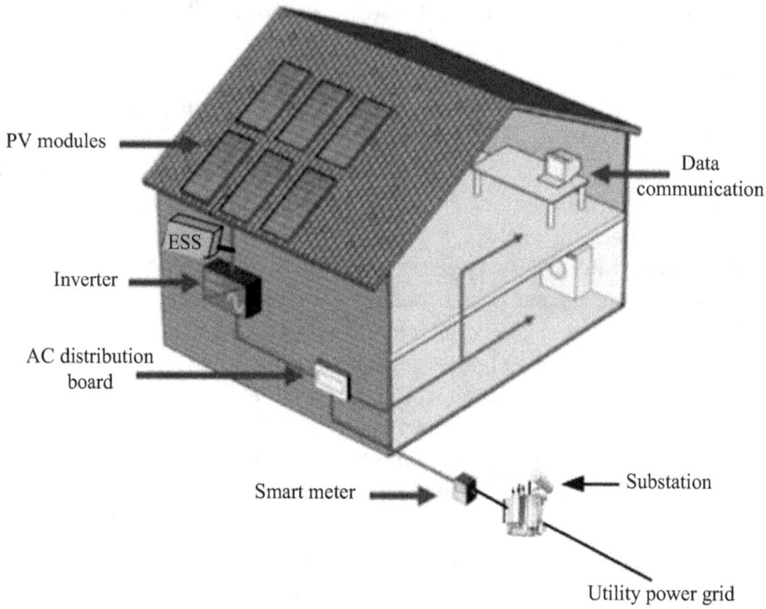

Figure 3.7 Combined PV-power grid powering a house

into the grid. When there is no sunlight to generate PV electricity at night, the ESS will supply the building's demand if it has enough energy to do that. Otherwise, ESS may share the power grid or, totally, the power grid supplies all demand. Of course, this operation needs monitoring and management systems that can be implemented using data communication channels and smart meters.

It is to be noted that PV arrays must be mounted on a stable, durable structure that can support the array and withstand wind, rain, hail and corrosion over decades. These structures tilt the PV array at a fixed angle determined by the local latitude, orientation of the structure and electrical load requirements. To obtain the highest annual energy output, modules in the northern hemisphere are pointed due south and inclined at an angle equal to the local latitude. Rack mounting is currently the most common method because it is robust, versatile and easy to construct and install. More sophisticated and less expensive methods continue to be developed [5].

For PV arrays mounted on the ground, tracking mechanisms automatically move panels to follow the sun across the sky, which provides more energy and high returns on investment. One-axis trackers are typically designed to track the sun from east to west. Two-axis trackers allow for modules to remain pointed directly at the sun throughout the day. Naturally, tracking involves more upfront costs and sophisticated systems are more expensive and require more maintenance. As systems have improved, the cost-benefit analysis increasingly favours tracking for ground-mounted systems.

However, solar systems have some advantages and disadvantages as follows:

Advantages of solar systems:

- Conversion from sunlight to electricity is direct, so that bulky mechanical generator systems are unnecessary.
- PV arrays can be installed quickly and in any size, required or allowed.
- The environmental impact is minimal as the PV system is inexhaustible, does not pollute the atmosphere and has great reliability.
- There is emission-free operation and no fossil fuel consumption.
- Solar systems have excellent modularity (nearly any building-related load can be matched well by multiple units).

Disadvantages of solar energy:

- The amount of sunlight that arrives at the earth's surface is not constant. It depends on location, time of day, time of year and weather conditions.
- Because the sun does not deliver that much energy to any one place at any one time, a large surface area is required to collect the energy at a useful rate.
- The price of delivered power exceeds other distributed generation resources.
- Temporal match of power produced to load is imperfect; batteries or other systems are often needed.
- The scheme may also fail to satisfy the emerging energy needs of the area and constrain industrial development and/or expansion.

3.2.2 Energy loss in a solar cell

The performance of a PV array is dependent upon sunlight. Climate conditions (e.g. clouds and fog) have a significant effect on the amount of solar energy received by a PV array and, in turn, its performance.

The amount of solar energy reaching the ground, as well as its spectral distribution, depends very much on how much atmosphere it has had to pass through to get there. Recall that the length of the path taken by the sun's rays through the atmosphere to reach a spot on the ground, divided by the path length corresponding to the sun directly overhead, is called the *air mass ratio*, AM. Thus, an air mass ratio of one (designated 'AM1') means that the sun is directly overhead. By convention, AM0 means no atmosphere; that is, it is the extraterrestrial solar spectrum. For most PV work, an air mass ratio of 1.5, corresponding to the sun being 42 degrees above the horizon, is assumed to be the standard. For an AM1.5 spectrum, 2% of the incoming solar energy is in the ultraviolet portion of the spectrum, 54% is in the visible and 44% is in the infrared. The solar spectrum at AM1.5 for a silicon PV cell, based on ERD/NASA (1977), is depicted in Figure 3.8(a). It illustrates that photons with wavelength longer than 1.11 μm do not have enough energy to excite electrons (20.2% of the incoming solar energy); those with shorter wavelengths cannot use all their energy, which counts for another 30.2% unavailable to a silicon PV cell.

Visible light is only part of the electromagnetic spectrum. Electromagnetic radiation is not monochromatic; it is made up of a range of different wavelengths,

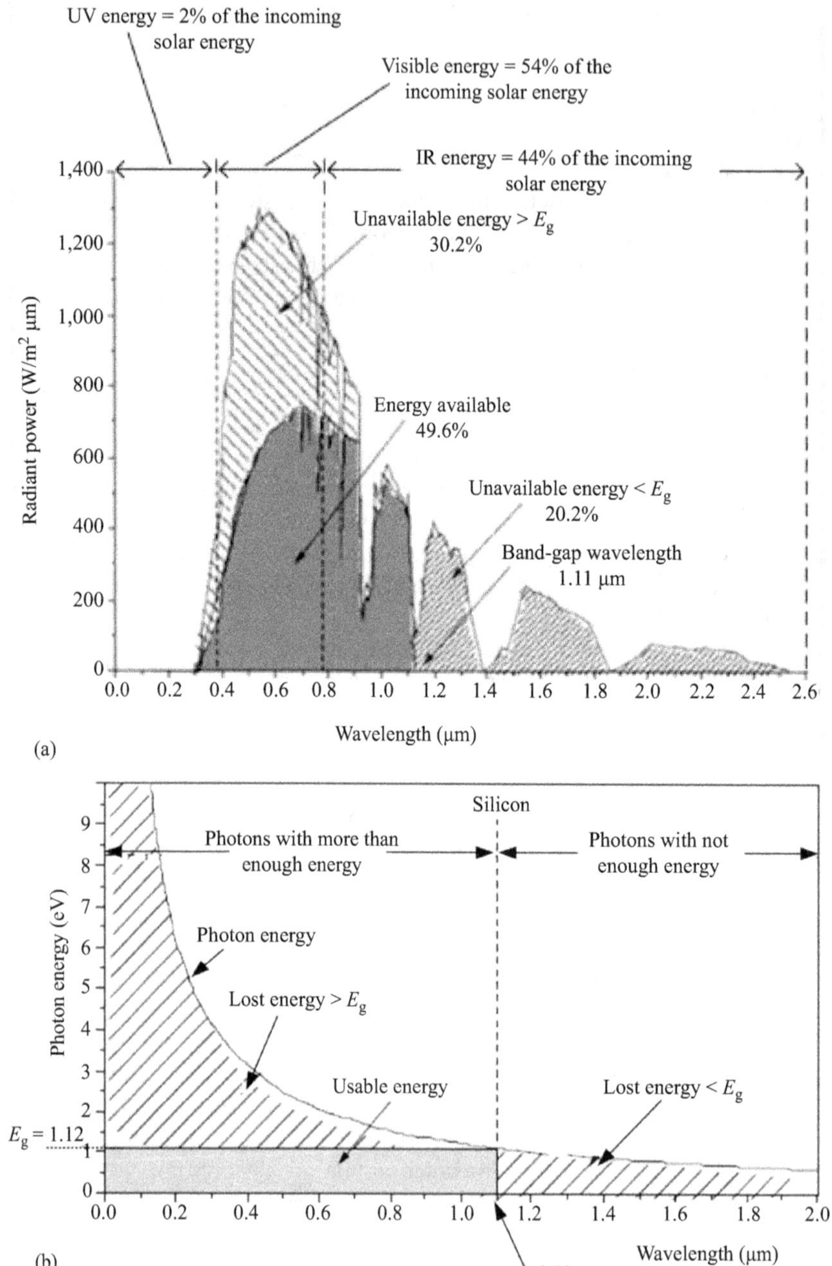

Figure 3.8 Solar spectrum at AM1.5 for a silicon PV cell: (a) radiant power versus wavelength indicating the percentage of available and unavailable photon energies; (b) solar spectrum illustrating usable and lost energies

and therefore energy levels. Since the light that hits the cell has photons of a wide range of energies, it turns out that some of them would not have enough energy to form an electron–hole pair. They will simply pass through the cell as if it were transparent. Still other photons have too much energy. Only a certain amount of energy, called 'band-gap energy', E_g, measured in electron volts (eV = 1.6×10^{-19} joule) and defined by the cell material (about 1.11 eV for crystalline silicon), is required to knock an electron loose. If a photon has more energy than the required amount, then the extra energy is lost (unless a photon has twice the required energy and can create more than one electron–hole pair, but this effect is not significant). These two effects alone account for the loss of around 70% of the radiation energy incident on the cell. Figure 3.8(b) shows that for a silicon PV cell, photons with wavelength greater than 1.11 μm have energy less than the 1.12-eV band-gap energy needed to excite an electron. None of those photons create hole–electron pairs capable of carrying current, so all their energy is wasted. It just heats the cell. However, photons with wavelengths shorter than 1.11 μm have more than enough energy to excite an electron. As one photon can excite only one electron, any extra energy above the 1.12 eV needed is also dissipated as waste heat in the cell [10].

A material with a low band gap could not be chosen, so more of the photons can be used. Unfortunately, the band gap also determines the strength (voltage) of the electric field, and if it is too low, then extra current by absorbing more photons is produced but at a small voltage which is not useful as power is product of voltage and current. The optimal band gap, balancing these two effects, is around 1.4 eV for a cell made from a single material.

There are other losses as well. Electrons must flow from one side of the cell to the other through an external circuit. The bottom can be covered with a metal, allowing for good conduction, but if the top is completely covered, then photons cannot get through the opaque conductor and all the current is lost (in some cells, transparent conductors are used on the top surface, but not in all). If the contacts are only placed at the sides of the cell, then the electrons must travel an extremely long distance (for an electron) to reach the contacts. It should be noted that silicon is a semiconductor; that is, it is not nearly as good as a metal for transporting current. Its internal resistance (called series resistance) is high, and high resistance means high losses. To minimize these losses, the cell is covered by a metallic contact grid that shortens the distance that electrons must travel while covering only a small part of the cell surface. Even so, some photons are blocked by the grid, which cannot be too small or else its own resistance will be too high.

3.2.3 Electricity generation

PV cells are made of a semiconductor material which can be described as a *p-n* junction diode. The conventional *p-n* junction diode has the voltage–current (V–I) characteristic as shown in Figure 3.9. Applying a voltage V_d across the diode terminals leads to flowing a forward current I_d through the diode from the *p* side to the *n* side; but in reverse direction, only a very small saturation current I_s will flow

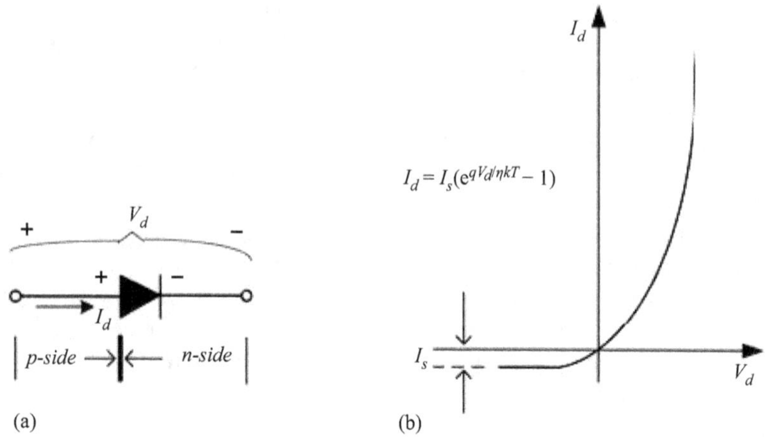

Figure 3.9 p-n junction diode: representation and characteristic curve. (a) p-n junction diode. (b) V–I characteristic curve

($\approx 10^{-12}$ A/cm^2). I_s is the result of thermally generated carriers with the holes being swept into the p side and the electrons into the n side. The voltage drop across the diode in the forward direction is only a few tenths of a volt.

V–I characteristic can mathematically be expressed by Shockley diode equation as

$$I_d = I_s\left(e^{qV_d/\eta kT} - 1\right) \tag{3.1}$$

where

$I_d \triangleq$ forward diode current (A);

$I_s \triangleq$ reverse saturation current of the diode (A), typically 100 pA for the silicon cell;

$k \triangleq$ Boltzmann constant (1.381×10^{-23} J/K);

$q \triangleq$ electron charge (1.6021×10^{-19} C);

$V_d \triangleq$ diode terminals voltage from p side to n side (V);

$T \triangleq$ absolute temperature (K);

$\eta \triangleq$ empirical constant to account for different mechanisms responsible for moving carriers across the junction (= 1 in PV application).

Substituting the values of q, k, η and T at 25 °C (= 25 + 273 + 25 = 298.15 K) into (3.1) obtains

$$I_d = I_s\left(e^{38.9V_d} - 1\right) \tag{3.2}$$

3.2.4 PV cell equivalent circuit

According to what is explained in Section 3.2.1 and depicted in Figures 3.2 and 3.3, the equivalent circuit of a PV cell can simply be deduced as a real diode in parallel with a current source driven by absorbed photons and directly proportional to the solar insolation as shown in Figure 3.10.

Based on two conditions, both the current flowing into the load, I, and the open-circuit voltage, V_{OC}, across the terminals can be determined. *First*, when the terminals are short-circuited, there is no current flowing in the diode, that is, $I_d = 0$, as the voltage $V_d = V = 0$ and all the ideal source current, I_{SC}, will flow through the short-circuited terminals. Consequently, the magnitude of the ideal current source must be equal to I_{SC}, and then, the current equation for the equivalent circuit can be written as

$$I = I_{SC} - I_d \tag{3.3}$$

Incorporate (3.2) and (3.3) to get

$$I = I_{SC} - I_s(e^{qV/\eta kT} - 1) \tag{3.4}$$

where $V = V_d$.

It can be seen from (3.4) that the current I is equal to I_{SC} added to the diode equation with negative sign. It is also noted that when $I_{SC} = 0$, i.e. there is no solar insolation (dark state), the current I is just the negative amount of diode current I_d. The plot of V–I relationship for both cases, with and without insolation, is shown in Figure 3.11.

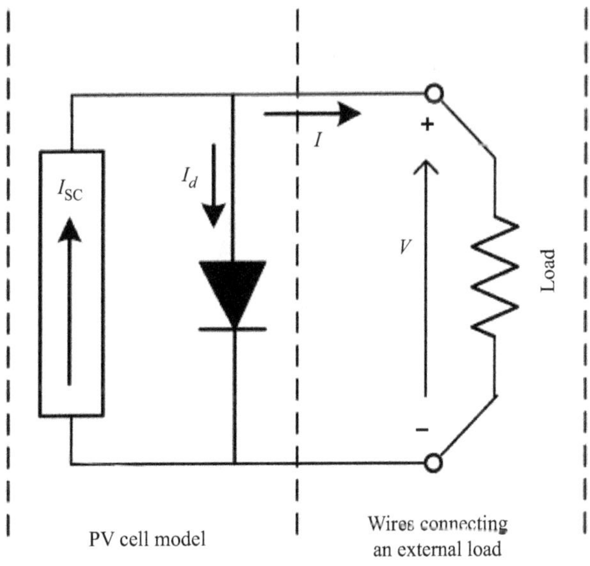

Figure 3.10 Equivalent circuit for a PV cell

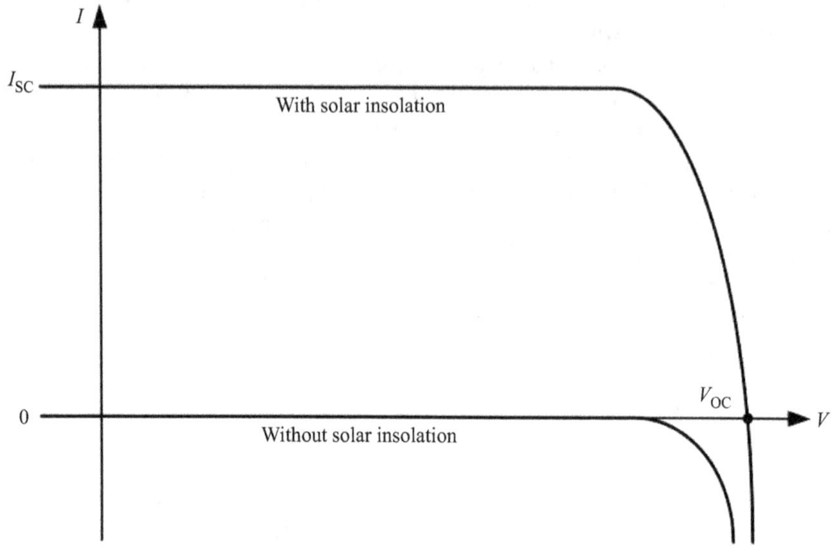

Figure 3.11 Current–voltage relationships for a PV cell

Second, if the diode terminals are opened, then the current I is equal to zero and the open-circuit voltage, V_{OC}, can be obtained by solving (3.4) as follows:

$$V_{OC} = \frac{kT}{q} \ln\left(\frac{I_{SC}}{I_S} + 1\right) \tag{3.5}$$

At 25 °C, (3.4) and (3.5) are written as

$$\left.\begin{aligned} I &= I_{SC} - I_S(e^{38.9V} - 1) \\ V_{OC} &= 0.0257 \ln\left(\frac{I_{SC}}{I_S} + 1\right) \end{aligned}\right\} \tag{3.6}$$

Example 3.1

(a) A *p-n* junction diode has a reverse saturation current of 10^{-10} A. Find the voltage drop across the diode when carrying (i) no current and (ii) currents of 1, 5 and 10 A, respectively.

(b) If the model of a 100 cm^2 PV cell consists of the diode described in (a) and the cell produces short-circuit current of 50 mA/cm^2, find the open-circuit voltage at full sunlight, and at 50% sunlight.

Solution

(a) Using (3.2), it can be found that

$$V_d = \frac{1}{38.9}\ln\left(\frac{I_d}{I_s}+1\right)$$

(i) At no current, $I_d = 0$ and accordingly $V_d = 0$
(ii) At diode current of 1 A, the voltage across the diode is

$$V_d = \frac{1}{38.9}\ln\left(\frac{1}{10^{-10}}+1\right) = 0.592 \text{ V}$$

Similarly, the voltage across the diode when $I_d = 5$ A is given by

$$V_d = \frac{1}{38.9}\ln\left(\frac{5}{10^{-10}}+1\right) = 0.633 \text{ V}$$

and at $I_d = 10$ A, V_d will be

$$V_d = \frac{1}{38.9}\ln\left(\frac{10}{10^{-10}}+1\right) = 0.651 \text{ V}$$

It is to be noted that, when the diode current changes from 1 to 5 A, the voltage across its terminals changes just by an amount of 0.041 V. Also, for current changing from 1 to 10 A, the terminal voltage changes by 0.059 V. This means that a little change of V_d corresponds to a large change of diode current, I_d. So, V_d is commonly assumed to be ≈ 0.6 V for circuit analysis.

(b) At full sunlight, $I_{SC} = 50 \times 10^{-3} \times 100 = 5$ A

Using (3.6),

$$V_{OC} = 0.0257 \ln\left(\frac{5}{10^{-10}}+1\right) = 0.633 \text{ V}$$

At 50% sunlight, $I_{SC} = 25 \times 10^{-3} \times 100 = 2.5$ A as it is directly proportional to solar intensity. Thus,

$$V_{OC} = 0.0257 \ln\left(\frac{2.5}{10^{-10}}+1\right) = 0.615 \text{ V}$$

So, *V–I* curves for the two cases, full sunlight and 50% sunlight, can be plotted as follows:

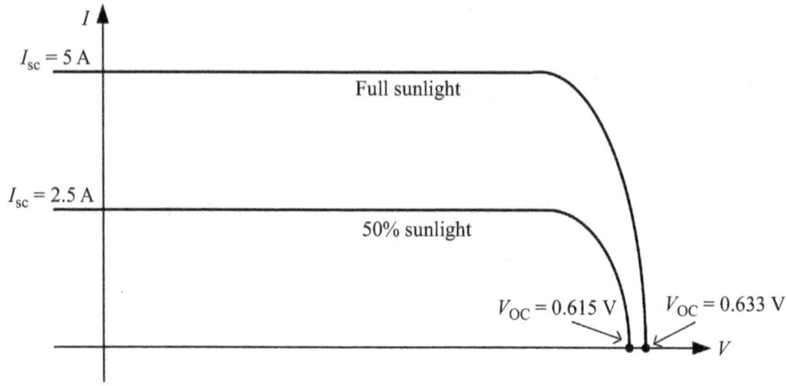

The equivalent circuit of a PV cell can be modified by adding a parallel resistance, R_P, and a series resistance, R_S (Figure 3.12). R_P can represent the effect of shading and internal losses or leakage current across the diode (its value ranges from 200 to 300 Ω). R_S is usually very small (0.05–0.10 Ω) and might be the contact resistance and the resistance of the semiconductor itself. However, R_S is reflected directly on the manufacturing quality of the PV cells.

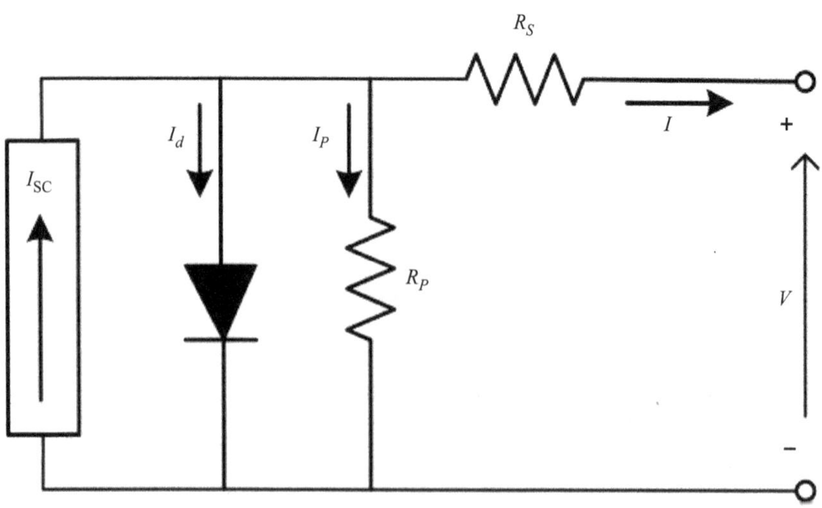

Figure 3.12 Modified equivalent circuit of a PV cell

As shown in Figure 3.12, the following relations (at 25 °C) can be derived.

$$I = I_{SC} - I_d - I_P \tag{3.7}$$

Using (3.2) and substituting $I_P = V_d/R_P$, (3.7) becomes

$$I = I_{SC} - I_s\left(e^{38.9V_d} - 1\right) - \frac{V_d}{R_P} \tag{3.8}$$

$$V_d = V + IR_S \tag{3.9}$$

Incorporating (3.8) and (3.9) obtains

$$I = I_{SC} - I_s(e^{38.9(V+IR_S)-1) - \frac{V+IR_S}{R_P}} \quad (3.10)$$

The open-circuit voltage can be obtained as

$$V_{OC} = V_d|_{I=0} = (I_{SC} - I_d)R_P = \left[I_{SC} - I_s\left(e^{38.9V_{oc}} - 1\right)\right]R_P \tag{3.11}$$

3.2.5 *Solar modules*

An individual PV cell produces only ≈0.6 V. So, it is impractical to find an application using just a single cell. Instead, a group of PV cells in series, known as a module, is used, and consequently, a schematic line diagram of a module of *n*-PV cells can be drawn as shown in Figure 3.13(a). Assuming that the cells are identical and have the same characteristic, all cells carry the same current whereas the

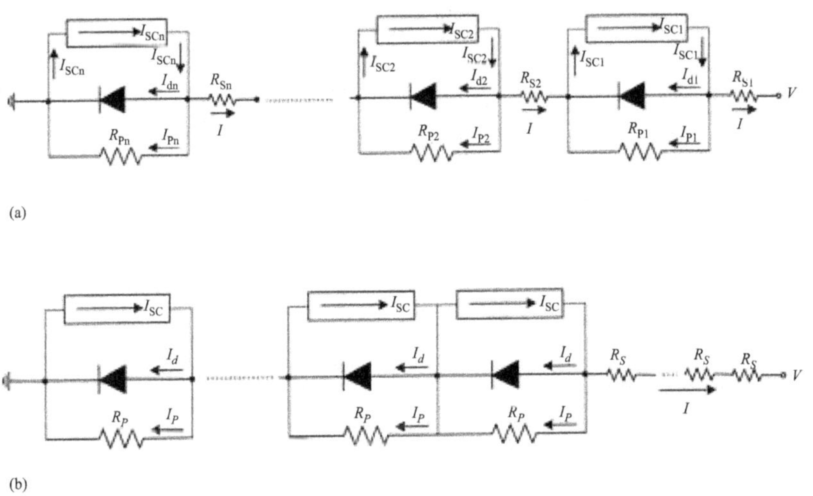

Figure 3.13 Line diagram of a module consisting of n-PV cells: (a) Module of n-PV cells. (b) Module of similar n-PV cells

corresponding voltages are added. In addition, there are no circulation currents among the cells, all series resistances are equal and parallel resistances are equal too. Accordingly, the schematic line diagram can be redrawn as shown in Figure 3.13(b) [11].

To obtain the equivalent circuit of a module consisting of n-PV cells, the line diagram in Figure 3.14(a) can be formed as shown in Figure 3.14(b). It contains n-series current sources having the same current I_{SC}, n-parallel resistances connected in series and n-series resistances connected in series as well. Thus, the equivalent circuit is a current source carrying a current of I_{SC} connected in parallel with a diode of voltage nV_d and a resistance nR_P and all in series with a resistance nR_S as shown in Figure 3.14(b).

From the equivalent circuit, it is noted that

$$V = n(V_d - R_s I) \tag{3.12}$$

where V_d is the diode terminal voltage for an individual cell. It can be obtained from (3.1) as

$$V_d = \frac{kT}{q}\ln\left(\frac{I_d}{I_s} + 1\right) \tag{3.13}$$

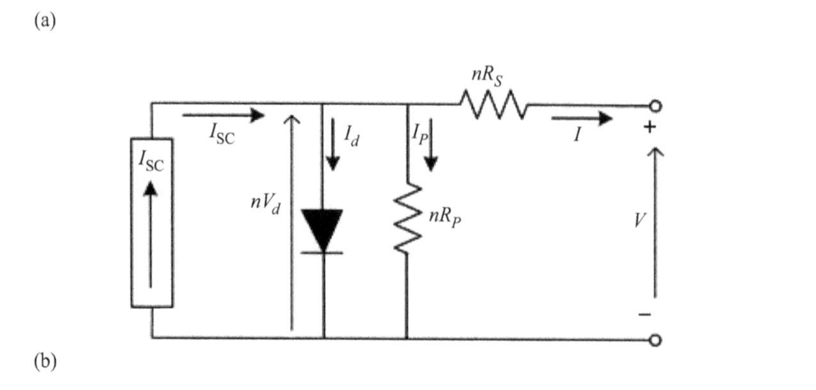

Figure 3.14 Equivalent circuit of a PV module consisting of n-PV cells. (a) Equivalent line diagram of n series PV cells. (b) Compact equivalent circuit for a PV module

Thus, (3.12) becomes

$$V = \frac{nkT}{q} \ln\left(\frac{I_d}{I_s} + 1\right) - nR_S I \tag{3.14}$$

and the current through the diodes can be obtained by Shockley's equation as

$$I_d = I_S \left\{ \exp\left[\frac{q(V + nR_S I)}{nkT}\right] - 1 \right\} \tag{3.15}$$

3.2.6 Solar arrays

Arrays are made of some combination of series and parallel modules to increase both voltage and current, which in turn increases the power. First, as analogous to the PV cells connected in series, it is required to calculate the voltage and currents by determining the equivalent circuit for two cells connected in parallel.

In case of connecting *m*-PV cells in parallel across common terminal, there is no direct interaction between current sources or individual voltages because they are considered the same. Therefore, the equivalent source current is *m* times the individual source current whereas the voltage across the terminal is the same as the voltage across the terminals of the individual cells and the equivalent resistance (series or parallel) is equal to its value for the individual cell divided by the number of cells *m*. Thus, the following relations can be written, and the corresponding compact equivalent circuit is as shown in Figure 3.15 [11].

$$\left.\begin{array}{c} I_{SC} = mI_{SCi},\ V = V_i,\ I = mI_i,\ R_P = R_P/m,\ R_S = R_S/m \\ I_d = mI_{di},\ I_P = mI_{Pi} \end{array}\right\} \tag{3.16}$$

where *i* denotes the individual cell.

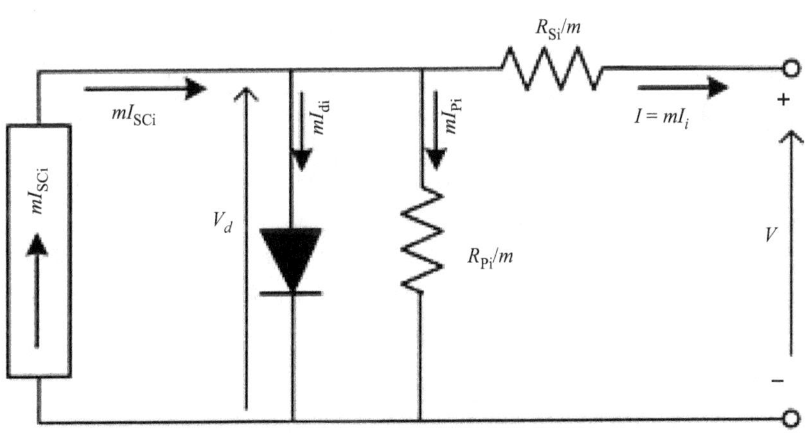

Figure 3.15 Compact equivalent circuit for m-PV parallel cells

The array is composed of a combination of series and parallel modules. For modules in series, *V–I* curves are simply added along the voltage axis, whereas they are added along the current axis for parallel modules. For instance, consider an array consisting of two parallel strings and each string is composed of three modules in series as shown in Figure 3.16. The total *V–I* curve is the sum of the individual module curves (Figure 3.17). The number of modules in series and the number of strings are determined according to the power needed.

3.2.7 PV cell power curve

Referring to *V–I* curve of a PV module, it is seen that the short-circuit current, I_{SC}, will flow when the cell terminals are short-circuited, and the output voltage will be zero. Also, when the cell is exposed to the sun and its terminals are opened, an open-circuit voltage is produced with no flowing current. So, in both cases there is no power delivered as the power is the product of current and voltage, and consequently, no power is received by the load. With connecting a load, some combination of current and voltage will result, and power is delivered. Therefore, at each point on the *V–I* curve the power delivered by the module can be obtained by calculating the voltage times its corresponding current. The module power curve will be as shown in Figure 3.18. It is noted that the maximum power point (MPP) is that point near the knee of the *V–I* curve where the product of current, I_m, and voltage, V_m, reaches its maximum. It can be located by finding the biggest possible rectangle fitted beneath the *V–I* curve where its sides will be I_m and V_m or analytically MPP can be obtained as follows.

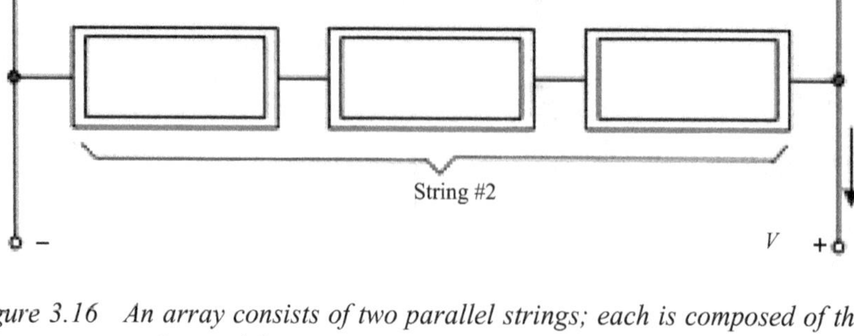

Figure 3.16 An array consists of two parallel strings; each is composed of three series modules

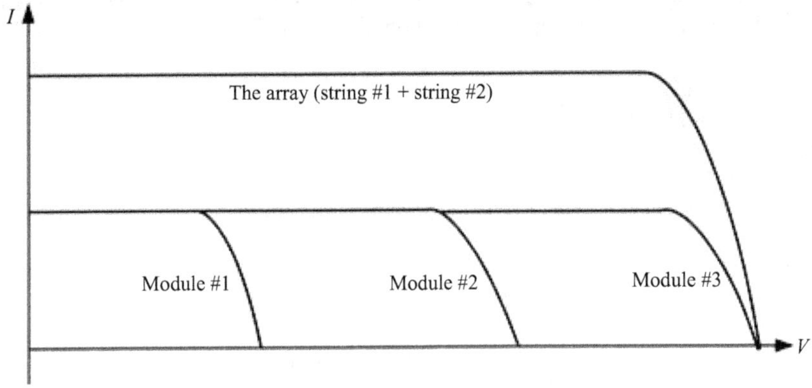

Figure 3.17 Current–voltage curve for the array in Figure 3.16

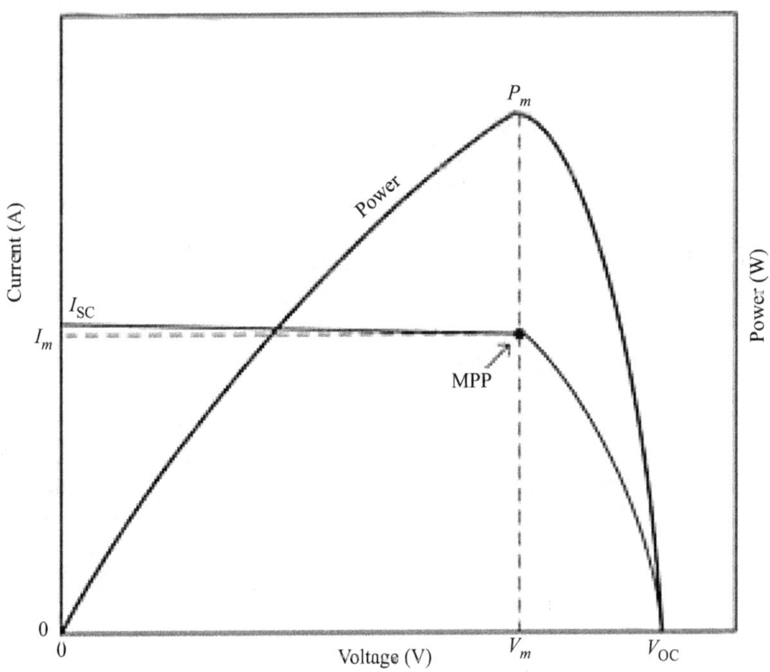

Figure 3.18 I–V and power output curves for a PV module

Assuming a load with resistance R_L is connected to the cell terminals shown in Figure 3.12, the output current, I, can be obtained from

$$I = \frac{R_P}{R_P + R_S + R_L}\left[I_{SC} - I_S\left(e^{38.9V_d} - 1\right)\right] \tag{3.17}$$

where

$$V_d = I(R_S + R_L) = V(1 + R_S/R_L) \tag{3.18}$$

The output power of the cell, P, is the product of the output current, I, and the terminal voltage, V, i.e.

$$P = \frac{VR_P}{R_P + R_S + R_L}\left[I_{SC} - I_S\left(e^{38.9V(1\ +\ R_S/R_L)} - 1\right)\right] \tag{3.19}$$

Differentiating (3.19) with respect to V and equating the derivative to zero, the terminal load voltage at maximum power of the cell, V_m, must satisfy

$$I_S e^{38.9V_m(1\ +\ R_S/R_L)} = \frac{I_{SC} + I_S}{1 + V_m^{38.9(1\ +\ R_S/R_L)}} \tag{3.20}$$

and the maximum power, P_m, may be given by

$$P_m = \frac{V_m(I_{SC} + I_S)}{1 + 0.0257/V_m(1\ +\ R_S/R_L)} \tag{3.21}$$

The conversion efficiency for the maximum power, η_{mc}, can be obtained in terms of the incident solar power, P_{SP}, as

$$\eta_{\mathrm{mc}} = P_m/P_{\mathrm{SP}} \tag{3.22}$$

In addition to MPP, another quantity is also used to characterize the module performance [8]. It is known as the fill factor (FF), which is the ratio of the power at MPP to the product of V_{OC} and I_{SC}, i.e.

$$FF = \frac{\text{Power at MPP}}{V_{\mathrm{OC}}I_{\mathrm{SC}}} = \frac{V_m I_m}{V_{\mathrm{OC}}I_{\mathrm{SC}}} \tag{3.23}$$

The deduction of output power curve depends on the form of *V–I* curve. So, it is important to investigate the factors having prominent effects on the characteristic curve (*V–I* curve) of a PV module. These factors are mainly cell temperature, insolation intensity and shading.

3.2.7.1 Temperature and insolation impact

V–I curves shift as cell temperature and insolation change. Typical such changes are shown in Figure 3.19. With increasing cell temperature, the short-circuit current is slightly increased whereas the open-circuit voltage is substantially decreased. Therefore, the MPP slides slightly upward and towards the left yielding a decrease in the amount of maximum power (e.g. for crystalline silicon cells I_{SC} increases by

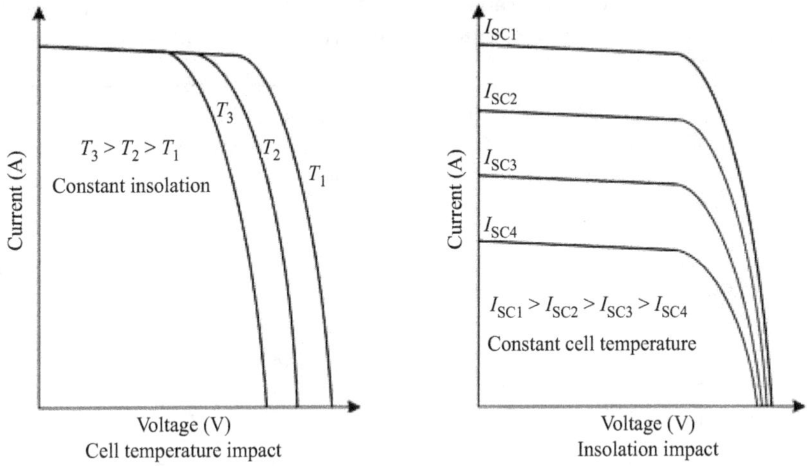

Figure 3.19 I–V curves at different cell temperatures (in the left) and insolation levels (in the right)

$\approx$0.05%/°C, V_{OC} drops by $\approx$0.37%/°C and P_m decreases by $\approx$0.5%/°C). Hence, the effect of module temperature should be considered when estimating the performance because of its significant impact.

It is to be noted that cell or module temperature differs from the ambient temperature. Cell temperature varies not only because of ambient temperature change, but also because of the change of insolation on the cell. A small fraction of insolation is converted into electricity and the rest is absorbed and converted into heat. In most cases, manufacturers provide an indicator known as nominal operating cell temperature (NOCT) to help designers in cell performance estimation with temperature. NOCT is the cell temperature in a module at ambient temperature of 20 °C, irradiation of 0.8 kW/m^2 and wind speed of 1 m/s. For other ambient temperatures, T_{amb} (°C), the cell temperature, T_{cell} (°C), may be obtained by the expression given as follows [8]:

$$T_{cell} = T_{amb} + S\left(\frac{NOCT - 20^{\circ}}{0.8}\right) \tag{3.24}$$

where S is solar insolation (kW/m^2).

Another expression for estimating cell temperature can be used, if NOCT is not given, as follows [8]:

$$T_{cell} = T_{amb} + \gamma\left(\frac{\text{Insolation}}{1\ \text{kW/m}^2}\right) \tag{3.25}$$

where γ is a proportionality factor. It depends on wind speed and the state of module ventilation when installed. Commonly, it ranges from 25 to 35 °C.

On the other hand, as the insolation drops, short-circuit current drops in direct proportion whereas the open-circuit voltage is reduced by modest changes.

Example 3.2 In a 36-cell PV module, all cells are identical. Each cell with 1-sun insolation (1 kW/m^2) has a short-circuit current $I_{SC} = 7.45$ A, reverse saturation current $I_S = 4 \times 10^{-10}$, parallel resistance $R_P = 7.2\ \Omega$ and series resistance $R_S = 0.006\ \Omega$.

(a) If the junction voltage of each cell, V_d, is 0.5 V, find the output voltage, current and power delivered by the module.
(b) Considering the open-circuit voltage, V_{OC}, of the module at the standard temperature of 25 °C is 21.5 V, the maximum power delivered, $P_m = 120$ W, and the ambient temperature = 30 °C. Estimate cell temperature, V_{OC}, and the maximum power delivered.

Solution

(a) From (3.8), the current can be obtained as

$$I = I_{SC} - I_s\left(e^{38.9V_d} - 1\right) - \frac{V_d}{R_P}$$

$$= 7.45 - 4 \times 10^{-10}\left(e^{38.9\times 0.5} - 1\right) - \frac{0.5}{7.2} = 7.27 \text{ A}$$

The voltage produced by 36-cell PV module as in (3.12) is

$$V_{\text{module}} = n(V_d - R_S I) = 36(0.5 - 0.006 \times 7.27) = 16.43 \text{ V}$$

Thus, the power delivered, P, will be

$$P = IV_{\text{module}} = 7.27 \times 16.43 = 119.44 \text{ W}$$

(b) Using (3.25) and assuming $\gamma = 30$ °C, the cell temperature is

$$T_{\text{cell}} = T_{\text{amb}} + \gamma\left(\frac{\text{Insolation}}{1\ \text{kW/m}^2}\right) = 30 + 30 = 60°\text{C}$$

That means the cells are getting hotter than the ambient by 30 °C.

As explained in Section 3.7.2.1, the open-circuit voltage drops by about 0.37% per degree centigrade. Thus, V_{OC} drops, at 60 °C, to

$$V_{OC}\ (\text{at } 60\,°\text{C}) = V_{OC}\ (\text{at } 60\,°\text{C})[1 - 0.0037(T_{\text{cell}} - T_{\text{std}})]$$

$$= 21.5[1 - 0.0037(60 - 25)] = 18.716 \text{ V}$$

and the percentage of open-circuit decrease is about 12.9%.

Similarly, the power delivered drops by about 0.5% per degree centigrade. Then,

$$P_m \text{ (at } 60\,^\circ\text{C)} = P_m \text{ (at } 25\,^\circ\text{C)}[1 - 0.005(T_{\text{cell}} - T_{\text{std}})]$$
$$= 120[1 - 0.005(60 - 25)] = 99 \text{ W}$$

that is, the maximum power delivered by the module is decreased by about 17.5%.

3.2.7.2 Shading impact

The output voltage of a PV module can reduce extremely when one or more cells are shaded. To illustrate that, consider a module of n-PV cells and one of these cells is shaded (e.g. the cell on the top), whereas the other $n - 1$ cells are in the sun as shown in Figure 3.20. The current source of the shaded cell is reduced to zero. The voltage drop across R_P due to current flow through it causes the diode to be reverse biased, and thus, the diode current will also be zero. Hence, the entire current flowing through the module must flow through both R_P and R_S in the shaded cell to complete its way to the load. That means the shaded cell reduces the output voltage rather than adding to it. The output voltage of the module with one cell shaded will drop to become

$$V_{\text{sh}} = V_{n-1} - I(R_P + R_S) \tag{3.26}$$

Assuming the output voltage in case of all cells in the sun is V and carrying a current I, then the voltage of the $n - 1$ cells can be obtained by

$$V_{n-1} = \left(\frac{n-1}{n}\right)V \tag{3.27}$$

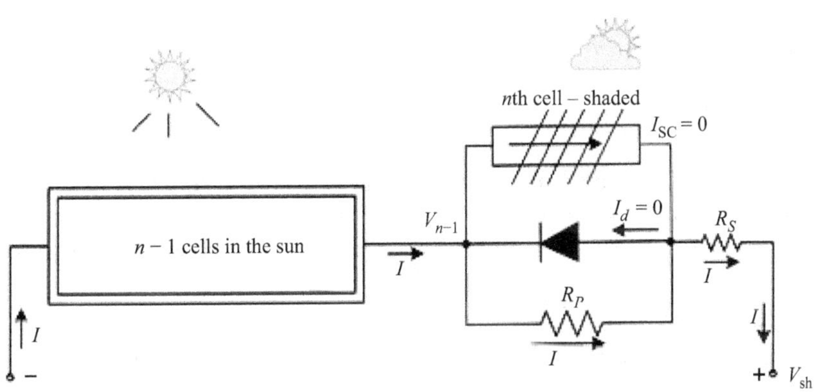

Figure 3.20 A PV module with (n – 1) cells in the sun and nth cell in the shade

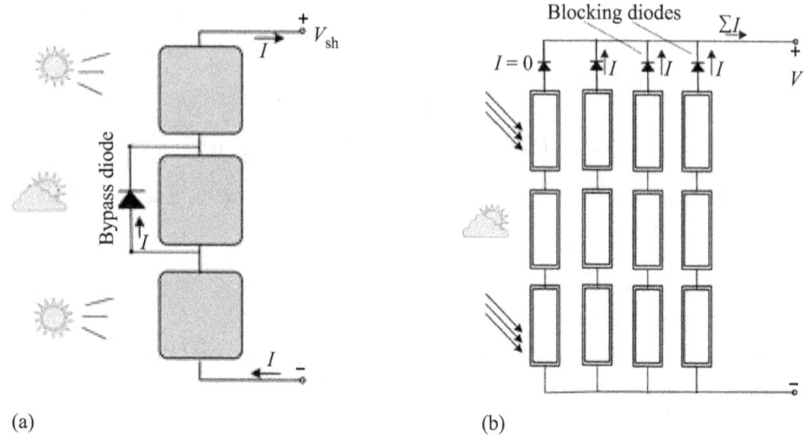

Figure 3.21 Mitigation of shading impact: (a) shaded cell in a module with bypass diode and (b) shaded string in an array with blocking diodes

Incorporate (3.26) and (3.27) to give

$$V_{sh} = \left(\frac{n-1}{n}\right)V - I(R_P + R_S) \tag{3.28}$$

Thus, the voltage drop caused by the shaded cell, ΔV, at any given current I is given by

$$\Delta V = V - V_{sh} = V - \left(\frac{n-1}{n}\right)V + I(R_P + R_S) = \frac{V}{n} + I(R_P + R_S) \tag{3.29}$$

Considering $R_P \gg R_S$, (3.29) can be approximated and rewritten as

$$\Delta V \cong \frac{V}{n} + IR_P \tag{3.30}$$

Different efforts are made by manufacturers or designers to mitigate shade problems such as using bypass diodes for a partial shaded PV module or block diodes to isolate the shaded string in an array (Figure 3.21). More details about that can be found in [10].

Example 3.3 Suppose that a 36-cell module has the following data for the standard conditions (1 kW/m^2, 25 °C, AM 1.5): effective illumination area = 0.136 m^2, and the voltage and current at the maximum power are 15.6 V and 1.4 A, respectively.

The parallel and series resistance equal 6.6 and 0.005 Ω, respectively. It is required to find the following:

(a) Module efficiency for the maximum electrical power.
(b) If one cell is shaded, calculate the module output voltage, voltage across the shaded cell and the power dissipated in this cell.

Solution

(a) Module efficiency for the maximum power, η, will be

$$\eta = \frac{P_{\text{electrical}}}{P_{\text{solar}}} = \frac{15.6(1.4)}{1,000(0.136)} = \frac{21.84}{136} = 16.06\%$$

(b) Using (3.29), the voltage drop across the module will be

$$\Delta V = \frac{V}{n} + IR_p = \frac{15.6}{36} + 1.4(6.6) = 9.673 \text{ V}$$

Thus, the output voltage, $V_{\text{sh}} = 15.6 - 9.673 = 5.927$ V and the corresponding power, $P_{\text{sh}} = V_{\text{sh}}I = 5.927(1.4) = 8.298$ W.

As well, the efficiency is reduced and becomes

$$\eta = \frac{8.298}{136} = 6.1\%$$

That means, the decrease in efficiency $T = \frac{16.06 - 6.1}{16.06} = 62.02\%$.

The voltage drop across the shaded cell, $V_{\text{cell-sh}} = I(R_p + R_s) = 1.4(6.605) = 9.247$ V. The shaded cell subtracts more than 9 V from the module voltage, 15.6 V. Power dissipated in the shaded cell, $P_{\text{cell-sh}} = V_{\text{cell-sh}}I = 9.247(1.4) = 12.945$ W. It is converted to heat causing a local hot spot that may cause damage in the plastic layer enclosing the cell.

3.3 Concentrating solar power

CSP technology is used primarily for utility-scale projects. It uses mirrors or lenses and tracking systems to reflect and focus a large area of sunlight onto a receiver that converts the solar energy to thermal energy through heating a working fluid. Then, the working fluid (heat transfer fluid) can be used as a heat source for conventional power plant or energy storage.

Therefore, the CSP process, as shown in Figure 3.22, mainly includes (i) concentrators supplemented with tracking system to track the sun and focus it onto thermal receivers, (ii) receivers to absorb the solar energy and convert it to heat, (iii) working fluid to be heated and used as a heat source and (iv) power plant or ESS to use the heat source for producing electricity.

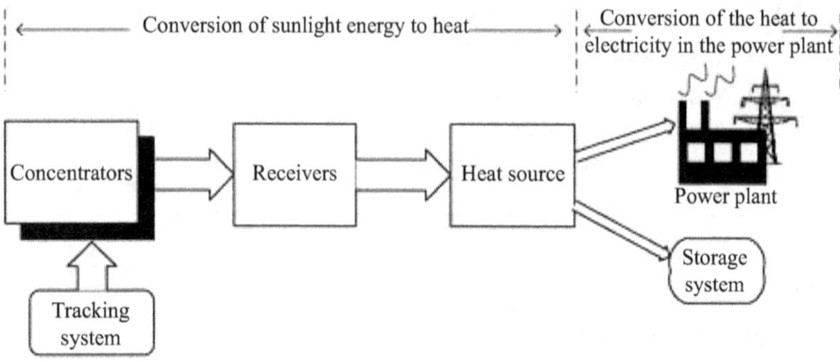

Figure 3.22 Concentrating solar energy process

A wide range of concentrating energy technologies are applied. Three main most developed are (i) linear concentrator technology, which produces high-pressure superheated steam, (ii) dish/engine technology, which can directly generate electricity in isolated locations and (iii) power tower technology, which produces air above 1,000 °C or synthesis gas turbine operation [12].

3.3.1 Linear concentrator systems

The linear concentration is the technology where the concentration can be done on a line. There is not a precise point of focusing, but a line. Linear concentrator systems collect the sun's energy using mirrors. Only one coordinate needs to follow the sun with an optimized collector adjustment. The mirrors are tilted towards the sun to reflect and concentrate sunlight onto linear receivers (or tubes) located along the focal line of the mirrors. The reflected sunlight heats a fluid flowing through the tubes. The hot fluid then is used to boil water in a conventional steam-turbine generator to produce electricity. There are two major types of linear concentrator systems: parabolic trough systems and linear Fresnel reflector systems.

For parabolic trough systems, the collectors are installed in parallel rows (each row consists of a number of concatenated parabolic-shaped mirrors) aligned on a north-south horizontal axis. Using a tracking system, they rotate from east to west to track the sun throughout the day. The receiver tubes are located along the foci of the parabolas (Figure 3.23).

A typical example is shown in Figure 3.24 to illustrate the process of generating electricity. The solar collector field consists of an array of parabolic trough solar concentrators to reflect sunlight onto the receiver (absorber tube). The fluid flowing through the absorber tube, usually synthetic oil or water/steam, transfers the heat to a conventional steam-turbine power cycle.

For linear Fresnel reflector systems, a developed collector with segmented flat mirrors following the principle of Fresnel is used. One receiver tube is positioned above several mirrors to allow the mirrors greater mobility in tracking the sun

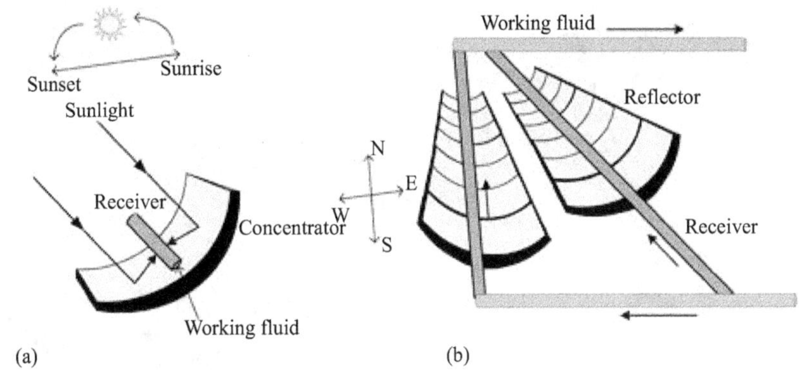

Figure 3.23 Structure of parabolic trough solar concentrating system: (a) singular parabolic trough concentrator; (b) array of parabolic trough collector system

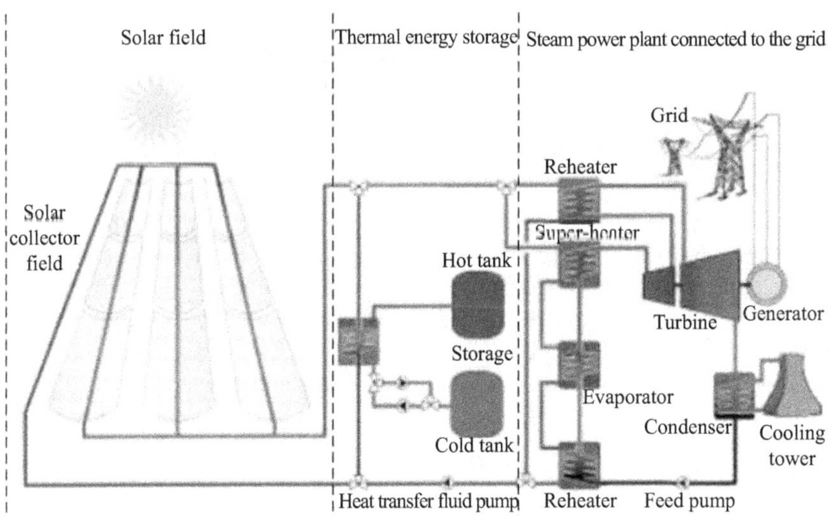

Figure 3.24 Schematic diagram of a steam cycle power plant with parabolic trough collector and energy storage [12]

(Figure 3.25). The linear Fresnel system also shows a good potential for low cost steam generation and provides a semi-shaded space below, which may be particularly useful in desert climates. However, further modelling and experimental work is still required to determine under what conditions it may be more cost-effective than the parabolic trough system with direct steam generation [13].

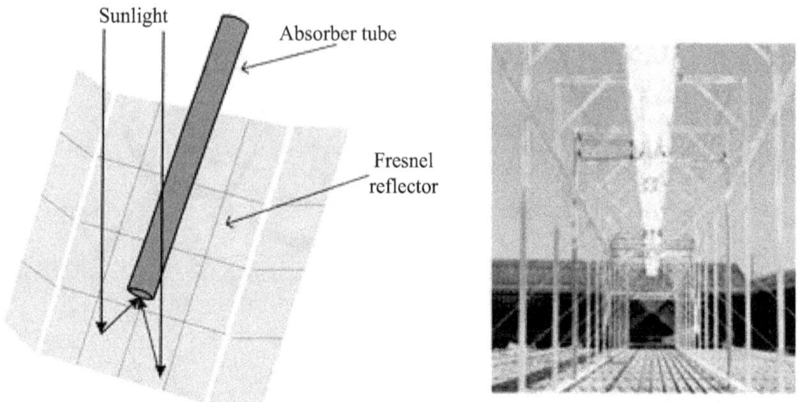

Figure 3.25 Fresnel concentrator structure (on the left) and a typical example (on the right)

Linear Fresnel reflectors are optically analogous to parabolic troughs. They are concentrating reflectors with linear focus, where the parabolic reflective surface is obtained by an array of linear mirror strips which independently move and collectively focus on absorber lines suspended from elevated towers [14,15].

3.3.2 Dish-engine systems

Dish-engine systems use dish-shaped parabolic mirrors, like a very large satellite dish, as reflectors to direct and concentrate sun's rays onto a thermal receiver. To minimize costs, the mirrored dish is usually composed of many smaller flat mirrors formed into a dish shape (Figure 3.26). The receiver is mounted above the dish at the dish focal point. It absorbs the sunlight energy and converts it into thermal energy that can be either transported to a central generator for conversion or converted directly into electricity at a local generator coupled to the receiver.

Dish-engine systems are the most efficient collector systems using dual-axis tracking system to keep the collector pointing at the sun. Different types of engines have been considered for using dish-engine system such as Rankine cycle engines, Brayton cycle engines and sodium-heat engines, but greatest attention has been paid to Stirling engine system. This system uses the fluid heated by the receiver to move pistons and create mechanical power which can be used to run a generator to produce electricity. In addition, a dish-Stirling system can be stand-alone power plant. It does not need access to fuel lines or sources of cooling water as well as it is ideal for producing electricity in sunny desert areas. To increase the reliability and guarantee the service continuity of dish-Stirling systems, some units have been designed to operate in a hybrid mode in which fuel is burned to heat the engine at the time of solar inadequacy.

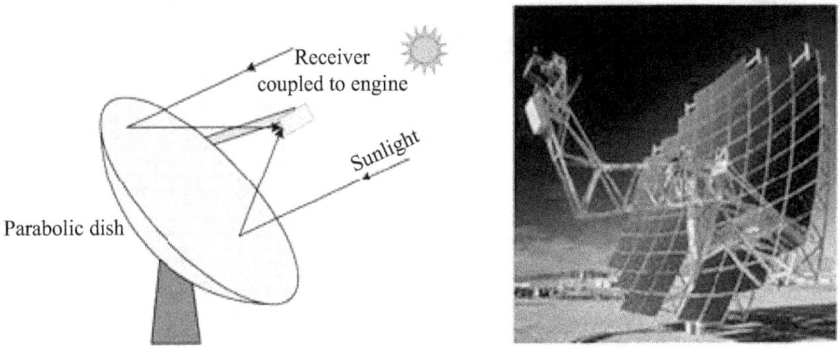

Figure 3.26 Dish-Stirling engine structure; parabolic mirror (on the left) and multi-flat mirrors formed into a dish shape (on the right) [http://www.csiok.cz/assets/docs/esap-CSP-02.pdf]

3.3.3 Power tower systems

Power tower systems use a large field of flat, sun-tracking, computer-controlled mirrors known as heliostats to concentrate sunlight onto a receiver located on the top of a tall tower (Figure 3.27). A heat transfer fluid in the receiver passes optionally to storage and finally to generate steam, which in turn, is used in a conventional turbine-generator to produce electricity and supplies it to the grid (Figure 3.28). Water/steam may be used as a heat transfer fluid for some towers, whereas other advanced designs use molten nitrate salt because of its superior heat transfer and energy storage capabilities. The energy storage capability adds an advantage of allowing the system to continue to dispatch electricity during cloudy weather or at night in addition to the simplicity, scalability, quick start-up, high efficiency due to high achievable temperatures. Therefore, the power tower system that can also be named as a central receiver system is composed of five main components: (i) heliostats including their tracking system, (ii) receiver, (iii) heat transport and exchange, (iv) thermal storage and (v) control system.

The heliostat can be constructed as a group of several distributed flat mirrors positioned in a manner that the reflected rays are collected at the receiver on the top of the tower (Figure 3.29).

An overview of some of the technical parameters of the different CSP technologies is given in Table 3.2. Parameters in the fifth and sixth columns are defined as follows:

$$\text{Solar efficiency} = \frac{\text{Net power generation}}{\text{Incident beam radiation}} \tag{3.31}$$

$$\text{Capacity factor} = \frac{\text{Solar operating hours per year}}{\text{8,760 hours per year}} \tag{3.32}$$

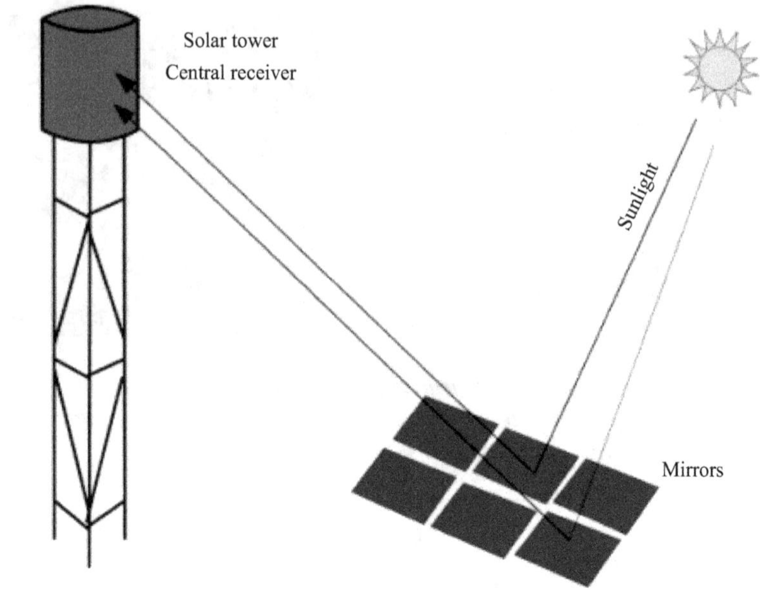

Figure 3.27 Power tower system basics

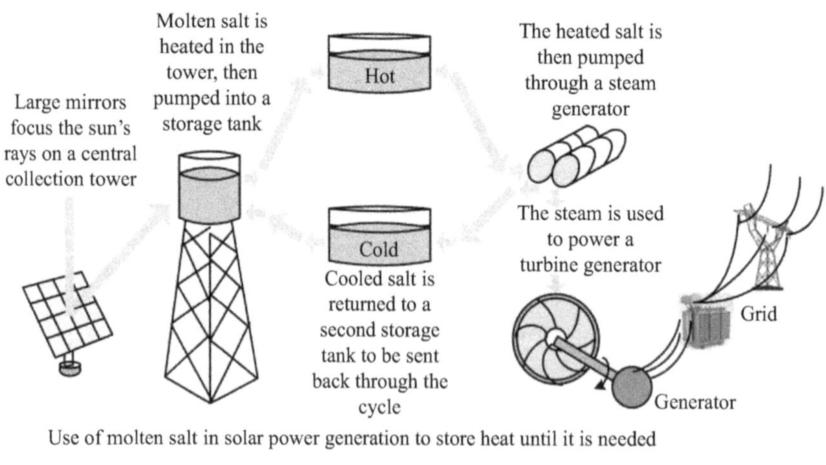

Figure 3.28 Power tower system connected to the grid [16]

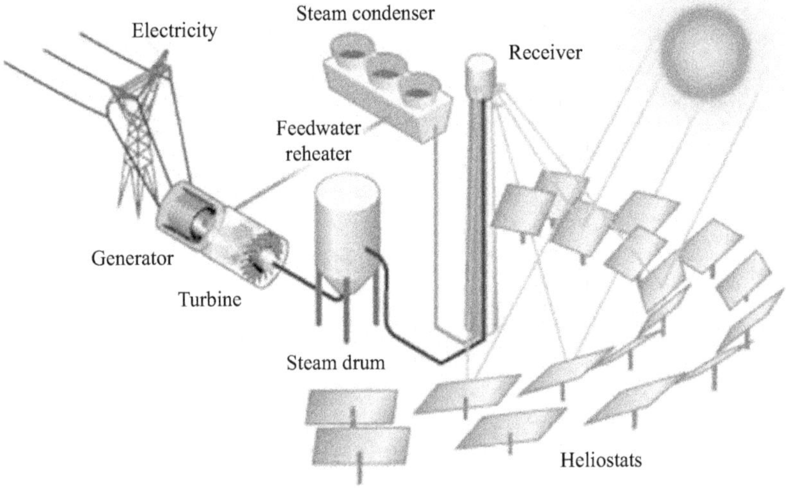

Figure 3.29 Power tower system with distributed flat mirrors [17]

Parabolic troughs, linear Fresnel systems and power towers can be coupled to steam cycles of 10–200 MW of electric capacity, with thermal cycle efficiencies of 30%–40%. The values for parabolic troughs, by far the most mature technology, have been demonstrated in the field. Today, these systems achieve annual solar-to-electricity efficiencies of about 10%–15% and may reach about 18%. The values for other systems are, in general, projections based on component and prototype system test data, and the assumption of mature development of current technology. Overall solar-electric efficiencies are lower than the conversion efficiencies of conventional steam or combined cycles, as they include the conversion of solar radiation energy to heat within the collector and the conversion of the heat to electricity in the power block. The conversion efficiency of the power block remains essentially the same as in fuel-fired power plants [12].

Because of their thermal nature, each of these technologies can be 'hybridized', or operated with fossil fuel as well as solar energy. Hybridization has the potential to improve dramatically the value of CSP technology by increasing its power availability and dispatch ability, decreasing its cost (by making more effective use of the power block equipment), and reducing the technological risk by allowing conventional fuel use if, for example, the collector must be repaired. Solar heat collected during the daytime can be stored in concrete, molten salt, ceramics or phase-change media. It can be extracted at night from storage to run

Table 3.2 Performance data for CSP technologies

Class	Type of concentrator	Capacity (MW)	Concentration	Peak solar efficiency (%)	Annual solar efficiency (%)	Thermal cycle efficiency (%)	Capacity factor (solar) (%)
Line concentration	Trough	10–200	70–80	21 (d)	10–15 (d)	30–40ST	24 (d)
					17–18 (p)		25–70 (p)
	Fresnel	10–200	25–100	20 (p)	9–11 (d)	30–40ST	25–70 (p)
Central point concentration	Dish-Stirling	0.01–0.4	1,000–3,000	29 (d)	16–1 (d)	30–40ST	25–70 (p)
					18–23 (p)	20–30GT	
	Power tower	10–150	300–1,000	20 (d) 35 (p)	8–10 (d)	30–40ST	25 (p)
					15–25 (p)	45–55CC	

(d) = demonstrated; (p) = projected; ST = steam turbine; GT = gas turbine; CC = combined cycle.

the power block. Fossil and renewable fuels such as oil, gas, coal and biomass can be used for co-firing the plant, thus providing power capacity whenever required.

Moreover, solar energy can be used for co-generation of electricity and heat. In this case, the high-value solar energy input is used with the best possible efficiencies of up to 85%. Possible applications include the combined production of electricity, industrial process heat, district cooling and seawater desalination.

References

[1] Sayigh A. A. M. *Solar energy engineering*. 2nd edn. Orlando, FL: Academic Press; 2013.

[2] Chwieduk D. *Solar energy in buildings*. San Diego, CA: Academic Press; 2014.

[3] National Renewable Energy Laboratory. *Solar energy basics* [online]. 2016. Available from https://www.nrel.gov/research/re-solar.html [Accessed 20 June 2016].

[4] Wikipedia, the free encyclopedia. *Solar energy* [online]. 2019. Available from https://en.wikipedia.org/wiki/Solar_energy [Accessed 31 January 2020].

[5] Gevorkian P. *Large scale solar power systems – Construction and economics*. Cambridge, UK: Cambridge University Press; 2012.

[6] Kiatreungwattana K., VanGeet O. and Stoltenberg B. *Facility scale solar photovoltaic guide book*. Report NREL/Bureau of reclamation, 2016.

[7] Ehrlich R. *Renewable energy: A first course*. Boca Raton, FL: CRC Press, Taylor & Francis Group; 2018.

[8] Masters G. M. *Renewable and efficient electric power systems*. New York: Wiley-Interscience, John Wiley & Sons; 2004.

[9] Department of Energy's Office of Energy Efficiency & Renewable Energy. *Solar photovoltaic technology basics* [online]. 2013. Available from http://energy.gov/eere/energybasics/articles/photovoltaic-technology-basics [Accessed 31 January 2020].

[10] Krauter S. C. W. *Solar electric power generation – Photovoltaic energy systems*. Berlin Heidelberg: Springer Verlag; 2006.

[11] Felix A. and Farret F. A. *Integration of alternative sources of energy*. New York: Wiley-IEEE Press; 2006.

[12] Steinhagen H. M. and Trieb F. *Concentrating solar power – A review of the technology* [online]. Stittgart, German: Institute of Technical Thermodynamics, German, Aerospace centre; 2003. Available from https://www.researchgate.net/publication/224797493_Concentrating_solar_power_-_A_review_of_the_technology [Accessed 31 January 2020].

[13] Mills D. R. and Morrison G. L. 'Compact linear Fresnel reflector solar thermal power plants'. *Solar Energy*. 2000;68(3):263–83.

[14] Cau G. and Cocco D. 'Comparison of medium-size concentrating solar power plants based on parabolic trough and linear Fresnel collectors'. *68th Conference of the Italian Thermal Machines Engineering Association, ATI2013*; Bologna, Italy, Sep 2013. *Energy Procedia, Elsevier*. 2014;45: 101–10.

[15] El-Gharbi N., Derbal H., Bouaichaoui S. and Said N. 'A comparative study between parabolic trough collector and linear Fresnel reflector technologies'. *Energy Procedia, Elsevier*. 2011;6:565–72. Available from www.sciencedirect.com [Accessed 31 January 2020].

[16] Vignarooban K., Xu X., Arvay A., Hsu K. and Kannan A. M. 'Heat transfer fluids for concentrating solar power systems – A review'. *Applied Energy, Elsevier*. 2015;146:383–96.

[17] U.S. Department of Energy. 'Power tower system concentrating solar power basics'. Solar Energy Technologies Office'; 2013. Available from https://www.energy.gov/eere/solar/articles/power-tower-system-concentrating-solar-power-basics [Accessed 31 July 2020].

Chapter 4
Ocean energy: tidal energy

4.1 Introduction

Ocean space has a huge amount of energy. Useful electric energy, generally known as 'ocean energy', can be derived from the ocean. It is an alternative renewable energy source (RES), like solar and wind energy. The important feature of ocean energy resources is that they have the highest density among the other sources of renewable energy.

Ocean energy comes from three distinct sources of water energy with different origins and conversion technologies as in the following:

- Tidal energy: It is derived from gravitational forces of earth–moon–sun system as explained in this chapter.
- Waves and swells energy: It is derived from the wind energy and is the subject of Chapter 5.
- Ocean thermal energy conversion: It is derived from the solar energy that is stored in a form of heat in ocean surface layers or hydrothermal energy at submarine volcanic centres. Chapter 5 focuses on this kind of energy.

Large amounts of renewable energy can be harnessed by these three sources because of the oceans' large area.

4.2 Physics of tidal phenomena

Tidal phenomena are characterized by the currents and variations in the sea level caused by the gravitational forces of both the moon and the sun on the earth. This movement of vast quantities of water around the oceans and seas producing tides every day can be classed as an alternative RES. The large amount of energy in the tides may be considered as a form of hydropower that can be converted into electrical energy. The sun does not always shine, and the wind does not always blow. However, what the tides will be doing at any instant in time can be reliably calculated for millennia in advance. Thus, tidal energy is more predictable compared to wind and solar energy, but its cost is relatively high as well as the availability of sites with sufficient flow velocities is limited. However, much attention is currently being devoted to technological developments aimed at getting both higher availability and lower cost.

As the earth, moon and sun rotate around each other in space, the gravitational movement of the moon and the sun with respect to the earth causes millions of litres of water to flow around the earth's oceans creating periodic shifts in these moving molecules of water. These vertical shifts of water are called 'tides'.

The attractive force exercised by the moon has the larger impact on the water molecules (two or three times greater) than that by the sun because of the shorter distance between the earth and moon. Depending on how the earth, moon and sun are positioned relative to each other, the illuminated part of the moon as seen from the earth varies and passes through different phases. So, the phases of the moon are the different appearances of the illuminated moon from a point on the earth. The moon itself does not change, but the part of it that is reflecting sunlight is visible to a different degree. For example, when the earth is located between the moon and the sun, one side of the moon can be seen fully illuminated by the sun. It looks perfectly round and is called the 'full moon'. When the moon is situated between the earth and the sun, the illuminated side of the moon faces away from the earth and this is what is called the 'new moon'. It cannot be seen, but a day or two after the new moon, a skinny crescent-shaped moon appears. It gets bigger and bigger until it reaches the full moon, then it wanes again, becomes a crescent and finally goes back to be a new moon. This cycle (new moon, full moon, new moon) is called 'lunar cycle' and takes approximately 29 days from one new moon to another.

The lunar cycle mostly coincides with a calendar month, but not completely. This means that a new moon will typically be seen once a month, but sometimes there will be two in a month. The second new moon in a month is called a 'black moon'. Like a blue moon (the second full moon in a month), a black moon occurs every two and a half years or so. The black moon does not get as much attention as a blue moon, which makes sense as the black moon cannot be seen.

The moon's position relative to the earth and the sun as viewed from the outer space, above the solar system, as well as the moon as seen from earth are shown as inner and outer sequence, respectively, in Figure 4.1 [1]. For instance, the waxing gibbous (upper left) as a large part of the illuminated moon is seen on the right side. A waning crescent (lower right) as a small slice of the moon illuminated would have the left side of the moon illuminated. The physical pictures of the moon phases illustrating the gradual change of the illuminated moon are shown in Figure 4.2.

At any instant, there are two high tides and two low tides around the earth because of the moon's gravitational attraction. One high tide is between the moon and the earth and closest to the moon (the sublunar point), and another on the opposite side of the earth and furthest from the moon (the antipodal point). The points in between are the resulting low tides (Figure 4.3). The difference in height between a high tide and low tide is defined as the tidal range [2].

The height variation of these tides is an effective factor to consider when producing tidal power. The height is affected by the relative positions of the moon and the sun. When the moon is in perfect alignment with the earth and sun, the combined effects of the lunar and solar tides result in a greater tidal bulge.

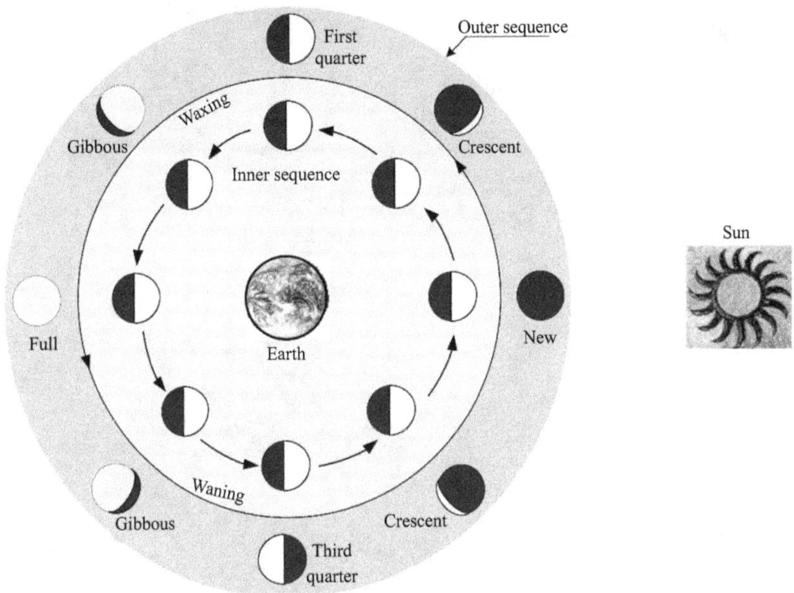

Figure 4.1 Moon phases diagram (the lunar cycle) [2]

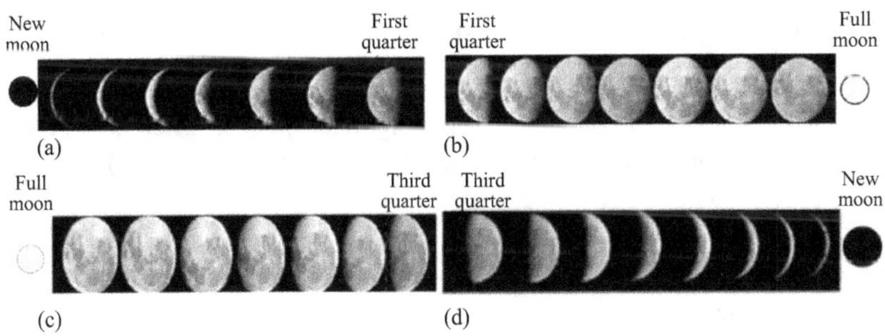

Figure 4.2 Physical pictures of the moon appearance in phase; (a) from new moon to first quarter, (b) from first quarter to full moon, (c) from full moon to third quarter, and (d) from third quarter to new moon [3]

Meanwhile, the tides have a larger tidal range (maximum) and become much stronger than normal. The high tides become very high and the low tides become very low during each tidal cycle. These tides occur during the full or new moon phase and are known as 'spring tides'.

Other neap tides having minimum height arise when the sun, moon and earth are not in alignment during first and third quarter moon phases (the pull of the moon is at 90° to the sun). The lunar and solar tides, therefore, occur in different places, so the overall gravitational effect on the water is less. The net result is a

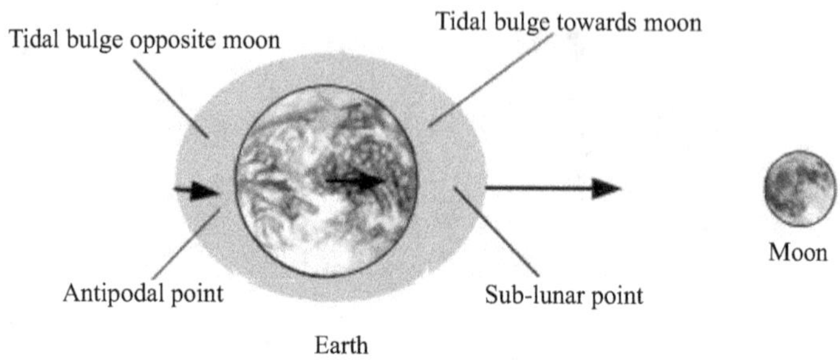

Figure 4.3 Basics of tide formation [2]

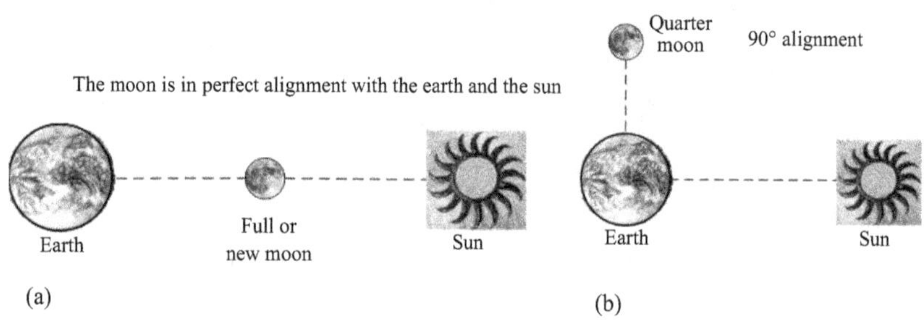

Figure 4.4 Alignment of moon and sun on (a) spring tides and (b) neap tides

smaller pulling action on the seawater creating much smaller differences between the high and low tides, thereby producing very weak tides (Figure 4.4).

4.2.1 Types of tides

Tidal phenomena are, therefore, periodic, though the exact nature of the periodic response varies depending, mainly, on the combined effect of the lunar and solar gravitation, and the relative positions of the earth, sun and moon in addition to other geographical considerations. There are three main types of tides at different locations around the globe as described in the following [4]:

1. Semidiurnal tides: These tides are regular resulting in two high and two low tides each day with the level of the oceans constantly moving between a high tide and a low tide. The rotational movement of the earth is 180° around its axis each 12 h, whereas for the moon around the earth is 6° in the same period, 12 h. So, the time taken for a tidal cycle to happen is about 12 h and 24 min (called the 'diurnal cycle') between two consecutive high tides. The lunar behaviour dominates this type of tide, which is observed on the non-Mediterranean stretches of the French coastline.

   The amplitude of the tide varies through the lunar month, with the tidal range being greatest at the full moon or new moon – that is, when earth, sun

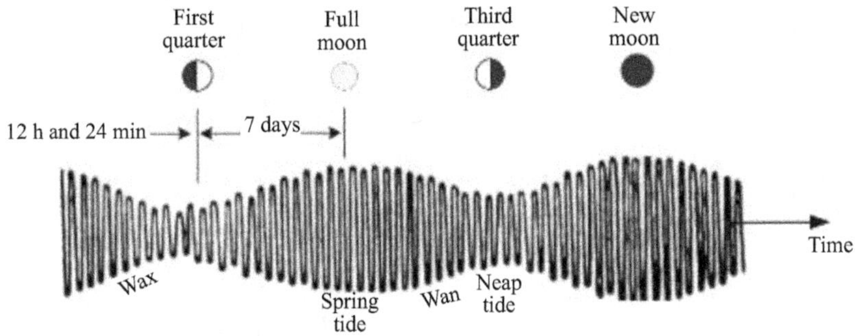

Figure 4.5 Semidiurnal tides monthly cycle

and moon are aligned and the pulls of both gravitational forces combine to produce a spring tide. The tidal range is at its lowest at the first and third quarters of the moon – that is, when the pull of the moon is orthogonal to the sun producing neap tides (Figure 4.5) [4].

2. Diurnal tides with monthly variation: The tidal period of this type corresponds to a full revolution (24 h and 50 min). The 'day-side' gravitational attraction outweighs the 'night-side' attraction because of the variation in the distance from the axis of rotation, and also the tidal range varies with the monthly lunar cycle. This type of tide is found in the China Sea and Tahiti.
3. Mixed tides: This type combines the characteristics of semidiurnal and diurnal tides, either of which may display monthly or bimonthly variation. The laws governing water level variation in mixed tide areas can be highly complex. Examples of mixed tides are those observed in the Mediterranean and at Saigon.

4.3 Tidal energy estimation

Throughout the year, the position of the earth and the moon with respect to the sun changes. So, the potential energy of the water contained in the daily movement of the rising and falling sea levels can be exploited to generate electricity. Electricity generation from tides is similar in many ways to hydroelectric generation. The difference is that the water flows in and out of the turbines in both directions rather than in just one forward direction.

Tidal energy, just like hydro energy, transforms water in motion into a clean energy. The motion of the tidal water, driven by the pull of gravity, contains large amounts of kinetic energy in the form of strong tidal currents called 'tidal streams'. The daily ebbing and flowing, back and forth of the oceans' tides along a coastline, and into and out of small inlets, bays or coastal basins, is little different to the water flowing down a river or stream.

The movement of the seawater is harnessed in a similar way using waterwheels and turbines to that used to generate hydroelectricity. Because of the bi-direction of seawater flow in a tidal energy system, electricity can be generated when the water is flowing in and also when it is ebbing out. It means that the tidal generators are

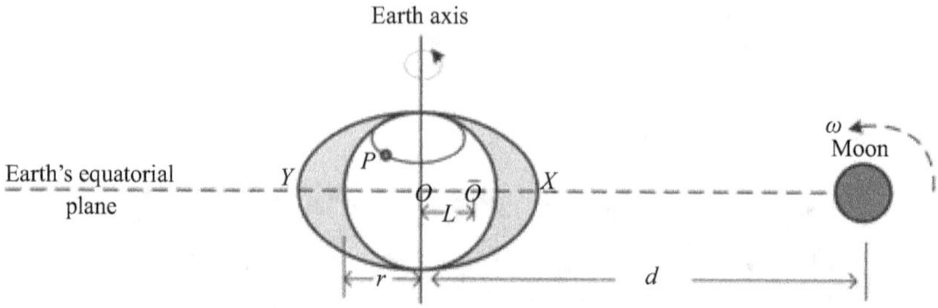

Figure 4.6 Physical explanation of the semidiurnal tide

designed to produce power when rotor blades are turning in either direction and, consequently, the reversible electrical generators cost more than the single direction generators.

4.3.1 Tide generating forces

Tidal energy is derived from the gravitational forces of attraction, which operate between the earth and the sun and between the earth and the moon. The gravitational force, F, which mutually attracts any two bodies of mass m_1 and m_2, respectively, is directly proportional to the product of their masses and inversely proportional to the square of the distance, d, separating them, that is, $F \propto m_1 m_2/d^2$ [5].

Thus, the attractive force exercised by the sun and the moon, respectively, on a water molecule at the surface of the earth, F_S and F_{lu}, can be given by

$$F_S = GM_S M_E/d_S^2 \quad \text{and} \quad F_{lu} = GM_{lu} M_E/d_{lu}^2 \tag{4.1}$$

where

$G \triangleq$ gravitational constant;
$M_S \triangleq$ mass of the sun;
$M_E \triangleq$ mass of the earth;
$M_{lu} \triangleq$ mass of the moon;
$d_S \triangleq$ distance between the centres of mass of the earth and the sun;
$d_{lu} \triangleq$ distance between the centres of mass of the earth and the moon.

Accordingly, and as shown in Figure 4.6, the gravitational forces F_{SX} and F_{luX}, at point #X, the nearest to the sun or the moon, are

$$F_{SX} = GM_S M_E/(d_S - r)^2 \quad \text{and} \quad F_{luX} = GM_{lu} M_E/(d_{lu} - r)^2 \tag{4.2}$$

where r is the equatorial radius of the earth.

Similarly, at point #Y, the furthest from the sun or the moon, the gravitational forces, F_{SY} and F_{luY}, are

$$F_{SY} = GM_S M_E/(d_S + r)^2 \quad \text{and} \quad F_{luY} = GM_{lu} M_E/(d_{lu} + r)^2 \tag{4.3}$$

and at the centre of mass of the earth, O, the gravitational forces, F_{SO} and F_{luO}, are

$$F_{SO} = GM_SM_E/d_S^2 \quad \text{and} \quad F_{luO} = GM_{lu}M_E/d_{lu}^2 \tag{4.4}$$

The tide-generating force is equal to the difference between the gravitational force at the centre of mass of the earth and that at other points on the earth. So, its value at point #X and point #Y, $F_{SX\text{-tide}}$ and $F_{SY\text{-tide}}$, the nearest to and furthest from the sun, respectively, are

$$\begin{aligned} F_{SX\text{-tide}} &= F_{SX} - F_{SO} = GM_SM_E/(d_S - r)^2 - GM_SM_E/d_S^2 \\ &= GM_SM_E\{[2rd_S(1 - r^2/2rd_S)]/r^4\} \quad \text{(ignoring the small value } r/d_S) \\ &= G(2M_SM_Er/d_S^3) \quad \text{(where } r/2d_S \text{ is very small)} \end{aligned} \tag{4.5}$$

and

$$F_{SY\text{-tide}} = F_{SO} - F_{SY} = G(-2M_SM_Er/d_S^3) \tag{4.6}$$

From (4.5) and (4.6), the tide-generating force on the earth due to the sun, $F_{S\text{-tide}}$, can generally be expressed as

$$F_{S\text{-tide}} = \pm G(2M_SM_Er/d_S^3) \tag{4.7}$$

Similarly, the tide-generating force at the surface of the earth (at the points #X and #Y) nearest to and furthest from the moon, $F_{luX\text{-tide}}$ and $F_{luY\text{-tide}}$, respectively, can be given by

$$F_{luX\text{-tide}} = F_{luX} - F_{luO} = G(2M_{lu}M_Er/d_{lu}^3) \tag{4.8}$$

$$F_{luY\text{-tide}} = F_{luO} - F_{luY} = G(-2M_{lu}M_Er/d_{lu}^3) \tag{4.9}$$

Therefore, the general form of the tide-generating force on the earth due to the moon can be written as

$$F_{lu\text{-tide}} = \pm G(2M_{lu}M_Er/d_{lu}^3) \tag{4.10}$$

From (4.7) and (4.10), the tide-generating force ratio (TGFR), which is the ratio of the tide-generating force by the moon to that by the sun, can be given by

$$TGFR = \pm\left(\frac{2M_{lu}M_Er}{d_{lu}^3}\right)\left(\frac{d_S^3}{2M_SM_Er}\right) = (M_{lu}/M_S)(d_S/d_{lu})^3 \tag{4.11}$$

The sun is 27×10^6 times greater in mass than the moon, and the moon is 389 times nearer to the earth than the sun. So, as given by (4.11), the moon is 2.16 times more influential in causing the tide than the sun.

It is to be noted that although the gravitational force varies inversely as the square of the distance between the two bodies, the tide-generating force varies inversely as the cube of the distance. This is very important as it accounts for the moon being 2.16 times more influential in causing the tide than the sun. Furthermore, the ranges of the lunar tide R_{lu} and solar tide R_S are proportional to the difference in force, that is,

$$\frac{R_{lu}}{R_S} = TGFR = 2.16 \tag{4.12}$$

As explained in Section 4.2.1, the solar tide moves in and out of phase with the lunar tide. When the sun, earth and moon are aligned, the lunar and solar tides are in phase and produce tides of maximum range, named as ‘spring tides’. These tides occur twice per lunar month at times of both full and new moons, Figure 4.5. When the sun/earth and moon/earth directions are perpendicular, the tides of least range, named as ‘neap tides’ are produced and occur twice per lunar month.

Considering the spring tide results from the sum of lunar and solar tides, and the neap tide from their difference, the ratio of spring to neap ranges can be calculated as

$$\frac{R_{\text{spring}}}{R_{\text{neap}}} = \frac{1 + (1/2.16)}{1 - (1/2.16)} = 2.72 \tag{4.13}$$

As an alternative development of using centrifugal force, the procedure is the same because the centrifugal force everywhere is equal (but of opposite sign) to the gravitational force at the centre of mass of the earth to maintain the earth-moon separation. Assuming the centre of revolution is at point $\#\overline{O}$ at L from the centre of the earth, as shown in Figure 4.6:

$$\frac{GM_{lu}M_E}{d^2} = M_{lu}(d_{lu} - L)\omega^2 = M_E L\omega^2 \tag{4.14}$$

where $\omega \triangleq$ the lunar frequency.

The force causing the tides can be estimated as the resultant of two forces: (i) the centrifugal force about $\overline{O}$ at the lunar frequency ω and (ii) the gravitational lunar force. So, for a water of mass m at the positions X and Y, the following relations can be written:

$$F_X = \frac{GmM_{lu}}{(d_{lu} - r)^2} + m(r - L)\omega^2 \tag{4.15}$$

$$F_Y = m(r + L)\omega^2 - \frac{GmM_{lu}}{(d_{lu} + r)^2} \tag{4.16}$$

where F_X and F_Y are the force causing tides at points X and Y, respectively.

At position O with definition of L, the relation in the following can be obtained:

$$\frac{GmM_{lu}}{d_{lu}^2} = mL\omega^2 \tag{4.17}$$

It is to be noted that $r << d$, and so,

$$\frac{1}{(d_{lu} \pm r)^2} = \frac{1}{d_{lu}^2}\left(1 \pm \frac{2r}{d_{lu}}\right) \tag{4.18}$$

Incorporating (4.15) to (4.18) gives

$$F_X = F_Y = mr\omega^2\left(1 + \frac{2L}{d_{lu}}\right) \tag{4.19}$$

Therefore, two lunar tidal ranges of equal amplitude are expected each day when the moon is in equatorial plane of the earth. For instance, and as shown in Figure 4.6, point #P experiences two equal tides each day (semidiurnal tide).

At some point #P with polar coordinates of r_e and θ, as shown in Figure 4.7 and assuming, for simplicity, that the earth is covered by water, the gravitational potential (ψ) for a unit mass of water due to the moon is given by [6]

$$\psi = -G(M_{lu}/d) \tag{4.20}$$

where d is the distance from P to the centre of the moon, and $1/d$ can be expanded as follows:

$$\begin{aligned}\frac{1}{d} &= \frac{1}{\sqrt{d_{lu}^2 + r_e^2 - 2d_{lu}r_e\cos\theta}} \\ &= \frac{1}{d_{lu}}\left[1 + \left(-\frac{2r_e}{d_{lu}}\cos\theta + \frac{r_e^2}{d_{lu}^2}\right)\right]^{-\frac{1}{2}} \\ &= \frac{1}{d_{lu}}\left[1 + \frac{r_e}{d_{lu}}\cos\theta + \frac{r_e^2}{d_{lu}^2}\left(\frac{3}{2}\cos^2\theta - \frac{1}{2}\right) + \cdots\right]\end{aligned} \tag{4.21}$$

Substituting (4.21) in (4.20) gives

$$\psi = \frac{GM_{lu}}{d_{lu}}\left[1 + \frac{r_e}{d_{lu}}\cos\theta + \frac{r_e^2}{d_{lu}^2}\left(\frac{3}{2}\cos^2\theta - \frac{1}{2}\right)\right] \tag{4.22}$$

It is to be noted from (4.22) that the first term can be ignored as it does not yield a force. The second term represents a constant force, $GM_{lu}r_e\cos\theta/d_{lu}^2$,

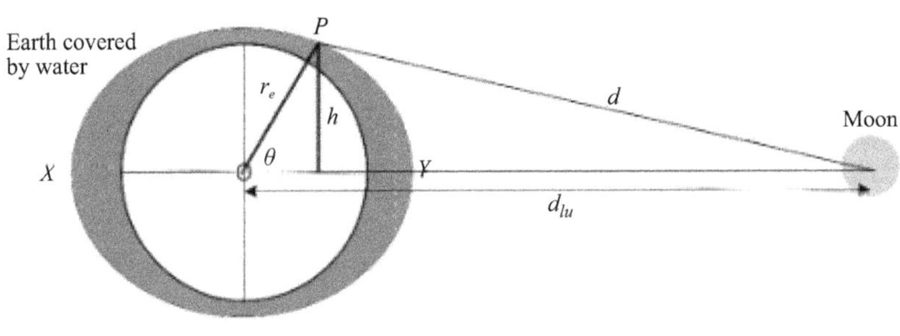

Figure 4.7 Lunar-tidal effect at some point on the surface of the earth

directed towards the centre of the moon. It acts on the earth and is balanced by the centrifugal force due to the rotation of the earth–moon system. The third term describes the variation of the moon's potential around the earth, $\frac{GM_{lu}r_e^2}{d_{lu}^3}\left(\frac{3}{2}\cos^2\theta - \frac{1}{2}\right)$. The surface profile of the water is an equipotential surface due to the combined effects of the moon and the earth. The potential of the unit mass of water to the earth's gravitation is gh, where h is the height of the water above its equilibrium level and $g = GM_E/r^2$ is the acceleration due to gravity at the earth's surface. Hence, the height of the tide $h(\theta)$ can be obtained from

$$gh(\theta) - \frac{GM_{lu}r_e^2}{d_{lu}^3}\left(\frac{3}{2}\cos^2\theta - \frac{1}{2}\right) = 0$$

That is,

$$h(\theta) = \frac{M_{lu}}{M_E}\frac{r_e^4}{d_{lu}^3}\left(\frac{3}{2}\cos^2\theta - \frac{1}{2}\right) = h_{\max}\left(\frac{3}{2}\cos^2\theta - \frac{1}{2}\right) \tag{4.23}$$

The maximum height of the tide, $h_{\max}$, occurs at points $\#X$ and $\#Y$, where $\theta = 0$ and π, respectively. For $M_{lu}/M_E = 0.0123, d = 384{,}400$ km and $r = 6{,}378$ km, the value of $h_{\max} \approx 0.36$ m, which is around the observed mean tidal height. More analysis and details about the rotation of celestial bodies and their orbits can be found in [7–9].

4.3.2 Tidal wave

Tidal wave is the form of a particular type of the ordinary tidal movement of the sea. The wavelength λ is much greater than the mean depth of the sea, h. It is assumed also that the amplitude of the wave is small compared to the depth, in which case the vertical acceleration is small compared to the acceleration due to gravity, g. A slice of water (column of water) of thickness Δx, depth h, width b and surface level above the mean position z is considered. Thus, the change in surface level over the thickness Δx is $(\partial z/\partial x)\Delta x$, which is small compared to z. The side area of the water slice is $A = hb$ (Figure 4.8).

Considering Newton's equation of motion for the slice and the requirement for conservation of water mass, the form of the wave is obtained as follows [5].

The pressure difference across each side of the slice, Δp, arising from the small change in height of the surface is

$$\Delta p = -\rho g\left(\frac{\partial z}{\partial x}\right)\Delta x \tag{4.24}$$

Considering force = mass × acceleration, the equation of motion of the slice of velocity u is

$$\left[-\rho g\left(\frac{\partial z}{\partial x}\right)\Delta x\right]hb = \rho hb\Delta x\frac{\partial u}{\partial t} \tag{4.25}$$

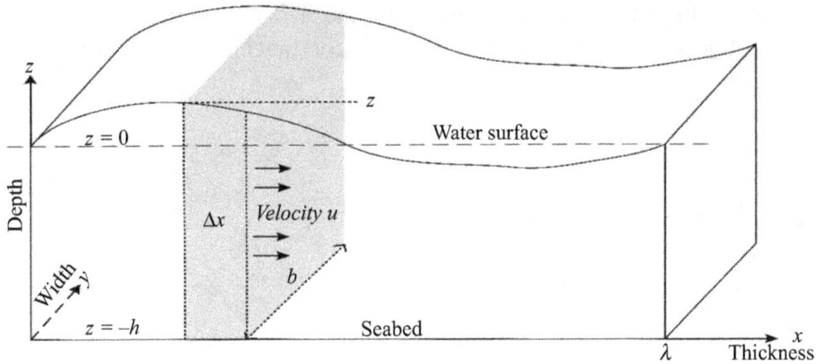

Figure 4.8 Water motion in a tidal wave

Hence,

$$\frac{\partial u}{\partial t} = -g\frac{\partial z}{\partial x} \tag{4.26}$$

The difference between the flow of water into and out of the slice must be accounted for by a change in volume V of the slice with time, and this conservation of water mass leads to the equation of continuity:

$$-\left[\frac{\partial}{\partial x}(A + bz)u\right]\Delta x = \frac{\partial V}{\partial t} = \frac{\partial[(A + bz)\Delta x]}{\partial t} \tag{4.27}$$

As $A = hb$ is constant and much greater than bz,

$$-A\frac{\partial u}{\partial x} = b\frac{\partial z}{\partial t}$$

Or

$$\frac{\partial u}{\partial x} = -\frac{1}{h}\frac{\partial z}{\partial t} \tag{4.28}$$

From (4.26) and (4.28),

$$\frac{\partial^2 u}{\partial t\partial x} = -g\frac{\partial^2 z}{\partial x^2} = \frac{\partial^2 u}{\partial x\partial t} = -\frac{1}{h}\frac{\partial^2 z}{\partial t^2} \tag{4.29}$$

Thus,

$$\frac{\partial^2 z}{\partial t^2} = gh\frac{\partial^2 z}{\partial x^2} = c^2\frac{\partial^2 z}{\partial x^2} \tag{4.30}$$

So, (4.30) is the equation of a tidal wave speed c, where

$$c = \sqrt{(gh)} \tag{4.31}$$

Another alternative proposed formula for the tidal wave speed as a function of the spring tide speed, neap tide speed and tide coefficient (C) is given by [10]

$$c_{\text{tide}} = c_{\text{nt}} + \frac{(C_s - 45) + (c_{\text{st}} - c_{\text{nt}})}{95 - 45} \tag{4.32}$$

where

$c_{\text{tide}} \triangleq$ tidal current speed;
$c_{\text{nt}} \triangleq$ neap tide speed;
$c_{\text{st}} \triangleq$ spring tide speed;
$C_s \triangleq$ constant (= 95 for spring, 45 for neap tide).

4.3.3 Tidal resonance

The only possibility for enhanced motion is for the natural tidal motion to be in resonance with the solar and the lunar forces. For instance, the Atlantic Ocean has a width of about 4,000 km and an average depth of about 4,000 m [6]. The speed of the tide wave as given by (4.31) is

$$c = \sqrt{(gh)} \approx \sqrt{10 \times 4{,}000} \approx 200\ \text{m/s}$$

The tidal frequency f_{tide} is about $2 \times 10^{-5}\ \text{s}^{-1}$, so the wavelength λ can be calculated by

$$\begin{aligned} \lambda &= \frac{c}{f_{\text{tide}}} \\ &\approx \frac{200}{2 \times 10^{-5}}\ \text{m} \approx 10^4\ \text{km} \end{aligned} \tag{4.33}$$

That is, the wavelength is about twice the width of the Atlantic and so resonance occurs. In addition, the amplitude builds up as the time taken for the water wave to make the round trip, reflecting off both shores, is about the same as the tidal period.

The wave amplitude also increases on the continental shelf and can reach about 3 m at the shores. River estuaries can also exhibit large tidal resonance if the length and depth of the estuary are favourable. From (4.31), the time taken for a wave to propagate the length of the channel, L, and back to the inlet is given by

$$t = \frac{2L}{c} = 2L/\sqrt{gh} \tag{4.34}$$

If this time is equal to half the time between successive tides then the tidal range is doubled.

4.4 Tidal energy extraction

Two primary methods to extract energy from the tides are (i) estuaries into which large amounts of ocean water flows in due to high tidal range, water is captured behind barrages and the turbines are rotated by exploiting the potential energy of the stored water, and (ii) the kinetic energy of moving water can be used to extract energy like the principle of extraction of wind energy. Each method has its own advantages and disadvantages. It may be also possible to employ pumping strategies for barrages to obtain better efficiency and to match electricity demand.

Accordingly, the devices used in energy generation are classified into three types: (i) tidal barrages that store tidal flow and generate power through discharge, (ii) tidal stream devices such as (a) current devices, which are fixed or moored within a tidal stream, which is formed by the horizontal flowing volumes of water or (b) fences that block a passage and extract energy in either or both directions of tidal flow.

4.4.1 Tidal barrage

A tidal barrage is a type of tidal power generation that involves the construction of a low dam wall across the mouth of the estuary through which the water flows in and out of the basin. This dam has many underwater tunnels cut into its width allowing seawater to flow through them in a controllable way using sluice gates. The water flows into the bay during high tide and is retained by closing the sluice gates at the beginning of low tide. The barrage gates are controlled according to the tidal range of the location and operating it at right times of the tidal cycle. As the tide ebbs and flows, seawater can flow in or out of the basin through a one-way underwater tunnel system. This flow of tidal water back and forth causes the water turbine generators located at the sluice gates to rotate producing tidal energy with special generators used to produce electricity on both the incoming and the outgoing tides (Figure 4.9).

The main disadvantage of tidal barrage generation is that it can only generate electricity when the tide is flowing either 'in' or 'out' as during high and low tide times the tidal water is stationary. However, because tides are totally predictable, other power stations can compensate for this stationary period when there is no

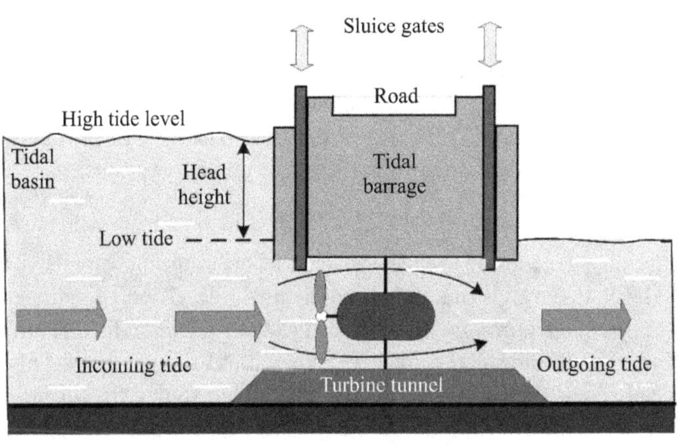

Figure 4.9 Schematic diagram of the tidal barrage

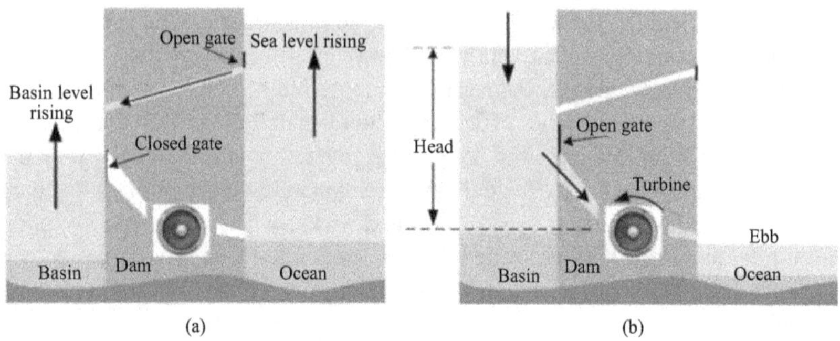

Figure 4.10 Ebb generation: (a) filling the tidal basin and (b) generating electricity

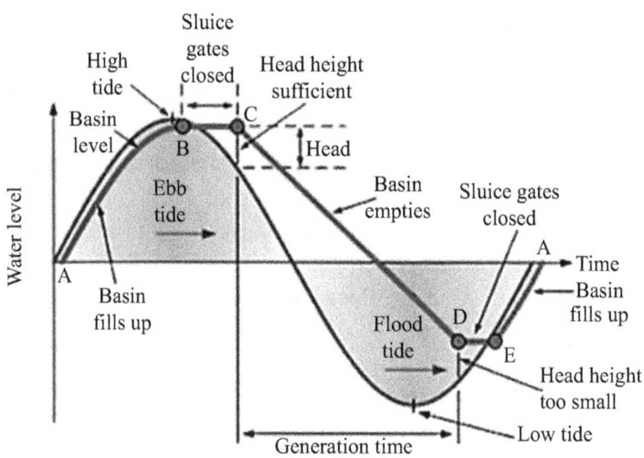

Figure 4.11 Ebb generation cycle

tidal energy being produced. Another disadvantage of the tidal barrage system is the environmental and ecological effects that long concrete dams may have on the estuaries they span. Different schemes of the tidal barrage system can be applied for power generation as explained in the forthcoming subsections.

4.4.1.1 Ebb generation

Tidal barrage ebb generation uses the energy of an outgoing or falling tide. The tidal basin is filled through the sluices until high tide. Then the sluice gates are closed and at this stage pumping may be used to raise the level further. As shown in Figure 4.10(a), the turbine gates are kept closed until the sea level falls to create sufficient head across the barrage, and then are opened so that the turbines generate until the head is again low (Figure 4.10(b)). Then the sluices are opened, turbines are disconnected and the basin is filled again, and the cycle repeats itself as shown in Figure 4.11. When there is sufficient head height between two sides of the barrage, the sluice gates are opened at point *C* releasing the trapped water back to the sea and power generation starts. The tidal basin continues to empty driving the tidal generators until the level of water in the basin reaches point *D*. At point *D*, the head height of the seawater across the barrage is no longer adequate to drive the turbines so the sluice gates are closed, and generation stops until the level of the seawater reaches point *E* again and the whole process starts over.

Ebb generation is also known as 'outflow generation' as the generation takes place when the tidal direction changes [11].

4.4.1.2 Flood generation

Tidal barrage flood generation uses the energy of an incoming rising tide as it moves towards the land. The water is filled into the basin through the turbines which will generate once at tide flood. The tidal basin is emptied through sluice gates or lock gates located along a section of the barrage, and at low tide, the basin is affectively empty. As the tide turns and starts to come in, the sluice gates are closed and the barrage holds back the rising sea level, creating a difference in height between the levels of water on either side of the barrage dam.

The sluice gates to the entrances to the dam's tunnels can either be closed as the seawater rises to allow for a sufficient head of water to develop between the sea level and the basin level before being opened generating more kinetic energy as the water rushes through, turning the turbines as it passes. Or may remain fully open, filling up the basin more slowly and maintaining the same water level inside the basin as out in the sea. The tidal basin is therefore filled up through the turbine tunnels which spin the turbines generating tidal electricity on the flood tide and is then emptied through the opened sluice or lock gates on the ebb tide. Then a flood tidal barrage scheme is a one-way tidal generation scheme on the incoming tide with tidal generation restricted to about 6 h per tidal cycle as the basin fills up. Once the tidal basin is full of water at high tide, all the sluice gates are opened allowing all the trapped water behind the dam to return to the ocean or sea as it ebbs away.

The movement of the water through the tunnels as the tidal basin fills up can be a slow process, so low-speed turbines are used to generate the electrical power.

Generally, this is not efficient if compared to that of ebb generation as the basin's upper half has more volume than the lower half, during the flood generation, if filled thoroughly. The utmost importance for the turbine to produce power is the level difference that is available between the barrage's sea side and the basin side, and it will reduce faster. Rather than ebb generation enhancement, water from the rivers flowing into the basin will reduce the potential of the energy [12].

4.4.1.3 Pumping at high tide

Pumping at or soon after high tide is like pump-storage generation for ebb-generation scheme. When the sluice gates are being closed, the turbines are operated in reverse so that they act as pumps moving water from the sea into the basin. Pumping stops when the sea has ebbed enough to increase the head across the turbines to an uneconomical height. With more water stored in the basin, generation will start earlier than would be the case for simple ebb generation scheme. The additional water in the basin is released through the turbines at the higher head than that through which it was pumped, so that there should be a net gain in energy. The turbine operated as a pump is supplied by the grid energy to raise the basin's level of water. The energy produced during generation is more than is returned; this is because the output power is strongly related to the head, as well as the correlation between the potential energy is not in a linear relationship, but rather is related to the square of the tidal height variation.

4.4.1.4 Two-way generation scheme

It is seen that both ebb tidal barrage and flood tidal barrage installations are 'one-way' tidal generation schemes. So, to increase the power generation time and thus improve the efficiency, special double effect turbines that generate power in both directions, tide ebbs and flows, should be used. Two-way electrical generation requires a more accurate control of the sluice gates keeping them closed until the differential head height is sufficient in either direction before being opened. As the tide ebbs and flows, seawater flows in or out of the tidal reservoir through the same gate system. This flow of tidal water back and forth causes the turbine generators located within the tunnel to rotate in both directions producing electricity.

However, this two-way generation is in general less efficient than one-way flood or ebb generation as the required head height is much smaller which reduces the period over which normal one-way generation might have otherwise occurred. Also, bi-directional tidal turbine generators designed to operate in both directions are generally more expensive and less efficient than dedicated unidirectional tidal generators.

One way of improving the operating time and efficiency of a two-way tidal barrage scheme is to use individual one-way unidirectional tidal turbines installed along the barrage. By controlling their individual sluice gates, one set of turbines can be made to operate on the flood tide between points E and B and the other set operate on the ebb tide between points C and D in Figure 4.11. While this method increases the total number of tidal turbines located along the tidal barrage, it has the advantage that the generation period is greatly extended.

In practice, this scheme that is only used for spring tides, affords only a slight increase in output, though it does make it possible to alter the pattern of distribution throughout the day. In addition, double action mode does not have any effect on the output variations produced by the fortnightly tide cycle [4].

Investment in double-action capability would only appear to be economically warranted in situations where there is a high tidal range and where accurate long-term forecasts are available as to relative energy values at different times of day. Efficient use of two-way scheme operation requires accurate information, several days in advance, on energy values throughout the day.

It is to be noted that the different schemes of tidal barrage system can use double-basin concept instead of one basin. Here the turbines are located on a dyke separating two basins: (i) upper or high-level basin 'A', which fills at high water, and (ii) lower or low-level basin 'B', which empties at low water as shown in

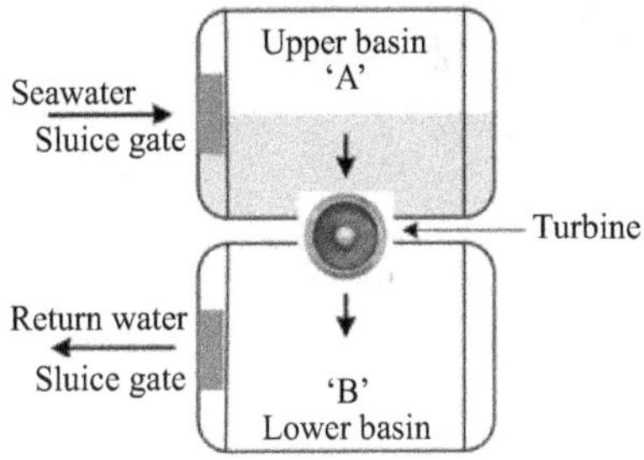

Figure 4.12 Two-basin system

Figure 4.12. At the beginning of the flood tide, the turbines are shut down. Basin A fills and B remains empty. As soon as the head difference of A and B is large enough the water flows from A to B and the turbines start. The separation can be controlled to continuously operate from upper to lower basin. However, as water head between basin varies during each tidal cycle, power generation can improve situation in using off peak power from tidal generator or from alternate system to pump water from low to high basin, and increased head is available from tidal power at peak times.

4.4.1.5 Energy calculations

The energy per tide can be obtained by considering the following: (i) the water is trapped at high tide in the basin and allowed to run out through the turbine, (ii) the basin has a constant surface area A, which remains covered in water at low tide, and (iii) the trapped water has a mass of ρAR at a centre of gravity $R/2$ above the low-tide level. Therefore, the potential energy available per tide when all the water falls through $R/2$ is

$$\text{Energy per tide} = (\rho AR)g\frac{R}{2} \tag{4.35}$$

neglecting the small changes in density from the value of the seawater and taking $\rho = 1.025 \times 10^3\ \text{kg/m}^3$.

The average potential power $(\overline{P})$ for one tidal period, τ, becomes

$$\overline{P} = \frac{\rho Ag}{2\tau}R^2 \tag{4.36}$$

The range R through the month varies sinusoidally from the highest value R_{spring} for the spring tides to the lowest for neap tides, R_{neap}, with a period of half the lunar month as explained in Section 4.1.2.1 (Figure 4.13).

At any time, t, after a mean high tide within the lunar month of period $T = 29.53$ days, the range is given by

$$\frac{R}{2} = \left(\frac{R_{\text{spring}} + R_{\text{neap}}}{4}\right) + \left(\frac{R_{\text{spring}} - R_{\text{neap}}}{4}\right)\sin(4\pi t/T) \tag{4.37}$$

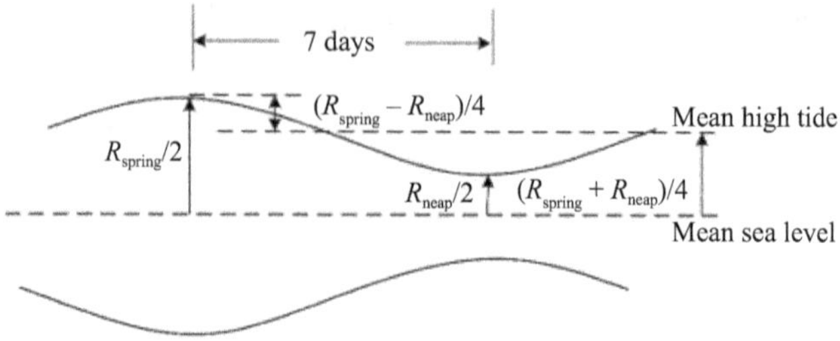

Figure 4.13 Sinusoidal variation of tidal range

Hence,

$$R = \frac{R_{\text{spring}}}{2}\left[(1+\alpha) + (1-\alpha)\sin(4\pi t/T)\right] \tag{4.38}$$

where $R_{\text{neap}} = \alpha R_{\text{spring}}$.

The power can be obtained from the mean square range, $\overline{R^2}$, as

$$\overline{R^2} = \frac{R^2_{\text{spring}}}{4}\frac{\int_{t=0}^{T}\left[(1+\alpha) + (1-\alpha)\sin(4\pi t/T)\right]^2 dt}{\int_{t=0}^{T} dt} \tag{4.39}$$

Thus,

$$\overline{R^2} = \frac{R^2_{\text{spring}}}{8}\left(3 + 2a + 3a^2\right) \tag{4.40}$$

The mean power produced over the month is

$$\overline{P}_{\text{month}} = \frac{\rho A g}{2\tau}\frac{R^2_{\text{spring}}}{8}\left(3 + 2a + 3a^2\right) \tag{4.41}$$

where τ is the intertidal period.

The value of α is about 0.5, and so, (4.41) differs little from the approximations that are commonly used in the literature by either

$$\overline{P} \approx \frac{\rho A g}{2\tau}\left(\overline{R}\right)^2 \tag{4.42}$$

where $\overline{R}$ is the mean range of all tides, or

$$\overline{P} \approx \frac{\rho A g}{2\tau}\left(\frac{R^2_{\max} + R^2_{\min}}{2}\right) \tag{4.43}$$

where $R_{\max}$ and $R_{\min}$ are the maximum and minimum ranges.

Example 4.1 Typical values of mean tidal power are given as $R_{spring} = 5\ \text{m} = R_{max}$, $R_{neap} = 2.5\ \text{m} = R_{min}, A = 10\ \text{km}^2,\ \rho = 1.025 \times 10^3\ \text{kg/m}^3,\ g = 9.81\ \text{m/s}^2$ and $\tau = 4.47 \times 10^4$ s ($\approx$ 12 h and 25 min). Therefore, $\alpha = 0.5$ and $\overline{R} = \frac{5+2.5}{2} = 3.75$ m.

Applying (7.40) gives $\overline{R^2} = 1.84, \frac{\rho A g}{2\tau} = 1.125 \times 10^6$.

Then, $\overline{P}$ can be calculated as given by (4.41), (4.42) and (4.43) to see the difference that results from the approximation as follows:

From (4.41),

$$\overline{P} = 16.70\ \text{MW}$$

From (4.42),

$$\overline{P} = 15.82\ \text{MW}$$

From (4.43),

$$\overline{P} = 17.58\ \text{MW}$$

It is to be noted that the value resulted from (4.41) is the average value of those resulted from (4.42) and (4.43) as approximated values.

4.4.2 Tidal stream

The tidal stream generation (TSG) system exploits the major tidal flows and ocean currents (e.g. gulf stream) to extract its tidal energy using turbine generators beneath the surface of the water. In principal, TSG looks like wind power generation, except that the water currents flow across turbine rotor blades rather than wind currents turning the blades for wind power turbines. So, TSG areas on the sea can look just like underwater wind farms.

TSG system includes a turbine-generator set as shown in Figure 4.14(a). The ocean's currents turn the turbine blades, powering the generator, and then electricity can be transmitted to the shore through submarine cables. Tidal turbines can range from about 5 to 30 m in diameter and are designed to swivel 180 degrees or reverse rotation as tides ebb and flood. The stream generator, often referred to as a tidal energy converter (TEC), is the machine that converts the kinetic energy provided by the turbine into electrical energy. TSG system may include several turbine-generator sets as shown in Figure 4.14(b). The set can be supported on the ocean floor by either tripod or concrete block or sunken pile.

Unlike offshore wind power, which can suffer from storms or heavy sea damage, tidal stream turbines operate below the sea surface or are fixed to the seabed. Tidal streams are formed by the horizontal fast-flowing volumes of water caused by the ebb and flow of the tide as the profile of the seabed causes the water to speed up as it approaches the shoreline.

Tidal streams' turbines have much smaller diameters and higher tip speed rates than those of wind turbine because the water is denser than air (1,000 times greater than air) and has a much slower flow rate. They generate tidal power on both the ebb and

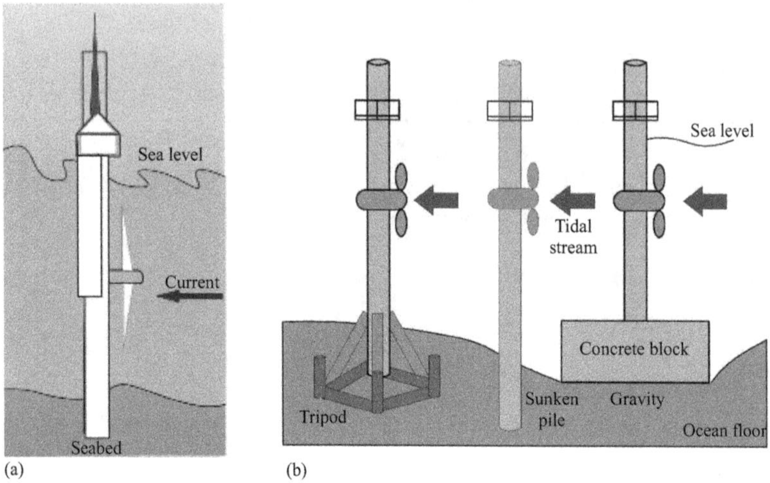

Figure 4.14 Tidal stream generators: (a) single turbine generator and (b) set of multi-turbine generator

flow of the tide. One of the disadvantages of TSG is that as the turbines are submerged under the surface of the water they create hazards to navigation and shipping.

Other forms of tidal energy include tidal fences, which use individual vertical-axis turbines mounted within a fence structure, known as the caisson, to completely block a channel and force water through them. An alternative way of harnessing tidal power is using an oscillating tidal turbine by which the tidal stream flowing past it causes its giant wing to oscillate up and down with the movement of the tidal current. This motion is used to generate electricity.

4.4.2.1 Tidal energy calculation

Tidal energy is the energy due to the tidal stream (kinetic energy). It is also a fact that the increase in tidal stream energy yields, to a large extent, an increase in energy extraction [13,14].

Various turbine designs have varying efficiencies and, therefore, varying power output. If the efficiency of the turbine, ξ, is known, the power output of a turbine is given by

$$P = \xi \frac{\rho A v^3}{2} \text{ kW} \tag{4.44}$$

where ρ = the density of the water (1,025 kg/m^3), A = the sweep area of the turbine (m^2) and v = the velocity of the flow (m/s).

Tidal velocities vary with time and can approximately be determined by [5]

$$= v_{\max} \sin(2\pi t/\tau) \tag{4.45}$$

where $v_{\max}$ is the maximum speed of the current and τ is the period of the natural tide (12 h and 25 min for a semidiurnal tide).

In practice, ξ may approach a maximum of 40%. Thus, the average electrical power output is

$$\bar{P} \approx \frac{0.4}{2}\rho A v_{\max}^3 \frac{\int_{t=0}^{\tau/4} \sin^3(2\pi t/\tau)dt}{\int_{t=0}^{\tau/4} dt} \approx 0.1\rho A v_{\max}^3 \text{ kW} \tag{4.46}$$

Example 4.2 For a tidal current maximum velocity of 4 m/s and a sweep area of a circle with a radius of 5.6 m, the power output of the wind turbine can be calculated as follows:

The sweep area is

$$A = \pi(5.6)^2 \approx 100 \text{ m}^2$$

From (4.44), the power is

$$P = 0.4\frac{1{,}025 \times 10^2 \times (4)^2}{2} \approx 1.3 \text{ MW}$$

From (4.46), the average power is $\overline{P} \approx 0.1 \times 1{,}025 \times 10^2 \times 4^3 \approx 6.6$ MW.

4.4.2.2 Resource assessment

It is important to recognize how best to estimate the resource of a site. The real amount of power generated will depend on the hydrodynamic and other characteristics of the turbine used. It is worthy to make an upper estimate of the resource having a detailed knowledge of tidal design.

It is found that early assessment of tidal energy in a channel focuses on calculations using kinetic energy flux model. It is, of course, a useful measure but has been shown not to be related to the maximum extractable power. Physically, significant complicated limitations of tidal power generation exist. For instance, the maximum possible energy extraction from a strait connecting two large basins is given to within 10% by [11]

$$P = 0.22\rho g \Delta H_{\max} Q_{\max} \tag{4.47}$$

where $\Delta H_{\max}$ is the maximum differential water surface elevation across the channel and $Q_{\max}$ is the maximum volumetric flow rate through the channel.

To understand this limit, considering the absence of turbines retarding the flow in a site, the current is a maximum, but no power can be generated as no work is done. Meanwhile, for a site so blocked by turbines that the flow is small, the extracted power is going to be low. This yields the existence of an optimal condition in between, at which the power is maximum. In practice, the maximum extractable power is likely to be significantly higher than that can be realized from

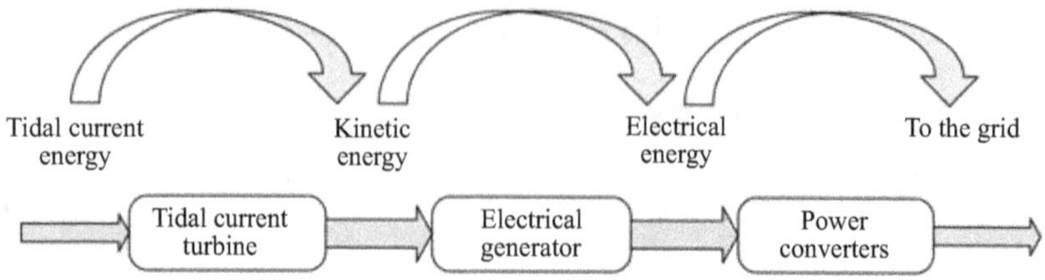

Figure 4.15 Typical structure of tidal current energy conversion system

a site because of two key reasons: (i) large losses from the wake mixing behind turbines, from the turbine support structures, in hydrodynamic inefficiency in the turbine, and in generator and transmission losses; (ii) in case more rows of turbines are placed at a site, there is a diminishing return – that is, when adding a second row of turbines in addition to the first one, it will not double the power but will lead to both rows of turbines producing less than the original one row [15].

4.5 Tidal current energy conversion devices

A typical structure of tidal energy conversion system is shown in Figure 4.15 as a chain of the devices: turbine, generator and converters. A brief description of both is in the next subsections, and more details are explained in Chapter 2, Section 2.3, for turbines and Section 2.4 for generators.

4.5.1 Tidal current turbines

The tidal current turbine is a device used to harness the kinetic energy in a tidal form and then transforming the energy into mechanical power, which can then drive a generator. The technology of tidal current turbine design to harness tidal energy is the same as that used to design many wind turbines to harness the wind power as both tidal current and wind are in the form of fluid. Most tidal turbines can be characterized as belonging to three fundamental types [16–19].

4.5.1.1 Horizontal axis turbines

This type of turbine rotates around a horizontal axis which is parallel to the current stream and is the most common type used to date. To generate greater starting torque and reduce balancing problems encountered with single-blade turbines, multi-bladed turbines are favoured. The number of blades should be decided with taking into consideration that hydrodynamic losses are increased with greater number of blades. Fixed pitch or variable pitch blades can be used depending on the turbine design.

4.5.1.2 Vertical axis turbines

The rotation speed of both vertical axis turbines and horizontal axis is very low ($\approx$15 rpm). However, vertical axis turbines can transmit the torque directly to the water surface without the need of complex transmission systems or underwater nacelle and they permit the harnessing of tidal flow from any direction.

4.5.1.3 Oscillating hydrofoil turbines

This kind of turbine induces hydrodynamic lift and drag forces due to a pressure difference on the foil section caused by motion of the tidal current. Consequently, a resultant tangential force to hydraulic fluid to drive an electrical generator is provided. It has been found that linear generators are the superior power take-off choice for this kind of turbine, which is normally used for wave energy applications.

4.5.2 Tidal current generators

The generator is the device that converts the input mechanical power delivered by the turbine through its shaft into electric power. As the tidal current turbine topologies are like wind turbine, many wind turbine generator topologies can be used for tidal current turbines.

4.5.2.1 Squirrel cage and wound rotor induction generators

The two types of induction generators, squirrel cage (SCIG) and wound rotor (WRIG), are widely used as they are relatively inexpensive, robust and require low maintenance. SCIG is mounted on the same shaft of the turbine through a multiple-stage gearbox and is connected to the grid through a full-scale power converter and a transformer. WRIG has a similar topology but with controllable rotor resistance.

4.5.2.2 Doubly fed induction generator

It has a stator directly connected to the grid through a transformer, whereas the wound rotor is connected through a power electronic converter. The rating of the converter is in the range of 25%–30% of the generator capacity and the generator can operate in a wide range of speed variation (e.g. ±30%) around the synchronous speed. So, doubly fed induction generator (DFIG) is promising and widely used from the economic point of view for wind integration but it is probably not the case in tidal turbine applications in comparison with the permanent magnet synchronous generator (PMSG).

4.5.2.3 Permanent magnet synchronous generator

It is commonly used in direct drive train with full-scale power converter connected to the grid. PMSG has high potential for the tidal current turbines because of its reduced failure, increased energy yield and reliability. It is favourable for oscillating hydrofoil systems with horizontal turbines.

4.5.2.4 Direct drive permanent magnet generator

The direct drive synchronous generator with permanent magnet (DDPM) is the most attractive because it has the highest yield of energy even if it is more expensive than the generator systems with gearbox. There are many different alternatives for the design of DDPM as suggested for tidal current turbine and wind turbine as well. Basically, DDPM can be divided into radial-flux machine, axial-flux machine and transverse flux machine according to the flux direction in the air gap [20].

4.5.3 Drawbacks of tidal stream generators

Because of high start-up and construction costs due to limited experience of installing, operating and maintaining plants, contractors' perceptions of risk are likely to be reflected in higher costs.

Like the wind, waves and tidal streams are variable RESs. Their intermittent generation has implications for large-scale grid integration.

Because the amount of energy of the tidal streams naturally varies over time, the power output of wave energy converters and tidal stream generators will also vary. This has implications for grid integration, particularly balancing supply and demand.

4.6 Tidal power plants connected to utility grid

Traditionally, to facilitate power flow from the generating stations to the load centres through the transmission system, the electrical networks are designed and operated in such a way that both frequency and voltage are fixed. Generation technologies such as hydro, thermal and nuclear are supplied with synchronous generators to produce bulk power at constant frequency and constant voltage by monitoring and controlling the active power and reactive power, respectively, in a reliable and cost-effective manner.

Renewable energy plants are deployed in areas in which the network availability is limited (e.g. remote areas) or to assist the power network for facing the demand increase in addition to the other benefits of green energy sources. However, tidal power plants can be developed in areas with network accessibility as most of the high-energy tidal channels are in naturally articulated channels already being used by shipping, navigation and neighbouring populated areas.

In the case of grid integration, some issues should be investigated, such as capacity limits, power quality and system analysis [21].

4.6.1 Capacity limits

For areas with limited network strength, reinforcement of transmission line thermal rating and fault protection mechanisms may be needed, as well as limitations in transmission capacity because of the technical and cost issues associated with it. However, high currents may flow into the distribution network and its feeders in fault conditions, and that makes the planners to pay more attention to the switchgear components.

4.6.2 Power quality

Tidal generation source behaves as a time-varying generation source, which in turn leads to a complex voltage profile in the steady state, and flicker because of resource intermittency, start-up and shutdown conditions.

Tidal power generation systems with induction machines may consume significant amount of reactive power. Therefore, compensation techniques may be essential to reduce system losses and enhance voltage stability. In switching conditions such as sudden disconnection of generators when fault occurs or turn on/off actions of intermittent energy converters coupled with induction machines and transformers, it may cause large, sudden and sustained alteration of network voltage.

Another important aspect of power quality is '*Harmonics*', which may be injected into the network in some cases such as use of PM generators through direct-drive or gear-coupled arrangements and power electronic systems (full-sized for PM generators and partially rated for DFIGs) used in grid-interfacing schemes.

4.6.3 System analysis

To analyse a system incorporating tidal energy sources, load flow studies are required prior to integration to a network. That necessitates the availability of adequate and accurate system performance data for applying numerical modelling. Typical with any other generation or load condition, tidal energy systems are exposed to faults or failures in system elements (e.g. internal connection circuits, site-to-shore cables and onshore network supplying power to the grid) due to harsh sea conditions, human activities (e.g. fishing, anchors and leisure activities) and short-circuit cables. For short-circuit calculations with a goal of protection design and service personnel safety, grounding is an important issue.

For tidal energy systems planning, some studies are essential such as (i) stability (transient stability, small signal stability, voltage stability and frequency stability), (ii) scheduling and unit commitment to maintain the balance between electricity demand and supply and (iii) load forecasting [22]. Therefore, dynamic model development and statistical data must be determined.

Intermittency and variability of tidal current systems can be mitigated through long-term storage techniques, which have high cost at present.

4.7 Environmental impacts

Tidal energy as a renewable resource does not result in any emissions. So, it is beneficial to the environment because its use reduces the demand on fuels (coal and oil) and it is attractive to many because of its reliance on existing technology and knowledge. In contrast, some of the non-monetary costs associated with barrage systems include destruction of habitat, interruption of organisms' travel routes and potential both electromagnetic interference and acoustic pollution. Moreover, hydroelectric systems contribute to killing fish, and the waste heat that ends up in the water reduces its capacity to store dissolved oxygen, harming not only fish but all organisms in the affected area. Tidal energy systems also lead to a reduction in the area of intertidal foreshore, which affects feeding habits of the large numbers of wading birds that use the estuary [23].

4.8 Tidal power pros and cons

Though the development of the tidal power is at the nascent stage, it is estimated to have a vast potential globally. It is generated from converting energy from the force of tides into power. Tidal energy production is more predictable compared to wind energy and solar power. Despite this, tidal power is still not exploited widely even though the world's first large-scale tidal power plant became operational in 1966. However, increased focus on generating power from renewable sources is expected to accelerate the development of new methods to exploit tidal energy.

4.8.1 Pros of tidal power

Renewable: As it is derived from tidal waves, which result from the gravitational pulls from the sun and the moon, and the earth's rotation around its axis, the energy is classified as a renewable source. Compared to limitations in power generation from finite fossil fuel reserves, tidal energy does not have any such constraints. Taking this into account, the tidal energy generation may last for thousands of years as long as the tides remain.

Environmentally friendly: As the tidal power generation emits no harmful gases into the atmosphere, it is an environmentally friendly energy source. Further, the tidal power projects do not need much space compared to solar farms which require a large area of land for installation of modules. Tidal power installations also occupy less space compared to other modes of power generation.

Predictability: As it is easy to know the occurrence of tidal currents, the scope of a tidal energy project can be measured accurately. The construction of the projects can also be carried out with correct dimensions based on the formation of high and low tides in certain cycles. Assessment of these factors can help easily determine the size and capacity of a tidal plant at an area.

Cost-effective: Given the longer lifespan of tidal power plants compared to those that depend on fossil fuels, they are considered as cost-effective. Also, the tidal plants can be effective at low tidal speeds and produce power to bring down the overall cost.

4.8.2 Cons of tidal power

Site availability constraints: Limited availability of sites has remained a major constraint in the development of tidal power projects. The limitation has restricted the growth of tidal power projects worldwide. However, recent innovations in design and turbine technology for tidal power are expected to offer a potential to increase the installed capacity of tidal plants. The technological developments are also expected to result in reducing economic and environmental costs of the projects, thereby increasing their availability.

Sites close to land: Tidal power plants must be constructed near to the land, as high tidal currents are needed to produce power. A huge variation in tides and tidal velocities can significantly increase the potential of a site for tidal electricity generation. Though the places where there is variation in tides offer greater potential for tidal projects, scientists are trying to develop technology which can generate enough energy even in low-tide environment.

Environmental effects: The impact of tidal power projects on the environment is yet to be ascertained precisely. Tidal power can damage marine life, as tidal turbines with their rotating blades may lead to death of living creatures in a sea. Noise from the rotation of the turbines may also impact fish habitations in the tidal power locations. Tidal energy can also impact the quality of water and sediment processes.

Tidal barrage: Installing a tidal barrage may result in a change to shoreline within a bay, damaging an ecosystem that relies on tidal flats. Flushing of the bay may also be affected by the prevention of the flow of water in and out of the bay. As a result, there may be reduction in saltwater, impacting marine life.

References

[1] Schatz D., Allan P. and Cooper D. *The moon and its phases: Modeling moon phases* [online]. 2003. The Pacific Science Center. Available from https://www.southampton.ac.uk/~sj2y10/moonPhasesSheetA3.pdf [Accessed 2 Feb 2020].

[2] Richardson C. J. *The 'ins and outs' of tides and their use as a renewable energy source* [online]. 2015. Institute of Physics – Environmental Physics Essay Competition. Available from https://www.iop.org/activity/groups/subject/env/prize/winners/file_65752.pdf [Accessed 2 Feb 2020].

[3] Scherrer D. *Teaching moon phases*. 2015. Stanford Solar Centre. Available from http://solar-center.stanford.edu/activities/MoonPhases/Teaching-Moon-Phases.pdf [Accessed 2 Feb 2020].

[4] Frau J. P. 'Tidal energy: Promising projects'. *IEEE Transactions on Energy Conversion*. 1993;8(3):552–8.

[5] Blokdyk G. *Tidal power* [online]. E-book, Chapter 13, pp. 429–52. Available from http://documents.routledge-interactive.s3. amazonaws.com/9780415584388/S12-1_(RER2e-ch13-tidal_power).pdf [Accessed 2 Feb 2020].

[6] Bhatia S. C. *Advanced renewable energy systems*. New Delhi: Woodhead Publishing India PVT Ltd.; 2014.

[7] Kawalik Z. and Luick J. *The oceanography of tides: Tidal forces*. Alaska: University of Alaska Fairbanks; 2013, Chapter 1, pp. 5–40.

[8] Micks S. D. *Understanding tides*. US Dept. of Commerce: National Oceanic and Atmospheric Administration (NOAA); 2006, Chapters 5–9, pp. 21–50.

[9] Gjevik B. *Lectures on tides*. Oslo: University of Oslo, UNIS Longyearbyn; 2011.

[10] Aly H. H. H. 'Dynamic modeling and control of the tidal current turbine using DFIG and DDPMSG for power system stability analysis'. *Electrical Power and Energy Systems*. 2016;83:525–40.

[11] Tousif S. R. and Taslim B. S. 'Tidal power: An effective method of generating power'. *International Journal of Scientific and Engineering Research*. 2011;2(5):1–5.

[12] Bohari Z. H., Sulaima M. F., Baharom, M. F. and Bukhari W. M. 'Are tidal power generation as the future generation for Malaysian climate and

location: A technical review?'. *International Journal of Applied Engineering Research.* 2016;11(10):7095–9.

[13] Vikas M., Ran S. and Seelam J. K. (eds.). 'Tidal energy: A review'. *Proceedings of International Conference on Hydraulic, Water Resources and Coastal Engineering (Hudro2016)*; Pune, India, Dec 2016, pp. 2320–8.

[14] Al Yusuf S., Gupta S. D., Ali T., Arnab I. Z. and Begum J. 'An introduction to tidal power generation and analyzing the potentiality in Bangladesh'. *Sustainable Energy Technologies (ICSET), 2012 IEEE 3rd Int. Conf. on*; Kathmandu, Sep 2012, pp. 316–20.

[15] Adcock T. A. A., Draper S. and Nishino T. 'Tidal power generation – A review of hydrodynamic modelling'. *Proc. IMechE Part A: Journal of Power and Energy.* 2015;229(7):755–71.

[16] Rourke F. O., Boyle F. and Reynolds A. 'Marine current devices: Current status and possible future applications in Ireland'. *Renewable and Sustainable Energy Reviews.* 2010;14(3):1026–36.

[17] Zhang J., Moreau L., Machmoum M. and Guillern P. E. 'State of the art in tidal current energy extracting technologies'. *First International Conference on Green Energy ICGE.* Sfax, Tunisia, March 2014, pp. 1–7.

[18] Mohammadi S. and Jam S. S. 'Usage of tidal power as power station'. *Electrical Power Distribution Networks (EPDC), 2012, Proc. of 17th Conf. on.* Tehran, Iran, May 2012, pp. 1–7.

[19] Chen H., Aït-Ahmed N., Zaïm E. H. and Machmoum M. 'Marine tidal current systems: State of the art'. *Industrial Electronics (ISIE), 2012 IEEE Int. Symposium on.* Hangzhou, China, May 2012, pp. 1431–7.

[20] Chen Y., Pillay P. and Khan A. 'PM wind generator topologies'. *IEEE Trans. on Industry Applications.* 2005;41(6):1619–26.

[21] Khan J., Bhuyan G., Moshref A., Morison K., Pease Jr. J. H. and Gurney J. 'Ocean wave and tidal current conversion technologies and their interaction with electrical networks'. *PES General Meeting, 2008 IEEE-Conversion and Delivery of Electrical Energy in the 21st Century.* Pittsburgh, PA, July 2008, pp. 1–8.

[22] Jones A. T. and Westwood A. 'Recent progress in offshore renewable energy technology development'. *PES General Meeting, 2005 IEEE.* San Francisco, USA, June 2005. IEEE 2017–22.

[23] Baker A. C. 'Tidal power'. *IEE Proceedings Pt. A.* 1987;134(5):392–8.

Chapter 5

Ocean energy: wave and thermal energy

Useful electric energy can be derived from the ocean waves and is known as 'ocean wave energy'. Physics of waves and their different forms as well as estimation of energy and its conversion into electrical energy are explained in the forthcoming sections.

5.1 Wave energy concept

Wave energy is the energy contained in the waves that are caused by winds blowing across the surface of the sea. Friction transfers some of the wind energy to the water thus forming waves. This energy is stored in the waves as (i) potential energy in the mass of water displaced from the mean sea level and (ii) kinetic energy in the motion of water particles. Thus, the stronger the wind is, the more energy in the waves and the larger the distance the waves travel. The wave energy is proportional to the square of the height of the wave, which is inversely proportional to the depth of the water. This means that the waves near the coast carry the most energy.

Wave energy is not tidal energy even though both fall under the category of ocean energy. While the tidal energy consists of long period oscillations, sea waves of short period are generated by the action of the wind. The total theoretical wave energy resource is very high, roughly twice the global electrical energy consumption as the oceans cover about 70% of the surface of the earth and are considered to be one of the greatest sources of renewable energy. Therefore, it is worthy to capture the wave energy to be transferred into useful energy, such as electricity. The devices used to capture wave energy and convert it into electrical energy are known as 'wave energy converters' (WECs).

5.2 Ocean waves creation

Radiation from the sun to the earth and ocean surface creates air circulation, which is wind. When wind blows over the surface of the oceans, waves will be generated by the effect of both gravity and friction (Figure 5.1(a)). Energy in the wind is transferred to the surface water of the sea or ocean producing waves of size that depends on the magnitude of that energy, which is a function of the wind speed, the duration the wind blows and the area affected by the wind (known as 'fetch'). The

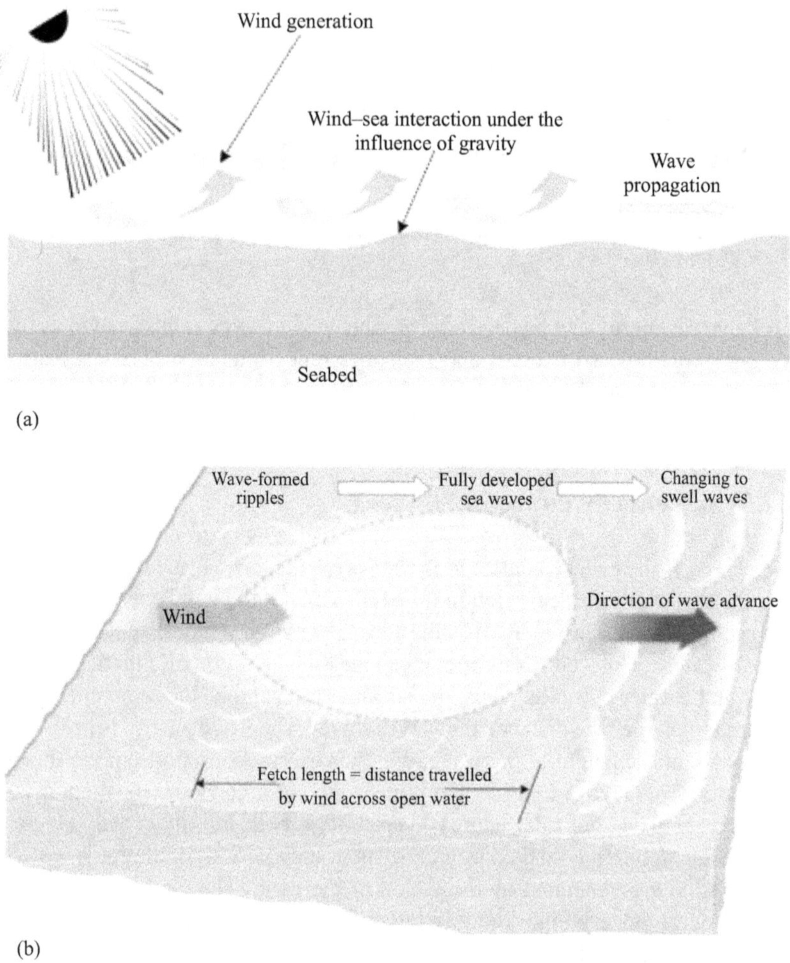

Figure 5.1 Ocean waves: (a) wave creation and (b) fetch area created by wind blows

fetch is specified by its length, where the fetch length, is expressed as the distance travelled by the wind across open water. When the wind blows across the water surface, wave-formed ripples or capillary waves may take place as well as continuous flows in the direction of wave advance. Thus, the wave is fully developed and can change to swell (Figure 5.1(b)) [1].

5.2.1 Ripples

In sedimentology, wave-formed ripples or wave-formed ripple marks are a feature of sediments (sandstones, limestone, siltstones) and dunes. These ripple marks are often characterized (and thus distinguished from current ripples) by symmetric

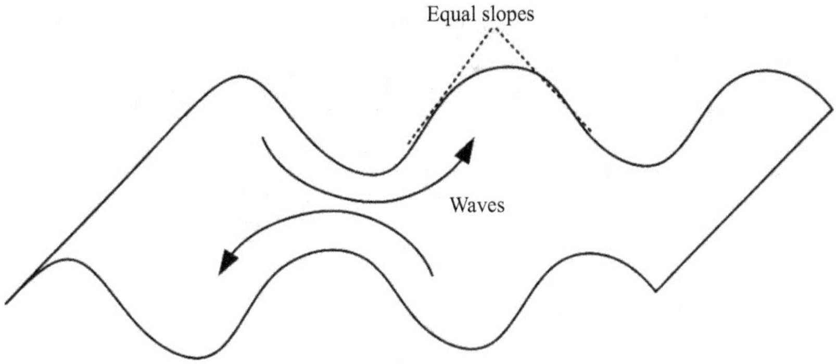

Figure 5.2 Symmetrical ripple

cross sections and long relatively straight crests, which may commonly bifurcate. Commonly, these crests can be truncated by subsequent flows. Their wavelength (periodicity) depends on the sediment grain size, water depth and water-particle orbits in the waves. Symmetrical ripples are commonly found in shallow waters. Beaches are a good place to find these ripples. A typical example is shown in Figure 5.2.

Though wave-formed ripples are traditionally described as symmetrical, asymmetric wave ripples are common in shallow waters along sandy shores. They are produced by bottom oscillations generated by passing breaker waves, which have unequal intensity in opposite directions [2].

Wave-formed ripples indicate an environment with weak currents where water motion is dominated by wave oscillations [3]. However, ripples are relatively small elongated ridges that form on bed surfaces perpendicular to current flow. With continuous current flow in one direction, asymmetrical ripples form. Asymmetrical ripples contain a steeper slope downstream. With an alternation in current flow from one direction to the opposite, symmetrical ripples form. Symmetrical ripples tend to have the same slope on both sides of the crest [4].

Capillary wave. It is a wave travelling along the phase boundary of a fluid whose dynamics and phase velocity are dominated by the effects of surface tension. The wavelength of capillary waves on water is typically less than a few centimetres, with a phase speed in excess of 0.2–0.3 m/s [5].

This type of wave is known as a gravity wave. In addition, the group velocity is times the phase velocity as: (i) the wavelength is short and (ii) the phase velocity is dominated by the surface tension effects. This type of wave is known as a capillary wave (Figure 5.3).

Swell waves. A swell, in the context of an ocean, sea or lake, is a series of mechanical waves that propagate along the interface between water and air and thus are often referred to as surface gravity waves. Occasionally, swells which are longer than 700 m occur as a result of severe storms (wind, tides and rainfall). A swell has a period in the order of a fraction of a minute to several minutes. Due to

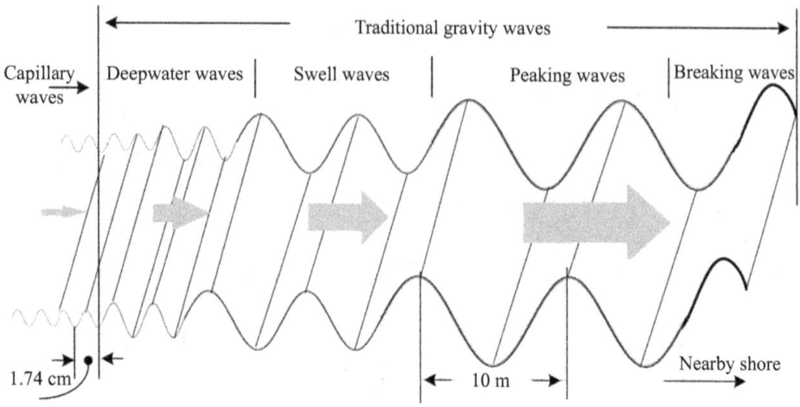

Figure 5.3 Capillary wave formation

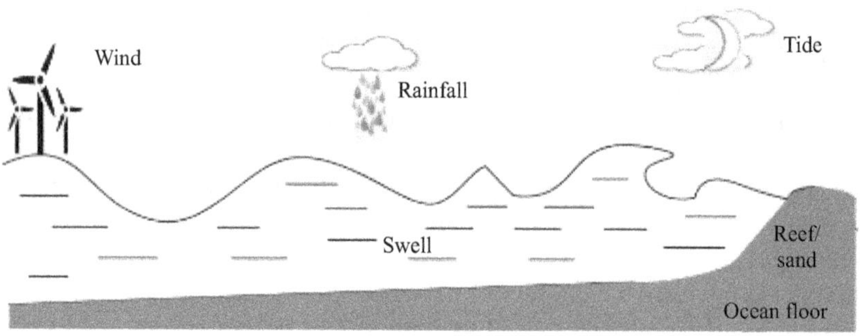

Figure 5.4 Swell wave formation

its long wavelength, this water wave travels very fast. It is also very rounded and most people when first encountering it may not recognize this as a wave. Some call it the hills and valleys of the sea. Swells need a large body of water to develop and are usually, but not always, associated with the ocean. A typical graph of swell is depicted in Figure 5.4.

A swell is a wave; so, often to prevent confusion, the shorter wavelength variety is called chop or choppy waves. These waves have periods in the order of a few seconds. Most waves are wind generated. As chop and swell are waves, they can interact and pass through each other. Most waves are wind generated. Swells move larger amounts of water, so it takes longer for the energy to build up and dissipate. Swells also travel fast and for a long distance. For that reason, there is often a swell present in many places of the ocean. Chop can whip up pretty quick and die off just as fast. A variant of chop is a whitecap. A whitecap is when the wind transfers more energy to a water wave than the wave can transmit, the top may curl and blow in the wind making the top appear white. Ocean sea spray is the result of whitecaps.

5.2.2 Shallow waves or shore waves

They are formulated when the wave energy interacts with the bottom or land (Figure 5.5). However, there is an ocean wave called a breaker. Breakers strike fear and concern in the heart of any wise seafarer. A breaker is a swell interfering with a shallow bottom in normally deep water. The result is a localized, out of place, chop that does not seem to travel on its way. It is an indication of shallow rocks, reefs or a shallow submerged object.

Breaking waves. They are formed out in the open ocean and can travel vast distances before breaking on a distant coastline (Figure 5.6). The energy carried by these waves and the way they break against the shoreline has a dramatic impact on erosion and how shorelines are shaped over time.

As waves approach the shore, the bottom of the wave meets the ocean floor. As they drag across the bottom, the front waves slow down, and the wavelength is reduced. The following waves start to build up behind the slow ones, and as the wavelengths get shorter, the wave energy gets transferred upwards, increasing wave height. The friction along the bottom slows down the base of the wave while the water at the surface continues forward. When the wave steepness (wave height to wavelength ratio) exceeds a ratio of 1:7, it becomes unstable and breaks at the surf zone (Figure 5.7). Surf is the wave activity in the area between the shoreline and the outer limit of breakers. It may refer to a breaking wave in shallow water, upon the shore or in the area in which waves break.

Therefore, the slope of the sea floor greatly influences how waves break. A physical graph of breaking waves is shown in Figure 5.8.

5.3 Wave propagation

As explained in Section 5.2, it can be concluded that along the wind direction towards the shoreline, the form of water waves may change from one to another depending on the depth of the seabed *D*. A typical example is shown in Figure 5.5. At the full depth of water, the deepwater waves may swell and propagate as transitional waves. Decreasing water depth results in an increase of the wave peak

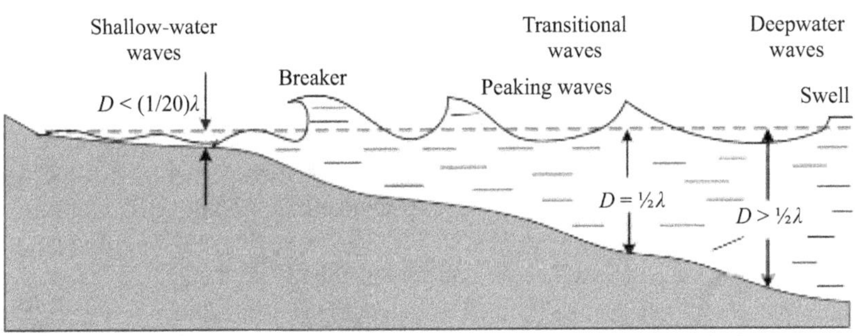

Figure 5.5 Types of ocean waves in terms of water depth

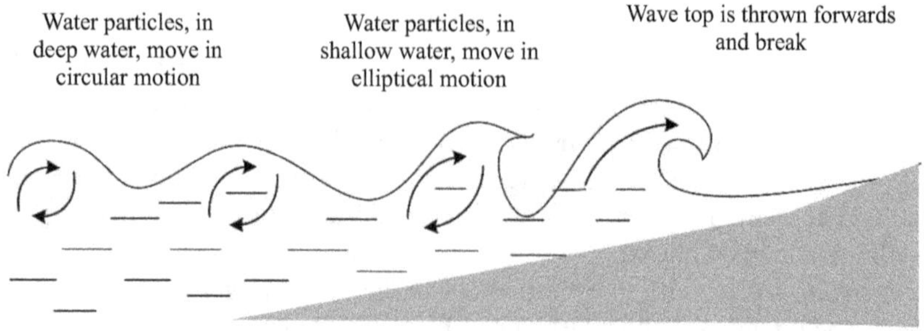

Figure 5.6 Breaking waves near the shore

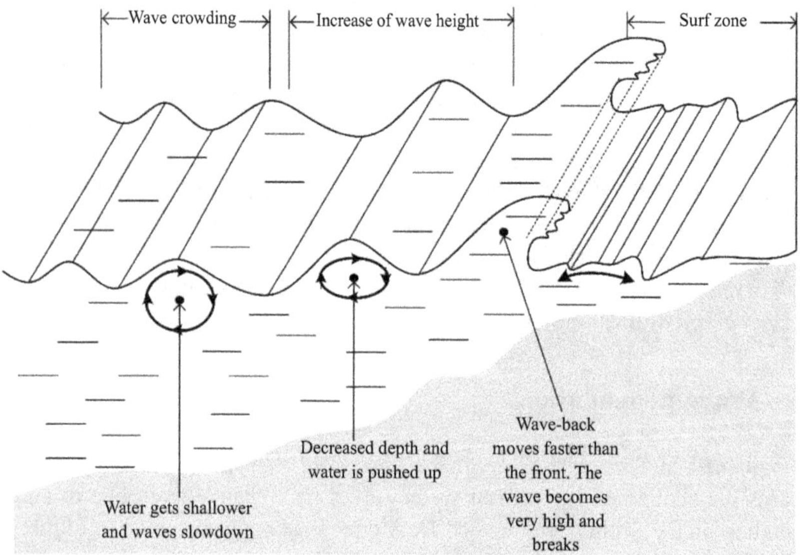

Figure 5.7 Breaking mechanism

(peaking waves). When approaching the shoreline, the increase of wave peak continues and the wave form changes to breaking wave form, and then shallow-water wave.

The deepwater wave (also called 'surface wave') is the most common form for which the most wave energy devices are designed to extract energy. For the water waves to behave as a deepwater wave, the mean depth of the seabed should be more than about half the wavelength, λ (i.e. $D > \lambda/2$). The deepwater wave form can be characterized by the parameters: crest and trough amplitudes, height (H), mean depth of the water or still-water level (D) and wavelength (λ) as illustrated in Figure 5.9.

As depicted in Figures 5.6 and 5.7, in a deepwater wave, the motion of particles is circular with an amplitude that decreases exponentially with depth and

Figure 5.8 Typical physical graph of breaking waves. Source: Earth and Ocean Sciences (EOSC)

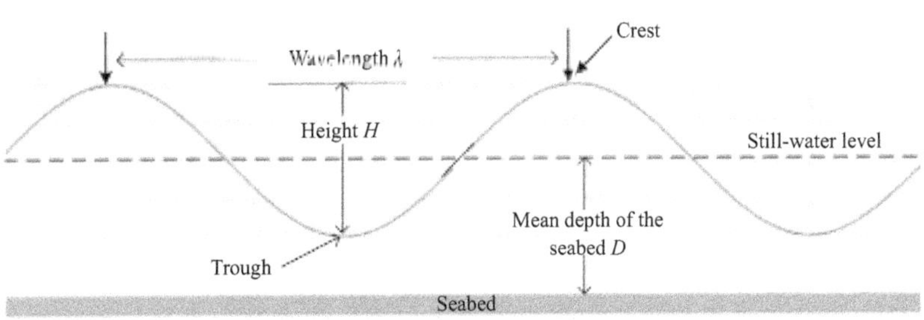

Figure 5.9 Parameters of deepwater wave

becomes ignored at $D < \lambda/2$. The particle motion for shallow-water wave is changed to be elliptical and the water movement occurs against the sea bottom producing energy dissipation (Figure 5.10(a) and (b)).

Deepwater waves have their own properties to be considered for the mathematical representation of these waves. They can be summarized as follows:

- Water on the surface remains on the surface.
- Surface waves are sets of continuous sine waves of varying wavelength, phase and direction.

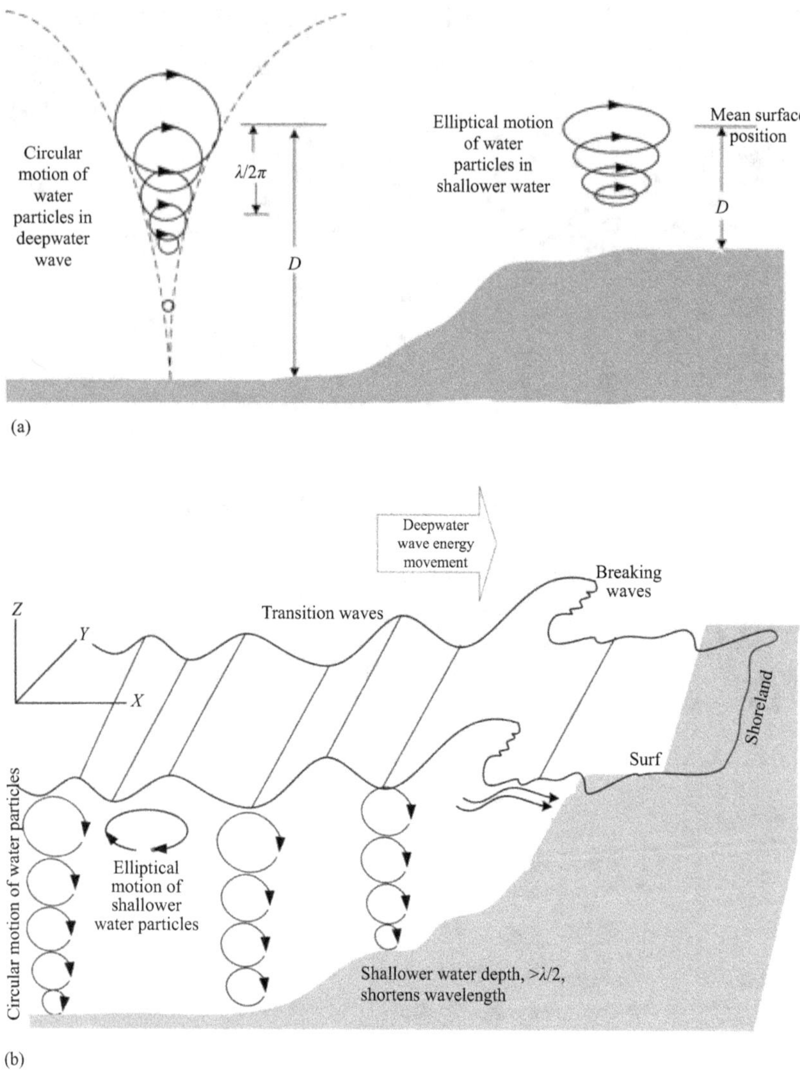

Figure 5.10 Particle motion in water waves (a) in deepwater and shallow-water waves and (b) representation of particle motion for different wave forms in three-dimension ($X =$ direction of wave energy movement, $Y =$ wave width, $Z =$ wave height)

- The motion of water particles is circular. The surface waves have a shape that progresses as a moving wave, although the particles themselves do not progress and have no net motion.
- The amplitude of the surface wave depends mainly on the history of the wind regimes above the surface and is slightly dependent on the wavelength, velocity and duration.

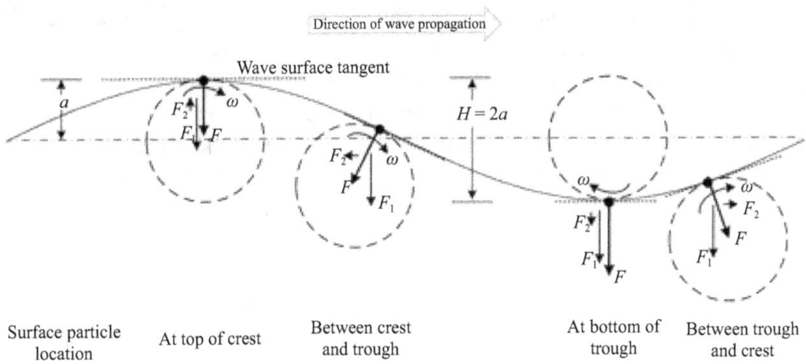

Figure 5.11 Forces affecting surface water particles at different locations

- The amplitude of water particle motion decreases exponentially with the water depth below the mean surface position; that is, it reduces to λ/e at a depth of $\lambda/2\pi$ (e = the base of natural logarithms = 2.72) as shown in Figure 5.10(a).
- The particle motion is negligible at depth of $\lambda/2$ as its amplitude is less than 5% of that of the surface motion.

5.4 Fundamentals of deepwater waves

For deepwater waves, the water particles in the surface are affected by friction, surface tension and inertial forces as well as forces of gravity (F_1) and circular motion (F_2), which are the dominant forces as the other forces are relatively small. The amplitude of the wave equals the radius of the surface particle circular motion, a, and therefore, the height H is twice the amplitude, that is, $H = 2a$, and ω (radian per second) is the angular velocity of the water particles.

As shown in Figure 5.11, along the direction of the wave motion the moving shape results from the phase differences in the motion of successive particles of water. As one particle in the crest drops to a lower position, another particle in a forward position circles up to continue the crest shape and the forward motion of the wave.

The surface water particle of mass m is subjected to a resultant force F of two components: the gravitational force F_1 and the centrifugal force F_2. The water surface takes up the position produced by this resultant force, that being perpendicular to the tangent to the surface. For instance, F_2 moves a particle at the top of a crest upwards. The particle is dropping a moment later and the position in the crest is taken by a neighbouring particle rotating with a delayed phase. At a particle position between crest and trough nearby the still-water level, the surface orientates perpendicular to the resultant force F. Similarly, for a water particle at the bottom of the trough, the centrifugal force is downward and the resultant force is maximum; whereas at a position between trough and crest, the water particle is going to complete a full cycle of its motion [6].

Consider a general position of a surface particle at a height of h above the mean water level and the wave moving along X-direction as shown in Figure 5.12. The forces affecting that particle can be expressed as $\vec{F}_1 = ma\omega^2$, $\vec{F}_2 = mg$ and, therefore, $\vec{F} = \vec{F}_1 + \vec{F}_2$ in a direction perpendicular to the wave surface tangent.

Assuming that the particle is initially at time $t = 0$, at the average water level, then,

$$\varphi = \left(\frac{\pi}{2}\right) - \omega t$$

The resultant acceleration γ, as in Figure 5.12, can be calculated as follows:

$$\gamma^2 = \left(g + a\omega^2 \cos\varphi\right)^2 + \left(a\omega^2 \sin\varphi\right)^2$$

Hence,

$$\gamma = g\sqrt{1 + \frac{2a\omega^2}{g}\cos\varphi} \qquad \left(\text{practically}, a\omega^2 \ll g\right) \tag{5.1}$$

The vertical component of particle velocity equals the rate of change of the height with respect to the time, that is, dh/dt, where

$$\frac{dh}{dt} = -a\omega \sin\varphi = -a\omega \cos\omega t \tag{5.2}$$

The negative sign in (5.2) indicates the opposition of rate of height change to the positive direction of Z-axis.

$$\text{The slope of surface tangent} = \tan\theta = \frac{dh}{dx} = \frac{a\omega^2 \sin\varphi}{g + a\omega^2\cos\varphi} = \frac{1}{g}\left(a\omega^2 \sin\varphi\right) \tag{5.3}$$

Incorporating (5.2) and (5.3) can get

$$h = a\sin\left(\frac{\omega^2 x}{g} - \omega t\right) \tag{5.4}$$

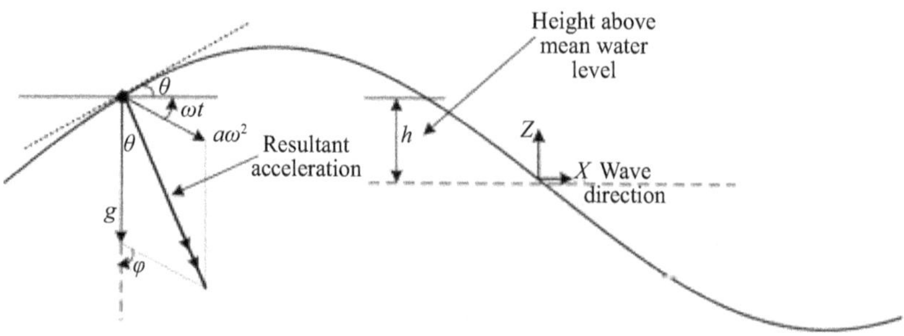

Figure 5.12 Vector diagram for forces affecting a surface-water particle

But the general equation of wave motion having wavelength λ and velocity c can be written as

$$h = a \sin\left[\frac{2\pi}{\lambda}(x - ct)\right] = a \sin\left(\frac{2\pi}{\lambda}x - \omega t\right) = a \sin(kx - \omega t) \tag{5.5}$$

where $k = 2\pi\lambda$ (it is defined as the wave number) and $c =$ the phase velocity of the travelling wave.

From (5.4) and (5.5), it can be seen that

$$\lambda = \frac{2\pi g}{\omega^2} \tag{5.6}$$

The period of motion or the period of the wave, T, is the time taken for one full wave or one wavelength to pass a fixed position.

$$T = \frac{2\pi}{\omega} = \sqrt{\frac{2\pi\lambda}{g}} \tag{5.7}$$

The velocity of the particle at the wave crest, v, is given by

$$v = a\omega = a\sqrt{\frac{2\pi g}{\lambda}} \tag{5.8}$$

The wave surface velocity, c, which is called the phase velocity of the travelling wave (made by the surface motion) in the X-direction is given by

$$c = \frac{\lambda}{T} = \frac{\omega}{k} = \frac{\omega\lambda}{2\pi} = g\sqrt{\frac{\lambda}{2\pi g}} = \frac{gT}{2\pi} \tag{5.9}$$

Hence, longer waves travel faster than the shorter ones. In addition, from (5.8) and (5.9), it is to be noted that the particle velocity, v, is a function of the amplitude a, whereas the phase velocity, c, does not depend on the amplitude. This means that c does not pertain to v.

5.5 Estimation of wave energy

Consider a single regular deepwater wave with surface particles moving in circular orbits of amplitude a, at varying phase, in the direction of propagation X (Figure 5.13). In a vertical column, the amplitude decreases exponentially with the increase of depth below the mean surface water level and becomes negligible at the sea bottom. It means that the amplitude of oscillation of particles decreases as moving into the depth of the sea (Figure 5.14). Therefore, the particle at mean

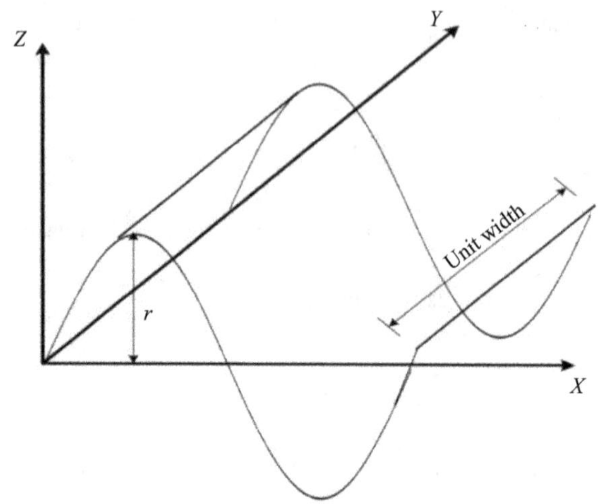

Figure 5.13 A wave of unit width and amplitude r moving along X-direction

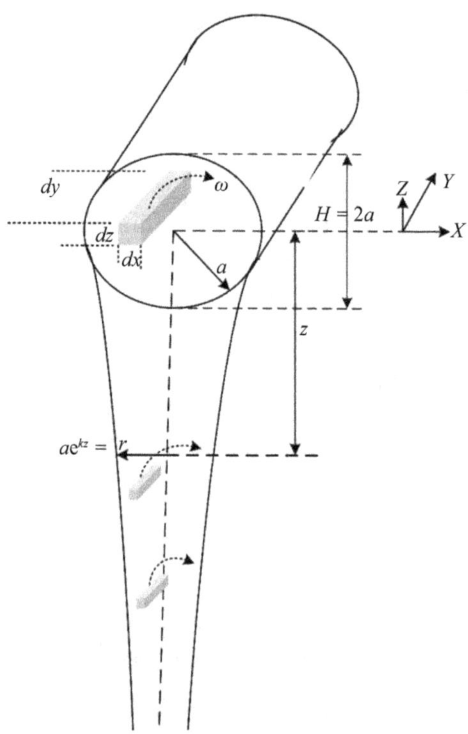

Figure 5.14 Elemental strip of water in a deep-water wave

position of z below the surface (i.e. in the negative direction of Z) moves in a circle of radius r that is given by [6]

$$r = ae^{kz} \tag{5.10}$$

where k is the wave number, $2\pi/\lambda$. It is to be noted that the particle motion remains circular if the seabed depth $D > 0.5\lambda$.

As shown in Figure 5.14, a vertical column as an elemental strip of water across the wavefront of dimension $dx \times dy \times dz$ (length×width×height) is considered. The volume of the strip is

$$dV = dx\,dy\,dz$$

Thus, the volume per unit width of the wave is given by

$$dV = dx\,dz \tag{5.11}$$

and the mass is

$$dm = \rho\,dV = \rho\,dx\,dz \tag{5.12}$$

where ρ is the density of seawater.

Assuming E_K is the kinetic energy of the total wave motion per unit length in X-direction, the kinetic energy of particle of width dx is

$$dE_K\,dx = ½mv^2 = ½(\rho\,dz\,dx)r^2\omega^2$$

That is,

$$dE_K = ½\rho r^2\omega^2\,dz \tag{5.13}$$

where each elemental strip of water is in circular motion at constant angular velocity ω, radius of circular orbit r and angular velocity $v = r\omega$.

Consider a moment in time when the element is at its mean position, and all other elements in the column are moving vertically with the same phase in the Z-direction. Incorporating (5.10) and (5.13) gets

$$dE_K = ½\rho\left(a^2e^{2kz}\right)\omega^2\,dz$$

Integrating both sides and multiplying by dx, the total kinetic energy in the column is

$$E_K\,dx = \int_{z=-\infty}^{z=0} ½\rho a^2\omega^2 e^{2kz}\,dz\,dx = \frac{1}{4}\rho\frac{a^2\omega^2}{k}dx$$

By substituting the value of $k = 2\pi/\lambda$ and the value of ω as in (5.6) can obtain the kinetic energy per unit length of wave per unit width of wavefront as

$$E_K = \frac{1}{4}\rho a^2\frac{2\pi g}{\lambda}\frac{\lambda}{2\pi} = \frac{1}{4}\rho a^2 g \tag{5.14}$$

For harmonic motions, the average kinetic and potential energies are equal, and so the potential energy per unit length of deepwater wave per unit width E_P is given by

$$E_P = \frac{1}{4}\rho a^2 g \tag{5.15}$$

Summation of (5.14) and (5.15) gives the total energy per unit length per unit width of wavefront, which is also the total energy per unit area of surface, as

$$E = E_K + E_P = \tfrac{1}{2}\rho a^2 g \tag{5.16}$$

Equation (5.16) indicates that the energy within a wave is contained near the surface and falls off sharply with depth.

The energy per unit width of front wave, per unit wavelength in the direction of wave motion, E_λ, is

$$E_\lambda = E\lambda = \tfrac{1}{2}\rho a^2 g\lambda \tag{5.17}$$

Substituting the value of λ given by (5.6) gets

$$E_\lambda = \pi\rho\left(\frac{ag}{\omega}\right)^2 \tag{5.18}$$

and in terms of period of motion T given by (5.7), (5.18) can be rewritten as

$$E_\lambda = \frac{\rho}{4\pi}(agT)^2 \tag{5.19}$$

5.5.1 Derivation of power formulae

As written in (5.9), from the phase velocity, $c = \frac{\omega}{k}$, and by the value of k and (5.6), it can be found that

$$\omega^2 = kg \tag{5.20}$$

The corresponding group velocity, c_g, in deepwater wave is

$$c_g = \frac{\partial\omega}{\partial k} = \frac{g}{2\omega} = \frac{c}{2} \tag{5.21}$$

This velocity is defined as the velocity at which the unseen energy in the group of waves is carried forward. The total energy (kinetic + potential) per unit length per unit width of wavefront is given by (5.16). Accordingly, the power carried in such waves across a vertical plane, per unit width of wavefront, can be expressed as

$$P = E \times c_g = \frac{1}{2}\rho g a^2 \frac{gT}{4\pi} = \rho g^2 T H^2/32\pi \text{ W/m} \tag{5.22}$$

In nature, waves are irregular and can be described by statistical models. If the wave conditions are measured, over 20 min for example, the mean wave height,

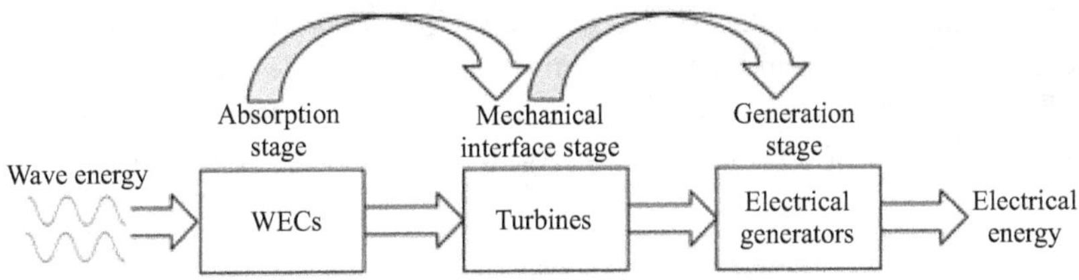

Figure 5.15 Stages of wave power conversion

H_m, and the significant wave height, H_s, can be calculated. The significant wave height is defined as the average of the highest 33% of the waves. Under such circumstances, the wave power can be stated to be

$$P = 550H_s^2T_z \text{ W/m} \tag{5.23}$$

where T_z, the mean zero crossing period is the duration of the record divided by n, where $(n + 1)$ is the number of upward crossings of the mean water level.

5.6 Wave energy converters

Wave energy should be extracted in a form adequate for transformation to electrical energy. WECs are devices used for this purpose, that is, to convert the wave power into electrical power. The conversion passes through three steps in cascade as shown in Figure 5.15. First, absorption stage at which the wave energy is transformed to energy in working fluids (different working fluids are used depending on the type of the device) and that is the function of WECs. Second, at mechanical interface stage, the working fluid energy is transformed to mechanical energy using a motor or a turbine. Third and finally, which is the electricity generation stage, the mechanical energy rotates an electric generator to produce electrical energy [7–9].

The efficiency of a WEC can be estimated as the ratio of power output, P_{out}, to power input, P_{in}. Input power per unit width can be calculated using either (5.22) or (5.23). Therefore, the output power from the device must be divided by the wave width, w, to get a dimensionless ratio that expresses the efficiency, that is

$$\eta = \frac{P_{out}}{wP_{in}} \tag{5.24}$$

In practice, estimation of capture wave width (may be called 'absorption width') meets difficulties as the device absorbs power from a wavefront wider than the device itself. So, it is better to define the device efficiency as the delivered power from the device to the power incident immediately on the device, P_{inc}. Thus,

$$\eta = \frac{P_{out}}{P_{inc}} \tag{5.25}$$

It will be noted that the efficiency depends greatly on the wavelength, amplitude of the oncoming waves and the characteristics of the device.

Design of WECs is based on a large variety of energy conversion techniques. However, WECs are generally categorized by location, type and modes of operation.

5.6.1 Location of WECs

There are three locations for WECs: onshore, nearshore and offshore.

Onshore. It means coastal areas in which the water depth is 10–15 m and the maximum wave height is 7.8 m. Shoreline devices have the advantage of being close to the utility network, are easy to maintain and as waves are attenuated as they travel through shallow water, WECs have a reduced likelihood of being damaged in extreme conditions. On the other hand, this leads to one of the disadvantages of shore-mounted devices, as shallow water leads to lower wave power.

Nearshore. It means shallow water areas in which the water depth is 15–25 m and the maximum wave height is 15.6 m. Nearshore devices are defined as devices that are in relatively shallow water of depth less than one-quarter wavelength. Devices in this location are often attached to the seabed, which gives a suitable stationary base against oscillations produced by an oscillating bodywork. Like shoreline devices, a disadvantage is that shallow water leads to waves with reduced power, limiting the energy harvesting potential.

Offshore. It means deepwater areas in which there is no natural limitation for both water depth and wave height. The water depth is always more than 50 m and wave height can exceed 30 m. In some literature, the depth is defined by 'greater than 40 m' and in others by 'exceeding one third of the wavelength'. The advantage of siting a WEC in this location is the capability of harvesting greater amounts of energy because of the higher energy content in deepwater waves. On the contrary, more difficulties for construction and maintenance of devices are present as well as the likely extreme climatic conditions require a proper design of the device construction with additional cost.

5.6.2 Types of WECs

There is a large variation in designs of WECs. However, there are three predominant types into which WECs are classified. Type classification is based on one or more of variations that arise when a wave passes a stationary position: (i) change of wave height at the surface, (ii) change of water pressure under the surface and (iii) rotation of small volumes of water near the surface.

Point absorber. It is a device that possesses small dimensions where both width and length are much less than the incident wavelength. So, wave direction is not important as the device size is relatively small and it absorbs the energy from all wave directions because of their small size compared to the wave length. This type of device is a floating structure that will move up and down on thc water surface or is submerged below the surface relying on pressure differential. Such devices have large capture width to extract power from the oncoming waves. They are basically offshore oscillating bodies, either floating or fully submerged.

The point absorbers usually use translational motion, such as surge, sway and heave, to produce electricity. Indeed, the energy will come from the motion of the buoyant top relative to the base. If one considers a floating buoy which has a generator at the bottom of the sea, all translational movements make a source for power generation. The larger the buoy is, the more energy can be absorbed as the exposed area to wave increases. However a larger buoy will cause an increase in diffraction forces. So, to design a point absorber the key parameter is the desired amount of produced electricity. Hence, it is required to configure the entire system considering the dimensions, the mass and inertia of the system. An optimization algorithm is used for a selected site with the given met-ocean data. There are many examples of point absorbers, the most well known is the ocean power technology's (OPT) power buoy as shown in Figure 5.16.

Floating point absorbers are divided into two categories, depending on whether the bottom end of the structure is fixed to the seabed or to a bottom-fixed structure, or if it is allowed to be in motion. Currently, the major point absorbers with a non-fixed bottom end include the power buoy as shown in Figure 5.17. This device floats on the surface of the water, and is held in place by cables connected to the seabed. The construction involves a cylindrical structure with one component relatively immobile as the bottom end, and a second component with movement driven by wave motion as the top end floating buoy inside a fixed cylinder [10].

The relative motion of the two components that is caused by the rising and falling of the waves is used to drive electromechanical generators or hydraulic energy converters. The electric power generated is transmitted to shore over a submerged transmission line. A 150 kW buoy has a diameter of 11 m and is 44 m tall, with approximately 10 m of the unit rising above the ocean surface.

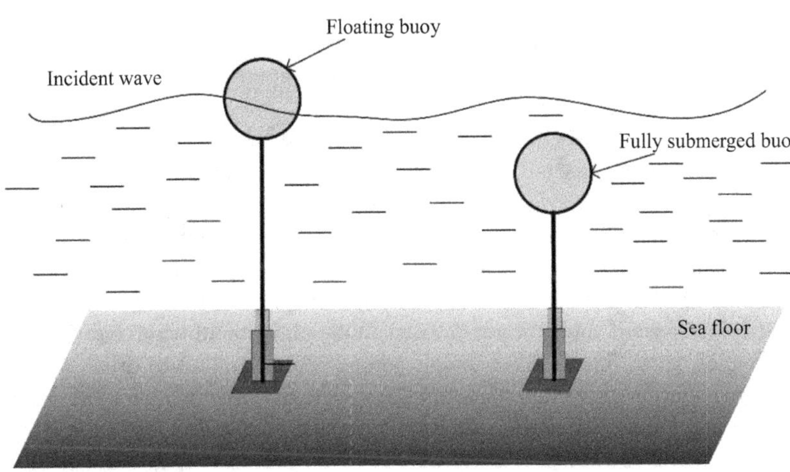

Figure 5.16 Point absorber device (OPT power buoy)

Figure 5.17 Power buoy with non-fixed bottom

Attenuators (line absorbers). This device is a floating device that operates parallel to the wave direction and literally rides the waves. An example of this type is the well-known 'Pelamis wave power', which is developed by ocean power delivery. These devices capture energy from the relative motion of the two arms as the wave passes them. More particularly, cylindrical steel sections are connected to each other by two degrees of freedom. Each joint contains the complete power conversion unit. Those joints are going to face both horizontal and vertical motions that are accommodated by four hydraulic cylinders. They act as pumps, which drive the fluid through a hydraulic motor, which in turn drives an electrical generator. In addition, accumulators are used to decouple not only the primary circuit which is the pump but also the secondary circuit which is the motor. This allows to regulate the flow of fluid and thus produce a more constant output. These devices have to be at around 50 m off the coast.

A Pelamis machine which has four 35 m long by 3.5 m diameter floating cylindrical sections connected by three hinged joints is shown in Figure 5.18. The Pelamis technology is intended for general deployment and operation offshore with water depths of 50–70 m and with nominal wave power of 55 kW/m. The four sections of the device articulate with the movement of the waves, each resisting motion between this and the next section. In this way, as waves run down the length of the device, since it is positioned parallel to the direction of wave travel, flexing at the hinged joints due to wave action pumps pressurized oil to drive hydraulic pumps and motors built into the joints. Electricity generated in each joint is transmitted to shore by a subsea cable.

Terminators. Terminator devices have their principal axis parallel to the wavefront, extend perpendicular to the predominant wave direction and capture or

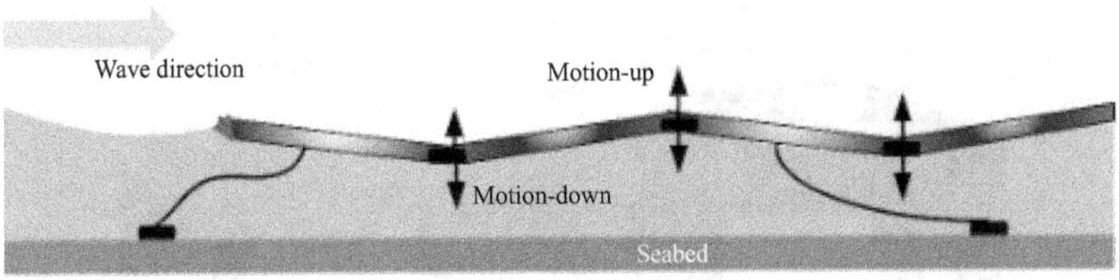

Figure 5.18 Pelamis wave attenuator

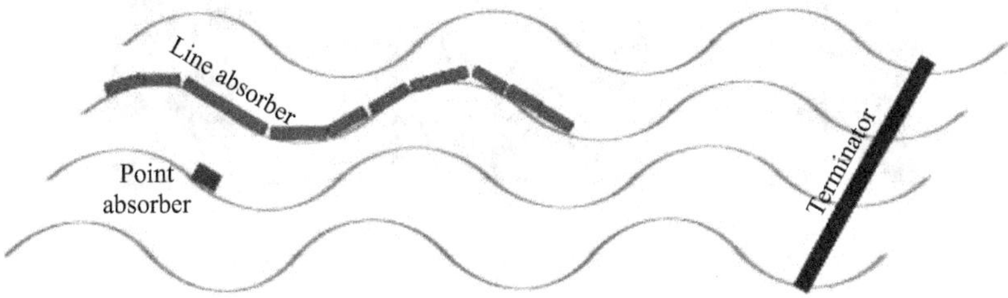

Figure 5.19 Point absorber, line absorber and terminator devices

reflect the power of the wave. These devices are typically installed onshore or nearshore, and the device structure is typically fixed on the seabed or on the shore. Placement of point absorber, line absorber and terminator devices with respect to the water waves is shown in Figure 5.19.

5.6.3 Modes of operation

WECs can also be categorized by their mode of operation such as submerged pressure differential device, oscillating wave surge converter, oscillating water column (OWC) and overtopping device.

Submerged pressure differential. The device is a submerged point absorber. The pressure difference between the crests and troughs of the wave passing above the device is used to actuate an air-filled cylindrical chamber that contains two parts: a seabed lower fixed chamber (called 'basement') and a moveable upper cylinder (called 'floater'). As a crest passes over the device, the water pressure above the device compresses the air within the cylinder, moving the upper cylinder down. As a trough passes over, the water pressure on the device reduces and the upper cylinder rises (Figure 5.20). Consequently, this up and down motion can be resisted by a linear electrical generator, with the interior air pressure acting as a spring. Because of the fixed part being attached to the seabed, the device is located nearshore. A typical example of this device is the Archimedes wave swing.

Oscillating wave surge converter. An oscillating wave surge converter is generally comprised of a hinged deflector, positioned perpendicular to the wave direction (a terminator), that moves back and forth exploiting the horizontal particle velocity of the wave. An example is the Aquamarine Power Oyster, a nearshore

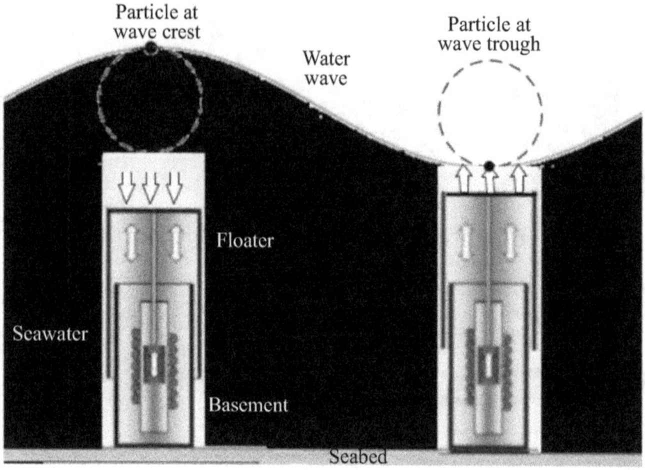

Figure 5.20 Submerged pressure differential device

device, where the top of the deflector is above the water surface and is hinged from the seabed. A prototype of this device has been constructed as illustrated in Figure 5.21.

Oscillating water column. An OWC consists of a chamber with an opening to the sea below the waterline. As waves approach the device, water is forced into the chamber, applying pressure on the air within the chamber. This air escapes to atmosphere through a turbine. As the water retreats, air is then drawn in through the turbine. A low-pressure Wells turbine is often used in this application as it rotates in the same direction irrespective of the flow direction, removing the need to rectify the air flow. It has been suggested that one of the advantages of the OWC concept is its simplicity and robustness [12].

There are examples of OWCs as point absorbers, as well as that being built into the shoreline, acting as a terminator. An example of a shoreline mounted device is the Wavegen Limpet. The device is installed on the island of Islay, Western Scotland, and produces power for the national grid. The design of the Limpet is shown in Figure 5.22. The OWC concept has also been proposed by Oceanlinx, an Australian wave energy developer, in a nearshore tethered device [13].

Overtopping device. An overtopping device captures seawater of incident waves in a reservoir above the sea level, then releases the water back to sea through turbines. An example of such a device is the Wave Dragon, a schematic diagram of which is shown in Figure 5.23(a). This device uses a pair of large curved reflectors, as shown in Figure 5.23(b), to gather waves into the central receiving part, where they flow up a ramp and over the top into a raised reservoir, from which the water can return to the sea via a number of low-head turbines.

The various categories of WECs described above can be summarized as illustrated by the block diagram shown in Figure 5.24.

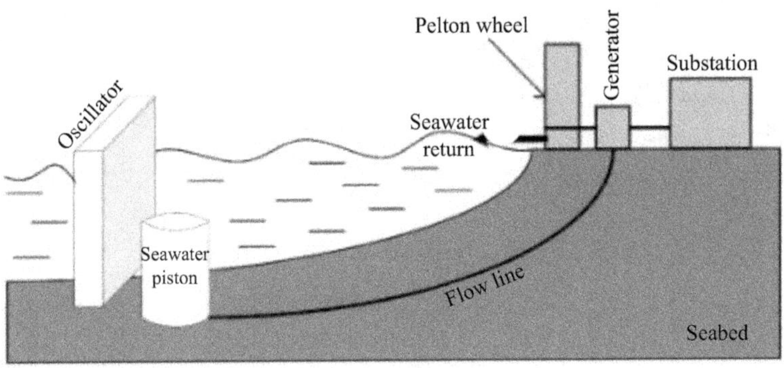

Figure 5.21 Oscillating wave surge converter: Aquamarine Power Oyster [11]

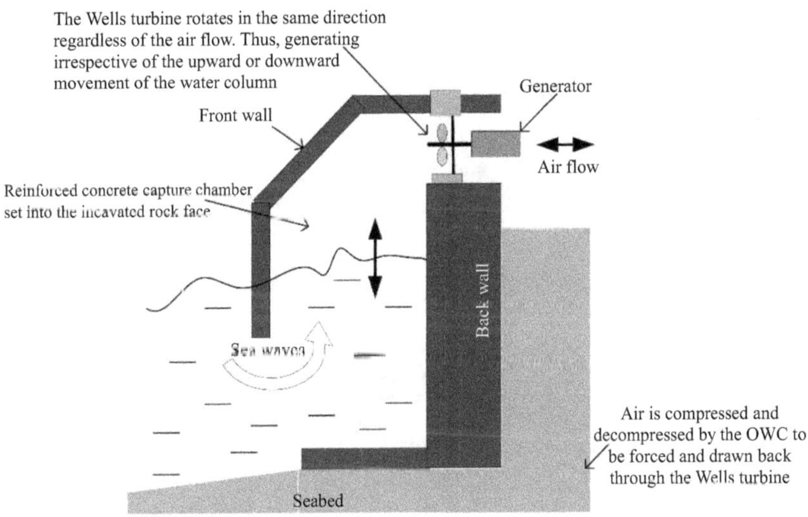

Figure 5.22 OWC: the Limpet

5.7 Mechanical interfaces

Wave energy captured using a WEC should be transformed into an adequate form that can drive a rotating generator to produce electricity. So, intermediate mechanical interfaces such as turbines are needed to make this adequate formation. Three major interfaces can be used: (i) self-rectifying air turbine, (ii) high- or low-head hydraulic turbine and (iii) high-pressure oil-driven hydraulic motor.

5.7.1 Air turbines

This type of turbine allows a single-directional air flow, and so it is typically used in all terminator or OWC devices. The self-rectifying air turbines have the

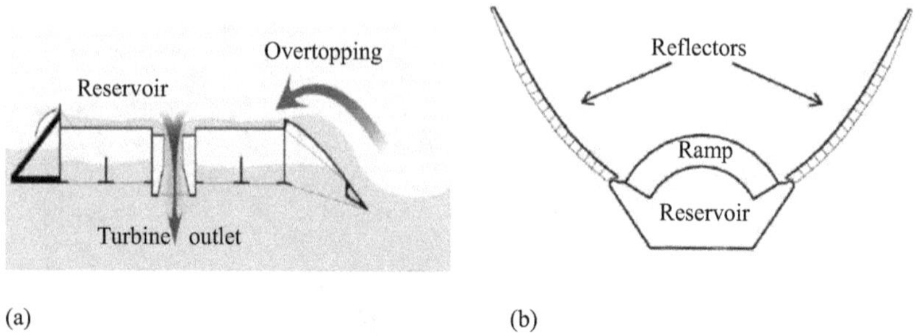

Figure 5.23 Overtopping WEC: the Wave Dragon. (a) A schematic diagram. (b) The large curved reflector

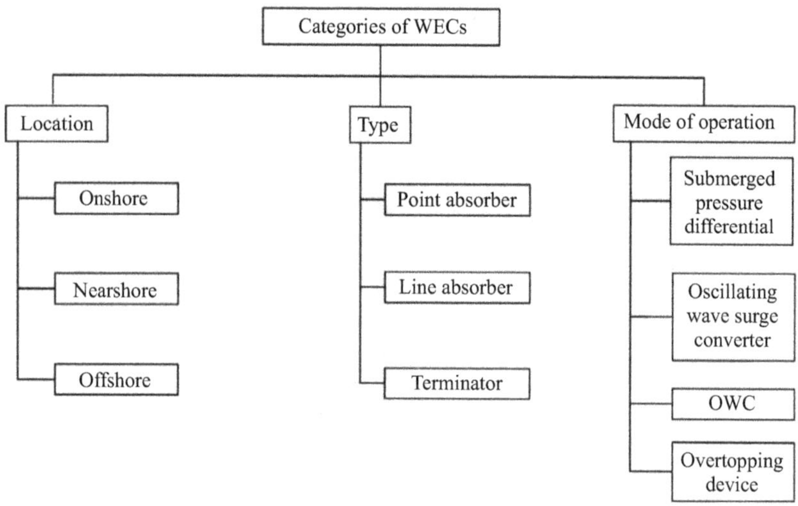

Figure 5.24 WEC categories

capability of controlling the rotational speed to match the current sea state. Wells turbine, as a typical example, is the most commonly used as an air turbine. It has (i) the capability to rotate at high speed for a low velocity of air flowing through the turbine, (ii) an efficiency of 70%–80% and (iii) simplicity and high reliability. On the other hand, it has a negative torque at low flow rates and aerodynamic losses at flow rates exceeding the stall-free critical value, i.e. resulting in output power drop. Kaplan turbines are the modified and enhanced version of Wells as the efficiency is increased by controlling the rotor blade pitch angle [14].

A block diagram of the air turbine is shown in Figure 5.25. The input of the turbine model is the air velocity, v_{air}, and the output is the mechanical torque, T_m [15].

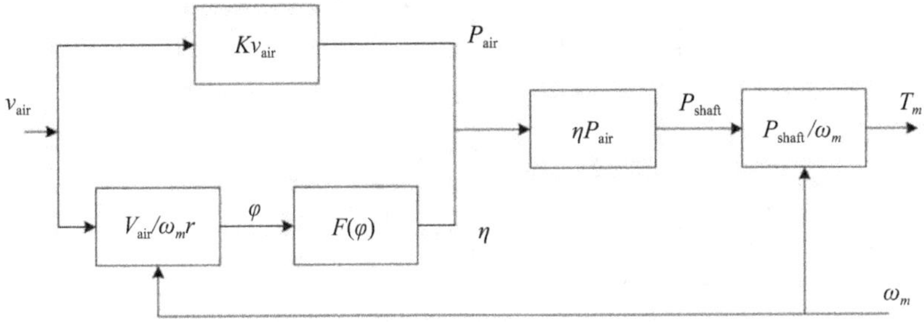

Figure 5.25 Block diagram of air turbine model

The output mechanical torque is given by

$$T_m = P_{\text{shaft}}/\omega_m \tag{5.26}$$

where ω_m is the shaft angular speed and P_{shaft} is the turbine shaft power determined by

$$P_{\text{shaft}} = \eta P_{\text{air}} \tag{5.27}$$

where η is the turbine efficiency and P_{air} is the air power. Turbine efficiency is a function of flow coefficient φ as given by Figure 5.26 [15]. The air power is proportional to the air velocity as

$$P_{\text{air}} = Kv_{\text{air}} \tag{5.28}$$

where K is a constant. The flow coefficient φ is determined by the turbine air velocity and the shaft speed as

$$\varphi = v_{\text{air}}/(\omega_m r) \tag{5.29}$$

where r is the turbine radius. The output turbine torque has a pulsating nature because of the variation of air velocity at the turbine input with the variation of wavelength and wave height. The pulsation of the output torque can be smoothed by coupling a large flywheel along the shaft between the turbine and the generator.

5.7.2 Hydraulic turbines

The operation of this type of turbine is based on converting a head of water, created between the reservoir of an overtopping device and the mean sea level, into mechanical rotational power to drive an electrical generator. Hydraulic turbines are classified into two categories: low-head (typically 3–4 m) and high-head (typically tens to hundreds of metres) turbines. The flow can be controlled by adjusting the inlet guide vanes or by a servomechanism, whereas the blades of the runners are also adjusted to improve the efficiency over a wide range of flow. This is costly and commonly not used in the small turbines. On the other hand, high-head hydraulic turbines (e.g. impulse Pelton type) are employed in some line and point absorber devices. They can also be considered as alternative to hydraulic motors. In general,

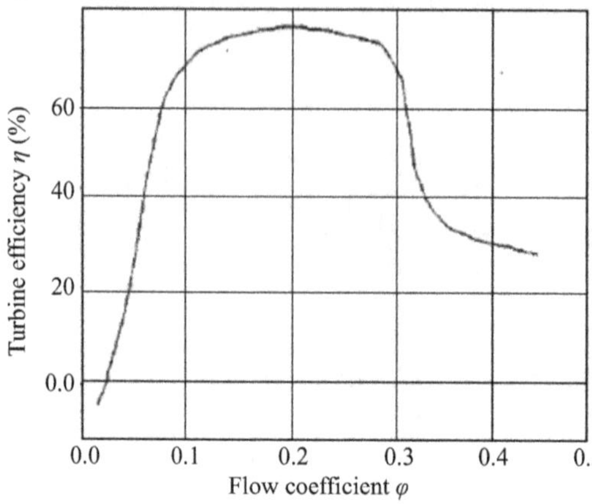

Figure 5.26 Turbine efficiency versus flow coefficient

the efficiency of the two types of hydraulic turbines (low and high head) is very sensitive to the head-to-rotational speed ratio and may reach 90%. So, the use of variable speed electrical generator is highly favourable.

The functional block diagram of a hydraulic turbine-generator set as a typical example is shown in Figure 5.27. More details including the mathematical representation of each function can be found in [16].

5.7.3 High-pressure oil hydraulic motors

The oil hydraulic motor acts as a prime mover of the electrical generator by converting its received hydraulic input energy that comes from the motion of an oscillating body through a hydraulic cylinder into rotational energy. As an example, a typical application of this oil hydraulic system to power buoy point absorbing device is illustrated in Figure 5.28. The hydraulic system is equipped with a gas accumulator system capable of storing energy over a few wave periods with a goal of smoothing out the irregular power absorbed from the wave. The oil-hydraulic mechanical interface can be used for point absorber or attenuator devices.

Direct mechanical interface. A floating cylinder (called 'Bristol cylinder') collects the wave's movement and is mechanically connected to the energy unit by flexible joints and rods. The rods move slowly with cylinder and the reciprocating motion is transferred to the axles in converter unit. This converter unit is the heart of the system and converts the reciprocating motion to a rotating shaft that is directly connected to a suitable slow speed electric generator for producing electricity at high efficiency. When transferring converter movements with mechanical arms and rotation to the generator, the efficiency should be kept as high as possible. The floating cylinder operates under sea level as shown in Figure 5.29.

The pendulum system can also be installed in the shoreline. It consists of a parallelepiped concrete box, which is open to the sea at one end, as shown in Figure 5.30. A pendulum flap is hinged over this opening, so that the actions of the

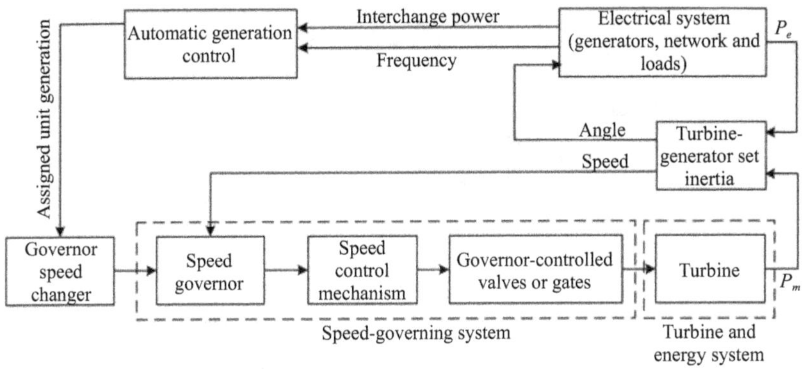

Figure 5.27 Functional block diagram of hydraulic turbine-generator set. P_e = electrical power; P_m = mechanical power

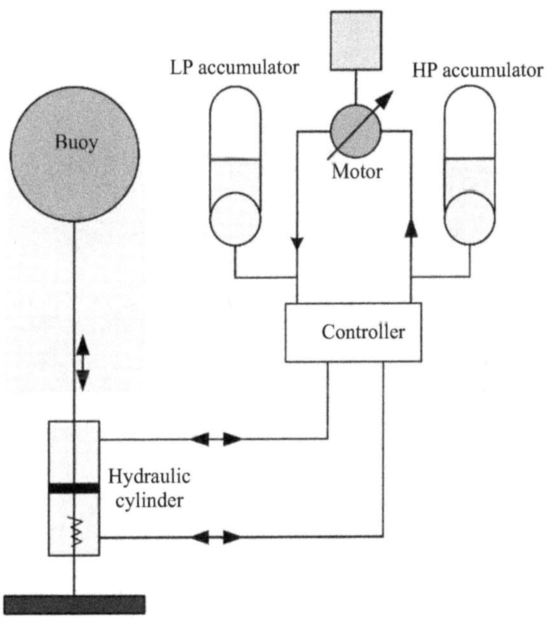

Figure 5.28 Hydraulic mechanism with point absorber device. HP = high pressure, LP = low pressure

waves cause it to swing back and forth. This motion is then used to power a hydraulic pump and an electric generator.

5.8 Electricity generation

The generator delivers power into the grid with constant frequency and rms voltage. Because the turbine rotates with a variable speed, a synchronous machine is

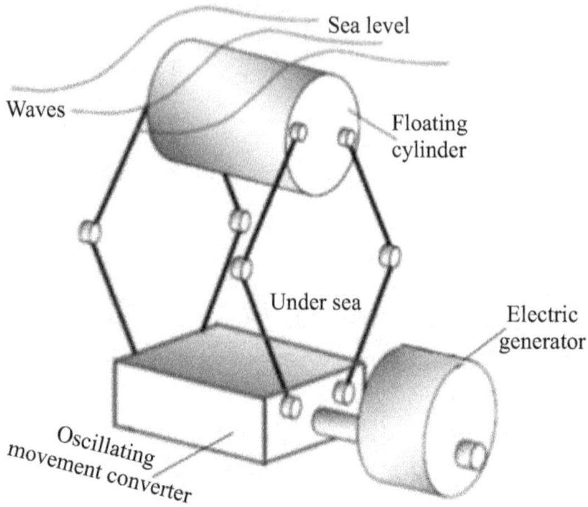

Figure 5.29 Direct mechanical interface for wave energy extraction

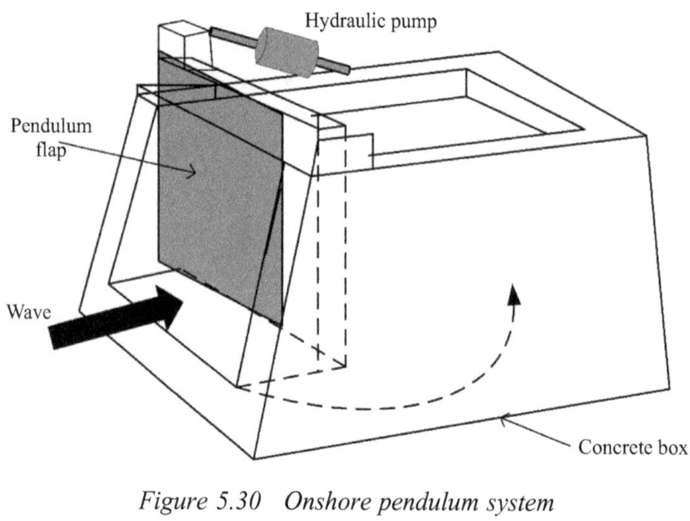

Figure 5.30 Onshore pendulum system

not appropriate. Instead, asynchronous machine such as a doubly fed wound rotor induction generator (DFIG) is used, as shown diagrammatically in Figure 5.31. The wound rotor is fed by the stator using a converter, and with this arrangement, the frequency and voltage are kept constant for a large range of turbine speed variation. More details about DFIGs are explained in Chapter 2, Section 2.4.

5.8.1 Direct drive wave energy

It is seen that the rotary generator is driven by one of the intermediate mechanical interfaces such as air turbine or hydraulic turbine. This leads to the use of large

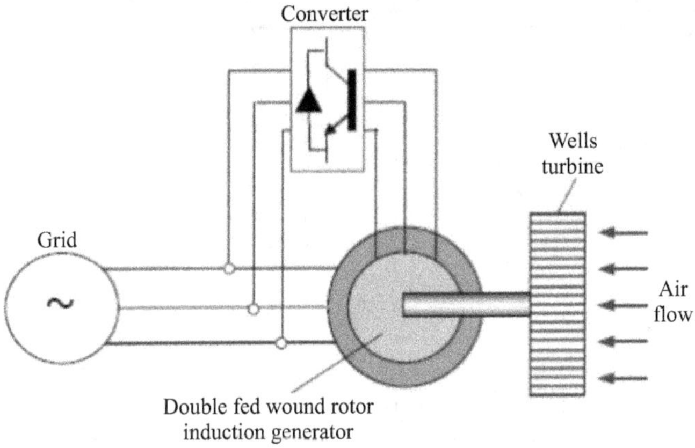

Figure 5.31 DFIG with a prime mover of Wells turbine

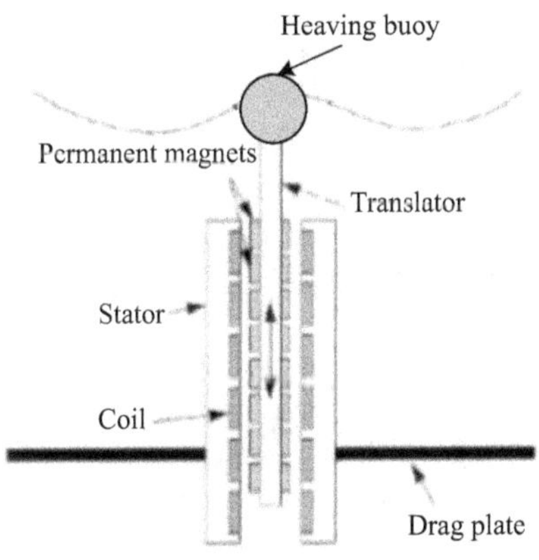

Figure 5.32 Schematic diagram of a linear permanent magnet electric generator

expensive equipment, reduced efficiency, limited operating conditions and reduced reliability of the WEC. Recently, wave power generation using a linear generator has received more attention to be applied as it can convert the reciprocating motion of a wave directly into electrical energy without any intermediate mechanism. Thus, the linear generator is required to convert the slow and large forces of the wave into electrical energy.

The linear generator has a stator which is a number of coils fixed to the seabed as well as a linear rotor (called 'translator') which is the moving part. For wave energy applications, the linear permanent magnet generator (LPMG) is commonly used (Figure 5.32). When the buoy is lifted by the wave, the buoy sets the translator in motion. It is the relative motion between the stator and the translator in the generator, which induces a voltage in the stator windings. But when a generator moves with varying speed and direction, it results in an induced voltage with

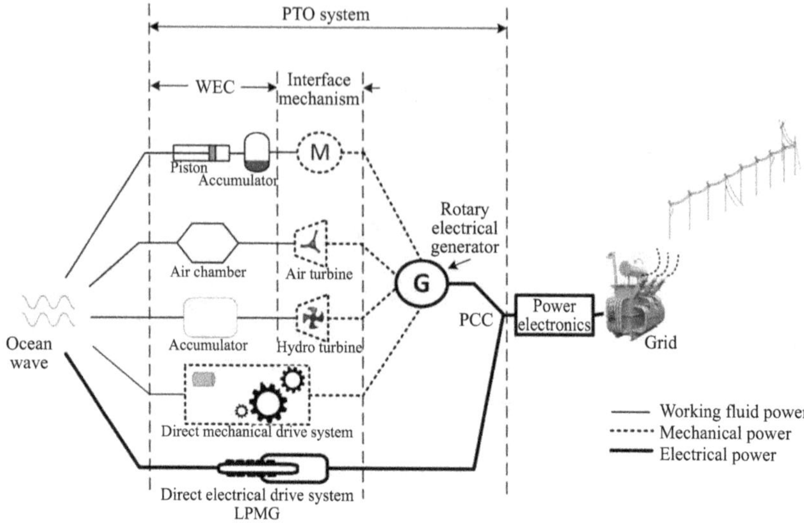

Figure 5.33 PTO systems

irregular amplitude and frequency. The peak value of the output power will be several times higher than the average power production. The generator and the electrical system have to be dimensioned for these peaks in power. More details about the principal of operation and modelling of LPMG, integrated with interface mechanisms, can be found in [17,18].

The power produced cannot be delivered directly to the grid without conversion. This is done in several steps. First, the voltage of each generator is rectified to DC and the resulting output DC voltage is filtered through capacitive filters. The filters smooth out the voltage from the generators and create a stable DC voltage. Outputs of various generators are connected to a bus, during short periods of time, and the power after the filter will also be constant. If the system is studied during hour scales, or more, there will be variations in the produced power and these variations are due to changes in the sea state. To compensate for voltage variations of the output that occur due to sea state variations, a DC/AC converter and a tap-changing transformer can be used.

5.9 PTO systems

As shown in Figure 5.33, the function of the power take-off (PTO) systems is to harvest the ocean wave energy and convert this energy into an adequate electric form to be delivered to the grid. Consequently, PTO systems can be classified as follows:

- Indirect PTO systems which use mechanical interface mechanism and deliver power to the grid through a chain of equipment as the following:
  - piston and accumulator-hydraulic motor-rotary generator;
  - air chamber-air turbine-rotary generator;

- accumulator-hydro turbine-rotary generator;
- direct mechanical interface-rotary generator.

- Direct PTO systems which directly deliver the ocean wave power to the grid through an LPMG and without any interface mechanism.

It is to be noted that for all PTO classes power electronics is used at the point of common coupling for power adoption-based power quality [19–21].

5.10 Pros and cons

Advantages of wave energy

1. Renewable: Waves are created by the winds and their energy can be captured. The wave energy will never run out and can be delivered with all benefits that renewable energy sources (RESs) would have.
2. Green: Power extraction from ocean waves does not emit harmful by-products such as gas, waste and pollution like there is with fossil fuels. So, it is contributing to the reduction of CO_2 emissions. The wave energy can be taken directly to electrical generators to produce electricity.
3. Energy potential: A huge amount of power can be extracted from the waves. It still needs more research and development to exploit this power in addition to increasing the efficiency of WECs.
4. Predictable: One of the major advantages of wave energy is its predictability as the waves are consistent throughout the day. Thus, it is easy not only to calculate the amount of power that can be produced but also to properly manage it.
5. Power density: It is greater for wave energy than that of wind energy and thus it will be more efficient.
6. Reliability: Compared to wind and solar as RESs, the generation of electrical power from wave energy is more reliable as the waves are hardly interrupted and are always in motion.
7. Onshore and offshore wave power: For onshore wave energy plants, the capturing of wave energy leads to reducing the power with which the waves hit the shoreline, thus causing less damage to the shore. On the other hand, offshore wave energy plants are installed in locations at which the waves have larger energy potential as well as they could minimize the negative environmental impacts.

Disadvantages of wave energy

1. Cost: Cost of harnessing wave power is generally very high because of equipment installation such as WECs and interface mechanisms in the sea area and in many cases offshore. Thus, the equipment may be exposed to storms, strong ocean waves and seawater. Therefore, monitoring and maintenance, which is not easy to implement, are required and of course this will increase the cost.

2. Location: Wave energy is not a viable power source everywhere. Its location is restricted to be onshore or offshore. Cities or lands near the ocean will benefit directly from it, whereas those far away have to find alternate sources.
3. Source of disturbance for private and commercial vessels: Another downside is that it disturbs commercial and private vessels. Power plants that gather wave energy have to be placed by the coastline to do their job, and they have to be near cities and other populated areas to be of much use to anybody. But these are the places that are major thoroughfares for cargo ships, cruise ships, recreational vehicles and beach goers. All of these people and vessels will be disrupted by the installation of a wave energy gathering source. This means that government officials and private companies that want to invest in wave energy sources have to take into account and consider the needs of those that may be disturbed.
4. Wavelength: Wind power is highly dependent on wavelength, that is, wave speed, wavelength and water density. They require a consistent flow of powerful waves to generate significant amount of wave power. Some areas experience unreliable wave behaviour and it becomes unpredictable to forecast accurate wave power and, therefore, cannot be trusted as a reliable energy source.
5. Wave energy generators may be unpleasant for some who live close to coastal regions. They look like large machines working in the middle of the ocean and destroy the beauty of the ocean. They also generate noise pollution but the noise is often shrouded by the noise of waves, which is much more than that of wave energy generators.

5.11 Principle of OTEC operation

Oceans of the globe constitute nearly 70% of the earth's surface and can be considered as large reservoirs for storing solar energy. The concept of ocean thermal energy conversion (OTEC) has been explored for years with the goal of utilizing this energy as a RES. It can be extracted by exploiting the thermal gradient that exists in the tropical ocean waters. OTEC is a process based on the temperature difference between the warm surface layer and deep cold layer of ocean water. The temperature difference is transformed to power to drive a heat engine-generator set for producing electricity.

OTEC plant operation is dependent on the natural thermal gradient found in tropical ocean waters, that is, $\pm 20°$ latitude in the Atlantic and Pacific oceans. Thus, the plant must be located at a site having (i) abundant warm surface water (upper layer) to provide the heat source for a heat engine and (ii) equally abundant cold water supply (lower layer) to be used as a heat sink for the engine.

Typical thermal gradient is shown in Figure 5.34. The temperature at the upper layer (~100 m from the water surface) is about 27 °C. It first decreases fast with increasing depth and then slowly to reach 5 °C at a depth of 1,000 m. Therefore, the thermal potential of $27 - 5\ °C = 22\ °C$ can be used to operate the OTEC system

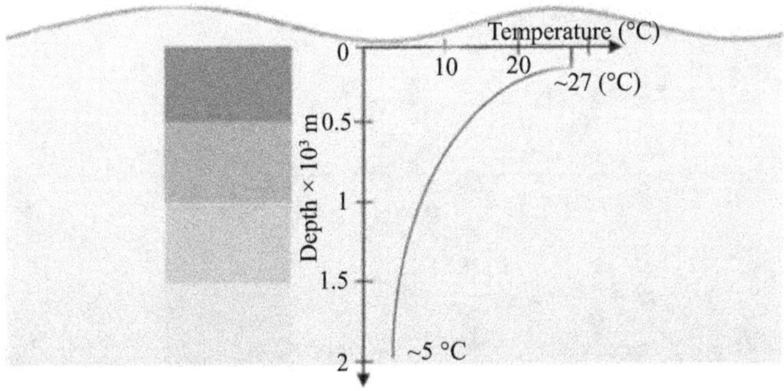

Figure 5.34 Typical temperature gradient in the tropics

[22]. Thermodynamically, OTEC plant can be designed to operate in either an open-cycle or a closed-cycle operation modes [23].

5.11.1 Open-cycle OTEC

Employing Claude cycle, seawater is used as a working fluid in the open-cycle system, which consists of the following steps: (i) flash evaporation of a fraction of the warm seawater by reduction of pressure below the saturation value corresponding to its temperature, (ii) expansion of the vapour through a turbine to generate power, (iii) heat transfer to the cold seawater thermal sink resulting in condensation of the working fluid and (iv) compression of the non-condensable gases (air released from the seawater streams at the low operating pressure) to pressures required to discharge them from the system. These steps are depicted in Figure 5.35. In case of a surface condenser, the condensate (desalinated water) must be compressed to pressures required to discharge it from the power-generating system. Other useful outcomes from the OTEC system can be obtained such as (i) generate hydrogen for use as a clean fuel source, (ii) generate fertilizer from biological nutrients that are drawn up from the ocean floor in the cold water intake and (iii) source of ocean water to be used as drinking water via desalination (taking out the salt).

The evaporator, turbine and condenser operate in partial vacuum ranging from 3% to 1% atmospheric pressure. This poses a number of practical concerns that must be addressed. First, the system must be carefully sealed to prevent in-leakage of atmospheric air that can severely degrade or shut down operation. Second, the specific volume of the low-pressure steam is very large compared to that of the pressurized working fluid used in closed-cycle OTEC. This means that components must have large flow areas to ensure that steam velocities do not attain excessively high values. Finally, gases such as oxygen, nitrogen and carbon dioxide that are dissolved in seawater (essentially air) come out of solution in a vacuum. These gases are uncondensable and must be exhausted from the system. Despite the complications, the Claude cycle enjoys certain benefits from the selection of water

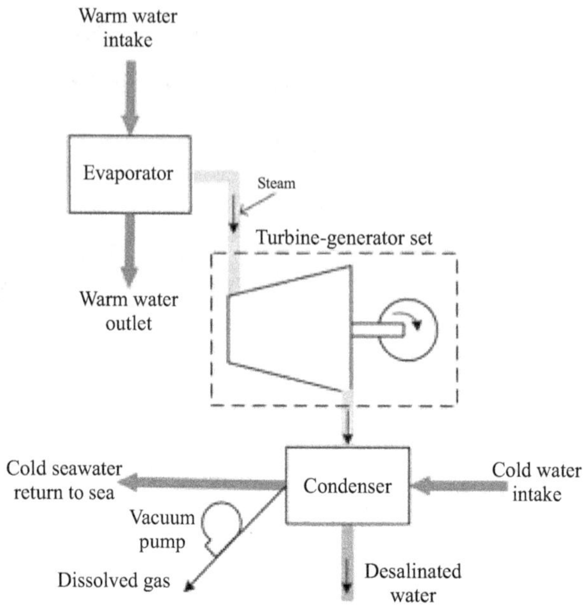

Figure 5.35 Open-cycle OTEC flow diagram

as the working fluid. Water, unlike ammonia, is non-toxic and environmentally benign. Moreover, since the evaporator produces desalinated steam, the condenser can be designed to yield fresh water. In many potential sites in the tropics, potable water is a highly desired commodity that can be marketed to offset the price of OTEC-generated electricity.

5.11.2 Closed-cycle OTEC

The fundamental thermodynamic description of a closed-cycle OTEC plant is similar to that of other thermal plants operating in closed-cycle mode, which follows a Rankine cycle. The common working fluid is water, as used in coal and nuclear power stations, but the small temperature differences and near-ambient conditions for OTEC lead to using ammonia as a working fluid. The cycle comprises four processes as shown in Figure 5.36(a) and (b), indicating the temperature-entropy (*T-S*) diagram and the corresponding schematic diagram of an OTEC system, respectively [24–26]. According to these figures and applying the first law of thermodynamics, the four processes can be described as follows.

Process #1, isobaric heat supply (evaporator) from State 1 to State 2: During this process, the heat from the high-temperature source is added to the pressurized working fluid to convert it into high pressure vapour at State 2. The heat supplied to

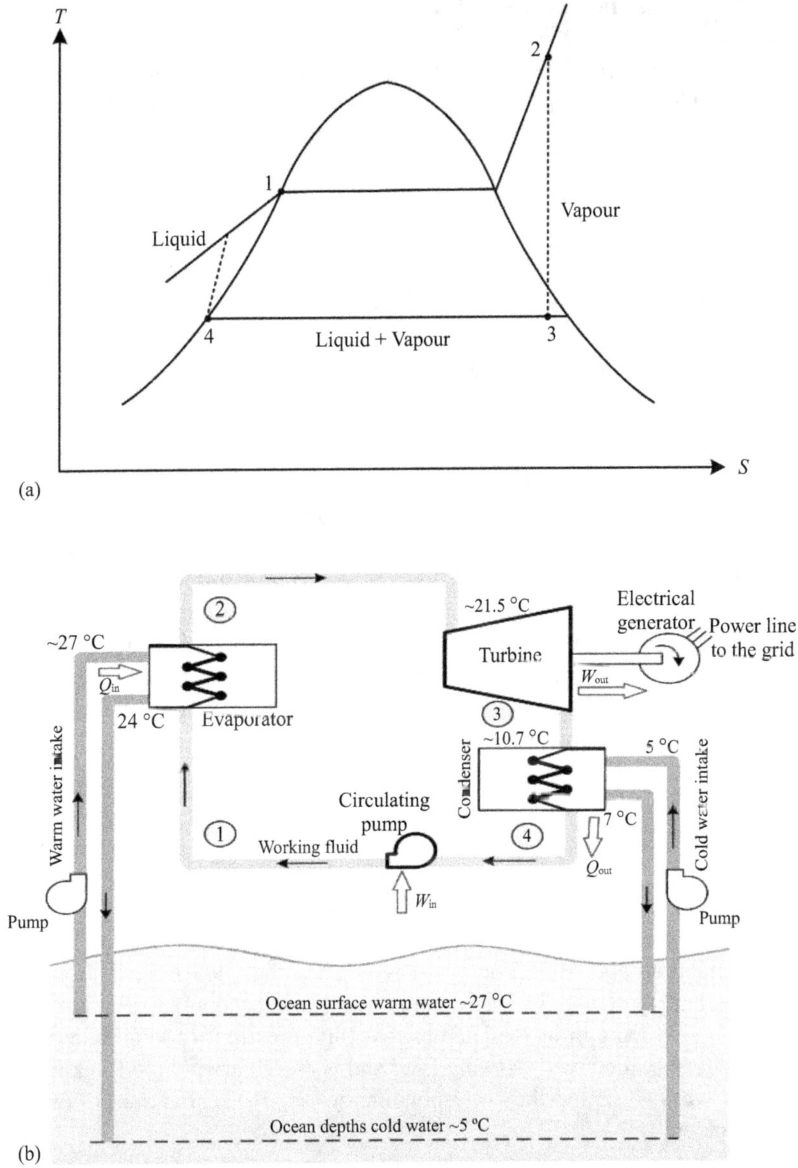

Figure 5.36 Rankine cycle: (a) T-S diagram and (b) OTEC system with typical values of temperature at each process

the cycle (evaporator) is given by

$$Q_{in} = m(h_2 - h_1) \text{ kJ} \tag{5.30}$$

where m is the mass flow of the cycle and h_i is the enthalpy of the working fluid at the ith state.

Process #2, isentropic expansion (turbine) from State 2 to State 3: an isentropic process, in which the entropy of working fluid remains constant. The vapour flows through the turbine where its internal energy is converted into mechanical work to drive the electrical generator. Thus, the work output of the working fluid during this process can be expressed as

$$W_{out} = (h_2 - h_3) \text{ kJ/kg} \tag{5.31}$$

Hence, the power delivered by the turbine to the electrical generator, P_{out}, is given by

$$P_{out} = mW_{out} = m(h_2 - h_3) \text{ kW} \tag{5.32}$$

$$P_{out} = mW_{out} = m(h_2 - h_3) \text{ kW} \tag{5.33}$$

Process #3, isobaric heat rejection (condenser) from State 3 to State 4: an isobaric process is one in which the pressure of working fluid remains constant. The condenser is a large shell-and-tube-type heat exchanger. This is positioned next to the turbine in order to receive a large flow rate of low pressure vapour. This vapour in the condenser goes under a phase change from vapour to liquid. External cooling water is pumped through thousands of tubes in the condenser to transport the heat of condensation of the vapour away from the plant. Upon leaving the condenser, the condensate is at a low temperature and pressure. Therefore, the heat rejected from the cycle, Q_{out}, is given by

$$Q_{out} = m(h_4 - h_3) \text{ kJ} \tag{5.34}$$

The value of Q_{out} is negative as $h_4 < h_3$. Thus, consistent with the sign convention, Q_{out} represents an outflow of heat from the condensed vapour.

Process #4, isentropic compression (pump) from State 4 to State 1: During the isentropic compression process, external work is done on the working fluid by means of pumping operation. The pump is a device that moves liquid from a low pressure to high pressure. In the Rankine cycle, the condensate is raised to the pressure of the evaporator by feed pumps. The high-pressure working fluid entering the evaporator is called feed-working fluid and is reused around the Rankine cycle again. The work, W_{in}, and the corresponding power, P_{pump}, required to drive the pump are given by

$$W_{in} = (h_4 - h_1) \text{ kJ/kg} \tag{5.35}$$

$$P_{pump} = m(h_4 - h_1) \text{ kW} \tag{5.36}$$

The pump work has a negative value as $h_1 > h_4$. This is consistent with the thermodynamic sign convention, which indicates that work and power must be supplied to operate the pump.

A measure of the effectiveness of an energy conversion device is its thermal efficiency, η_{th}. This is defined as the ratio of the cycle network to the heat supplied from external sources. Thus, using (5.30), (5.32) and (5.36) the ideal Rankine-cycle thermal efficiency in terms of cycle enthalpies is given as the ratio of the net power delivered by the Rankine cycle, which is the difference between the turbine power and the magnitude of the pump power, to the total heat supplied to the cycle, that is,

$$\eta_{th} = \frac{P_{net}}{Q_{in}} = \frac{P_{out} - P_{pump}}{Q_{in}} = \frac{h_2 - h_3 + h_4 - h_1}{h_2 - h_1} \tag{5.37}$$

To determine the overall efficiency and simulate the OTEC system as a power source connected to a grid, the combined turbo-generator efficiency, η_{tg}, should be considered rather than the turbine efficiency. OTEC simulation is defined by a given value of the volume flow rate of OTEC deep seawater per unit (latitude–longitude) area, w_{cw}, and the OTEC net power area density P_{net} is calculated from the following formula [27,28]:

$$P_{net} = w_{cw}\rho c_p \eta_{tg} \left\{ \frac{9}{80} \frac{(\Delta T)^2}{T} - \frac{9}{200} \right\} \tag{5.38}$$

where T is the absolute temperature of the OTEC surface seawater intake, and ΔT is the temperature difference between surface and deep seawater intakes. The average seawater density ρ is taken as 1,025 kg/m^3, the seawater specific enthalpy c_p as 4,000 J/kg K and the OTEC combined turbo-generator efficiency η_{tg} as 0.75. The numerical coefficients account for a surface-to-deep seawater flow rate ratio of 1.5, and seawater pumping power losses equal to 30% of the turbo-generator output at standard conditions ($T = 300$ K, $\Delta T = 20$ K). It is noted that these somewhat arbitrarily defined standard conditions for the OTEC thermal resource are conservative, and correspond to the net power production of 1 MW with about 3 m^3/s of deep cold seawater. These figures would be much better in most of the OTEC region; that is, less seawater would produce the same amount of net power, or equivalently, pumping power losses would represent a smaller fraction of the turbo-generator output.

5.12 OTEC and environment

It is seen that the OTEC technology requires drawing seawater from the surface layer and the deep layer to feed the evaporator and condenser, respectively, and returning it to the sea. From the environmental point of view, it is noted the following [29]:

- The carbon dioxide outgassing from the seawater used for OTEC operation of either open cycle or closed cycle is much less than that released by fossil fuel plants.
- Protection of heat exchangers (evaporators and condensers) from biofouling is necessary because of their exposition to surface seawater. Several mechanical and chemical means can be used for this purpose that leads to reducing the

probable release of working fluid during operations as well as having effective heat transfer.

- Marine food web is minimally affected because of the absence of the sea surface temperature anomalies that may result from the sustained flow of cold seawater.
- The working fluid (ammonia is commonly used) and biocide emissions from the plant must be too low to avoid their potential hazard to the workers and persons in surrounding areas.
- The inlet flow velocity of intakes should be limited to minimize entrainment and impingement of organisms in addition to keeping benign fishing life.
- Similar to the other steam-electric power generation plants, OTEC is associated with risks of safety issues such as electrical hazards, use of compressed gases and so on. On the other hand, because OTEC power plant operates at low temperature and low pressure, the probable hazards to plant workers and local population are much less than those of fossil fuel plants.

5.13 Technical limitations and challenges

Thermodynamic fundamentals applied to determine the performance of OTEC power-generating cycles are the same as those of conventional steam power plants. However, large quantities of both warm and cold seawater are required for heat transfer processes. In addition, the pumping process, the pipe-fluid frictional losses and gravitational energy due to the density difference between the colder water inside the pipe and the surrounding water column, all consume about 20%–30% of the power generated by the turbine-generator set [29].

The actual energy conversion for 26 °C warm seawater and 4 °C cold seawater is from 3% to 4%, whereas the ideal is about 8%. Thus, the efficiency in both cases is very low, in particular, compared to the efficiency of conventional power plants. So, to improve the performance of OTEC system, it is required to increase the size of cold water pipe and use of large diameter pipes to transport the deep seawater to the evaporators, which are presently limited and represent a big challenge of significant magnitude [30].

Design and installation of OTEC plants present engineering challenges. For instance, position keeping of the components of OTEC floating plants such as submarine power cable, mooring systems designed for water depth of more than 1,000 m [31,32].

5.14 Site selection

The OTEC concept is based on the difference in temperature, ΔT, between the warm (T_w of about 22–29 °C) tropical surface waters and the cold deep ocean waters (T_c of about 4–5 °C at 1,000-m depth), as the source of the thermal energy required.

T_c at a given depth, below 500 m, does not vary much throughout the regions of interest for OTEC. It is also a weak function of depth, with a typical gradient of 1 °C per 150 m between 500 and 1,000 m. These considerations may lead one to regard T_c as nearly constant, with a value of 4–5 °C at 1,000 m.

A desirable OTEC thermal resource of at least 20 °C requires typical values of T_w of 25 °C. Generally, regions between latitudes 20°N and 20°S are adequate. Some definite exceptions exist due to strong cold currents: (i) along the west coast of South America, tropical coastal water temperatures remain below 20 °C, and are often 15 °C; and (ii) a similar situation prevails to a lesser extent for the west coast of Southern Africa. Moreover, T_w varies throughout the year, and sometimes exhibits a significant seasonal drop due to the upwelling of deeper water induced by the action of the wind. A careful OTEC site selection requires comprehensive knowledge of local climate features as they may affect T_w seasonally [30].

The following summarizes the availability of the OTEC thermal resource throughout the world:

- Equatorial waters, defined as lying between 10°N and 10°S, are adequate except for the west coasts of South America and Southern Africa.
- Tropical waters, defined as extending from the equatorial region boundary to, 20°N and 20°S are adequate, except for the west coasts of South America and of Southern Africa; moreover, seasonal upwelling phenomena would require significant temperature enhancement for the west coast of Northern Africa, the Horn of Africa and off of the Arabian Peninsula.

5.15 Advantages and disadvantages

Oceans surround many of the world's countries, and these provide an abundant source of tapping this form of ocean thermal energy. But, in order to benefit from such green energy, it is crucial to understand the advantages and disadvantages of OTEC. In addition, this will help in understanding why ocean thermal energy is not widely used despite most countries having a coastal region.

Advantages

- OTEC technology feeds off an unchanging condition, the solar energy stored in the world's oceans. Therefore, it can run virtually continuously, unlike other RESs such as solar and wind power. On an average day, the world's oceans absorb an amount of energy equivalent to 250 million barrels of oil, approximately 4,000 times the population's current energy needs. Once the generators and water pipes are in place, only minimal upkeep is required to keep the flow of electricity running and no harmful by-products result from the process.
- OTEC can also support numerous spin-off industries. Chilled water that has already been used can be pumped out of the plant and used in air conditioning, industrial cooling and chill-soil agriculture (where pipes containing chilled water are used to chill soil so it can support temperate crops in tropical climates). Additionally, desalinized water can be produced through OTEC

systems by implementing surface condensers to turn evaporated seawater into potable (fresh) water. A 2 MW plant, for instance, could produce about 4,300 m^3 of potable water.

Disadvantages

- At present, government subsidies are required to make OTEC energy economically viable. Electricity can be produced at about \$0.07/kWh, as opposed to subsidized wind power systems that can produce energy for as low as \$0.05/kWh. Moreover, OTEC requires expensive, large-diameter pipes submerged more than 1 km below the ocean's surface. Many of the countries within the viable geographical belt (between the Tropic of Cancer and Tropic of Capricorn) lack the economic resources to build this infrastructure.
- Because OTEC facilities are stationary surface platforms, they are essentially considered artificial islands and, therefore, their exact location affects their legal status under the United Nations Convention on the Law of the Sea treaty (UNCLOS). According to UNCLOS, coastal nations are given 5-, 20- and 320-km zones of varying legal authority. The amount of political autonomy among these zones varies greatly. Consequently, jurisdictional conflicts could arise based on international boundary disputes between nations.

References

[1] Rodrigues L. 'Wave power conversion systems for electrical energy production'. *RE&PQJ.* 2008;1(6):601–7.

[2] Evans O. F. 'The classification of wave-formed ripple marks'. *Journal of Sedimentary Research.* 1941;11(1):37–41.

[3] Monroe J. and Wicander R. 'The changing Earth: Exploring geology and evolution'. 7th edn. *Oceans, shorelines, and shoreline processes*. Stamford, CT, USA: Cengage Learning; 2015, pp. 385–403.

[4] Marshak S. 'Earth portrait of a planet'. 5th edn. *Sedimentary rocks.* New York: W.W. Norton & Company; 2015, pp. 202–30.

[5] Hung L. and Tsai W. 'The formation of parasitic capillary ripples on gravity-capillary waves and the underlying vertical structures'. *Journal of Physical Oceanography.* 2009;39:263–89.

[6] Drew B., Plummer A. R. and Sahinkaya N. N. 'A review of wave energy converter technology'. *Proceedings of the Institution of Mechanical Engineers, Part A: Journal of Power and Energy, IMechE.* 2009; 223(8):887–902.

[7] Chenari B. A., Saadatian S. S. and Ferreira A. D. 'Wave energy systems: An overview of different wave energy converters and recommendation for future improvements'. *8th International Technology, Education and Development Conference*; Valencia, Spain, March 2014, pp. 1–7.

[8] Kempener R. and Neumann F. *Wave energy technology brief.* International Renewable Energy Agency (IRENA) Report, June 2014.

[9] Twidell J. and Weir T. *Renewable energy resources*. 3rd edn. USA: Taylor & Francis Group, Routledge (imprint of); 2015.

[10] Poullikkas A. 'Technology prospects of wave power systems'. *Electronic Journal of Energy & Environment*. 2014;2(1):47–69.

[11] Aquamarine Power Oyster. Available from http://tinyurl.com/cn9k4k/ [Accessed 2 July 2018].

[12] Duckers L. 'Wave energy' in Boyle G. (ed.). *Renewable energy*. 2nd edn. Oxford, UK: Oxford University Press; 2004, Chapter 8.

[13] Prudell J. H., Schacher A. and Rhinefrank K. 'Direct drive ocean wave energy electric plant design methodology'. *Proceedings of 2012 Oceans*; Hampton Roads, VA, USA, Oct 2012. IEEE; 2012, pp. 1–7.

[14] Huang F. and Beattie W. C. 'Modeling and simulation of the Islay wave power conversion system'. *1997 IEEE Electric Machines and Drive Conference Record*; Milwaukee, WI, USA, May 1997, pp. MB2/2.1–MB2/2.4.

[15] Wu F., Ju P., Zhang X., *et al.* 'Modeling, control strategy, and power conditioning for direct-drive wave energy conversion to operate with power grid'. *Proceedings of the IEEE*. 2013;101(4):925–41.

[16] IEEE Committee. 'Dynamic models for system and hydro turbines in power system studies'. *IEEE Transactions on Power Apparatus and Systems*. 1973;92(6):1904–15.

[17] Sanada M., Morimoto S. and Inoue Y. 'Generator design and characteristics in direct-link wave power generating system considering appearance probability of waves'. *Renewable Energy Research and Applications (ICRERA), 2012 International Conference on*; Nagasaki, Japan, Nov 2012, pp. 1–6.

[18] Okano K., Sanada M., Morimoto S. and Inoue Y. 'Output characteristics in direct-link wave power generating systems considering place of installation'. *Power Electronics and Drive Systems (PEDS), 2013 IEEE 10th International Conference on*; Kitakyushu, Japan, April 2013. IEEE; 2013, pp. 187–93.

[19] Qing K., Xi X., Zanxiang N., Lipei H. and Kai S. 'Design of grid-connected directly driven wave power generation system with optimal control of output power'. *Power Electronics and Applications, 2013 15th European Conference on*; Lille, France, Sep 2013, pp. 1–8.

[20] Müller N., Kouro S., Glarfa J. and Malinowski M. 'Medium-voltage power converter interface for wave dragon wave energy conversion system'. *Energy Conversion Congress and Exposition (ECCE)*; Denver, CO, USA, Sep 2013. IEEE; 2013, pp. 352–8.

[21] Szabo L., Oprea C., Festilă C. and Dulf É. 'Study on a wave energy-based power system'. *Proceedings of Electrical machines, 2008. ICEM. 18th International Conference on*; Vilamoura, Portugal, Sep 2008. IEEE; 2008, pp. 1–6.

[22] Lavi A. 'Ocean thermal energy conversion: A general introduction'. *Journal of Energy*. 1980;5(6):469–80.

[23] Nihous G. C. 'Ocean thermal energy conversion' in Crawley G. M. (ed.). *The world scientific handbook of energy*. Singapore: World Scientific Publishing Co.; 2013, pp. 359–72.

[24] Kapooria R. K., Kumar S. and Kasana K. S. 'An analysis of a thermal power plant working on a Rankine cycle: A theoretical investigation'. *Journal of Energy in Southern Africa*. 2008;19(1): 77–83.

[25] Chan M. 'Ocean thermal energy conversion'. *Penn Sustainability Review, University of Pennsylvania*. 2015;1(7):29–31.

[26] Irianto D., Dibyo S., Salimy D. H. and Pane J. S. 'Thermodynamic analysis on Rankine cycle steam for cogeneration systems RGTT200K'. *Prosiding Seminar Nasional Teknologi Energi Nuklir* 2016; Batam, Aug 2016, pp. 865–72.

[27] Rajagopalan K. and Nihous G. C. 'Estimates of global ocean thermal energy conversion (OTEC) resources using an ocean general circulation model'. *An International Journal, Renewable Energy*. 2013;50:532–40.

[28] Jia Y., Nihous G. C. and Rajagopalan K. 'An evaluation of the large-scale implementation of ocean thermal energy conversion (OTEC) using an ocean general circulation model with low-complexity atmospheric feedback effects'. *Journal of Marine Science and Engineering*. 2018;6(1):1–28.

[29] Vega L. A. 'Ocean thermal energy conversion'. *Encyclopedia of sustainability and technology*. New York: Springer; 2012, pp. 7296–328.

[30] Lohani B. and Vega L. A. *Wave energy conversion and ocean thermal energy conversion potential in developing member countries*. Asian Development Bank, ADB, Report, 2014.

[31] Tanner D. 'Ocean thermal energy conversion: Current overview and future outlook'. *An International Journal, Renewable Energy*. 1995;6(3):367–73.

[32] Martins M., Sinama F. and Lucas F. 'Equivalent Gibbs systems for modelling an onshore OTEC experimental plant on Reunion Island'. *International Journal of Energy Research*. 2013;37:1112–21.

Chapter 6
Biomass energy

6.1 What is biomass?

Biomass is defined as the organic material that comes from living plants and animals. It has the ability of storing the energy absorbed from the sun (called 'bio-energy') internally. As with most living things, energy is stored in the form of carbohydrates including celluloses, starches and sugars that are produced as part of the process of photosynthesis. In the process of photosynthesis, plants convert radiant energy from the sun into chemical energy in the form of glucose or sugar, as shown in Figure 6.1, which can be expressed as

$$\text{Water} + \text{Carbon dioxide} + \text{Sunlight} \rightarrow \text{Glucose} + \text{Oxygen}$$

or

$$6H_2O + 6CO_2 + \text{Radiant energy} \rightarrow C_6H_{12}O_6 + 6O_2$$

The process of photosynthesis enables green plants to capture the electromagnetic radiation from the sun and transform it into chemical energy, with the biomass energy obtained by reversing this photosynthesis process. The stored energy within the living matter is released when the chemical compounds within the organic materials are broken down due to decomposition. The organic residue left over from this process is called 'biomass'. When biomass is burnt, the chemical energy in biomass is released as heat. Biomass can be burnt directly or converted to liquid biofuels or biogas that can be burnt as fuels [1].

Therefore, biomass energy can be produced from a total natural process and source, and consequently it is regarded as a green and climate-friendly form of energy, making biomass energy a renewable energy resource unlike fossil fuels (Figure 6.2).

6.2 Types of biomass

Biomass sources can potentially comprise a wide variety of different natural materials from liquids to solids to gas. Biomass can be made out of all natural products, for example, agricultural and forestry products, some of the energy crops growing everywhere including trees, wood, grass, plants and animal wastes

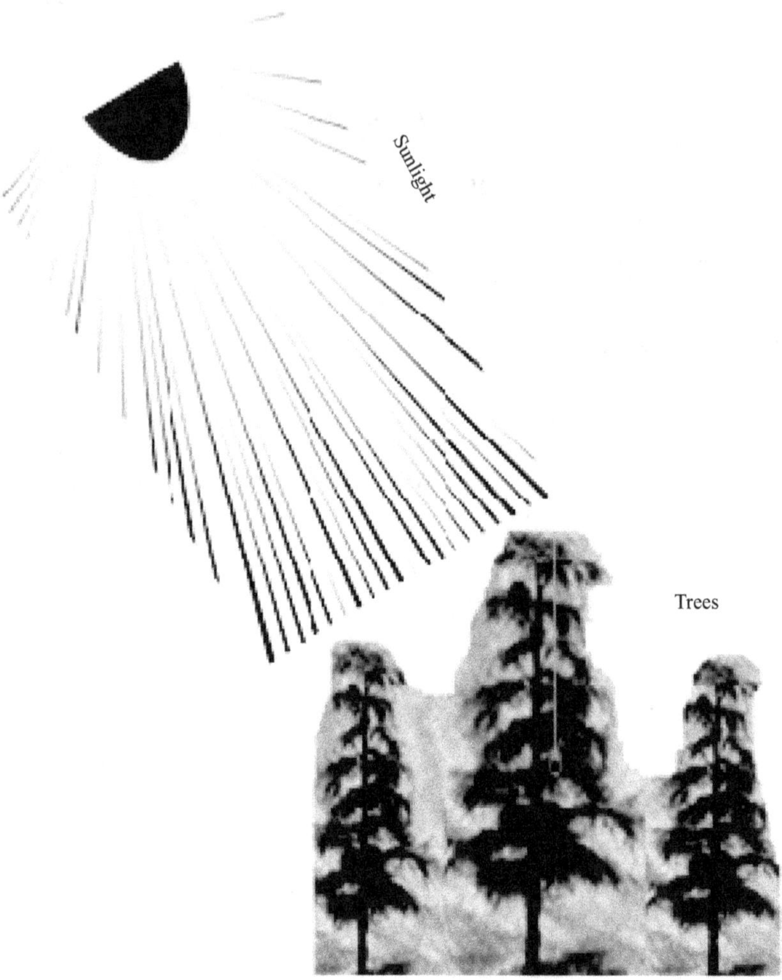

Figure 6.1 Photosynthesis process [National Energy Education Project (public domain)]

and slurries. In addition, humans are an integral part of the systems making life, and so the waste from human beings also can be a source of biomass (e.g. rubbish). All these can be used to produce biomass and bioenergy as shown in Figure 6.3 [2,3].

Biomass can be used in solid form (the most common example is that of wood) and it is very advantageous because of having a wide range of conversion and refining processes through which it can be made into both liquids and gases. These biomass solids, liquids and gases can then be used to produce electrical power, heat, new chemicals and even combustible liquid fuels.

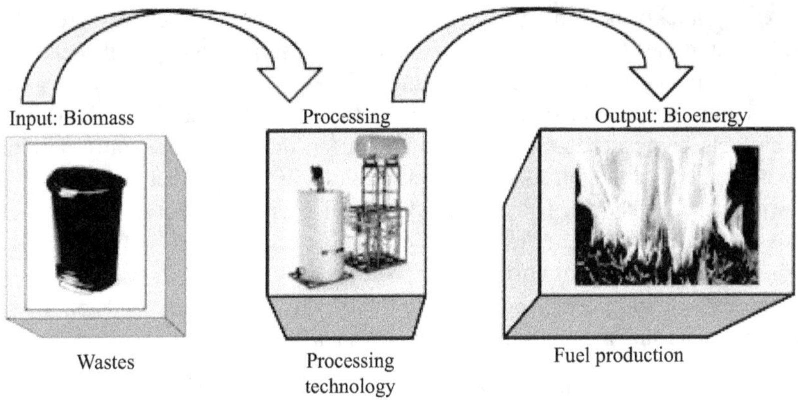

Figure 6.2 Biomass as a renewable source of energy

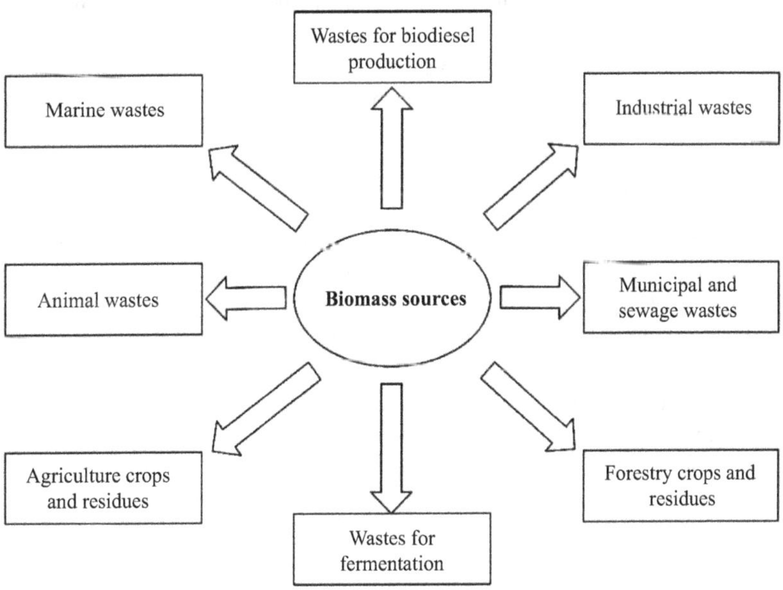

Figure 6.3 Biomass sources

6.3 Biomass energy

Energy can generally be recovered through several ways such as burning of fossil fuels, utilization of sunshine to generate solar energy or by making use of flow of water for hydroelectric generation. All these are ways of creating energy. Even the Earth's core can be used to collect energy, e.g. geothermal energy. However, today biomass is quickly becoming a growing source of renewable energy around the world. It continues to grow and help reduce the consumption of fossil fuels.

As mentioned above, there are many different types of biomass energy and biomass fuel sources available with many related terms, such as biomass, bioenergy and biofuels being used as though they all mean the same. Generally speaking, biomass is the organic non-fossil material of biological origin that can be used as a renewable energy source. Bioenergy is the final energy product or useful energy that is made available from the conversion of biomass, whereas biofuels are liquid or gaseous fuels of biomass origin.

The conversion process to obtain the final energy product depends on the type of organic materials included in the biomass such as follows:

1. Solid biomass, such as wood and garbage, can be burnt directly to produce heat for building, industry and generating electrical energy. Biomass can also be converted into a gas called biogas or into liquid biofuels such as ethanol and biodiesel. These fuels can then be burnt to produce energy.
2. Biomass materials, such as paper, food scraps and yard waste, that normally decompose in landfills can be burnt to generate electricity or converted into biogas in landfills. Similarly, sewage and animal manure can be processed in special vessels called digesters to produce biogas. Biogas can be burnt as a fuel.
3. Ethanol can be made by fermentation from crops such as corn and sugarcane. Biodiesel is produced from vegetable oils and animal fats. These liquid fuels (ethanol or biodiesel) can then be used as a source of energy in vehicles and/or as process heat.

Biomass energy production involves converting biological materials or wastes into substances that can be used as carbon negative or at the very least carbon neutral fuel [4]. Carbon neutral means that the carbon released during the use of the fuel is reabsorbed and balanced by the carbon absorbed by new plant growth during photosynthesis. The biomass crops are then harvested to make the next batch of fuel, thus maintaining the cycle of carbon within the Earth and its atmosphere without adding to the problem. Although the environmental impact of burning biomass is less harmful than those associated with fossil fuel, it can be more harmful than the impact associated with burning natural gas.

A typical example is shown in Figure 6.4 to illustrate the carbon cycle that starts by the harvesting of the crops (e.g. corn) to be separated into their component sugars. Then, the sugars are distilled to produce ethanol, which in turn can be used as an alternative fuel to power vehicles or engines that produce electricity. The released carbon dioxide, CO_2, is reabsorbed by the original crops to complete the cycle that is repeated again.

6.4 Biomass properties

The choice of biomass is affected by the form of energy required, and its properties determine the choice of conversion process. Therefore, the properties of concern

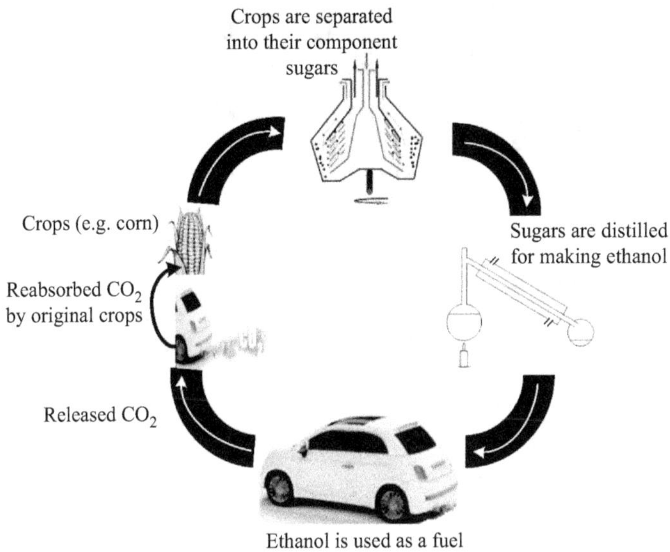

Figure 6.4 Carbon cycle from crops to a liquid fuel

that should be investigated to specify the adequacy of the biomass materials and consequently select the conversion process are based on [5] the following:

- Calorific value to define how much energy content in the material can be released during burning in the air.
- Alkali metal content, which is important for the thermochemical conversion processes.
- Ash residue content produced by combustion in air, which affects the cost of bioenergy cycle.
- Moisture content because of its impact on the conversion process, in particular, the thermal conversion which requires low moisture content feedstock.
- Proportions of fixed carbon and volatile matter which provide an indication of how the biomass is to be exploited as an energy source.
- Bulk density which determines the sizing of the materials handling system and consequently the costs of transport and storage.

6.5 Biomass energy conversion technologies

Key biomass energy conversion technologies include thermal, thermochemical, biological and biochemical as briefly presented next and depicted in Figure 6.5 [6,7].

Thermal technology. Direct combustion converts forestry, agricultural and municipal residues into heat and power under environmentally sound conditions. The biomass feedstock after handling and storage goes through different

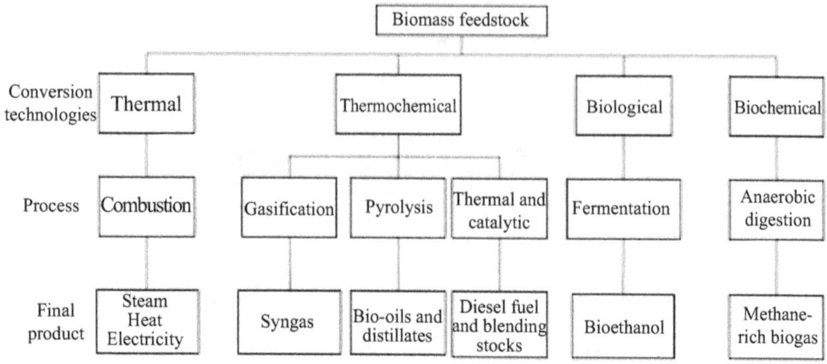

Figure 6.5 Conversion of biomass feedstock into final bio-products

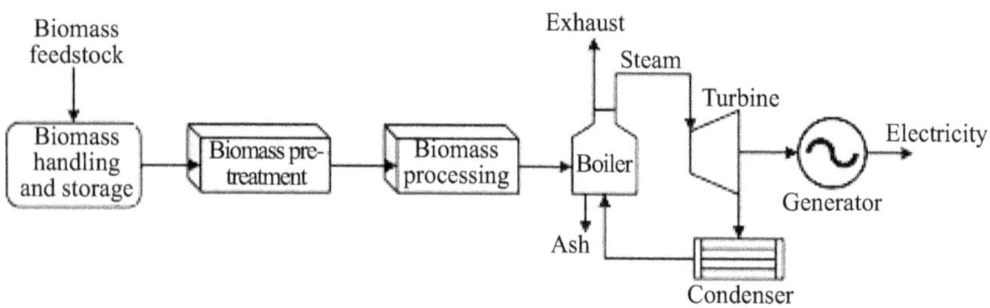

Figure 6.6 Thermal technology by direct combustion

stages: preparation and treatment, combustion to produce heat, using the heat energy to provide steam, and finally driving a turbine-generator set to generate electricity (Figure 6.6).

Thermochemical technology. It includes the following:

- Gasification: It converts forestry, agricultural and municipal residues into syngas.
- Pyrolysis: It converts forestry and agricultural residues into bio-oils and value-added products [8,9].
- Thermal and catalytic processing: It converts a variety of new and used vegetable oils, liquid rosin (obtained from the kraft pulping process), waste greases and animal fats into renewable diesel fuels and diesel blending stocks.

An example of gasification technology is illustrated by the block diagram of Figure 6.7. The gasification process is carried out on the collected biomass, and then by combustion to drive the gas-turbine as a prime mover of a generator to produce electricity.

Biological technology. Fermentation to convert the starch and cellulose components of biomass to bio-ethanol. The liquid fuel, ethanol, as an example, is the final product of processing the biomass by fermentation and distillation [10,11].

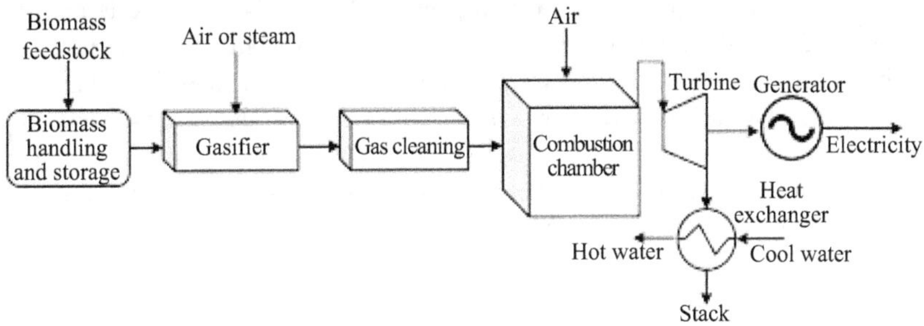

Figure 6.7 Thermochemical technology by gasification

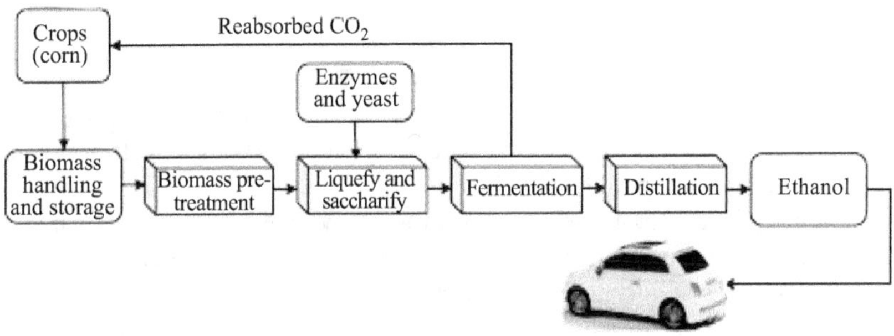

Figure 6.8 Biological conversion technology to produce ethanol

Ethanol is a corn alcohol that can be used to power cars or engines that produce electricity (Figure 6.8).

Biochemical technology. Anaerobic digestion to convert manures and food processing and municipal wastes into methane-rich biogas.

6.6 Pros and cons of biomass energy

Using biomass for energy has positive and negative effects [12,13]. Biomass and biofuels made from biomass are alternative energy sources to fossil fuels: coal, petroleum and natural gas. Burning either fossil fuels or biomass releases CO_2, a greenhouse gas. However, the plants that are the source of biomass capture a nearly equivalent amount of CO_2 through photosynthesis while they are growing, which can make biomass a carbon-neutral energy source.

6.6.1 Burning wood

Using wood, wood pellets and charcoal for heating and cooking can replace fossil fuels and may result in lower CO_2 emissions overall. Wood can be harvested from forests, woodlots that have to be thinned or urban trees that fall down or have to be cut down [14].

Wood smoke contains harmful pollutants such as carbon monoxide and particulate matter. Modern wood-burning stoves, pellet stoves and fireplace inserts can reduce the amount of particulates from burning wood. Wood and charcoal are major cooking and heating fuels in poor countries, but if people harvest the wood faster than the time required for the trees to grow, it causes deforestation. Planting fast-growing trees for fuel and using fuel-efficient cooking stoves can help slow deforestation and improve the environment.

6.6.2 *Burning municipal solid waste or wood waste*

Burning municipal solid waste or garbage to produce energy in waste-to-energy plants means that less waste is buried in landfills. However, burning garbage produces air pollution and releases the chemicals and substances from the waste into the air. Some of these chemicals can be hazardous to the people and the environment if the chemicals are not properly controlled.

The US Environmental Protection Agency (EPA) applies strict environmental rules to waste-to-energy plants and mandates that waste-to-energy plants use air pollution control devices such as scrubbers, fabric filters and electrostatic precipitators to capture air pollutants [15].

Scrubbers clean emissions from waste-to-energy facilities by spraying a liquid into the combustion gases to neutralize the acids present in the stream of emissions. Fabric filters and electrostatic precipitators also remove particles from the combustion gases. The particles called 'fly ash' are then mixed with the ash that is removed from the bottom of the waste-to-energy furnace.

A waste-to-energy furnace operates at high temperatures (1,800 °F to 2,000 °F), which breaks down the chemicals in municipal solid waste into simpler, less harmful compounds.

6.6.3 *Disposing ash from waste-to-energy plants*

Ash can contain high concentrations of various metals that were present in the original waste. Textile dyes, printing inks and ceramics, for example, may contain lead and cadmium.

Separating waste before burning can solve part of the problem. As batteries are the largest source of lead and cadmium in municipal waste, they should not be included in regular trash. Florescent light bulbs should also not be put in regular trash because they contain small amounts of mercury.

The EPA tests ash from waste-to-energy plants to make sure that it is not hazardous. The test looks for chemicals and metals that could contaminate ground water. Some municipal solid waste landfills use ash that is considered safe as a cover layer for their landfills, and some other ash is used to make concrete blocks and bricks.

6.6.4 *Collecting landfill gas or biogas*

Biogas forms as a result of biological processes in sewage treatment plants, waste landfills and livestock manure management systems. Biogas is composed mainly of methane (a greenhouse gas) and CO_2. Many facilities that produce biogas capture it and burn the methane for heat or to generate electricity. This electricity is considered renewable and, in many states, contributes to meeting state renewable portfolio standards. This electricity may replace electricity generation from fossil fuels and can result in a net reduction in CO_2 emissions. Burning methane produces CO_2, but because methane is a stronger greenhouse gas than CO_2, the overall greenhouse effect is lower.

6.6.5 Liquid biofuels: ethanol and biodiesel

Biofuels are transportation fuels such as ethanol and biodiesel. Ethanol is promoted to be used as a transportation fuel to help reduce oil imports and CO_2 emissions. As a result, nearly all gasoline now sold in some countries, for example, the United States, contain some ethanol [16].

Biofuels may be carbon-neutral because the plants that are used to make biofuels (such as corn and sugarcane for ethanol and soya beans and palm oil trees for biodiesel) absorb CO_2 as they grow and may offset the CO_2 emissions when biofuels are produced and burnt.

Growing plants for biofuels is controversial because the land, fertilizers and energy for growing biofuel crops could be used instead to grow food crops. In some parts of the world, large areas of natural vegetation and forests have been cut down to grow sugar cane for ethanol and soya beans and palm oil trees for biodiesel. The US government supports efforts to develop alternative sources of biomass that do not compete with food crops and that use less fertilizer and pesticides than corn and sugar cane. The US government also supports methods to produce ethanol that require less energy than conventional fermentation. Ethanol can also be made from waste paper, and biodiesel can be made from waste grease and oils and even algae.

Ethanol and gasoline–ethanol blends burn cleaner and have higher octane ratings than pure gasoline, but they have higher *evaporative emissions* from fuel tanks and dispensing equipment. These evaporative emissions contribute to the formation of harmful, ground-level ozone and smog. Gasoline requires extra processing to reduce evaporative emissions before it is blended with ethanol. Biodiesel combustion produces fewer sulphur oxides, less particulate matter, less carbon monoxide, and fewer unburnt and other hydrocarbons, but it does produce more nitrogen oxide than petroleum diesel.

It can be seen from the above that no particular approach is dealt with and each one has its own positive and negative aspects. While making a decision about which approach to adopt, it is necessary to weigh various alternatives and adopt one that offers the most positive outcome with respect to net environmental benefits.

References

[1] Mirza U. K., Ahmad N. and Majeed T. 'An overview of biomass energy utilization in Pakistan'. *Renewable and Sustainable Energy Reviews.* 2008;12:1988–96.

[2] Field C. B., Campbell J. E. and Lobell D. B. 'Biomass energy: The scale of the potential resource'. *Trends in Ecology and Evolution.* 2007;23(2):65–72.

[3] Doelle H. W. *Biomass and organic waste conversion to food, feed, fuel, fertilizer, energy, and commodity products* [online]. *Encyclopedia of Life Support Systems (EOLSS)*. 2003. Available from http://www.eolss.net/Eolss-sampleAllChapter.aspx [Accessed 8 Feb 2020].

[4] Tilman D., Hill J. and Lehman C. 'Carbon-negative biofuels from low-input high-diversity grassland biomass'. *Science Magazine.* 2006;314(5805): 1598–1600.

[5] McKendry P. 'Energy production from biomass (Part 1): Overview of biomass'. *Bioresource Technology*. 2002;83:37–46.

[6] McKendry P. 'Energy production from biomass (Part 2): Conversion technologies'. *Bioresource Technology*. 2002;83:47–54.

[7] McKendry P. 'Energy production from biomass (Part 3): Gasification technologies'. *Bioresource Technology.* 2002;83:55–63.

[8] Shuit S. H., Tan K. T., Lee K. T. and Kamaruddin A. H. 'Oil palm biomass as a sustainable energy source: A Malaysian case study'. *Elsevier, Energy.* 2009;34:1225–35.

[9] Mekhilef S., Saidur R., Safari A. and Mustafa W. E. S. B. 'Biomass energy in Malaysia: Current state and prospects'. *Renewable and Sustainable Energy Reviews*. 2011;15:3360–70.

[10] Abbasi T. and Abbasi S. A. 'Biomass energy and the environmental impacts associated with its production and utilization'. *Renewable and Sustainable Energy Reviews*. 2010;14:919–37.

[11] Farrell A. E., Plevin R. J., Turner B. T., Jones A. D., Hare M. O. and Kammen D. M. 'Ethanol can contribute to energy and environmental goals'. *Science Magazine.* 2006;311(5760):506–8.

[12] Herbert G. M. J. and Krishnan A. U. 'Quantifying environmental performance of biomass energy'. *Renewable and Sustainable Energy Reviews.* 2016;59:292–308.

[13] Reilly R. and Paltsev S. *Biomass energy and competition for land*. Science and Policy of Global Change Report no. 145, 2007.

[14] Dornburg V. and Faaij A. P. C. 'Efficiency and economy of wood-fired energy systems in relation to scale regarding heat and power generation using combustion and gasification technologies'. *Biomass and Bioenergy.* 2001;21:91–108.

[15] Goldemberg J. 'Ethanol for a sustainable energy future'. *Science Magazine.* 2007;315(5813):808–10.

[16] Heller M. C., Keoleian G. A., Mann M. K. and Volk T. A. 'Life scale energy and environmental benefits of generating electricity from willow biomass'. *Renewable Energy*. 2004;29:1023–42.

Part II

Energy storage systems

Chapter 7

Electrical energy storage

7.1 Introduction

The traditional power system comprised a chain of three segments in cascade: generation, transmission and distribution systems. Generation of electricity is derived from a source of energy to produce electrical power, for example, fossil fuels for thermal plants, water for hydraulic plants, wind, solar irradiation, etc. Traditional generation of electrical power is highly centralized and, often, a long distance away from its end users. The transmission system (second segment) is required to transmit the electrical power at a high-voltage level to the end users through the distribution system at medium voltage (third segment) in a unidirectional power flow (Figure 7.1). It is crucial to recognize that the centralized production of electricity leads to a complex power system configuration, high losses, low electric power reliability and increased risk of instability. To solve these problems, distributed energy resources (DERs) can offer an alternative solution.

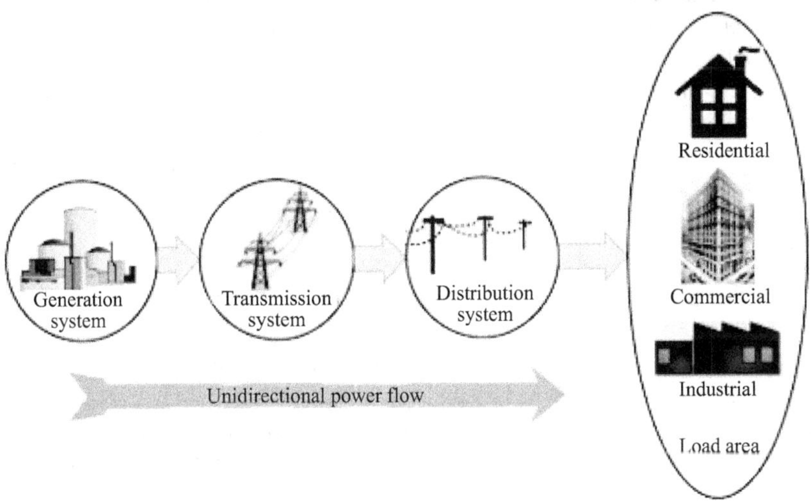

Figure 7.1 Schematic diagram for traditional power system

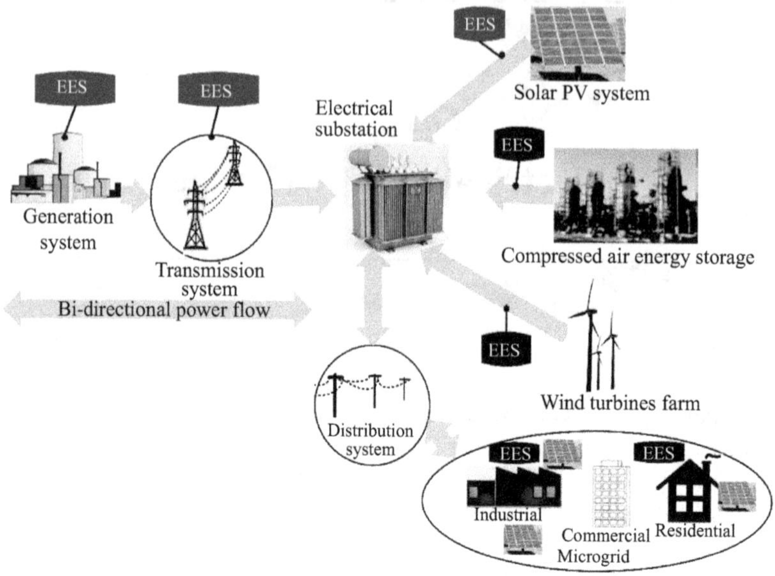

Figure 7.2 Future perspectives of power systems

DERs are small, modular, energy generation and storage technologies that provide electric energy installed on site and of a size meeting local need. Furthermore, DER systems may be either connected to the local electric grid or isolated from the grid in stand-alone applications. DER technologies include renewable energy sources (e.g. wind turbines and solar photovoltaics), micro-turbines, reciprocating engines, combustion turbines, cogeneration and electrical energy storage (EES) systems. Utilities can use DER technologies to delay, reduce or even eliminate the need to obtain additional central power generation, transmission and distribution equipment and infrastructure. Meanwhile, DER systems can provide voltage support and enhance local reliability.

Therefore, the perspective of the electrical power grid in future can be seen as an integration of conventional generation and DERs and consequently the power flow can be bi-directional as shown in Figure 7.2.

7.2 EES and renewable energy

Because of the rapid growth and development of renewable energy, sustainable energy systems will have to be based not strictly on traditional resources but make greater use of renewable energy. Furthermore, distributed electrical energy production from renewable energy sources leads to a cleaner supply with fewer environmental hazards for consumers [1].

However, the unpredictable character of renewable resources requires more attention from planners to keep the system operating in an optimal manner. In addition, the power supplied by these resources is fluctuating as it depends on unpredictable primary sources of energy, for example, wind speed for power generated by wind turbines and solar irradiation for solar photovoltaic systems. For the reasons of unpredictability and fluctuating renewable energy, use of energy storage systems is imperative to avert the problems, such as power imbalance between supply and demand and the difficulty of voltage and frequency regulation that may appear. This will help increase the penetration rate of renewable energy to the power networks [2].

7.3 EES concept

Electrical energy is non-storable in its original form. EES refers to a process of converting electrical energy from a power network into a storable form that can be converted back to electrical energy when needed. Such a process enables electrical energy to be captured at any time, for example, low demand, low generation cost or surplus from intermittent energy sources, and to be used later of high demand, high generation cost or in the presence of generation capacity shortage. The process of energy capture is regarded as 'charging,' whereas the process of releasing the energy to be used is regarded as 'discharging'. The energy is stored using different technologies and the storage medium (commonly known as carrier) is decided accordingly.

7.3.1 Elements of EES

Energy storage systems can be configured by three major components as shown in Figure 7.3 [3]:

- *Energy storage device:* It is designated to contain the carrier by which the energy is held until needed, for example, water reservoirs in pumped hydro storage or chemical/electrolyte used in battery, flywheel, etc.
- *Power conversion system:* It is the power electronics used to adjust the power characteristics of the storage (e.g. the voltage and current) based on the load requirements and ensures proper and safe charge/discharge of storage device.
- *Facilities* for housing the equipment, wiring, control and system protection from physical damage.

7.4 The need for EES

Major reasons for installing energy storage can be summarized as follows:

- *Renewable integration:* EES can store extra renewable energy to be used at the desirable time, eliminating power curtailment and oversized construction of hybrid power system capacity. For instance, for wind power generation, EES can contribute in relieving the fluctuations and provide voltage control support during low-voltage ride through, resulting in smooth power output.

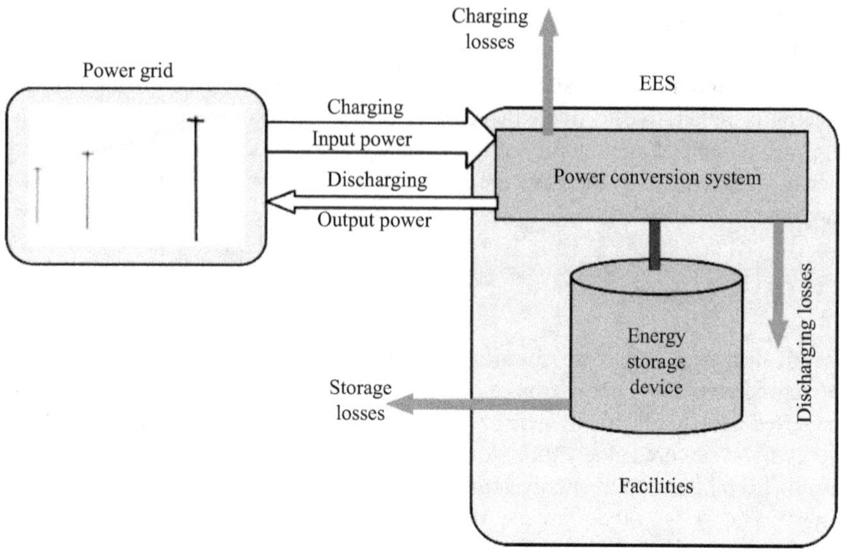

Figure 7.3 Elements of EES systems and energy losses

- *Transmission and distribution (T&D) upgrade deferral:* Storage provides buffering capability to enable high penetration of variable renewables and asset deferral for T&D systems (load management and ancillary services).
- *Power quality:* EES contributes in providing electrical service to customers without any oscillations or disruptions to the voltage waveform such as swells/ sags, spikes or harmonics.
- *Improved power reliability:* Bridging power can be supplied by the EES for consumers to ride through a power disruption.
- *Energy systems stability:* Energy storage helps in power system planning and matching demand with supply to maintain grid stability.

7.5 Critical challenges

There exist some critical challenges in making energy storage cost-competitive. These are the following:

- Critical challenges for energy storage are high system cost and cycle life with
  - existing storage solutions too expensive;
  - deep discharge and longer cycle life;
  - safe and reliable chemistry;
  - scalable technology to cover all markets.
- To make storage cost-competitive, advances across all major areas are required:
  - storage, power electronics and power conversion systems;
  - power quality and integration;

- engineered safety of large systems;
- codes and standards;
- optimal use of storage resources across the entire electricity infrastructure.

7.6 Functions of EES

An electric utility normally encompasses many electricity-generating plants. A characteristic of electricity is that it is consumed at the same time as it is generated. Thus, a proper amount of electricity must always be generated at any time to meet the varying demand, thus keeping the balance between power generation and consumption. Imbalance between supply and demand may result in undesirable quality (voltage and frequency) of power supply and even lead to instability.

Environmental concerns are leading to difficulties in installing new transmission lines to feed remote consumers. Depending on both network topology and power flow, there is the probability of congestion happening in a specific line causing a failure and consequently power interruption.

To continue supplying the load with enough energy and reconcile the differences between supply and demand during both normal operation and contingencies, power plants are built and operated with a variety of functions, as described in the following [4]:

- *Base-load plants*, such as large fossil fuel thermal power stations, provide centralized generation with a bulk capacity to meet the base (constant) demand for electricity. Such stations are normally run at full rating because technical constraints do not allow rapid load following.
- *Load-following plants*, also known as 'cycling' or 'intermediate load plants', have the capability to change the power output quickly and thus can follow the changing balance between supply and demand. Hydroelectric and natural gas fired plants are typically used in this category.
- *Peaking units*, designated to meet the peak demand, generally run less than a few hundred hours per year.

Taking a typical daily load curve, Figure 7.4 as an illustrative example, the base and intermediate loads are supplied, respectively, by base-load units and load following plants with more flexible but less cost-effective generation. The base-load and load following units are complemented by the peaking units to supply the load during the peak load conditions, #1 and #2.

To meet unforeseen increase in demand, loss of conventional units and transmission lines, additional generating units, known as '*operating reserves units*', are also made available by the utilities. Furthermore, some generating units, called *spinning reserves*, are required for (i) frequency regulation, i.e. providing the ability to respond to small, random fluctuations around normal load; (ii) providing the capability to respond to a greater or less than predicted change in demand due to load forecasting errors; and (iii) contingencies so as to be able to respond to, for example, unplanned generator or transmission line outage.

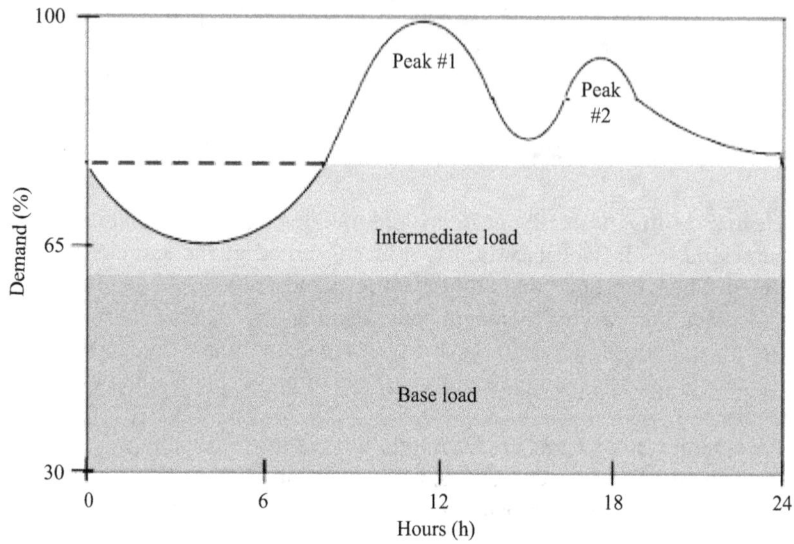

Figure 7.4 Typical daily load curve [4]

Excess energy available in a form that is difficult to store can be converted to a form in which it is more convenient or economic to store and use later. In case of excess electrical energy available from the electric grid or another source, particularly a renewable energy source, energy can be stored in different forms of energy such as electrochemical, electromagnetic, kinetic or potential energy, using a suitable energy storage device and make it available when needed. Some energy conversion technologies for energy storage provide short-term energy storage, whereas others can store for longer periods. Various technologies available for EES and the role played by them in the operation of the power system corresponds to the available generation capacity versus the varying load demand over a period, say 1 day.

To increase the reliability and resilience of the electric grid, EES technologies and devices can play a prominent role in a variety of the functions mentioned above as described in the following [5,6]:

Electric energy time shift: Because of the dependency of generation cost on the type of generating plant, base, load following and peaking, the price of electricity is higher at peak demand periods and lower at off-peak periods. Electric energy time shift is often referred to as arbitrage. Unlike arbitrage in the financial sense, energy arbitrage involves the purchase and sale of electricity at different times to benefit from a price discrepancy in the cost of generation. For example, electricity generated from wind at night or solar power during the day can be purchased at low cost during these off-peak hours and sold during peak hours. Storage devices, such as pumped hydro plants, compressed air energy storage facilities and large battery installations, used for electric energy time shift

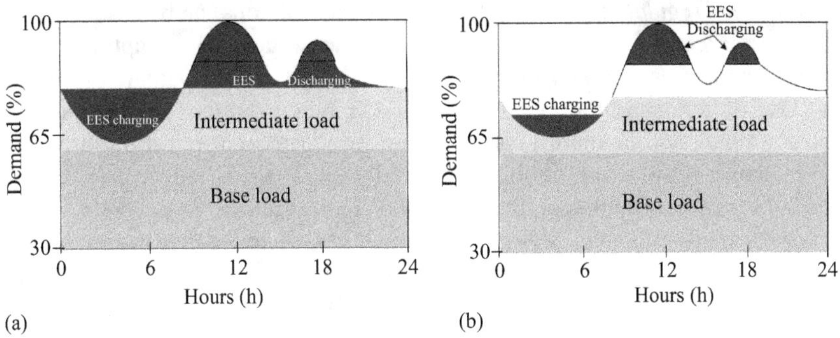

Figure 7.5 Load profile for large-scale EES system: (a) for load following and levelling; (b) for peak clipping

can store large amounts of electricity to optimize the gain from electricity price differentials between off-peak and peak demand times. This can also offset the disadvantages of intermittent renewable energy sources by shifting this energy to times when it is needed most.

Performance targets for electric energy time shift, a medium-duration energy management application, focus on system capital cost, operations and maintenance cost, discharge duration, efficiency and response time. The environmental impact of storage devices used for electric energy time shift is also an important factor to consider.

Load following and levelling: Energy storage technologies with very quick response are ideal for load following, a medium-duration energy management application (Figure 7.5(a)), and thus can isolate the rest of the grid from rapid and substantial changes in net supply in comparison to demand. Many of these technologies can operate at partial output/input capacity without compromising the system performance. In addition, they can quickly respond to increasing or decreasing load, respectively, by discharging or charging rapidly.

The need to quickly respond to fluctuating demands requires the storage device to have the ability to communicate and respond to the grid through the system operator. Performance targets for the storage technologies used for load following emphasize the importance of system capital cost, operations and maintenance cost, and discharge duration. An energy storage technology, which can meet these targets for the load following application, is well positioned to be adopted by the electric power industry to assist with abrupt fluctuations in electricity demand.

Peak clipping: It is more economical and desirable to build enough generating capacity to meet average demand rather than peak demands that may only last for a short time each year. Utility planners wish to reduce the load during peak demand periods as it defers the need for additional generation capacity. Therefore, large-scale EES can be used to implement peak clipping by providing the electrical energy required during peak periods (Figure 7.5(b)).

Frequency regulation: Sustained electricity interruptions can be costly and fast response time can minimize both momentary and sustained interruptions. Some storage technologies can vary output, from zero output to full output, within seconds. Use of an appropriate energy storage technology offers the potential for frequency regulation, a short-duration power management application that is far faster than by a gas or steam turbine and, thereby, significantly reduce the impact of power interruption of high-tech industrial and commercial facilities.

For optimum efficiency and response time, it is desirable that the energy storage technologies used for frequency regulation is able to communicate with the grid in a quick and efficient manner. The performance targets of energy storage technologies used for frequency regulation emphasize the importance of system cost, system lifetime, discharge duration, response time and roundtrip efficiency metrics.

T&D deferral and substitution: The continued increase in demand for electricity requires additional T&D infrastructure to transport electricity from the point of generation to where it is consumed. It requires building new transmission lines, substations and distribution infrastructure, a costly and time-consuming process. Also, as transmission lines are designed for high reliability at peak conditions, they normally operate at a capacity lower than the design capacity and thus low utilization.

The ability to store the power generated during times of low demand and use during the period of high demand can reduce the stress on power plants as well as on the T&D infrastructure. In addition, EES closer to the point of use lowers the congestion of the transmission system during peak demand. The target of the storage in this case, a long-duration energy management application, can be achieved using relatively small amount of storage and help to postpone or eliminate the need to build new T&D lines.

In addition to emphasizing the importance of system cost, discharge duration, capacity, reliability and system lifetime in the performance targets for energy storage technologies applied to T&D upgrade deferral and substitution, safety is also an important metric for storage systems used for this application. The storage locations throughout the grid must be safe and have cybersecurity protection consistent with the current utility practice. The targets mentioned above must be met to justify investment in energy storage technologies compared to investing in upgrades to the generation, transmission and distribution infrastructure.

Increased integration of renewables in the grid: Increasing the integration of renewable energy into the grid can reduce reliance on fossil fuels and thus emissions from electricity generation. However, when connected to the electric grid on a large scale, the intermittent nature of these alternative sources introduces generation variability that raises other operational and integration issues that can be categorized into the following two general areas.

First, output power fluctuations (e.g. ramp-up/ramp-down) of short duration may make system frequency control difficult. It may also lead to deterioration of the system operation under wide frequency deviation. Frequency control in conventional systems is mostly managed by the capability to control the output of conventional, hydro and thermal, generating units. This necessitates a partial capacity operation of these generators with the output margin being used to adjust frequency, even if it yields inefficient operation. Increased penetration of renewable generation will need a further increase of the output margin, and, consequently, may even further lower the efficiency of conventional generation. Proper use of EES can mitigate the output fluctuation of renewable generation sources and reduce the margins of conventional generators that can be operated at a higher efficiency. A coordinated and judicious operation of renewable generation sources and energy storage technologies can smooth the power generated from these sources. It can thus support the increased penetration of renewables-generated electricity and, thereby, ease the grid operation where large amounts of wind and solar generation are deployed. To optimize the effectiveness of these operations, it is necessary to provide the storage technologies the ability to communicate and respond to the operational requirements of the grid through the system operator.

Second, ESS can be used for mitigation of medium-duration output power fluctuation. For example, in the operation with a function of electric energy time shift, ESS can store large amounts of electricity at off-peak periods (low electricity price). Then, the energy can be discharged when it is needed to match energy production with demand.

Performance targets for storage technologies used for renewables grid integration address roundtrip efficiency, system lifetime, capacity and response time metrics.

7.7 Classification of EES technologies

Different technologies are used for storing the electrical energy. A widely used approach for classifying these technologies is according to the form of stored energy. It includes mechanical, chemical, electrochemical, electrical and thermal energy storage. The different classes of EES technologies are shown in Figure 7.6. More description about these matured technologies is presented in the forthcoming chapters of this part. There are other technologies which still are in the demonstration and development stage such as cryogenic and synthetic natural gas energy storage [7].

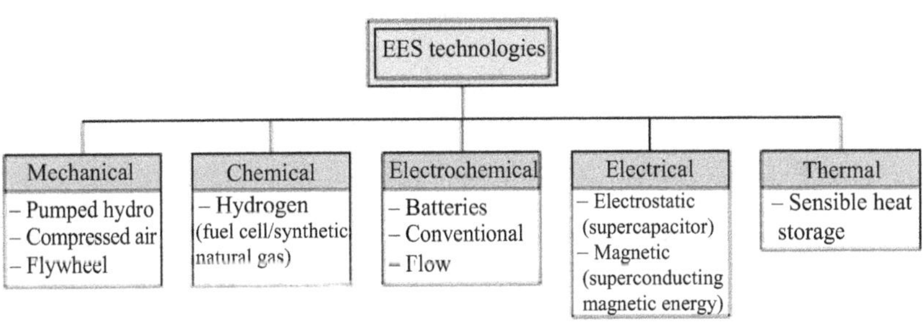

Figure 7.6 Classification of EES technologies

7.8 Characteristics of EES technologies

Each EES technology has its own characteristics (technical and economic), which may differ from the others. Therefore, it is critical to generally identify the main items of these characteristics to help the planners to properly select the storage system that meets the application needs. These items are the following:

- *Power availability:* It is determined by both the average and peak value of charge/discharge power (P_{av} and P_{max}, respectively).
- *Storage capacity:* It is defined as the total energy stored, E_{st} (Wh), which exceeds the actual retrieved stored energy, E_{act} (Wh), as the discharge is often incomplete. It is to be noted that the usable energy is limited by the depth of discharge (DoD) and so storage capacity represents the limit of discharge depth.
- *DoD:* Energy storage should have a quick energy release action on demand. The power discharge can be a limiting factor and called '*DoD*' or 'transmission rate' by which the time required to extract the stored energy is determined.
- *Discharge time:* It is the duration of maximum power discharge, t_{dis}, and is given by

$$t_{dis} = \frac{E_{st}}{P_{max}} \tag{7.1}$$

- It is a characteristic of storage system adequacy for some applications.
- *Efficiency:* On a single operating point, the efficiency, η, can be calculated as the ratio between released and stored energy, that is,

$$\eta = \frac{E_{act}}{E_{st}} \tag{7.2}$$

  It is a simplified definition as the systems have charging, no-load and discharging losses, which must be considered. In addition, in conditions of quick charge/discharge, the efficiency deteriorates as E_{act} is much less than E_{st}.
- *Durability:* It is also named as 'cycling capacity' and expressed as the maximum number of cycles where 'on' cycle corresponds to one charge and one discharge. So, durability refers to the times the storage unit can release the designed energy level after each recharge.
- *Autonomy:* It is the maximum time the storage unit can continuously release energy. It is calculated as the ratio of retrieved stored energy, E_{act}, to the maximum discharge power, P_{dis}, that is,

$$A_{ut} = \frac{E_{act}}{P_{dis}} \tag{7.3}$$

- *Self-discharge:* It is the portion of the energy that was initially stored and has dissipated over a given non-use time.
- *Energy density:* It represents the maximum amount of stored energy per unit mass or volume of the storage unit. It is an important indicator, especially for portable applications.

- *Operational constraints:* They are commonly related to safety and operating conditions such as temperature, pressure and so on, which affect the choice of a storage technology in terms of energy needs.
- *Reliability:* For energy storage systems, reliability is an important characteristic as it is a guarantee of on-demand service.
- *Costs:* As shown in Section 7.5, one of the main challenges is the cost of EES systems, which determines how much the storage system is competitive. Of course, investment of using a storage system is highly useful and interesting when the total gains exceed the total expenses. Like any other investment, the costs mainly include fixed and running costs. Fixed costs represent the cost of purchase, installation and delivery of system assets, whereas operation and maintenance (O&M) costs form the running costs. It is, therefore, crucial to study the cost of EES technologies [8]. However, as reported in the literature, two main approaches can be applied to estimate the cost as described in the following:
  - (i) *Total capital cost (TCC)* to evaluate the related costs of the system elements shown in Figure 7.3. *TCC* includes the following:
    - PCS costs, C_{PCS}, expressed in per unit of power capacity ($/kW) and represent the cost of purchase, installation and delivery.
    - Energy storage costs, C_{st}, in per unit of stored or delivered energy ($/kWh) and represent the cost of purchase, installation and delivery of reservoirs.
    - Balance of power costs, C_{BOP}, including costs of project engineering, grid connection interface, integration facilities, land and all other required services. C_{BOP} can be expressed in per unit of power ($/kW) or energy ($/kWh).

    Accordingly, *TCC* can be represented by the following formula [3]:

    $$TCC = C_{PCS} + C_{BOP} + C_{st} \times h \tag{7.4}$$

    where h is the charging/discharging time.
  - (ii) *Life cycle cost, LCC*, which is a more effective indicator from the ownership point of view. *LCC* can be calculated by adding to *TCC* the fixed and variable expenses of O&M, replacement, disposal and recycling. It is represented annually in per unit of power ($/kW/year). Consequently, *TCC*/year should be calculated by its present value of money as [9] follows:

    $$TCC/\text{year} = TCC \times CRF \tag{7.5}$$

    and

    $$CRF = \frac{i(1+i)^T}{(1+i)^T - 1} \tag{7.6}$$

where

CRF is the capital recovery factor;

i is the interest rate;

T is the lifetime.

The annual O&M costs ($C_{\text{O\&M/year}}$) can be expressed by adding the annualized costs of fixed O&M ($C_{\text{FO\&M/year}}$) and variable O&M ($C_{\text{VO}\times\text{M/year}}$) multiplied by yearly operating hours (h) as in the following.

$$C_{\text{O\&M/year}} = C_{\text{FO\&M/year}} + C_{\text{VO\&M/year}} \times h \tag{7.7}$$

Thus,

$$LCC = TCC/\text{year} + C_{\text{O\&M/year}} + C_R + C_{\text{DR}} \tag{7.8}$$

where C_R is the annual replacement cost, and C_{DR} is the annual disposal and recycling cost.

7.9 Fields of energy storage applications

The principle of energy storage can be applied to both the 'bulk power' transmission grid and the distribution system applications. Some of the technologies are available for both applications. Currently, distribution applications are receiving more attention due to the concept of micro-grid and distributed generation, and other power quality issues. In addition, the need to become less dependent on fossil fuels and imported oil is also driving research into energy storage devices for the transportation industry. There are many things in common between these energy storage applications for the transportation and backup power systems industries as they utilize similar technologies. The concept of energy storage in the transmission grid is not new either. Large pumped hydroelectric plants have been operating off this principle for many decades. Over the past two decades many other energy storage technologies have matured to the point that they have become viable options for selected applications in large-scale power systems. Some have already seen small-to-medium scale applications. The viable storage technologies for different applications are shown in Figure 7.7 [10].

It is to be noted that, for a specific application, different available EES systems can be used commercially, but no single storage system meets all the requirements that satisfy the best of all characteristics (high density, high efficiency, long lifetime, low cost, etc.). Each storage system has a suitable range; for instance, batteries, compressed air, flywheel and pumped hydroelectric are suitable for load following or load levelling and energy management applications. In contrast, supercapacitors and flywheel are suitable for power quality. So, more than one solution can be found. In this case, technical and economical comparison should be investigated to achieve the best solution [11,12].

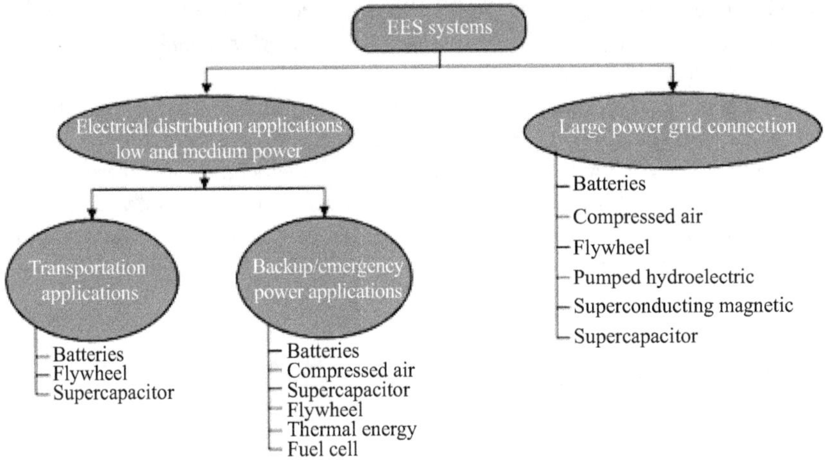

Figure 7.7 Applications of different storage technologies

References

[1] Ibrahim H., Ilinca A. and Perron J. 'Energy storage systems – Characteristics and comparison'. *Renewable & Sustainable Energy Reviews*. 2008;12: 1221–50.

[2] Beaudin M., Zareipour H., Schellenberglabe A. and Rosehart W. 'Energy storage for mitigating the variability of renewable electricity sources: An updated review'. *Energy for Sustainable Development*. 2010;14:302–14

[3] Zakeri B. and Syri S. 'Electrical energy storage systems: A comparative life cycle cost analysis'. *Renewable and Sustainable Energy Reviews*. 2015; 42:569–96.

[4] Denholm P., Ela E., Kirby B. and Milligan M. *The role of energy storage with renewable electricity generation*. National Renewable Energy Laboratory report NREL/TP-6A2-47187. January 2010. Available at http:// www.osti.gov/bridge [Accessed 2 August 2020].

[5] International Electrotechanical Commission, IEC. *Electrical energy storage*. White Paper, 2011.

[6] Lichtner S., Brindle R. and Pack L. *Electric power industry needs for grid-scale storage applications*. Sandia National Laboratories Report, 2010.

[7] Aneke M. and Wang M. 'Energy storage technologies and real-life applications – A state of the art review'. *Applied Energy*. 2016;179:350–77.

[8] Chen H., Cong T. N., Yang W., Tan C., Li Y. and Ding Y. 'Progress in electrical energy storage system: A critical review'. *Progress in Natural Science*. 2009;19:291–312.

[9] Brealey R. A., Myres S. C. and Allen F. *Principles of corporate finance*. New York, NY: McGraw-Hill/Irwin; 2011.

[10] Kondoh J., Ishii I., Yamaguchi H., *et al.* 'Electrical energy storage systems for energy networks'. *Energy Conversion & Management.* 2000;41(17): 1863–74.

[11] Schoenung S. M. and Burns C. 'Utility energy storage applications studies'. *IEEE Transactions on Energy Conversion*. 1996;11(3):658–65.

[12] Smith S. C., Sen P. K. and Kroposki B. 'Advancement of energy storage devices and applications in electrical power systems'. *Proceedings of the 2008 IEEE Power and Energy Society General Meeting – Conversion and Delivery of Electrical Energy in the 21st Century*; Pittsburgh, Pennsylvania, USA; 2008, pp. 1–8.

Chapter 8

Mechanical energy storage systems

Among the energy storage system (EES) types based on the form of energy stored (Chapter 7, Section 7.7), mechanical energy storage (MES) systems are one of these technologies. They include pumped hydroelectric storage (PHES), compressed air energy storage (CAES) and flywheels (FWs).

PHES technology is suitable for energy management applications that move the power over longer time scales and require continuous discharge ratings of several hours and more. CAES is adequate for short- and long-duration energy management applications, whereas FWs can be used for power quality applications that require rapid response (less than a second), transient stability and frequency regulation. Principles and operation of each technology are described in the forthcoming sections.

8.1 PHES system

8.1.1 Physical principles

The principle of PHES is to store electrical energy by utilizing the potential energy (PE) of water at a specific head, established between two reservoirs, one being highly elevated than the other.

At periods of low demand and high availability of electrical energy, water is pumped from the lower reservoir and stored in the upper reservoir. On demand, water is released back into the lower reservoir through the same piping system to generate electricity. The potential energy is converted into kinetic energy (KE) and enthalpy (h). If m_1 is a part of total mass m, which is moving with velocity v to produce the kinetic energy, then the remaining mass, $(m - m_1)$, multiplied by pressure, p, over water density, ρ, gives the enthalpy. Therefore, the energy balance is verified by the relation

$$\begin{aligned} \mathrm{PE} &= \mathrm{KE} + h \\ &= \frac{1}{2} m_1 v^2 + (m - m_1) p/\rho \end{aligned} \tag{8.1}$$

The generated output power, P, in watts (W) can be given by [1]

$$P = Q \times H \times \rho \times \eta \tag{8.2}$$

where Q is the fluid flow in cubic metres per second (m^3/s), H is the hydraulic head height in metres (m), ρ is the fluid density in kilograms per cubic metre (1,000 kg/m^3 for water), g is the gravitational acceleration (9.81 m/s^2) and η is the facility efficiency.

It is to be noted from (8.2) that the changeable variables are Q, H and η. The flow is inversely proportional to the head; that is, if the head (H) is maximized, the water utilization can be minimized and vice versa.

The cycle of operation is depicted in Figure 8.1. It is seen that the water discharged from the turbine is stored in the lower-level water tank. The corresponding components and their functions are given in Table 8.1 [2].

Recently, the total number of components has been decreased to reduce the capital cost. Pump, turbine, motor and generator can be arranged in different ways, which increases the overall efficiency and makes the system more compact. The system which is composed of four units, i.e. motor, generator, pump and turbine, makes the power house big in size. So, reversible motor/generator together with pump and turbine configuration having three units can be used instead. Additionally, reversible motor/generator and reversible pump/turbine systems that have two compact components are preferred, as the investment cost is highly reduced. A schematic diagram of a typical PHES system is shown in Figure 8.2.

Another aspect to increase the operation range of PHES is using asynchronous motor-generators. The so-called doubly fed induction machines (DFIMs) increase the flexibility, particularly during pumping mode. While the efficient pumping for synchronous motor-generators is in the range of 90%–100% of the nominal power (P_N), DFIM can operate in pumping mode between 65% and 100% of P_N.

Kaplan or Francis turbines are commonly used to maximize efficiency [3]. They are reversible and capable of handling both the pumping and the generating

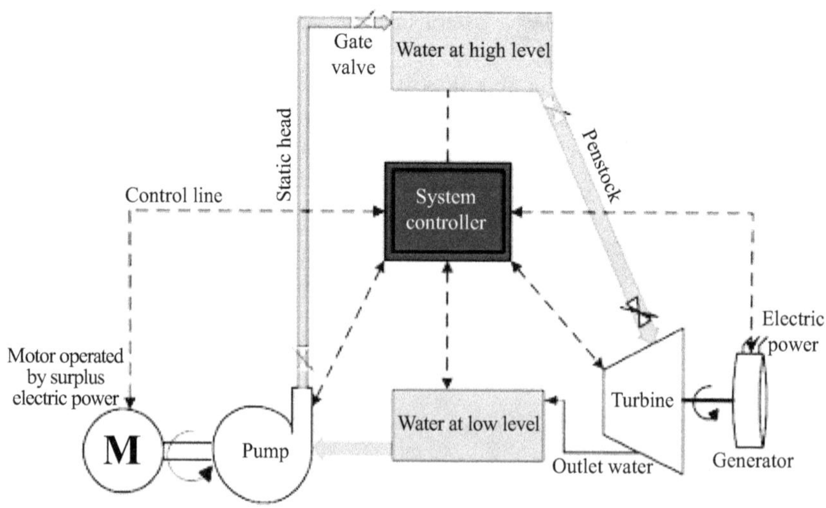

Figure 8.1 PHES operation cycle

Table 8.1 Functions of the main components of PHES system

Component name	Component function
A hydroelectric pumped storage system consists of two reservoirs (high and low), pumps, turbines, motors, generators, penstocks, flow control valves, system controller and auxiliary equipment	Both reservoirs are open tanks that feature an inlet/outlet to a pipe, located on the bottom. The elevation difference between the two reservoirs is called the 'head'
Penstocks	These are large diameter steel pipes, connecting the reservoirs to the pump and turbine
Pump	When the system must store energy, pump shaft is rotated with the force coming from an electric motor
Turbine	The centrifugal turbine converts stored energy to mechanical energy, rotating the shaft connected to the generator
Generator	The generator converts mechanical energy from the turbine output shaft to electrical energy and feds into the electric grid
Motor	The motor converts electrical energy (AC power from the grid) to mechanical energy to drive the pump when the system must store energy
Flow control valve	The flow control valve modulates the water flow in the turbine and pump mode. It serves as an emergency valve during operation. It is completely open during operation
System controller	The system controller decides the operating mode of the system, based on the power plant output and grid demand
Auxiliary equipment	Auxiliary equipment such as transformers and AC/DC panels are used for operation and protection of the system

process. Similarly, synchronous machine can be operated as motor during pumping and as a generator to generate electricity.

Developments in the machinery do not only decrease the number of units but also increase the round-trip efficiencies of the system. Global efficiency of a pumped-storage system is between 75% and 80%. Typical key performance data is given in Table 8.2.

8.1.2 Modelling of PHES dynamics

Dynamic models for PHES systems are necessary for studying the dynamic interactions between these systems and the grid incorporating renewable energy sources.

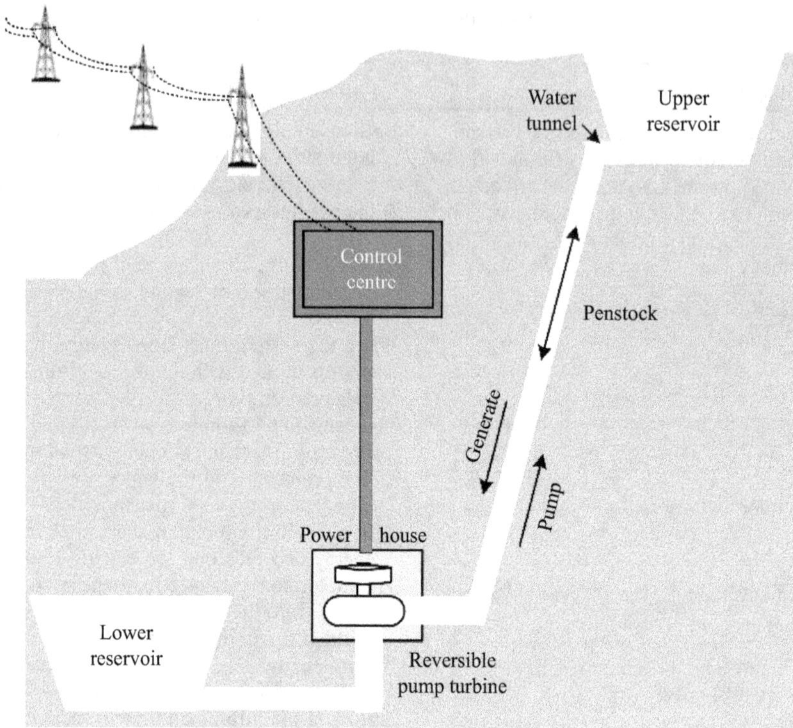

Figure 8.2 Developed PHES scheme

Table 8.2 Typical performance data

Power range (in general no limitation)	10 MW–30 GW	Life duration	>80 years
Energy range	Up to some 100 GWh	Reaction time	Some seconds to few minutes
Discharge time	Some minutes to 10 h	Cycle efficiency	70%–85%
Cycle life	Technically unlimited	Energy (power) density	0.5–3 Wh/kg

The PHES system shown in Figure 8.2 includes upper and lower reservoirs, reversible pump-turbine unit and reversible synchronous machine. The water flow direction and synchronous machine rotation direction in the generating mode are opposite to those in the pumping mode. So, the synchronous machine rotation direction should be reversed by changing the phase sequence at the machine terminals to get it operating as a motor rather than generator.

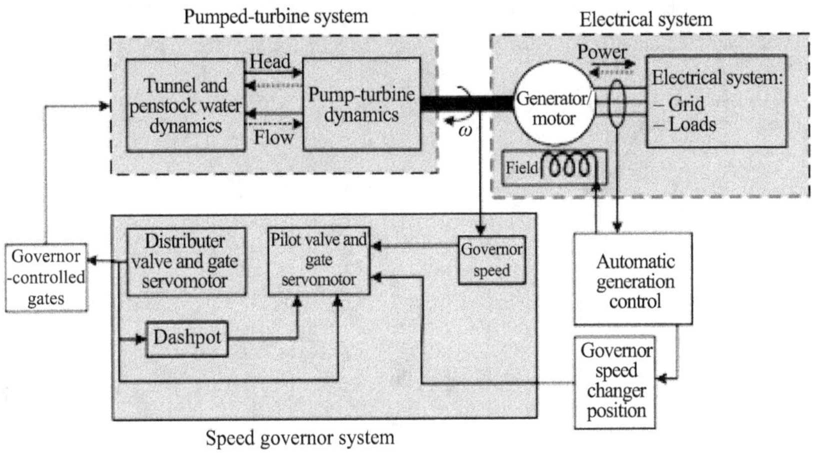

Figure 8.3 Functional block diagram of a PHES system (solid arrows for generating mode and doted arrows for pumping mode). The speed governor system only operates in the generating mode

The functional block diagram, which represents the main elements of a typical PHES system, is depicted in Figure 8.3. It is to be noted that the speed governor system only operates to control the valve opening during the generating mode. The valve opening is normally fixed at the maximum efficiency point in the pumping mode.

The water dynamics for the pumped-turbine system penstock during generating mode can be represented by a simplified relation as follows [4]:

$$Q = \frac{1}{T_w s}(H_T - H - H_{\text{loss}}) \tag{8.3}$$

where Q is the output water flow from the system, H_T is the total available static head, H is the dynamic head established by pump-turbine unit, H_{loss} is the frictional head loss, T_w is water starting time and s is the Laplace operator. The output mechanical power, P_m, can be expressed as

$$P_m = H(Q - Q_{\text{nl}}) \tag{8.4}$$

where Q_{nl} is the no load flow.

The water flow is defined as positive when it flows downwards to drive the pump turbine, otherwise, it is negative. The head also is positive in the direction from upper to lower reservoirs (Figure 8.4(a)).

According to (8.3) and (8.4), the block diagram of a pump-turbine unit in generating mode is as shown in Figure 8.4(b).

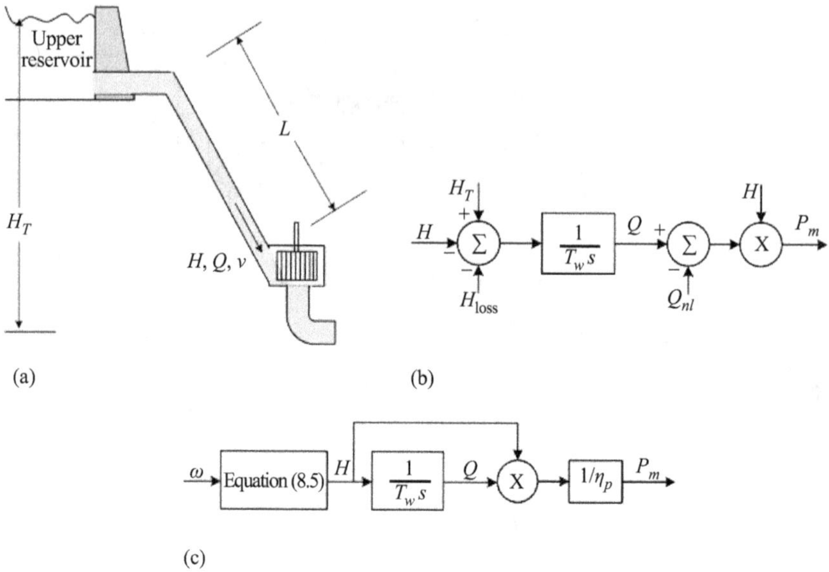

Figure 8.4 Water dynamics. (a) Penstock. (b) Block diagram of a pump-turbine unit in generation mode. (c) Block diagram of a pump-turbine unit in pumping mode

On the other hand, in the pumping mode, the water dynamics depend on the pump head-flow characteristic curve, which can be approximated by a quadratic polynomial relation as given in the following:

$$H = a_0\omega^2 + a_1\omega|Q| + a_2Q^2 \tag{8.5}$$

where a_0, a_1 and a_2 are coefficients for curve fitting and ω is the pump operating speed. Consequently, the block diagram of a pump-turbine unit in pumping mode is shown in Figure 8.4(c).

The water starting time, T_w (s), can be calculated by the relation [5]

$$T_w = \frac{Lv}{gH_T} \tag{8.6}$$

where L is the length of the penstock (m), v is water velocity (m/s) and g is the acceleration of gravity (m/s^2). Water velocity can be expressed in terms of power generated by the system as

$$P = 0.733vH_TA\eta \tag{8.7}$$

where P is the generated electrical power (W), A is the average penstock area (m^2) and η is the product of turbine efficiency and generator efficiency. Hence,

$$T_w = \frac{0.139PL}{H_T^2 A\eta} \tag{8.8}$$

8.1.3 Applications

PHES can be used for energy management applications that include moving power over longer timescales and generally require continuous discharge ratings of several hours or more. So, PHES technology plays an important role for load levelling, reducing the gap between off-peak and peak periods and power grid stabilization. Besides balancing the peak and off-peak periods, PHES provides ancillary services such as primary frequency and voltage control to the power grid. To fulfil the power system control, PHES can switch within seconds for different operation modes [6].

Because of their flexibility, large-scale storage possibilities and grid operations benefits, PHES systems will enable utilities to efficiently balance the grid and to develop their renewable energy portfolios. In fact, the installation of intermittent renewable generation has added a new degree of uncertainty to the dispatch of interconnected power system. Pumped storage is therefore set to play a key role in enabling renewables' grid integration while helping countries meet their ambitious targets of cutting greenhouse gas emissions and of building additional clean renewable energy capacity.

8.1.4 Types of PHES

In the literature, PHES is classified according to the structure and operation type. Structurally, pumped-hydro storage systems are classified as three types: pure pumped-storage (also named as off-stream PHES or closed-loop systems), pumped-back pumped storage and hybrid pumped storage plants [7].

1. Pure pumped storage plants shift water between two reservoirs, one of which is located off-stream and other is river, lake or sea.
2. Back pumped storage system approach (mixed pumped storage) is a combination of pumped storage and conventional hydroelectric plants that use natural stream flow.
3. Hybrid pumped storage systems seem like more conceptual when compared to the other two types. There are many examples of hybrid pumped storage power plant, which are associated with wind power plant (WPP) or photovoltaic (PV) solar farms. PHES plants are very dependent on geology and topography such as WPP and PV farms. Physically combining PHES and WPP units in the same location can be challenging due to their dependency on the natural conditions. More often, installed capacity of the wind farm, which is the pump capacity of PHES, is directly related with the energy need for pumping. That is why combining those two plants in the same location restricts the total installed capacity.

According to its reservoir capacity and operation policy, PHES can be classified as daily, weekly and seasonally storage power plants.

At the daily operating plants, electricity is generated during peak hours and water is stored at out off-peak hours. In the weekly cycle of water, some portion of the water used in the generation of electricity during peak hours of weekdays is pumped back to the upper reservoir within that day. Except during peak hours of weekends, water is stored in the upper reservoir, which becomes empty at the end of the weekdays. On the other hand, seasonally operating PHESs store water while the river flow and energy are in excess, and to increase its firm energy use reserved water while the river flow is low.

8.1.5 Advantages and disadvantages of PHES

Main advantages of PHES. Pumped storage plants have the biggest share among other storage technologies so far. Their advantages can be summarized as in the following:

- *Fast response:* Like other hydropower plants, PHES is faster in start-up and it takes short time to reach its full generating capacity when compared to the other types of power plants. Thus, it is possible to balance the varying electrical energy demand for consumers and some unplanned outages of other power plants in the grid system.
- *Load balancing:* Voltage and frequency are critical. Sudden changes in these parameters damage the equipment in the consumer premises. Broken equipment mean loss of value for domestic consumers and loss of production in the industry.
- *Black-start ability:* Other power plants need an external power during the start-up of the system. However, like hydropower plants, PHES does not require external energy to initiate generation. Thus, network restoration is possible if a power blackout occurs.
- *Energy storage:* Grid energy storage balances electricity supply and demand, which ensures stability and reliability in supply.
- *High global efficiency:* Round-trip efficiency can vary significantly for different energy storage technologies, depending on the number of cycles and duration of usage. Pumped-hydro storage systems are approximately 80% efficient and PHES has the highest global cycle efficiency when compared to other storage technologies.
- *Decrease in peak-hour electricity prices:* Pumped storage hydro plant provides the possibility of levelling the price of electricity in the market by being used in connection with daily peak shaving and load levelling as well as weekly and seasonal variations. PHES can lower electricity costs as it can store electricity bought at low off-peak prices and they can use it during peak periods in place of expensive power.
- *Decrease water wastage:* Pure pumped storage plants use the same water several times and prevent water consumption. Seasonally operating PHES facilities store excess water during the overflow season of the river, preventing flood and storing water before reaching to the sea.
- *Low operation and maintenance cost:* PHES operation and maintenance costs are low and these costs are directly related to the electricity prices. As PHES is used in peak hours, high prices are lowered significantly when pumped storage is in operation.

Disadvantages of PHES. There are several drawbacks of PHES facilities. The negative sides of pumped hydro storage are cost of the overall system and environmental issues, which are caused by the construction of the power plant, operation of pumps and turbines as explained in the following [8].

- *High capital cost:* Although the operation and maintenance costs are very low, there is a high upfront capital investment, which can be recouped only over a long time. Construction of reservoirs, dams and in some cases underground structures is expensive and resource-consuming.
- *Very dependent on location:* The two basic requirements for a PHES facility are head and water. Pumped storage systems require not only large volumes of water but also considerable amount of land with specific type of conditions. Finding suitable place for a PHES is not an easy task; that is why building the storage and generation facility may need some improvements in the geology, topography and so on, which increase the costs.
- *Environmental impacts:* Environmental impacts are also serious concerns for PHES projects as well as hydropower. In general, hydropower receives criticism due to flooding large areas, destroying terrestrial wildlife habitats and significantly changing the landscape. Blocking natural water flows and/or change in the path of the water disrupt the aquatic ecosystem and affect sediment carrying capacity of the river resulting in scouring downstream. Pumping may also increase the water temperature and stir up sediments at the bottom of the reservoirs and deteriorate the water quality. PHES operation may also trap and kill fish.

8.2 CAES system

Another type of commercialized MES technology with high technical maturity and large capacity is CAES. It can supply an output power of more than 100 MW per single unit in addition to storing bulk energy in the form of high-pressure compressed air. The energy can be stored during the periods of low electrical energy demand and released to generate electricity at the peak periods to fill the gap between the grid power and the high demand. CAES can be structured in different scales (from small to large) and different storage durations (from short to long). Moreover, CAES can efficiently operate at partial load with moderate response time compared to other EES technologies.

Different types of CAES technologies have been commercialized (e.g. conventional CAES systems), whereas others are more complicated, such as large-scale adiabatic CAES (A-CAES) and liquid air energy storage systems. At the discharge cycle, A-CAES is operated by integrating a thermal energy storage system for converting the energy stored in the compressed air into electrical power without involving a combustion process, which in turn results in a zero emission of CO_2 [9,10]. On the other hand, small-scale EES systems incorporating CAES have been also developed. It is also possible to use a reversible generator/motor unit to reduce the system cost and improve system efficiency.

8.2.1 Principles of CAES operation

The structure of CAES system includes an integration of different interacting components, devices and processes such as the following:

- A motor and a generator or reversible motor/generator unit with clutches on both ends (to engage/disengage it to/from the compressor train, the expander train, or both).
- Multi-stage air compressors with intercoolers to reduce the power requirements needed during the compression cycle, and with an aftercooler to reduce the storage volume requirements.
- An expander train consisting of high- and low-pressure turbo-expanders with combustors.
- Auxiliary equipment (fuel storage and handling, cooling system, mechanical systems, electrical systems and heat exchangers).
- Underground or aboveground compressed air storage, including piping and fittings.

A schematic diagram of a conventional cycle compressed-air storage system is shown in Figure 8.5 and the functional block diagram is depicted in Figure 8.6. The operation of CAES can be described as follows:

The charging cycle (compression cycle)

- Surplus electricity in the storage system or electricity from the grid during periods of low demand can be used to run compressors (low pressure (LP) and

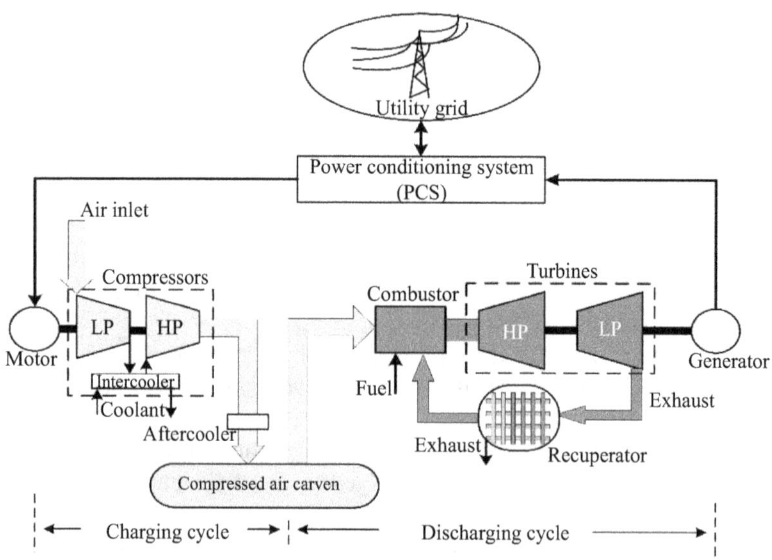

Figure 8.5 Schematic diagram of a conventional CAES system

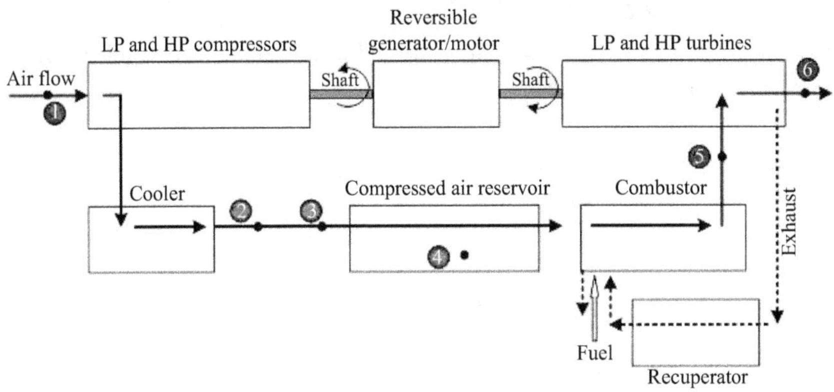

Figure 8.6 Functional block diagram of a CAES system with reversible motor/ generator model. Solid arrows represent the air flow direction whereas the exhaust is indicated by dotted arrows. State points are shown by numbered black circles

high pressure (HP) compressors) driven by a motor to inject the air into a cavern (normally, installed underground). The high-pressure compressed air is stored in the cavern at surrounding formation temperature.

- To improve the compression efficiency, coolers can be used to reduce the operating temperature of the injected air.

The discharging cycle (expansion cycle)

- The stored high-pressure compressed air is released and heated and consequently expanded through gas turbines (HP and LP turbines) and sometimes steam turbines.
- The combustor implements the combustion process with the mixed compressed air and fuel to rotate the turbines.
- The turbines are connected to an electrical generator to generate electricity.
- A recuperation device can be used to recycle the waste heat of the system exhaust before it is released into the atmosphere, thus resulting in reduction of both CO_2 emissions and fuel consumption.

The recuperation cycle

A recuperation cycle plant, the conventional CAES thermal cycle with an additional component (the recuperator), is shown in Figure 8.5. A recuperator recovers the low-pressure turbine waste heat to preheat the stored air before it goes into the high-pressure combustor. This reduces the fuel consumption of the plant (as compared to the conventional plant) by about 25%. This configuration is used in the Alabama McIntosh plant that was designed for primary operation as a source of peak power and as a load-management storage plant. As the recuperator reduces the plant heat rate during generation by about 25%, it reduces the cost of the plant's peak power supply.

8.2.2 CAES system modelling

Modelling of the system is based on identifying each component of behaviour and the following assumptions [11]:

- Ideal air with a constant heat capacity.
- Fixed maximum and minimum pressure in the tank.
- Constant heat transfer exchange coefficient for the tank.
- Accumulation (mass and energy) only occurs in the tank.

To identify the behaviour of each component, CAES circuit can be described by the stages of the major transformation of the air in the system. At these stages, represented by the state points (1–2, 2–3, 3–4, 4–5, 5–6) as shown in Figure 8.6, the thermodynamic variables of the air (temperature, pressure, specific volume, specific enthalpy, specific entropy and specific internal energy) are instantly calculated by the relationships of ideal gas law, and energy and mass balances [10–13].

At point #1, the air flows through the compressors and then to the cooler providing the compressor isentropic output at point #2. Assuming an isentropic transformation, from point #1 to point #2, the compressor can be modelled at point #3 by means of its isentropic efficiency that is defined by

$$\eta_{\mathrm{is,comp}} = \frac{h_2(t) - h_1(t)}{h_3(t) - h_1(t)} \tag{8.9}$$

In the compressed air reservoir, at point #4, the rate of mass change, $\frac{dm(t)}{dt}$, and the rate of increase in internal energy of the air, $\frac{dm_u}{dt}$, that is, energy balance equation can be written as

$$\left.\begin{aligned} \frac{dm(t)}{dt} &= \dot{m}_{\mathrm{in}}(t) - \dot{m}_{\mathrm{out}}(t) + N \\ \frac{dm_u}{dt} &= \dot{m}_{\mathrm{in}}(t)h_3(t) - \dot{m}_{\mathrm{out}}(t)h_4(t) + \dot{Q}_{\mathrm{tnk}} + \dot{W} \end{aligned}\right\} \tag{8.10}$$

where h stands for the specific enthalpy of unit mass of the fluid, $\dot{m}_{\mathrm{in}}(t)$ and $\dot{m}_{\mathrm{out}}(t)$ are the inlet and outlet mass flow rates, respectively, N represents the net production rate of mass, m, by the chemical reactions (equals zero when there is no chemical reaction inside the reservoir), $\dot{Q}_{\mathrm{tnk}}$ is the rate of change of heat transfer and $\dot{W}$ is the rate of change in energy transfer because of work done (equals zero for the energy balance of compressed air storage).

$\dot{Q}_{\mathrm{tnk}}$, $\dot{m}_{\mathrm{in}}(t)$ and the volumetric efficiency, $\eta_{V,\mathrm{comp}}$, are, respectively, given by

$$\left.\begin{aligned} \dot{Q}_{\mathrm{tnk}}(t) &= K_{\mathrm{rsv}}A_{\mathrm{rsv}}(T_4(t) - T_{\mathrm{ext}}(t)) \\ \dot{m}_{\mathrm{in}} &= \eta_{V,\mathrm{comp}}\frac{\dot{V}}{V_1(t)} \\ \eta_{V,\mathrm{comp}} &= 1 - \left[\left(0.05 \times \beta_{\mathrm{comp}}(t)\right)^{0.714} - 1\right] \end{aligned}\right\} \tag{8.11}$$

where K_{rsv} is the overall heat exchange coefficient between the reservoir and the environment, A_{rsv} is the external area of the reservoir, $\eta_{V,\text{comp}}$ is the compressor volumetric efficiency and β_{comp} is the compression ratio.

At points #5 and #6, the power required by the compressor and that produced by the turbine, respectively, can be calculated by the following relations:

$$\left.\begin{aligned} P_{\text{comp}}(t) &= \frac{1}{\eta_{m,\text{comp}}} \times \dot{m}_{\text{in}}(t) \times (h_3(t) - h_1(t)) \\ P_{\text{turb}}(t) &= \eta_{m,\text{turb}} \times \dot{m}_{\text{out}}(t) \times (h_6(t) - h_5(t)) \end{aligned}\right\} \tag{8.12}$$

where $\eta_{m,\text{comp}}$ and $\eta_{m,\text{turb}}$ are the mechanical efficiency of compressor and turbine, respectively.

Thus, the overall efficiency of the CAES system (η_{CAES}) and the coverage ratio (CR_{CAES}) can be expressed as

$$\eta_{\text{CAES}} = \frac{E_{\text{turb}}}{E_{\text{comp}}} \tag{8.13}$$

$$CR_{\text{CAES}} = \frac{E_{\text{turb}}}{E_{\text{Bldg}}} \tag{8.14}$$

where E_{turb}, E_{comp} and E_{Bldg} are, respectively, the turbine energy production, the compressor energy consumption and the total building demand over a specific period.

8.2.3 Advantages and disadvantages of CAES

Application of a CAES system provides the grid with ancillary services such as peak shaving, spinning reserve, reactive power (VAr) support and arbitrage. By exploiting CAES, the energy delivered from a variety of sources (e.g. wind, solar and the grid itself) can be temporarily stored to be recovered when needed.

Advantages. For specific applications, CAES plants are designed to provide economic benefit to owners and/or operators of power generation facilities. The benefits of using a CAES plant to support power generation include the following:

- Increased use of generation facilities during off-peak hours (i.e. during the storage plant charging cycle) and times of rapidly decreasing demand as the compressor absorbs excess generating capacity. This is particularly useful when base nuclear, hydro or fossil capacity is available at very low prices during off-peak time periods.
- Provide ramping, intermediate and peaking power during the day.
- Store daytime solar energy and night-time wind energy to be delivered during the higher priced time.
- Provide frequency regulation as CAES can provide much better frequency control than a base-load power plant.
- Provide VAr support (e.g. by operating the CAES plant to supply reactive power in the synchronous condenser mode).

- Provide off-peak to on-peak arbitrage.
- CAES technology is the only one that can provide significant energy storage (in the thousands of megawatt hours) at relatively low cost. The plant has practically unlimited flexibility for providing significant load management at the utility or regional levels.
- CAES plants are capable of black start.
- CAES plants have fast start-up time. If a CAES plant is operated as a hot spinning reserve, it can reach the maximum capacity within a few minutes.
- CAES plants have a ramp rate of about 30% of maximum load per minute.
- The nominal heat rate of a CAES plant at maximum load is about three times lower than the heat rate of a comparable combustion turbine plant using the same turbine expander. CAES plants also excel at part load. Their heat rate at 20% of maximum load is 80% of the nominal heat rate at maximum load. This is very good and unique, since all other oil, gas and coal power plants have poor efficiency at 20% load, making them uneconomical for operation at part load for normal duty. This characteristic of CAES plants makes them very useful (and efficient) for ramping, part load and regulation duty [14].

Disadvantages. A few of the potential reasons are listed as follows:

- The underground geology is likely perceived as a risk issue by utilities.
- Site selection is somewhat limited because of the need of installing underground tanks (air compressed reservoirs) that may require certain geological formations.
- Like any energy conversion technology, certain losses are inevitable.
- The requirement for additional heating in the expansion process is the most significant disadvantage. By some estimates, 1 kWh worth of natural gas will be needed for every 3 kWh generated from a CAES system. This is particularly problematic if fossil fuels are used for the heat generation. As natural gas prices increase, the economics of CAES, marginal at present, could fail [14].

8.3 FW energy storage

As an energy storage device, flywheel energy storage (FES) stores electrical energy in the form of kinetic energy in a rotating mass. It operates like any other storage technology in the charging and discharging mode. FES systems can be categorized into two classes: (i) conventional low-speed systems (up to 10,000 rpm) in which steel rotors are used to increase inertia with mechanical bearings that result in large and heavy system and relatively large standby losses (between 0.5% and 1% of rated power) and (ii) high-speed composite systems with speeds up to 100,000 rpm in which the rotor is made of composite materials with magnetic or superconducting bearings resulting in a compact system but because of the high speed, air drag forces cause losses of about 0.1% of the rated power [15].

However, the cost and maintenance of low-speed FWs is less than that of high-speed FWs. On the other hand, high-speed FWs have the advantage of low losses, low mass and high-energy storage capacity.

8.3.1 FES architecture and operation

An FES system, Figure 8.7, includes the following main components:

- An FW (rotor), which stores the kinetic energy in the form of a rotating mass. It is made of carbon-fibre composite in high-speed FWs to have higher tensile strength than steal in low-speed FWs. Hence, the overall configuration as well as the specific energy and specific power of high-speed FWs can be optimized.
- Mechanical or magnetic bearings to support the rotor. The type of bearings is the major element that the FW speed and performance are dependent upon. Mechanical bearings require frequent maintenance and produce friction, which can hinder rotation at high rates, and so, they are commonly used in the low-

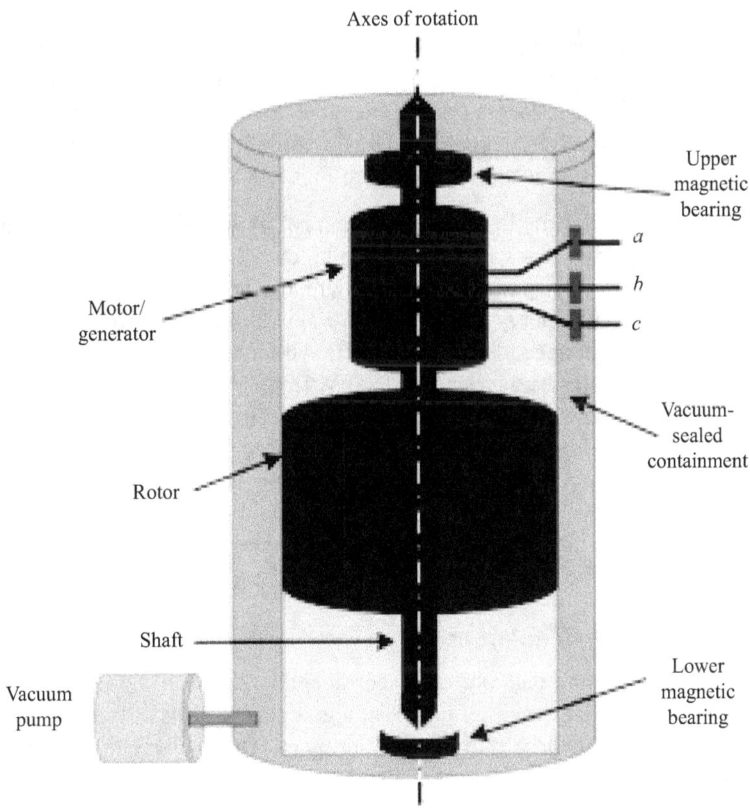

Figure 8.7 Typical flywheel main components

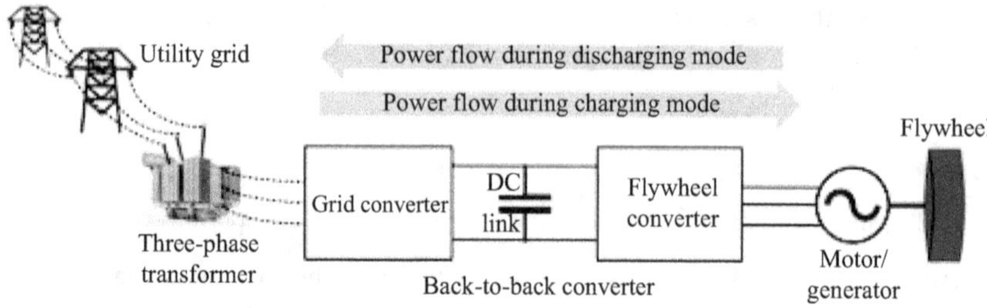

Figure 8.8 Schematic diagram of a flywheel connected to a grid

speed FWs. On the other hand, magnetic or superconducting bearings eliminate the friction between the rotor and its supports and have long life and low maintenance, and thus, they are preferred for high-speed FW systems [16].

- Vacuum enclosure for housing the system components at low pressure to help reduce the self-discharge losses.
- Electrical motor/generator coupled to the shaft of the FW. It acts as a motor during the charging mode and as a generator during the discharging mode.
- Power conditioning system, which includes two converters (back-to-back converter) connected by a common DC link, one for motor/generator and the other for the grid connection as shown in Figure 8.8 [17].

FES system, also named as 'flywheel battery' as it looks like a mechanical battery, uses the inertia of a spinning mass to kinetically store energy or generate power. Kinetic energy is quantified by one half the product of mass and speed squared. So, the stored energy for low-speed FWs is directly proportional to the mass, whereas for high-speed FW it is proportional to speed squared.

FES system operates by accelerating the FW (rotor) to a high speed and storing the energy in the system as rotational energy; that is, it works in a charging mode and the electrical machine acts as a motor. In the discharging mode, the energy is extracted from the system as the electrical machine works as a generator, and hence, the rotational speed of the FW is reduced according to energy conservation principle. This means that electricity is used to accelerate the rotor in the charging mode and decelerate the rotor in the discharging mode.

8.3.2 Mathematical relations

FW is a rotating, cylindrical object of considerable mass. It stores the kinetic energy or increases the moment of inertia of a given system. Its design is based on the storage capacity and the maximum rotating speed. The total energy stored in the FW as a rotating mass, E, can be calculated by

$$E = \tfrac{1}{2} I \omega^2 \tag{8.15}$$

where I is the moment of inertia and ω is the rotational speed. For simplicity, the FW is approximated to be a thin-walled cylinder. Thus, the moment of inertia can be expressed in terms of its mass, m, and radius, r, as

$$I = mr^2 \tag{8.16}$$

Substituting (8.16) into (8.15) gives

$$E = ½mr^2\omega^2 \tag{8.17}$$

The useful stored energy, intrinsically, depends on the ratio between the minimum and maximum rotating speed, ω_{min}, ω_{max}, respectively, as [15]

$$E = E_{\text{max}} \frac{\omega_{\text{max}}^2 - \omega_{\text{min}}^2}{\omega_{\text{min}}^2} \tag{8.18}$$

The FW is mainly affected by two forces: the circumferential and radial stresses. Therefore, the tensile strength, σ, of a specific material gives a measure of how much the FW can withstand these stresses during rotation. It can be calculated by

$$\sigma = \rho r^2\omega^2 \tag{8.19}$$

where ρ is the density of material.

From (8.19), the allowable rotational speed can be determined. In addition, substituting (8.19) into (8.17) gives the energy stored by the FW as

$$E = ½\frac{m\sigma}{\rho} \tag{8.20}$$

Hence, increasing the tensile strength and decreasing the material density will lead to an FW with a high ratio of energy to weight. This in turn necessitates a proper selection of the FW material according to the desired speed.

For FW dynamic analysis, the general form of equation of motion can be written as [18]

$$[M]\{\ddot{Z}\} + [C]\{\dot{Z}\} + [K]\{Z\} = \{f\} \tag{8.21}$$

where $[M]$ = symmetric mass matrix, $[C]$ = symmetric damping matrix, $[K]$ = symmetric stiffness matrix, $\{Z\}$ = generalized coordinate vector and $\{f\}$ = external force vector.

Several types of electrical machines can be used with FWs such as induction machine (IM), DFIM, permanent magnet synchronous machine (PMSM), brushless DC (BLDC) machine, and switched reluctance machine (SRM). Advantages and disadvantages of each type are summarized in Table 8.3 [15]. More details about the dynamics and equivalent circuits for the first three types are given in Chapter 2, Section 2.4, whereas the latter two types are explained in Chapter 12, Sections 12.4.3 and 12.4.4, respectively. The power conditioning systems used for interfacing are explained in Chapter 13.

Table 8.3 Pros and cons of types of motor/generator unit

Machine type	Advantages	Disadvantages
IM*	• High reliability • High robustness • Low cost	• Speed limitations
DFIM	• Very large inertia	• Low power-conversion rating • More losses in power electronics
PMSM	• Low rotor losses • High energy density • High overall efficiency	• Idle losses due to eddy currents in the stator • High cost • Low ruggedness
BLDC	• High energy density • High efficiency	• More power electronics are required, e.g. electronic switching controller • High cost
SRM	• Low idle losses • Wide speed range • Good robustness	• High torque, flux and current ripples • Difficulty of torque control at low speed

*The best choice for power smoothing in wind energy conversion system and power conditioning applications [15].

8.3.3 FES characteristics

FES systems can be mainly specified by the following characteristics:

- High specific energy (Wh/kg).
- Round-trip efficiency reaches 90%.
- Large maximum power output.
- Capacities ranging from small to large kilowatt hours are available.
- Time charging response of a system may be less than 15 min.
- Little or no maintenance.
- Very long lifetime as well as FWs have hundreds of thousands cycles of use.

8.3.4 Applications

Different fields of applications can use FES systems such as transportation, uninterruptible power supplies, aircraft launching systems, spacecraft energy storage, pulse power source and grid energy storage.

FES systems can be used for grid energy storage as they can perform the following functions [19]:

- Short-term spinning reserve for momentary frequency regulation of the grid and voltage sag correction [20,21].
- Balancing sudden changes between generation and demand.
- Storing the energy during off-peak hours rather than traditional sources, for example, natural gas turbines with no carbon emissions and faster response.

Table 8.4 Comparison between FWs and batteries

FWs	Batteries
• Temperature changes have no effect and FWs can operate at a wider range	• Operation is sensitive to the temperature changes
• Not subjected to failures of chemical reactions	• Subjected and affected by the failures and incorrect chemical reactions
• Less potentially damaging to the environment	• May cause environment damage
• Easy to know exactly the amount of stored energy by measuring the rotational speed	• Not easy to know the exact amount of stored energy
• Almost infinite working lifespan	• Operates for a finite period
• Minimal maintenance throughout the service life is required as the containment is completely sealed. For instance, magnetic bearing FWs in a vacuum enclosure do not need bearing maintenance, whereas with mechanical bearing, FWs have a limited lifespan because of wear	• Frequent maintenance is required
• In some applications, especially in those having constrained space, FWs may not be applicable	• Batteries can be arranged in a wide variety of configurations

- Supporting the intermittent renewable energy sources, for example, wind turbines [22].

FES operation is like the batteries in the same applications but may differ economically. A comparison between them is given in Table 8.4.

References

[1] Barnes F. S. and Levine J. G. *Large energy storage systems*. Boca Raton, FL, USA: CRC Press, Taylor & Francis Group; 2011, pp. 51–110.

[2] Çetinkaya S. *Capacity determination of pumped storage projects using market electricity prices*. MSc Thesis, Middle East Technical University (METU); 2014.

[3] Luo X., Wang J., Dooner M. and Clarka J. 'Overview of current development in electrical energy storage technologies and the application potential in power system operation'. *Applied Energy*. 2015;137:511–36.

[4] Liang J. and Harley R. G. (eds.). 'Pumped storage hydro-plant models for system transient and long-term dynamic studies'. *IEEE PES General Meeting*; RI, USA, July 2010, pp. 1–8.

[5] IEEE Committee. 'Dynamic models for steam and hydro turbines in power system studies'. *IEEE PES Winter Meeting*; New York, USA, Jan–Feb 1973, pp. 1904–15.

[6] Denholm P., Ela E. and Kirby B. *The role of energy storage with renewable electricity generation*. National Renewable Energy Laboratory (NREL) Report, 2010.

[7] Ming Z., Junjie F. and Song X. 'Development of China's pumped storage plant and related policy analysis'. *Energy Policy*. 2013;61:104–13.

[8] Baxter R. Energy storage—A nontechnical guide. Nashville, TN, USA, PennWell Corporation; 2007.

[9] Luo X., Wang J., Dooner M., Clarka J. and Krupke C. 'Overview of current development in compressed air energy storage technology'. *Int. Conf. on Sustainability in Energy and Buildings, SEB-14*; Warwick, UK. *Energy Procedia*. 2014;62:603–11. Available from www.sciencedirect.com [Accessed 6 Feb 2020].

[10] Luo X., Wang J., Krupke C., *et al.* 'Modeling study, efficiency analysis and optimization of large-scale adiabatic compressed air energy storage systems'. *Applied Energy*. 2016;162:589–600.

[11] Simpore S., Garde F., David M. and Marc O. 'Design and dynamic simulation of a compressed air energy storage system (CAES) coupled with a building an electric grid and a photovoltaic power plant'. *Proceedings of the 12th REHVA World Congress*; Heiselberg, Per Kvols, Denmark, May 2016, pp. 1–10.

[12] Haung Y., Chen H. S., Zhang X. J., *et al.* 'Techno-economic modelling of large-scale compressed air energy storage systems'. *Energy Procedia*. 2017;105:4034–9.

[13] Shamshirgaran S. R., Ameri M. and Ghadamian H. 'Design of compressed air energy storage (CAES) power plant using the genetic algorithm'. *4th International Conference on Sustainable Energy and environment (SEE 2011)*; Bangkok, Thailand, Feb 2012, pp. 401–8.

[14] McCalley J. D. *Compressed air energy storage*. Educational Chapter. USA: Trishna Das; 2012.

[15] Daoud M. I., Abdel-Khalik A. S., Massoud A., Ahmed S. and Abbasy N. H. 'On the development of flywheel storage systems for power system applications: A survey'. *Electrical Machines (ICEM), 2012 XXth International Conference on*; Marseille, Sep 2012, pp. 2119–25.

[16] Yamauchi Y., Uchiyama N., Suzuki E., Kubota M., Fujii M. and Ohsaki H. 'Development of 50 kWh-class superconducting flywheel energy storage system'. *International Symposium on Power Electronics, Electrical Drives, Automation and Motion, SPEEDAM 2006*; Taormina, Italy, May 2006, pp. 484–6.

[17] Chen Z., Zou X., Duan S. and Wei H. 'Power conditioning system of flywheel energy storage'. *8th International Conference on Power Electronics ECCE Asia*; The Shilla Jeju, Korea, May 2011, pp. 2763–8.

[18] Owusu-Ansah P., Yefa H. and Misbawu A. 'Rotor dynamic modeling and analysis of a flywheel rotor'. *Proceedings of the 2015 International Conference on Electronics and Automation Control (ESAC 2015)*; Zhengzhou, China, Aug 2015, pp. 35–9.

[19] Theubou T. A., Wamkeue T. R. and Kamwa I. 'Modeling and simulation of a flywheel energy storage system for microgrids power plant applications'. *EIC Climate Change Technology Conference (CCTC 2015)*; Montreal, QC, Canada, May 2015, pp. 1–12.

[20] Takahashi R. and Tamura J. 'Frequency stabilization of small power system with wind farm by using flywheel energy storage system'. *Diagnostic for Electric Machines, Power Electronics and Drives, 2007, SDEMPED 2007, IEEE International Symposium on*; Cracow, Poland, Sep 2007, pp. 393–8.

[21] Samineni S., Johnson B. K., Hess H. L. and Law J. D. 'Modeling and analysis of a flywheel energy storage system for voltage sag correction'. *IEEE Transactions on Industry and Applications*. 2006;42(1):42–52.

[22] Takahashi R. and Tamura J. 'Frequency control of isolated power system with wind farm by using flywheel energy storage system'. *Proceedings of the 2008 International Conference on Electrical Machines, ICEM'08*; Vilamoura, Portugal, Sep 2008, pp. 1–6.

Chapter 9

Chemical energy storage systems: fuel cells and power-to-gas

9.1 Introduction

A fuel cell (FC) is a static device having energy conversion function. Chemical energy of a fuel is supplied as an input to the FC, which converts it directly into electrical energy. Energy conversion results from a chemical reaction of positively charged hydrogen ions with oxygen or another oxidizing agent. FCs using hydrogen (H_2) may be called 'hydrogen FCs'. They are different from batteries in that a continuous source of fuel and oxygen (or air) are required to sustain the chemical reaction, whereas the chemicals in batteries react with each other to generate an electromotive force (EMF). FCs can produce electricity continuously if the proper inputs are supplied. They can be used, as an alternative energy source with low emission of pollutant gases, for primary and backup power for commercial, industrial and residential buildings as well as in remote or inaccessible areas.

Different types of FCs such as polymer electrolyte membrane fuel cell (PEMFC), solid oxide fuel cell (SOFC) and molten carbonate fuel cell (MCFC) exist. They have a wide variety of potential applications, particularly in distributed generation (DG) applications, for example, transportation, stationary power for buildings, auxiliary power and cogeneration applications.

Compared to conventional power plants, FC systems have many advantages. These include high efficiency, zero or low emission of pollutant gases and flexible modular structure. In addition, FCs can normally be located at any site at the distribution level in a power system resulting in grid reinforcement, deferring or eliminating the need for system upgrades, and improving system integrity, reliability and efficiency. However, the most common disadvantage of all types of FCs currently is the cost.

The major processes of a hydrogen FC energy system are depicted in Figure 9.1. They start with cleaning the fuel and converting it into a hydrogen-rich gas using the fuel processor. Power-to-gas (PtG) energy system can be used as a source of fuel to provide the FC with natural gas or hydrogen as discussed in Section 9.9. FC stacks normally convert the energy of hydrogen into electric energy in DC form. Then, power conditioner unit converts the electric power from DC into regulated DC or AC for consumer use. An energy storage system may also be included as a part of the FC system for energy management and mitigating load

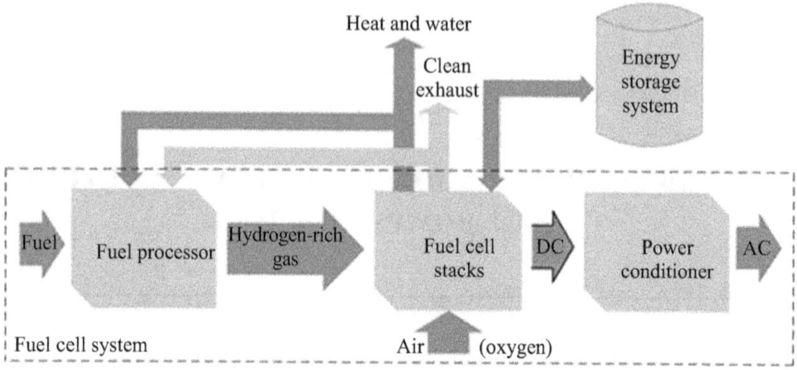

Figure 9.1 Processes of an FC energy system

transient impacts, where bi-directional energy flow is available between FC system and the storage system. FC stack by-products include heat, water and clean exhaust, which can be used for space heating or to produce additional electricity.

To produce the desired power and voltage for a specific application, the cells are bundled together in series and parallel combinations forming what is called 'FC stack'. Also, the output power and voltage can be controlled through proper control of the interfacing power electronic devices.

9.2 Principles of FC operation

The physical structure of all types of FCs is mainly composed of two electrodes (anode and cathode) separated by a thin layer of electrolyte that only allows positively charged hydrogen ions (or protons) to move between the two sides of the FC (Figure 9.2). Electrodes are usually made flat to achieve the maximum possible contact between the electrode, electrolyte and the gas. They contain catalysts that cause the fuel to undergo oxidation reactions generating positively charged hydrogen ions and electrons. Both the electrolyte from one side and the gas from the other side can penetrate each electrode [1].

To understand how the reaction between hydrogen and oxygen produces DC electricity, reactions that occur at each individual electrode are considered. An acid electrolyte is assumed to be used as it is the most common type.

At the anode, the hydrogen gas ionizes, releasing electrons and creating H^+ ions (or protons) according to

$$2H_2 \rightarrow 4H^+ + 4e^- \tag{9.1}$$

The hydrogen ions are drawn through the electrolyte after the reaction. Meanwhile, electrons are drawn from the anode to the cathode through an external circuit producing DC current.

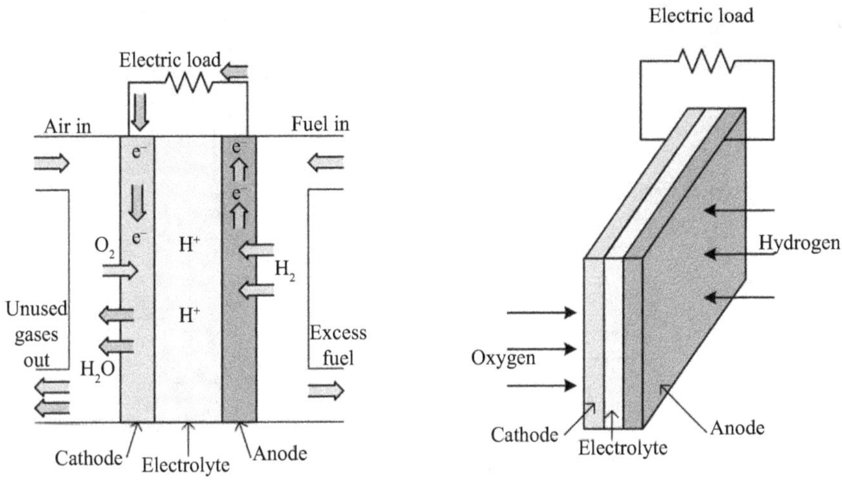

Figure 9.2 Physical structure of an FC

At the cathode, hydrogen ions, electrons and oxygen react to form water as the reaction is expressed by

$$O_2 + 4e^- + 4H^+ \rightarrow 2H_2O \tag{9.2}$$

Continuing the reactions (9.1) and (9.2), H^+ ions are produced at the anode and must pass through the electrolyte, whereas electrons must pass through an external electrical circuit to the cathode (Figure 9.3). It is to be noted that the cathode is the electrically positive terminal of the FC as the electrons are flowing from the anode to the cathode, that is, from − to +. This holds true for electrolysis, cells, valves, forward biased diodes and FCs where the cathode is always the electrode into which electrons flow. Similarly, the anode is always the electrode from which electrons flow and is the electrically negative terminal.

FCs are classified by the type of electrolyte used and by the difference in start-up time (from 1 s for PEMFC to 10 min for SOFC). Additional parameters, such as temperature and efficiency, are given in Section 9.8. For all types, the most important design features in an FC are the following:

- The anode catalyst breaks down the fuel into electrons and ions.
- The cathode catalyst turns the ions into the waste chemicals such as water or carbon dioxide.
- Reaction rate, an important factor for FC design. It should not proceed slowly. To deal with this problem, use of catalysts and raising the temperature can be applied to any chemical reaction. For FCs, an additional way, which is increasing the electrode area, is applied because the reaction rate is proportional to the area of the electrode. Increasing the electrode area is a vital issue to improve the FC performance that is determined in terms of the current per cm^2. For this reason, the electrode area can be increased not only by increasing

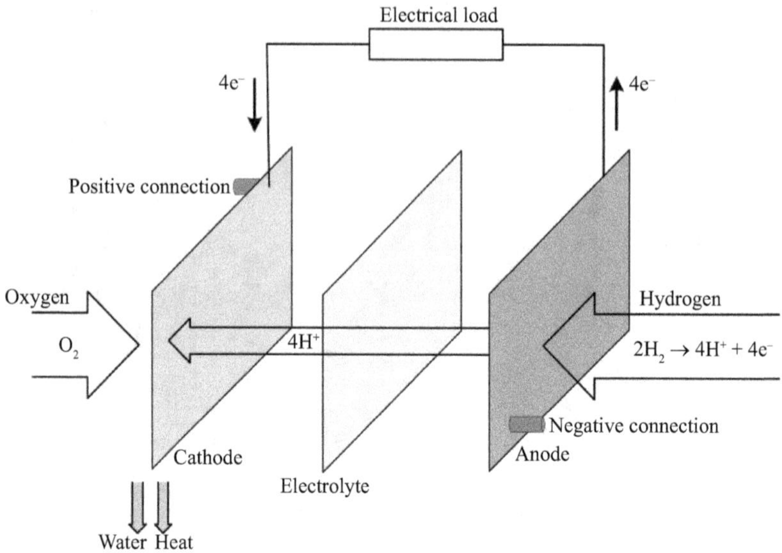

Figure 9.3 Chemical reactions in an FC with acid electrolyte

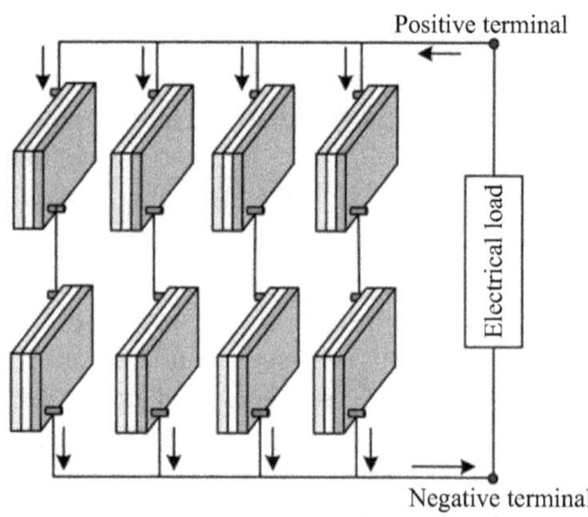

Figure 9.4 An FC stack, four parallel sets and two series FCs per set (arrows indicate the electron flow)

both width and length but also by using highly porous electrode, which has the effect of greatly increasing the effective surface area.

- Individual FCs produce relatively small electric potential, 0.6–0.7 V. So, FCs are stacked or placed in series and parallel combinations to create enough voltage and current to meet the specific requirements of an application (Figure 9.4). FCs produce not only electricity but also water, heat and, depending on the fuel source, very small amounts of nitrogen dioxide and other emissions. The energy efficiency of an FC is generally between 40% and 60%, or up to 85% efficient in cogeneration if waste heat is captured for use [2].

9.3 Reversible open-circuit voltage of FCs

The basic principles of operation of an FC are explained in Section 9.2. It is seen that hydrogen and oxygen provide input energy, whereas energy in the form of electricity, heat and water is the output (Figure 9.5).

The energy of chemical inputs (H_2 and O_2) and outputs (H_2O) is not simply defined, whereas the electrical energy output can be easily calculated as a multiplication of the voltage, current and time.

The concept 'Gibbs free energy' [1] is used to calculate the chemical energies in FCs. It is defined as the 'energy available to do external work, neglecting any work done by changes in pressure and/or volume'. Applying this definition to the FC, it is seen that 'external work' involves moving electrons around an external circuit, and any work done by a change in volume between input and output is not involved by the FC. Energy is all the extracted work including that due to some parameter changes (e.g. pressure, gas concentration and temperature).

To calculate FC efficiency and open-circuit voltage, the point of zero Gibbs free energy as well as the change of energy must be specified. When working with chemical reactions, the point of zero Gibbs free energy is defined as pure elements, in the normal state, at standard temperature and pressure (25 °C, 100 kPa). Also, when adopting this convention, the term 'Gibbs free energy of formation', G_f, instead of 'Gibbs free energy' is commonly used. Therefore, G_f of the input is equal to zero for an ordinary hydrogen FC operating at standard temperature and pressure.

On the other hand, the change in energy in an FC is the change in Gibbs free energy of formation, ΔG_f, which gives the energy released. It can be given by the difference between the Gibbs free energy of the products and that of reactants, that is,

$$\Delta G_f = G_f \text{ of products} - G_f \text{ of reactants} \tag{9.3}$$

It may be convenient to express the quantities in (9.3) in their 'per mole' form. Thus, (9.3) can be rewritten, replacing the upper-case letters by lower-case letters under $^-$ to denote quantities/mole, as

$$\Delta \overline{g_f} = \overline{g_f} \text{ of products} - \overline{g_f} \text{ of reactants} \tag{9.4}$$

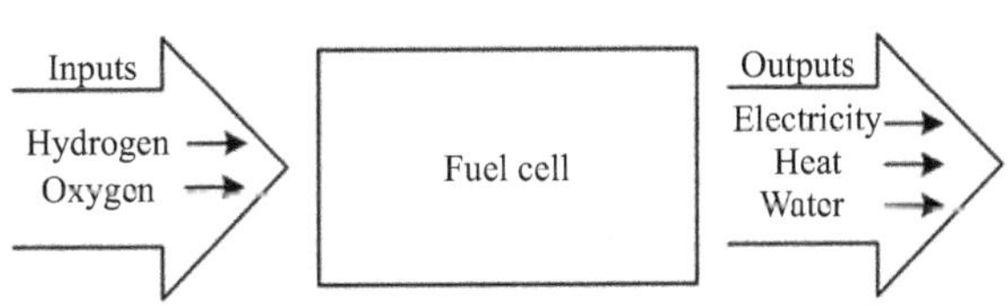

Figure 9.5 Inputs and outputs of an FC

As an illustrative example, the basic reaction of the hydrogen/oxygen FC is

$$2H_2 + O_2 \rightarrow 2H_2O,\ \text{which is equivalent to}\ H_2 + ½O_2 \rightarrow H_2O$$

It is seen that the reactants are one mole of H_2 and half a mole of O_2, whereas the product is one mole of H_2O. Applying (9.4) gives

$$\Delta\overline{g_f} = (\overline{g_f})_{H_2O} - (\overline{g_f})_{H_2} - ½(\overline{g_f})_{O_2} \tag{9.5}$$

The values of $\Delta\overline{g_f}$ in (9.5) at different conditions (temperature and state) are given in Table 9.1 indicating that the Gibbs free energy of formation is not constant. More detail about calculation of $\Delta\overline{g_f}$ is given in [1].

The negative values of $\Delta\overline{g_f}$ indicate that the energy is released. If the process is reversible, that is, it has no losses, Gibbs free energy is fully converted into electrical energy.

From the basics of chemistry, a mole of any substance always has the same number of entities, for example, molecules, which is 6.022×10^{23} and is called 'Avogadro's number', N_a.

Thus, a mole of electrons is 6.022×10^{23} electrons. Accordingly, its charge is N_a.e, where e is the charge on one electron, equal to 1.602×10^{-19} coulombs. (N_a.e) is called the 'Faraday constant, F' or the charge on one mole of electrons. Thus,

$$F = N_a.\text{e} = 96485\ \text{C}.$$

Referring to Section 9.2, it can be found that for the hydrogen FC, two electrons pass around the external circuit for each water molecule produced and each molecule of hydrogen used. So, for one mole of hydrogen used, $2N_a$ electrons pass around the external circuit, then the charge that flows is $-2N_a.\text{e} = -2F$ coulombs.

Assuming the open-circuit voltage of the FC is denoted by V_{OC}, the electrical work done moving the charge $-2F$ around the circuit is

$$\text{Electrical work done} = \text{charge} \times \text{voltage} = -2F.V_{OC}\ \text{J} \tag{9.6}$$

As the system is reversible, the electrical work done in (9.6) will be equal to the Gibbs free energy released, $\Delta\overline{g_f}$. Hence,

$$\Delta\overline{g_f} = -2F.V_{OC}\ \text{or}\ V_{OC} = \frac{-\Delta\overline{g_f}}{2F} \tag{9.7}$$

For instance, from Table 9.1, $\Delta\overline{g_f} = -210.3$ kJ for an FC operating at 400 °C, so

$$V_{OC} = \frac{210.3 \times 10^3}{2 \times 96{,}485} = 1.09\ \text{V}$$

It is to be noted that this value is higher than the actual as the system is assumed to be reversible as well as pure hydrogen and oxygen are at standard pressure (100 kPa). In practice, because some voltage drop occurs and reactions are irreversible, the voltage, V_{OC}, would be lower even when no current is drawn.

Table 9.1 Values of $\Delta\overline{g_f}$ for the reaction $H_2 + ½O_2 \rightarrow H_2O$ at different temperatures

Water state	Liquid	Liquid	Gas	Gas	Gas	Gas	Gas	Gas	Gas
Temperature (°C)	25	80	80	100	200	400	600	800	1,000
$\Delta\overline{g_f}$ (kJ/mol)	−237.2	−228.2	−226.1	−225.2	−220.4	−210.3	−199.6	−188.6	−177.4

The open-circuit voltage given in (9.7) can specifically be applied to hydrogen FCs as the number '2' in this equation represents the number of electrons per molecule of that fuel. Thus, to generalize (9.7) to be applied to any other fuel, the number '2' can be replaced by 'n' to obtain the general formula as

$$V_{OC} = \frac{-\Delta \overline{g_f}}{nF} \tag{9.8}$$

9.4 FC's efficiency

As explained in Section 9.3, for FCs, Gibbs free energy is converted into electrical energy. If the process of chemical reaction is a reversible process and Gibbs free energy is assumed to be totally converted into electrical energy, the efficiency of an FC would be 100% and may be considered as the efficiency limit as the efficiency, η, is defined as

$$\eta_{FC} = \frac{\text{Electrical output energy}}{\text{Gibbs free energy}} \tag{9.9}$$

This is not an accurate definition for the following reasons:

- Gibbs free energy changes not only with temperature change as seen in Table 9.1 but also with pressure and other factors as explained in the next sections. Consequently, efficiency limit is always 100% regardless of the conditions used.
- The definition of energy given in (9.9) depends on the energy produced by hydrogen as an input to the FC. In fact, there is a preceding process, which is fuel burning to release its energy. Therefore, it would be more useful to calculate the efficiency in terms of the heat produced by burning the fuel. This heat differs from one fuel to another and is evaluated by its calorific value or the change in 'enthalpy of information', $\overline{\Delta h_f}$ with a negative sign when the energy is released. Accordingly, the efficiency of an FC can be updated and rewritten as

$$\eta_{FC} = \frac{\text{Electrical output energy}}{-\overline{\Delta h_f}} \tag{9.10}$$

It is noted that for hydrogen FCs, $\overline{\Delta h_f}$ has two different values that can be used depending on the form of water (gas or liquid) produced by burning hydrogen. In case of chemical reaction: $H_2 + \frac{1}{2}O_2 \rightarrow H_2O$, the product water may be in the form of either steam or condensed back to water. For steam, $-\overline{\Delta h_f} = -241.83$ kJ/mol. and is called the 'lower heating value', LHV, whereas it equals to $-285{,}84$ kJ/mol. for water and is called the 'higher heating value', HHV. Thus, for any value of efficiency, the value (LHV or HHV) at which the efficiency is based on should be stated. LHV that gives higher efficiency is commonly used, particularly, when HHV is not known [1].

As given by (9.10), FC efficiency reaches its maximum value, efficiency limit, when the electrical output energy is maximum, that is, equal to Gibbs free energy (no waste of heat). So,

$$\eta_{FC(max)} = \frac{\Delta \overline{g_f}}{\Delta \overline{h_f}} \times 100\% \tag{9.11}$$

For instance, at FC temperature of 80 °C, $\Delta \overline{g_f} = -228.2$ kJ/mol. as given in Table 9.1. Using (9.7), the corresponding reversible open-circuit voltage, V_{OC}, is obtained as

$$V_{OC} = \frac{228.2 \times 10^3}{2 \times 96{,}485} = 1.18 \text{ V}$$

Applying (9.11), the efficiency limit will be

$$\eta_{FC(max)} = \frac{\Delta \overline{g_f}}{\Delta \overline{h_f}} \times 100 = \frac{228.2}{285.84} \times 100 = 79.8\% \text{ referred to HHV}$$

Similarly, at FC temperature of 400 °C, $\Delta \overline{g_f} = -210.3$ kJ/mol., then $V_{OC} = 1.09$ V and the efficiency limit, $\eta_{FC(max)} = 73.6\%$ referred to HHV.

From these values, it might be suggested that lower temperatures are better, but the voltage losses are less at higher temperatures. In practice, FC voltages are usually higher at higher temperatures and the waste heat is more useful as well.

The actual FC efficiency can be estimated in terms of the operating voltage, V_{FC}. If all the energy from the hydrogen fuel, $-\Delta \overline{h_f}$, is converted into electrical energy, the efficiency would be 100% and (9.7) can be adapted to give the corresponding reversible open-circuit voltage as

$$V_{OC} = \frac{-\Delta \overline{h_f}}{2F} = \begin{cases} 1.48 \text{ V referred to HHV} \\ 1.25 \text{ V referred to LHV} \end{cases}$$

Therefore, the actual efficiency of the cell, η_{FC}, can be obtained as the ratio of V_{FC} to V_{OC} value (1.48 V related to HHV or 1.25 V related to LHV), that is,

$$\eta_{FC} = \frac{V_{FC}}{V_{OC}} \times 100\% = \begin{cases} (V_{FC}/1.48) \times 100\% \text{ using HHV} \\ (V_{FC}/1.25) \times 100\% \text{ using LHV} \end{cases} \tag{9.12}$$

In practice, it is found that not all the fuel fed to the FC reacts as some of it passes through the electrolyte unreacted by diffusion (it is known as 'fuel crossover'). A fuel utilization coefficient, μ_f, is used when calculating the efficiency of the cell by rewriting (9.12) as

$$\eta_{FC} = \mu_f \frac{V_{FC}}{V_{OC}} \times 100\% \tag{9.13}$$

where $\mu_f = 0.95$ for a more accurate estimation.

9.5 Factors affecting V_{OC}

As given by (9.8), V_{OC} depends on Gibbs free energy, $\Delta\overline{g_f}$. It is noted that $\Delta\overline{g_f}$ changes in a reversible electrochemical reaction with temperature as shown in Table 9.1. Not only the change of $\Delta\overline{g_f}$ is due to temperature but also can result from variation of reactant partial pressure and concentration or system pressure as discussed in the following.

Consider the reversible reaction in a general form as

$$bX + cY \Rightarrow dM \tag{9.14}$$

where b and c are moles of reactants X and Y, respectively, and d is the moles of product M. Each of X, Y and M has an associated activity, a. Thus, a_x and a_y are the activities of reactants and a_m is the activity of the product. In the ideal gases, the activity a is defined as

$$a = \frac{P}{P^o} \tag{9.15}$$

where P is the pressure or partial pressure when the gases are part of a mixture and P^o is standard pressure, 100 kPa. In case of water produced in FCs, the activity depends on whether it is in a steam or liquid form. For steam, the activity is

$$a_{H_2O} = \frac{P_{H_2O}}{P^o_{H_2O}}$$

where $P^o_{H_2O}$ is the vapour pressure of the steam at the temperature concerned and can be obtained from steam tables. If the water produced is in a liquid form, $a_{H_2O} = 1$.

Therefore, Gibbs free energy change for (9.14) should be modified by the activities of the reactants and product to be as [3]

$$\Delta\overline{g_f} = \Delta\overline{g_f}^o - RT\ln\left(\frac{a_X^b . a_Y^c}{a_M^d}\right) \tag{9.16}$$

where $\Delta\overline{g_f}^o$ is the change in molar Gibbs free energy of formation at standard pressure, R is the molar gas constant, 8.314 J/(K mol.) and T is the temperature in kelvin.

For hydrogen FC reaction: $H_2 + ½O_2 \rightarrow H_2O$, (9.16) becomes

$$\Delta\overline{g_f} = \Delta\overline{g_f}^o - RT\ln\left(\frac{a_{H_2} . a_{O_2}^{0.5}}{a_{H_2O}}\right) \tag{9.17}$$

Typical values of $\Delta\overline{g_f}^o$ are given in Table 9.1. From (9.17), with increasing activity of reactants, $\Delta\overline{g_f}$ will be more negative and more energy is released, whereas $\Delta\overline{g_f}$ increases and becomes less negative when the activity of product increases.

Accordingly, (9.7) can be modified by substituting (9.17) to obtain V_{OC} as

$$V_{OC} = \frac{-\Delta \overline{g_f}^o}{2F} + \frac{RT}{2F}\ln\left(\frac{a_{H_2}.a_{O_2}^{0.5}}{a_{H_2O}}\right)$$

Or

$$V_{OC} = V_{OC}^o + \frac{RT}{2F}\ln\left(\frac{a_{H_2}.a_{O_2}^{0.5}}{a_{H_2O}}\right) \tag{9.18}$$

where V_{OC}^o is the open-circuit voltage at standard pressure.

The second term on the RHS of (9.18) determines how the voltage is affected by both temperature and activities of both reactants and product which are function of the pressure. The voltage calculated from (9.18) is known as 'Nernst voltage', which is the reversible open-circuit voltage of a cell at given temperature and pressure, and the equation is called 'Nernst equation'.

As an illustrative example, the reaction: $H_2 + ½O_2 \rightarrow H_2O$ (steam) is considered at high temperature and assuming that the steam behaves as an ideal gas, using (9.15), the activities will be as follows:

For reactants: $a_{H_2} = \frac{P_{H_2}}{P^o}$ and $a_{O_2} = \frac{P_{O_2}}{P^o}$

For products: $a_{H_2O} = \frac{P_{H_2O}}{P^o}$

Substituting in (9.18) gives

$$V_{OC} = V_{OC}^o + \frac{RT}{2F}\ln\left[\frac{\frac{P_{H_2}}{P^o}\left(\frac{P_{O_2}}{P^o}\right)^{0.5}}{\frac{P_{H_2O}}{P^o}}\right]$$

Substituting $P^o = 1$ when all pressures are given in bar, obtains

$$V_{OC} = V_{OC}^o + \frac{RT}{2F}\ln\left(\frac{P_{H_2}.P_{O_2}^{0.5}}{P_{H_2O}}\right) \tag{9.19}$$

The pressures in (9.19) are in most cases partial pressures, for example, the oxygen is mostly a part of air. Assuming the system pressure is P, the relations in the following can be written.

$$P_{H_2} - \alpha P,\ P_{O_2} = \beta P \text{ and } P_{H_2O} = \gamma P \tag{9.20}$$

where α, β and γ are constants depending on the molar masses and concentration of H_2, O_2 and H_2O, respectively.

Incorporating (9.19) and (9.20) gives

$$\begin{aligned} V_{OC} &= V_{OC}^{o} + \frac{RT}{2F}\ln\left(\frac{\alpha\beta^{0.5}}{\gamma}.P^{0.5}\right) \\ &= V_{OC}^{o} + \frac{RT}{2F}\ln\left(\frac{\alpha\beta^{0.5}}{\gamma}\right) + \frac{RT}{4F}\ln(P) \end{aligned} \tag{9.21}$$

Therefore, two forms of Nernst equation are presented by (9.19) and (9.21). It is seen that the second term on the RHS of (9.21) determines the effect of concentration, for example, if both α and β are decreased and γ increases, the second term is getting smaller, and so the cell voltage will fall. Similarly, the third term shows the effect of system pressure. If the system pressure changed from P_1 to P_2, the change in cell voltage will be

$$\Delta V_{OC} = \frac{RT}{4F}[\ln(P_2) - \ln(P_1)] = \frac{RT}{4F}\ln\left(\frac{P_2}{P_1}\right) \tag{9.22}$$

which shows that the FC voltage is increased by system pressure.

On the other hand, to discuss the effect of reactants partial pressure, for example, hydrogen that can be supplied pure or as part of a mixture, (9.19) can be rewritten by isolating the term of hydrogen pressure as

$$V_{OC} = V_{OC}^{o} + \frac{RT}{2F}\ln\left(\frac{P_{O_2}^{0.5}}{P_{H_2O}}\right) + \frac{RT}{2F}\ln(P_{H_2})$$

With a change in the hydrogen pressure from $P1_{H_2}$ to $P2_{H_2}$ at constant P_{O_2} and P_{H_2O}, the FC voltage will change by

$$\Delta V_{OC} = \frac{RT}{2F}[\ln(P2_{H_2}) - \ln(P1_{H_2})] = \frac{RT}{2F}\ln\left(\frac{P2_{H_2}}{P1_{H_2}}\right)$$

Similarly, for oxygen partial pressure change from $P1_{O_2}$ to $P2_{O_2}$, at constant P_{H_2} and P_{H_2O}, (9.19) can be rewritten as

$$V_{OC} = V_{OC}^{o} + \frac{RT}{2F}\ln\left(\frac{P_{H_2}}{P_{H_2O}}\right) + \frac{RT}{4F}\ln(P_{O_2})$$

and the FC voltage change will be

$$\Delta V_{OC} = \frac{RT}{4F}[\ln(P2_{O_2}) - \ln(P1_{O_2})] = \frac{RT}{4F}\ln\left(\frac{P2_{O_2}}{P1_{O_2}}\right)$$

Calculations of reversible open-circuit voltage, V_{OC}, of an FC are explained in Sections 9.3 and 9.5 and given by (9.8). V_{OC} may fall at higher temperatures as illustrated in Table 9.1 or when gas pressure decreases as in (9.22). It is practically found that losses due to irreversibility may take place causing voltage drops. Because of these drops, the voltage at the FC terminals (operating voltage) is lower than the voltage developed inside the cell; V_{OC}^{o}. Irreversibility losses can be classified as activation losses, ohmic losses and concentration losses. The voltage drops produced by these losses are functions of load current and FC temperature and/or pressure.

9.5.1 Activation losses

Activation losses are the most important irreversibility, particularly, in low and medium temperature FCs. These losses result from the slowness of chemical reactions on the surface of the electrodes that may cause a loss of a proportion of the generated voltage. The voltage drop is mainly at the cathode for hydrogen FCs whereas the FCs using fuel other than hydrogen may have a voltage drop at both electrodes (cathode and anode). At higher temperatures, the impact of activation losses becomes less important.

The voltage drop produced by the activation losses, ΔV_{act}, is considered as the sum of the voltage drop at both electrodes. It can be calculated as a function of the current density of the FC, i, using an empirical formula deduced from experiments and is known as 'Tafel equation', that is [1,3],

$$\Delta V_{\mathrm{act}} = A_a \ln\left(\frac{i}{i_{\mathrm{oa}}}\right) + A_c \ln\left(\frac{i}{i_{\mathrm{oc}}}\right) \text{ for } i > i_{\mathrm{oa}} \text{ and } i > i_{\mathrm{oc}} \tag{9.23}$$

where parameters subscripted by 'a' and those by 'c' are referred to anode and cathode, respectively, and

$A \triangleq$ constant with values higher for slow chemical reactions;

$i_o \triangleq$ constant and represents the current density at which the voltage drop begins to move from zero. It is higher for faster reactions and called 'exchange current density'.

The form of (9.23) can be rewritten as

$$\Delta V_{\mathrm{act}} = A \ln\left(\frac{i}{b}\right) \text{ for } i > b \tag{9.24}$$

where

$$A = A_a + A_c \quad \text{and} \quad b = i_{\mathrm{oa}}^{A_a/A} + i_{\mathrm{oc}}^{A_c/A}$$

So, it can be found that i_o has the most variation of voltage rather than A. It is also noted that for hydrogen FC, the terms referred to the anode in either (9.23) or (9.24) can be neglected compared to that of the cathode because i_{oc} is much smaller than i_{oa}. The constant A can be calculated as

$$A = A_c = \frac{RT}{2\zeta F}$$

where

$\zeta \triangleq$ charge transfer coefficient. Its value depends on the reaction and the material of the electrode and lies in the range from 0.1 to 0.5.

Therefore, to effectively reduce the activation voltage drop, it is required to increase the value of i_o which can be done by (i) raising the cell temperature,

(ii) using more effective catalysts, (iii) increasing the surface area of the electrode by increasing its roughness and (iv) increasing both reactant concentration and pressure.

On the other hand, the current density of an FC, i, in (9.24), may be modified when considering the energy loss resulting from (i) the small amount of fuel not passing through the electrolyte (fuel crossover), and (ii) the electron conduction through the electrolyte which causes a small internal current and produces no current from the cell. The modification includes a value of internal current density, i_{int}, being added to i to express this impact on the cell voltage. Accordingly, (9.24) becomes

$$\Delta V_{act} = A \ln\left(\frac{i + i_{int}}{b}\right) \text{ for } i > b \tag{9.25}$$

However, it is found that the effect of the internal current and diffusion of fuel through the electrolyte in an FC is usually not very important except in case of low temperature FCs as it has a noticeable effect on V_{OC}.

9.5.2 Ohmic losses

In FCs, positive hydrogen ions flow in the electrolyte and electrons go through the electrodes. So, the resistance to the flow of ions in the electrolyte and the electrical resistance of the electrodes will cause what is called 'ohmic losses', resulting in a voltage drop which is equal to the product of cell current and the equivalent resistance. To replace the current (mA) by the cell current density (mA/cm^2) to obtain a relation that is consistent with the others for voltage drop, the resistance (Ω) should be replaced by the area specific resistance (kΩ cm^2). Thus, the ohmic voltage drop, ΔV_{ohm}, can be calculated by

$$\Delta V_{ohm} = i.R_{ASR} \tag{9.26}$$

where

$i \triangleq$ the current density (mA/cm^2) and R_{ASR} = the area specific resistance, ASR (kΩ cm^2).

Therefore, to reduce this ohmic voltage drop, it is essential to select electrodes with highest conductivity and appropriate materials for cell interconnects and make the electrolyte as thin as possible.

9.5.3 Concentration losses

Because of a current being drawn from an FC, the gas pressure (or partial pressure) will be reduced causing a voltage drop. For example, if oxygen at the cathode is extracted from air, its concentration may change, which in turn causes a reduction in pressure, depending on the rate of air circulation around the cathode and how quickly the oxygen can be replenished. Also, at the anode a reduction of hydrogen pressure may result due to fluid resistance during hydrogen flow through tubes.

An empirical formula that is the most common approach used for calculating the voltage drop produced by concentration losses, ΔV_{conc}, is given by [1]

$$\Delta V_{conc} = -m \exp(ni) \tag{9.27}$$

where

$m \triangleq$ constant, typically about 3×10^{-5} V;
$n \triangleq$ constant, about 8×10^{-3} cm^2/mA;
$i \triangleq$ FC current density.

9.6 FC equivalent circuit

Due to the activation, ohmic and concentration, voltage drops inside FCs, and the voltage at the FC terminals is lower than the reversible open-circuit voltage, V_{OC}, which is the voltage developed inside the FC (it can be denoted by E as well). These voltage drops as given by (9.25), (9.26) and (9.27) are functions of load current and FC temperature and/or pressure. The ohmic voltage drop is a linear function of FC load current but the ohmic resistance, R_{ASR}, is a function of FC temperature. The activation and concentration voltage drops are nonlinear functions of load current, pressure and temperature inside the FC. Specifically, for the two types of FCs, PEMFC and SOFC, which have a great potential for use in distributed power generation applications, the equivalent circuit can be deduced as shown in Figure 9.6.

The output voltage, V, at FC terminals can be written as

$$\begin{aligned} V &= E - \Delta V_{act} - \Delta V_{ohm} - \Delta V_{conc} \\ &= E - A \ln\left(\frac{i + i_{int}}{b}\right) - i.R_{ASR} + m \exp(ni) \end{aligned} \tag{9.28}$$

where ΔV_{act}, ΔV_{ohm} and ΔV_{conc} are given by (9.25), (9.26) and (9.27), respectively.

In addition to the factors, load current, temperature and pressure, that affect the voltage, V, as included in (9.28), there is another factor, which is the electric storage capability of FCs and called 'capacitance of double-layer charge effect'. It is common to most FCs and affects their dynamic response as it makes them behave like a large capacitor in the order of a few Farads. In an FC, the two electrodes (anode and cathode) are separated by a thin layer of electrolyte forming two opposite polarity

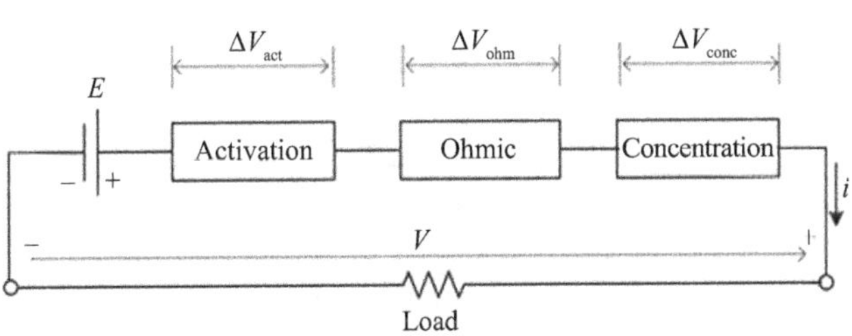

Figure 9.6 FC simple equivalent circuit

charged layers across the boundary between the electrodes and electrolyte; anode–electrolyte layer and electrolyte–cathode layer. These two layers (known as electrochemical double layers) can store electrical energy and behave much like a capacitor. The example shown in Figure 9.7 illustrates the situation that might arise at the electrode of an acid electrolyte FC. Electrons are collected at the surface of the porous electrode and the positive hydrogen ions will be attracted to the surface of the electrolyte forming the effect of a capacitor with capacitance, C, given by

$$C = \varepsilon \frac{A}{l} \tag{9.29}$$

where

$\varepsilon \triangleq$ electrolyte's electrical permittivity;
$A \triangleq$ effective surface area between the electrolyte and electrode;
$l \triangleq$ distance between the layers.

A is large because of the porous structure of electrodes, whereas l is very small (in the order of nanometers) resulting in a high value of capacitance.

Therefore, to represent the capacitance of double-layer charge effect the equivalent circuit of Figure 9.6 should be modified to be as shown in Figure 9.8. Resistances R_{ohm}, R_{act} and R_{conc} model the ohmic, activation and concentration losses, respectively. The voltage across the capacitance, V_C, can be obtained by

$$V_C = \left(i - C\frac{dV_C}{dt}\right)(R_{\text{act}} + R_{\text{conc}}) \tag{9.30}$$

and the output voltage, V is

$$V = E - V_C - V_{\text{ohm}} \tag{9.31}$$

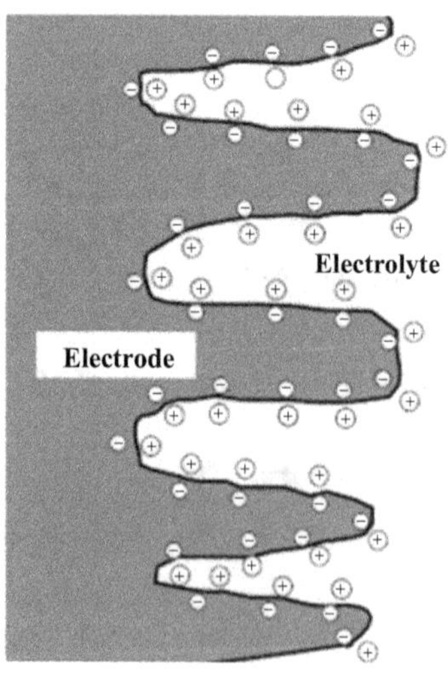

Figure 9.7 Double-layer charge effect at the electrode surface

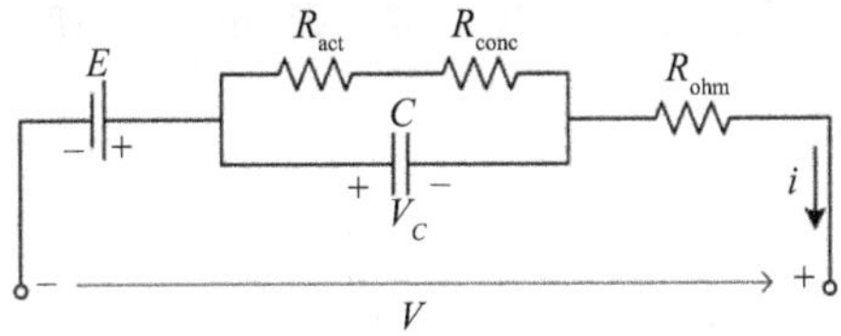

Figure 9.8 FC equivalent circuit with double-layer charge effect

It is found that the time constant, τ, of the circuit in Figure 9.8 only depends on R_{act} and R_{conc}, where

$$\tau = (R_{act} + R_{conc})C \tag{9.32}$$

The changes in these two resistances are only due to the electrochemical reactions inside FCs, whereas the ohmic resistance is the resistance due to the ohmic voltage drop; that is, a change in current gives an immediate change in the voltage drop across R_{ohm}. Thus, the output voltage given in (9.31) is a dynamic quantity as V_C is a function of the time constant, τ, given by (9.32).

9.7 PEMFC characteristics

PEMFC is one of the most common types used as a cogeneration unit. To predict its dynamic response, a computer model can be developed based on electrochemical and thermodynamic characteristics [4–8].

9.7.1 Calculation of resistances

As shown in Figure 9.8, the different voltage drops due to irreversibility losses are represented by resistances R_{act}, R_{ohm} and R_{conc}. Considering the PEMFC, the resistances can be calculated as follows.

Activation resistance: Activation losses result in a voltage drop as explained in Section 9.5.1 and can be calculated by (9.23). In case of PEMFC, from (9.23) and (9.24) the following relation can be obtained:

$$\begin{aligned} V_{act} &= A_c \ln\left(\frac{I}{I_{oc}}\right) = T.\left\{\frac{R}{2\zeta F}[\ln(I) - \ln(I_o)]\right\} = T.[a + B\ln(I)] \\ &= \Delta V_{act1} + \Delta V_{act2} \end{aligned} \tag{9.33}$$

where $A_c = \frac{RT}{2\zeta F}$, $I \triangleq$ cell current and $I_{oc} \triangleq$ constant and represents the cathode current at which the voltage drop begins to move from zero. ΔV_{act1} is the activation voltage drop component affected only by FC internal temperature and ΔV_{act2} is the component affected by both current and temperature. From (9.33), it can be found that

$$\Delta V_{act2} = T.B\ln(I) \tag{9.34}$$

where $B = \frac{R}{2\zeta F}$.

The equivalent resistance of activation is defined as: $\Delta V_{\mathrm{act2}}/I$. Thus,

$$R_{\mathrm{act}} = \frac{T.B \ln(I)}{I} \tag{9.35}$$

Ohmic resistance: It composes of polymer membrane resistance, conducting resistance between the membrane and electrodes and electrodes resistances. The ohmic resistance is current and temperature dependent and can be expressed as a sum of three components: constant, current-dependent and temperature-dependent components.

$$R_{\mathrm{ohm}} = R_{\mathrm{ohm0}} + k_{\mathrm{RI}}I + k_{\mathrm{RT}}T \tag{9.36}$$

where

$R_{\mathrm{ohm0}} \triangleq$ the constant component;

k_{RI} and $k_{\mathrm{RT}} \triangleq$ empirical constants for calculating R_{ohm} in Ω/A and Ω/K, respectively.

Concentration resistance: It can be obtained by calculating the ratio of concentration voltage drop, given by (9.27), and the FC current. Thus,

$$R_{\mathrm{conc}} = \frac{-m \exp(ni)}{I} \tag{9.37}$$

Resistances R_{act}, R_{ohm} and R_{conc} are defined by (9.35), (9.36) and (9.37). Accordingly, the equivalent circuit can be redrawn as shown in Figure 9.9. It is like that of Figure 9.8, but by adding a voltage source, ΔV_{act1}, with polarity opposite to E to satisfy (9.33).

9.7.2 PEMFC computer modelling

It is to be noted that the output voltage is a function of temperature and load current, whereas the cell temperature and the voltage across the capacitance of double-layer charge effect are a function of time during a transient state. So, the output voltage is a dynamic quantity.

A computer model based on the block diagram shown in Figure 9.10 can be developed. The inputs are anode pressure P_a, cathode pressure P_c, initial FC temperature T_{initial} and room temperature T_{room}. The overall fuel and oxidant delay effect are employed in addition to the other effects (irreversibility and double-layer charge effects) to determine the internal potential of the FC. This effect is represented by ΔV_d, which is subtracted from the right side of (9.19). It is a function of

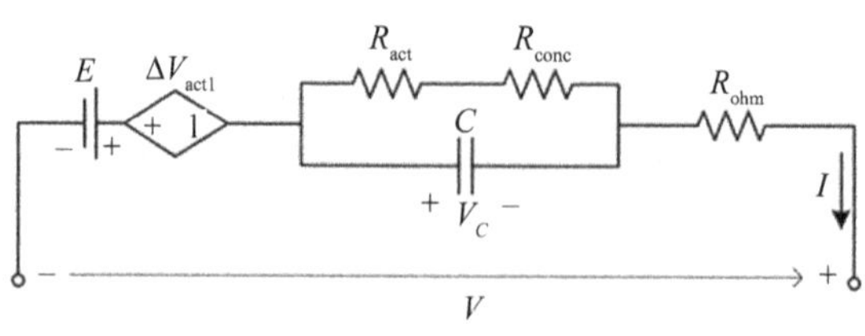

Figure 9.9 PEMFC equivalent circuit with double-layer charge effect

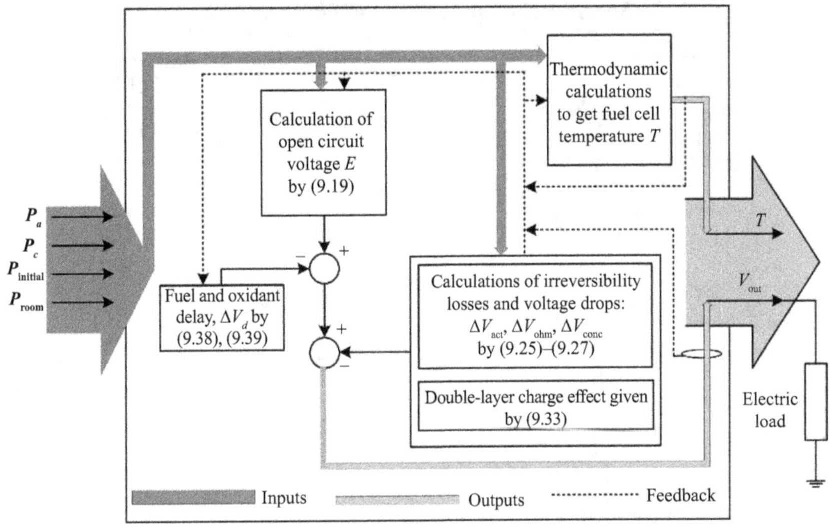

Figure 9.10 Block diagram for PENFC dynamic computer model

time and determines the influence of the FC and oxidant delay on the cell output voltage during load transients. It can be given in Laplace domain as [3]

$$\Delta V_d = \lambda_e I(s)\left(1 - \frac{1}{\tau_e s + 1}\right) = \lambda_e I(s)\frac{\tau_e s}{\tau_e s + 1} \tag{9.38}$$

where λ_e is a constant (Ω) and τ_e is the overall flow delay. In time domain, it becomes

$$\Delta V_d = \lambda_e[i(t) - i(t) \otimes \exp(t/\tau_e)] \tag{9.39}$$

where $\otimes$ is the convolution operator and t/τ_e is the time lag.

By simulation and based on the block diagram of Figure 9.10, a typical example for a PEMFC stack with specification given in Table 9.2 shows the steady-state voltage–current and power–current (*V–I* and *P–I*) characteristics as depicted in Figure 9.11.

It is shown that the *P–I* characteristic of the FC at given temperature and pressure has only one maximum power point (MPP) at which the efficiency is maximum. The output power and voltage can be controlled through proper control and MPP tracking algorithms associated with power electronic interface devices to improve FC efficiency [9–11].

9.8 Types of FCs

FCs have been classified into different types usually distinguished by the electrolyte that is used [12]. Some key figures can be referred to for identifying each type

Table 9.2 Specification of PEMFC stack

Maximum power (P_{max})	613.5 W
Voltage at P_{max} (V_{mp})	36.1 V
Current at P_{max} (I_{mp})	17 A
Short-circuit current (I_{SC})	20 A
Open-circuit voltage (V_{OC})	54 V

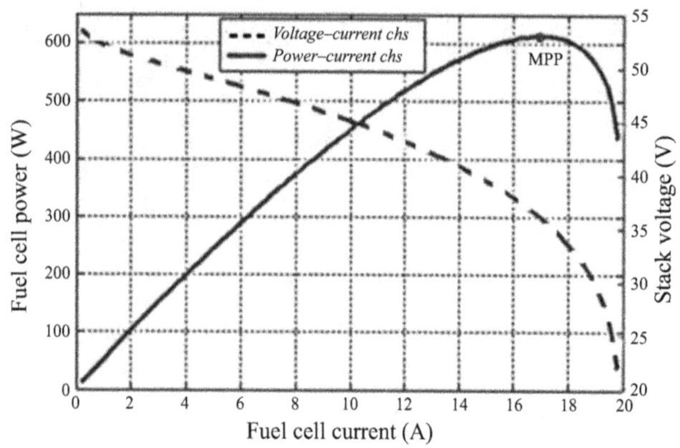

Figure 9.11 V–I and P–I characteristics for a 613.5 W PEMFC stack

such as operating temperature, produced power, cell efficiency and cost per kilowatt. These figures for some types of FCs are given in Table 9.3. Cell efficiency means the efficiency of the FC as an individual element, whereas system efficiency expresses the efficiency of the overall system including all the components shown in Figure 9.1.

9.9 Power-to-gas

The concept of PtG technology is about converting electrical energy into chemical energy (gaseous energy) in the form of hydrogen and/or methane as energy carriers. Therefore, linking the electricity and gas networks permits effective dealing with excess and lack of electrical power generated, especially by intermittent renewable sources (e.g. wind and solar) [13].

9.9.1 Principle of operation

To address the problem of electrical energy conversion into gaseous energy, the concept of PtG technology can be used as a solution. The core of this concept is water electrolysis to yield hydrogen and oxygen, then hydrogen conversion with an oxidation source to methane. The produced hydrogen and methane can be injected into a gas distribution grid or gas storages to be utilized in all other natural gas users

Table 9.3 Comparison of some FC types

FC type	Electrolyte	Power (W)	Operating temperature (°C)	Efficiency (%)	
				Cell	System
Direct methanol FC	Polymer membrane (ionomer)	100 W–1 kW	90–120	20–30	10–25
Reformed methanol FC	Polymer membrane (ionomer)	5 W–100 kW	250–300	50–60	25–40
Proton exchange membrane FC	Polymer membrane (ionomer)	1 W–500 kW	50–100	50–70	30–50
Phosphoric acid FC	Molten phosphoric acid (H_3PO_4)	<10 MW	150–200	55	40 and cogen. 90
Solid acid FC	H^+-conducting oxyanion salt (solid acid)	10 W–1 kW	200–300	55–60	40–45
Molten carbonate FC	Molten alkaline carbonate	100 MW	600–650	55	45–55
Tabular solid oxide FC (TSOFC)	O^{2-}-conducting ceramic oxide	<100 MW	850–1,100	60–65	55–60
Planar solid oxide FC	O^{2-}-conducting ceramic oxide	<100 MW	500–1,100	60–65	55–60

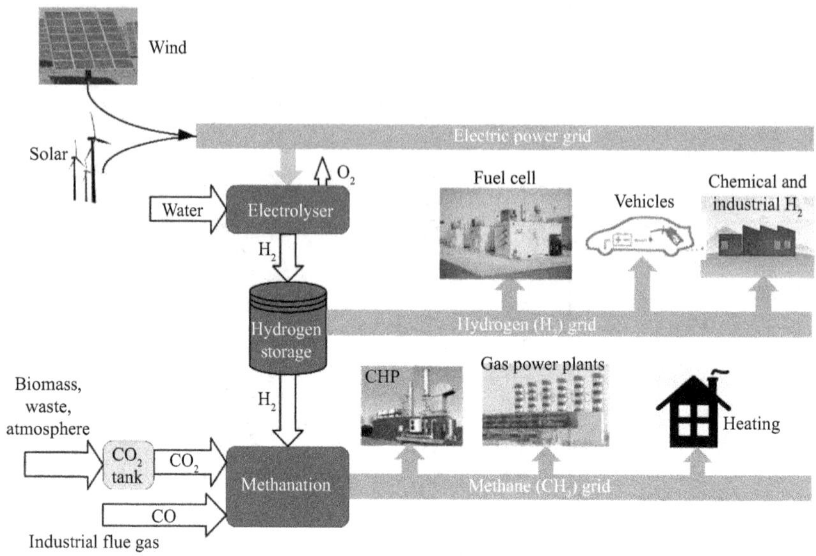

Figure 9.12 PtG concept

such as FC vehicles, chemical industries, combined heat and power (CHP) and heating as shown in Figure 9.12. Therefore, PtG systems include the main components as in the following [14].

Water electrolyser: The surplus electrical energy is delivered from the electrical power grid as input to the electrolysis device to split the water into its components, hydrogen and oxygen through the chemical reaction expressed by

$$H_2O \rightarrow H_2 + ½O_2$$

Three electrolysis technologies are commonly used in PtG process: alkaline (AEL), polymer electrolyte membrane (PEM) and solid oxide electrolysis (SOEC).

Hydrogen storage: Because of the fluctuation of power supplied by renewable sources, temporary hydrogen storage is of utmost importance to be employed to overcome the intermittent operation of the electrolyser. The storage capacity and its operating parameters depend on the amount of electrical power required to supply the demand. Different methods may be applied to establish hydrogen storage compressed gas tanks as follows:

1. Cryogenic compressed liquid hydrogen tanks, which have continuous gas leakage and expensive insulation. However, this type of tanks is a poor choice for PtG systems.
2. Physical storages, which are normally underground caverns and the cheapest option of all storage methods. In addition, this type is suitable and favourable for high capacity, long-term hydrogen storage, that is, not viable for small PtG systems.
3. High-pressure cylindrical gas tanks, which are the best option for temporary hydrogen storage.

4. Metallic hybrid tanks, which are an alternative method of the third method.

Methanation: It is a process to synthetically produce natural gas, that is, methane (CH_4) as the name is referred to. Carbon dioxide (CO_2) or carbon monoxide (CO) is required for methanation and can be obtained from either biogenic carbon source (e.g. biogas) or industrial processes [15]. Thermodynamically, the reaction can be expressed as

$$CO_2 + 4H_2 \rightarrow CH_4 + 2H_2O$$
$$\text{Or}$$
$$CO + 3H_2 \rightarrow CH_4 + H_2O$$

Methanation can be done in methanation reactors either biological or catalytic by use of catalyst, e.g. nickel. The gases produced by PtG technology (H_2 and CH_4) can be combined to form a combined natural gas (CNG), which can in turn be integrated into natural gas infrastructure. Figure 9.13 shows the sequence of processes of a typical PtG system [16–18].

9.9.2 *Applications*

The natural gas produced by PtG system can be utilized in stationary power applications such as CHP plants and small industries by respectively using gas turbines (GTs) and microturbines, which are described in the next two subsections. However, CHP covers a wide capacity range from 25 kW microturbines to 250 MW GTs [19].

9.9.2.1 Gas turbines

GTs are designed to produce high-quality thermal output for most CHP applications and district heating systems as they are a cost-effective alternative for commercial and industrial end users and reliable power generation device. GT size is available to supply the demand of many end users up to about 250 MW. GTs operate on either gaseous fuel (natural gas and synthetic gas) or stored liquid fuel (oils) for backup.

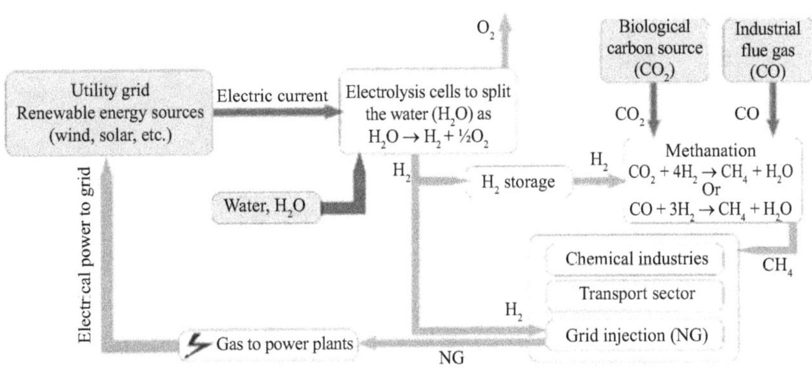

Figure 9.13 Typical flow diagram of PtG processes (inputs flow is denoted by black arrows and interior or output flows by grey arrows)

The technology of GTs is mainly based on Brayton thermodynamic cycle, which includes compression of atmospheric air, feeding and ignition of fuel and expansion of the heated combustion gases through the gas producer and power turbine. The provided power from the turbine as a prime mover can drive the compressor and rotate the electrical generator. Accordingly, the main components of a GT are shown in Figure 9.14. Simple-cycle industrial GTs electrical efficiency may vary within the range of 25% to 40%. Development of large GTs using combined cycles yields an increase in GT electrical efficiency to become approximately 60% and more economic installations.

Further improvement of electrical efficiency can be achieved by heat recovery. It is to be noted that there is no heat recovery in the exhaust gas of the simple cycle. Hot exhaust gas can directly be used in an additional process including a heat recovery steam generator (HRSG) in a combined cycle as depicted in Figure 9.15. The exhaust heat can generate steam to drive a steam turbine which in turn rotates an electrical generator.

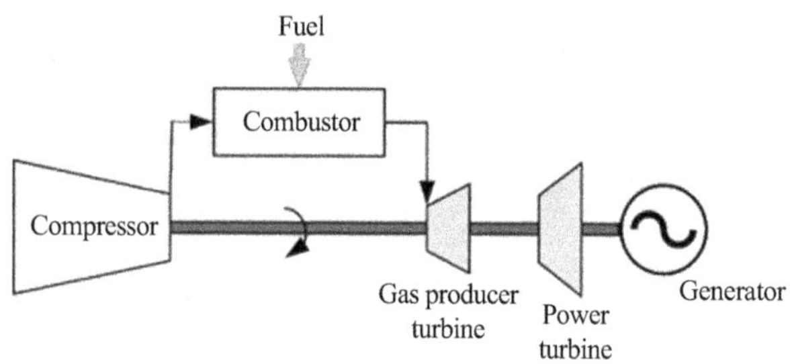

Figure 9.14 GT components

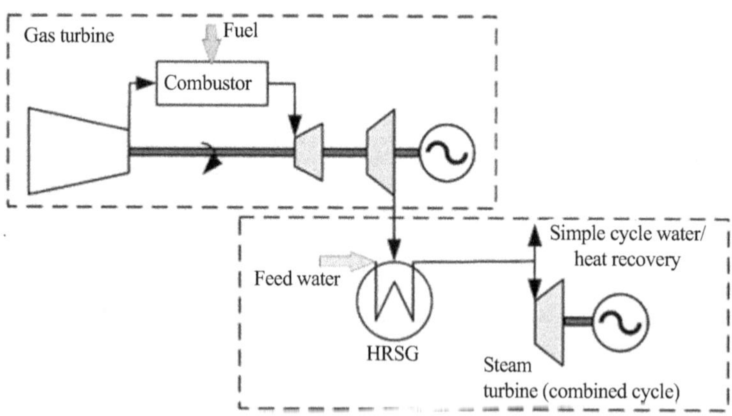

Figure 9.15 Heat recovery-GT system

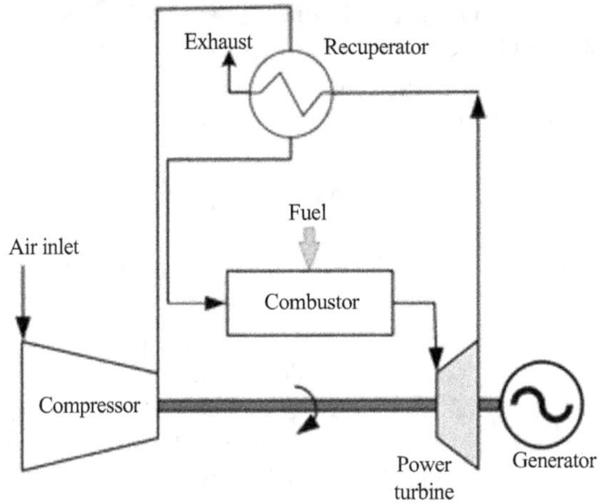

Figure 9.16 Microturbine components

9.9.2.2 Microturbines

Microturbine is a class of small GTs and operates with the same theory as that of large GTs but a difference in the design due to use of recuperator to recover part of the exhaust heat for preheating the combustion air. The air is drawn through a compressor and mixed with fuel (natural gas, gasoline, diesel and alcohol), then ignited to power the turbine and rotating an electrical generator (e.g. synchronous generator and permanent-magnet generator) at a speed up to 120,000 rpm as shown in Figure 9.16.

References

[1] Larminie J. and Dicks A. *Fuel cell systems explained*. 2nd edn. West Sussex, England: John Wiley & Sons Ltd.; 2003.
[2] Wikipedia. *Fuel cell* [online]. Available from https://en.wikipedia.org/wiki/Fuel_cell [Accessed 24 Aug 20018].
[3] Nehrir M. H. and Wang C. *Modeling and control of fuel cells: Distributed generation applications*. Hoboken, NJ, USA: IEEE-Wiley Press; 2009.
[4] Wang C., Nehrir M. H. and Shaw S. R. 'Dynamic models and model validation for PEM fuel cells using electrical circuits'. *IEEE Transaction on Energy Conversion*. 2005;20(2):442–51.
[5] Nehrir M. H., Wang C. and Shaw S. R. 'Fuel cells: Promising devices for distributed generation – Understanding their modeling and need for control'. *IEEE Power Engineering Magazine*. 2006;4(1):47–53.
[6] Maher A. R. and Al-Baghdadi S. 'Modelling of proton exchange membrane fuel cell performance based on semi-empirical equations'. *Renewable Energy*. 2005;30:1587–99.

[7] Ramos-Paja C. A., Giral R., Martinez-Salamero L., Romano J., Romero A. and Spagnuolo G. 'A PEM fuel-cell model featuring oxygen-excess-ratio estimation and power-electronics interaction'. *IEEE Transactions on Industrial Electronics.* 2010;57(6):1914–24.

[8] Lee J. H., Lalk T. R. and Appleby A. J. 'Modeling electrochemical performance in large scale proton exchange membrane fuel cell stacks'. *Journal of Power Sources.* 1998;70:258–68.

[9] Karami N., Outbib R. and Moubayed N. 'Maximum power point tracking with reactant flow optimization of proton exchange membrane fuel cell'. *Journal of Fuel Cell Science and Technology.* 2013;10:1–14.

[10] Ramos-Paja C. A., Spagnuolo G., Petrone G. and Giral R. 'Fuel cell MPPT for fuel consumption optimization'. *IEEE International Symposium on Circuits and Systems (ISCAS)*; Paris, France,.May/June 2010, pp. 2199–202.

[11] Karami N., Outbib R. and Moubayed N. 'Fuel flow control of a PEM fuel cell with MPPT'. *IEEE International Symposium on Intelligent Control (ISIC)*; Dubrovnik, Croatia, Oct 2012, pp. 289–94.

[12] Chambers A. *Renewable energy in nontechnical language.* Tulsa, Oklahoma: PennWell Corporation; 2004.

[13] Götz M., Lefebvre J., Mörs F., *et al.* 'Renewable power-to-gas: A technological and economic review'. *Renewable Energy.* 2016;85:1371–90.

[14] Liu W., Wen F. and Xue Y. 'Power-to-gas technology in energy systems: Current status and prospects of potential operation strategies'. *Journal of Modern Power Systems and Clean Energy.* 2017;5(3):439–50.

[15] Mostert C., Ostrander B., Bringezue S. and Kneiske T. M. 'Comparing energy storage technologies regarding their material and carbon footprint'. *Journal of Energies.* 2018;11(12):3386–411.

[16] Walker S. B., Lanen D. V., Fowler M. and Mukheriee U. 'Economic analysis with respect to power-to-gas energy storage with consideration of various market mechanisms'. *International Journal of Hydrogen Energy.* 2016; 41(19):7754–65.

[17] Schulze P., Holstein J., Noort A. V. and Knijp J. 'Power-to-gas in a decarbonized European energy system based on renewable energy sources'. White paper; *European POWERTOGAS.* Available from http://www.afhypac.org/documents/European%20Power%20to%20Gas_White%20Paper.pdf [Accessed 8 Feb 2020].

[18] Jentsch M., Trost T. and Sterner M. 'Optimal use of power-to-gas energy storage systems in an 85% renewable energy scenario'. *Energy Procedia.* 2014;46:254–61.

[19] Onsite Sycom Energy Corporation. 'Review of combined heat and power technologies'. Office of Industrial Technologies, US Department of Energy; October 1999, pp. 1–31.

Chapter 10
Electrochemical energy storage systems

Electrochemical energy storage (EcES) systems are technologically mature for practical use. The electricity is stored as chemical energy, which can be delivered in the form of electrical energy using electrochemical reactions. They include all types of battery energy storage (BES): (i) conventional secondary (rechargeable) BES such as lead-acid (PbA), nickel cadmium (NiCd), nickel metal hydrate (NiMH), lithium ion (Li-ion), metal air (Me-air), sodium sulphur (NaS) and sodium nickel chloride (NaNiCl) batteries and (ii) flow batteries (FBs) such as vanadium redox (VR) and hybrid FBs.

Batteries are built in different sizes with capacity ranging from <100 W to several megawatts and round efficiency from 60% up to 80% depending on the operational cycle and the type of electrochemistry. BES systems can be used in most stationary and mobile applications such as electric vehicles, marine and submarine missions, aerospace operation and so on. In addition, for grid-scale application, deep cycle-type batteries are the most commonly used with efficiency in the range of 70%–80%. A brief description of each type of batteries is illustrated in the forthcoming sections [1,2].

10.1 Conventional secondary BES technologies

10.1.1 PbA batteries

These types of batteries are the oldest and most widely used rechargeable electrochemical devices. It consists of electrodes of lead metal (Pb) with high surface area as negative electrode (i.e. cathode) and lead oxide (PbO_2) as anode, and both are immersed in an electrolytic solution of sulphuric acid (H_2SO_4).

A double sulphate reaction takes place inside the battery. In the discharging state, oxidation and reduction reactions at the anode and cathode, respectively, result in formation of lead sulphate ($PbSO_4$), and the electrolyte loses its dissolved sulphuric acid and becomes primarily water. Thus, the electrochemical reactions are

$$\text{At anode: } Pb + SO_4^{2-} \leftrightarrow PbSO_4 + 2e^-$$
$$\text{At cathode: } PbO_2 + SO_4^{2-} + 4H^+ + 2e^- \leftrightarrow PbSO_4 + 2H_2O$$
$$\text{Overall chemical reaction: } Pb + PbO_2 + 2H_2SO_4 \leftrightarrow 2PbSO_4 + 2H_2O$$

The reactions are reversible in the charging state. The two states, discharging and charging, are shown in Figure 10.1.

PbA batteries have the advantages of low cost and high reliability and efficiency. They are commonly used for power quality and some spinning reserve applications.

However, because of their short cycle life and low energy density their application to energy management is limited.

10.1.2 Nickel-based batteries

Depending on the type of metallic cathode, different types of nickel (Ni) batteries exist. The anode of all types is made of the active material 'nickel oxyhydroxide' and the electrolyte is an aqueous solution of potassium hydroxide (KOH). Whereas a metal such as cadmium (Cd), metal hydride (MH), iron (Fe), zinc (Zn) or hydrogen (H_2) is used as cathode, the electrochemical discharging reactions of nickel-based batteries are

$$\text{At anode: } Cd + 2OH^- \leftrightarrow Cd(OH)_2 + 2e^-$$
$$\text{At cathode: } NiO(OH) + H_2O + e^- \leftrightarrow OH^- + Ni(OH)_2$$
$$\text{Overall chemical reaction: } 2NiO(OH) + Cd + 2H_2O \leftrightarrow 2Ni(OH)_2 + Cd(OH)_2$$

NiCd and NiMH batteries are common, commercialized and widely used in applications of uninterruptible power supply (UPS), telecommunication and electronic devices. NiFe and NiZn batteries have limited and specific applications

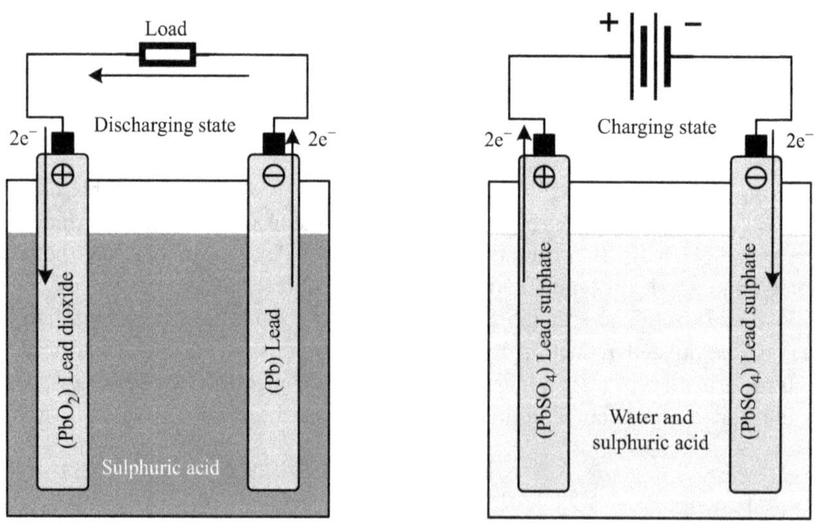

Figure 10.1 Discharging and charging of a typical lead acid battery

because of their high maintenance, low specific energy, short cycle life and high cost as well as the efficiency is about 75%. NiH_2 battery has long cycle life, large ratings and high overcharging/discharging tolerance, but on the other hand, it has low volumetric energy density and high self-discharge.

However, NiCd batteries have generally the following advantages and disadvantages.

Advantages

- Self-discharge rate at 10% per month.
- Efficiency of 70%–90%.
- Energy density of 40–60 Wh/kg.
- Large number of life cycles (200–2,500).
- Nominal cell voltage of 1.2 V.
- Capability of operation at temperatures of −20 to +40 °C.
- Reliable and do not require much maintenance.
- Depth of discharge (*DoD*), a measure of how much charge is drained from a battery in one cycle, has been developed to be only 10% and accordingly the life cycles reach 50,000.

Disadvantages

- The toxic heavy metal 'cadmium' causes environmental hazard and it has been prohibited for consumer use.
- Relatively high cost as the manufacturing process is expensive.
- Suffering from 'memory effect' as the full charge takes place only after a series of full discharges.

NiMH batteries have a nominal capacity much higher than that of equivalent NiCd and PbA batteries. Their design has been developed to have the advantages of NiCd batteries with the memory effect mitigated, operating at a wide temperature range, the environmental impact eliminated, increased specific energy (60–120 Wh/kg), increased life cycles (Σ3,000) and very low internal resistance. So, this type of batteries is very useful for hybrid electric vehicle applications and sensitive electronic devices. On the other hand, high self-discharge rate (5%–30% per month), *DoD* time dependence, life-cycle degradation and overcharging sensitivity are the limitations of NiMH batteries.

10.1.3 Li-ion batteries

In this kind of battery, the anode is made of layered graphitic carbon and the cathode is a lithiated metal oxide ($LiCoO_2$, $LiMO_2$ and $LiNiO_2$). The electrodes (anode and cathode) are separated by a foil to make electrical isolation immersed in an electrolyte of dissolved lithium salts in organic carbonates ($LiPF_6$). The nominal voltage of each cell of the battery is 3.7 V, i.e. three times the cell voltage of nickel batteries.

The operation of Li-ion battery is based on reversible extraction of ions on the porous electrodes. During charging process, the lithium atoms in the cathode become ions and travel through the electrolyte towards the anode to combine with external electrons and are deposited between the carbon layers as lithium atoms. The process is reversed during the discharging phase as shown in Figure 10.2. The overall chemical reactions is given as

$$LiC_6 + CoO_2 \leftrightarrow C_6 + LiCoO_2$$

Li-ion batteries are employed in both portable and grid-scale stationary applications. They have high energy density (~200 Wh/kg), long cycle life (3,000 cycles at 80% *DoD*), high efficiency (>90%), rapid response time (in milliseconds), fast charge/discharge capabilities and low self-discharge rate (<8% per month). So, Li-ion batteries are also applied in electric vehicles because of their reliable chemical and thermal characteristics. For large-scale applications, the main hurdle is the high cost as special packaging and internal overcharging protection circuits are required.

10.1.4 Me-air batteries

An Me-air electrochemical cell consists of the anode made from pure metals such as calcium (Ca), magnesium (Mg), Fe, aluminium (Al), lithium (Li) and Zn and a

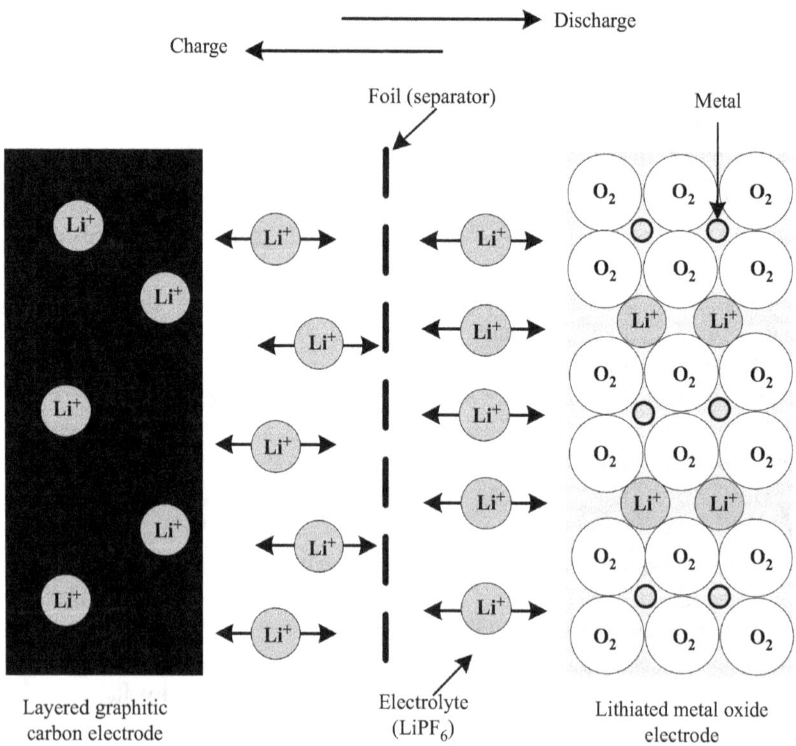

Figure 10.2 Basic components of Li-ion batteries

cathode connected to an inexhaustible air supply provided by open-cell structure. The oxygen in the air is only used for the electrochemical reaction. Among the various Me-air batteries, the Li-air is most attractive as its theoretical energy density excluding oxygen, which is not stored in the battery, is 11.14 kWh/kg (100 times more than that of other battery types as well as more than that of petrol (10.15 kWh/kg)). On the other hand, Li-air battery has a high safety risk because of the fire that can result from the high reactivity of lithium with air and humidity.

At present, only a rechargeable Zn-air battery with a theoretical specific energy excluding oxygen of 1.35 kWh/kg is technically feasible. It has some properties of fuel cells and conventional batteries: the zinc is the fuel; the reaction rate can be controlled by varying air flow and oxidized zinc/electrolyte paste can be replaced with fresh paste.

As shown in Figure 10.3, in discharging phase, the electrons are liberated through the electrochemical reaction of the zinc anode ($Zn \rightarrow Zn^{2+} + 2e^-$). In the presence of the electrolyte, an oxygen reduction reaction takes place at the air cathode resulting in the formation of hydroxide ions ($O_2 + 2H_2O + 4e^- \rightarrow 4OH^-$). These hydroxide ions migrate from cathode to anode and form the insoluble zinc oxide (ZnO). During charging phase, the reversible redox reaction takes place at the electrode–electrolyte interface, which stores the electrical energy in the Zn-air battery. The electrochemical reactions inside the battery are given as

$$
\begin{aligned}
&\text{At anode: } Zn + 4OH^- \rightarrow Zn(OH)_4^2 + 2e^- \\
&\text{In fluid: } Zn(OH)_4^2 \rightarrow ZnO + H_2O + 2OH^- \\
&\text{At cathode: } O_2 + H_2O + 4e^- \rightarrow 4OH^- \\
&\text{Overall chemical reaction: } 2Zn + O_2 \rightarrow 2ZnO
\end{aligned}
$$

10.1.5 NaS batteries

NaS battery has the capability of providing high density (151–170 Wh/m^3), is more than 85% efficient, has long cycle life (2,500 cycles at 90% *DoD*) and a long discharge period (~6 h). Furthermore, NaS can provide the response that makes it very effective in the application of power quality regulation of the utility grid. These properties are suitable for the utilities to use it as a commercial energy storage for distribution grid support, wind integration and high-value grid services.

The structure of NaS comprises liquid (molten) sulphur at the positive electrode and liquid (molten) sodium at the negative electrode as active materials separated by a solid beta alumina ceramic (β''-Al_2O_3) electrolyte as shown in Figure 10.4.

The electrolyte allows only the positive sodium ions (Na^+) to flow through it and combines with sulphur (S) to form sodium polysulphides (Na_2S_4) as the electrochemical reactions are

$$
\begin{aligned}
&\text{At cathode: } 2Zn \leftrightarrow 2Na^+ + 2e^- \\
&\text{At anode: } 4S + 2Na^+ + 2e^- \leftrightarrow Na_2S_4 \\
&\text{Overall chemical reaction: } 2Za + 4S_2 \rightarrow 2ZnO
\end{aligned}
$$

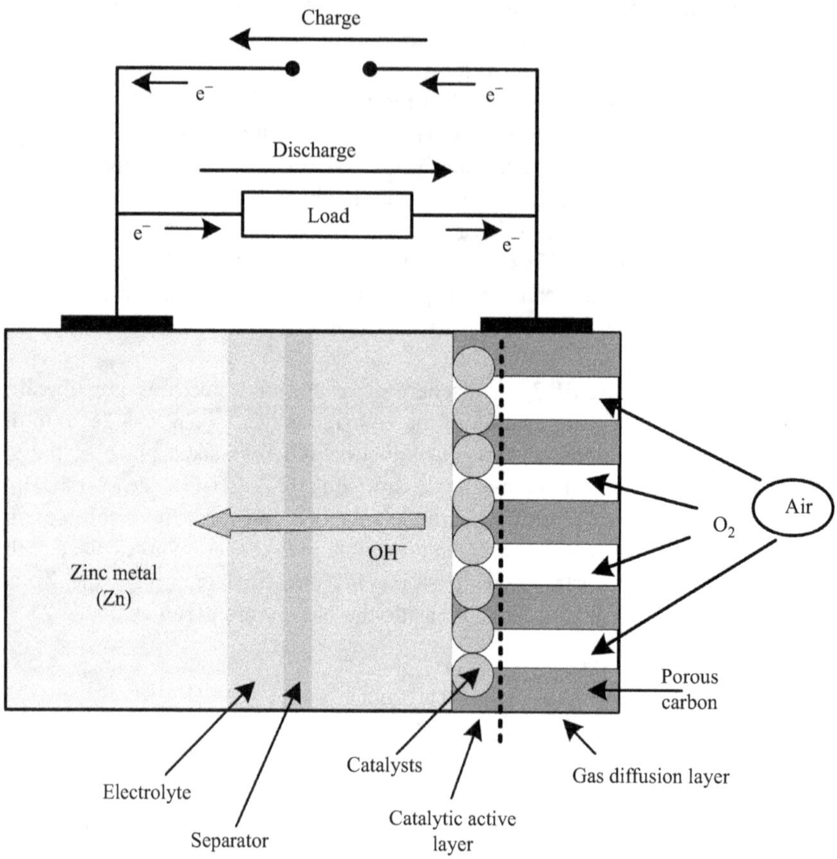

Figure 10.3 Zn-air battery representation

In discharging phase, the sodium is oxidized at the beta alumina ceramic interface forming Na^+ ions, which go through the electrolyte and electrons flow in the external circuit of the battery producing 2.0 V. The charging phase is the reverse where the Na^+ is released from sodium polysulphides back through the electrolyte to recombine as elemental sodium.

The operation temperature of NaS battery is kept at 300–350 °C, and thus a heat source is required, which uses the own stored energy of the battery resulting in partial reduction of its performance.

10.1.6 NaNiCl batteries

NaNiCl batteries are also known as ZEBRA batteries, which are high temperature (~300 °C) devices and can operate across a broad temperature range (−40 to +70 °C).

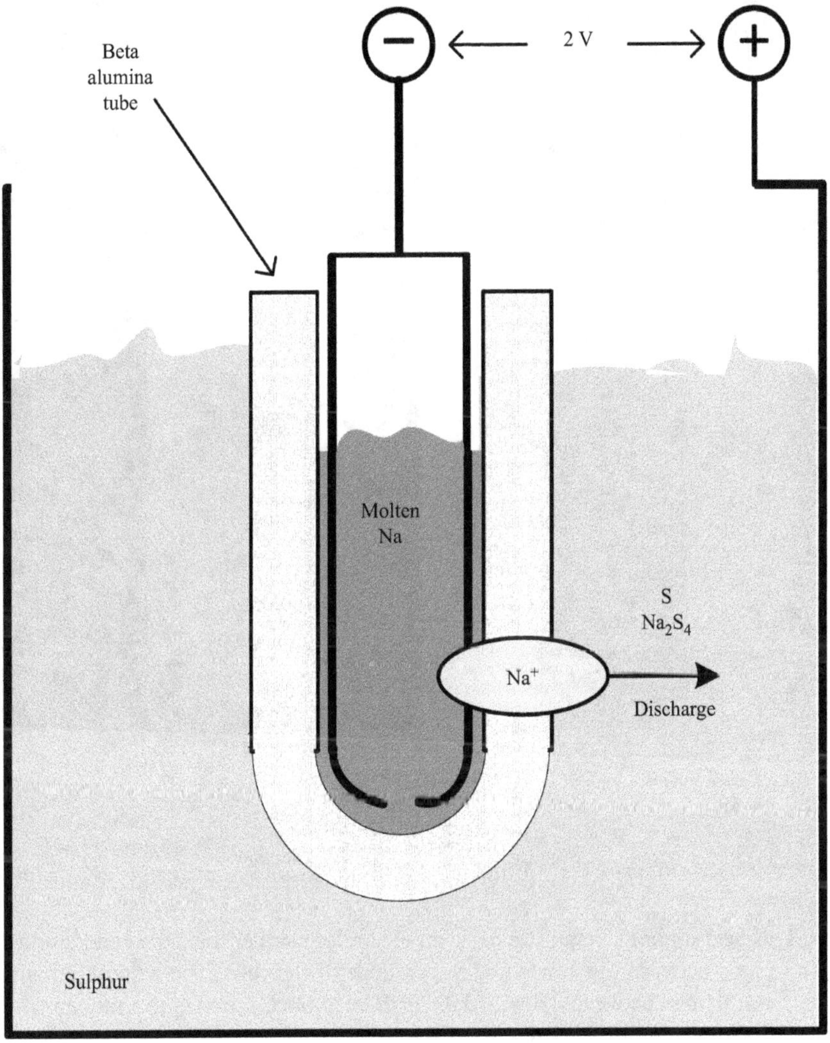

Figure 10.4 Main components of NaS battery

They are useful for such applications as time shifting, UPS, peak shaving, smart grid management, renewable generation optimization, frequency regulation and so on.

$NaNiCl_2$ secondary battery as an energy storage system is based on electrochemical charge/discharge reactions consisting mainly of a positive electrode (cathode) made of Ni and sodium chloride (NaCl) and a negative electrode (anode) that is typically made of sodium (Na). The electrodes are separated by a beta-alumina ceramic wall that is conductive for sodium ions but an isolator for electrons. A schematic diagram of the main components is depicted in Figure 10.5.

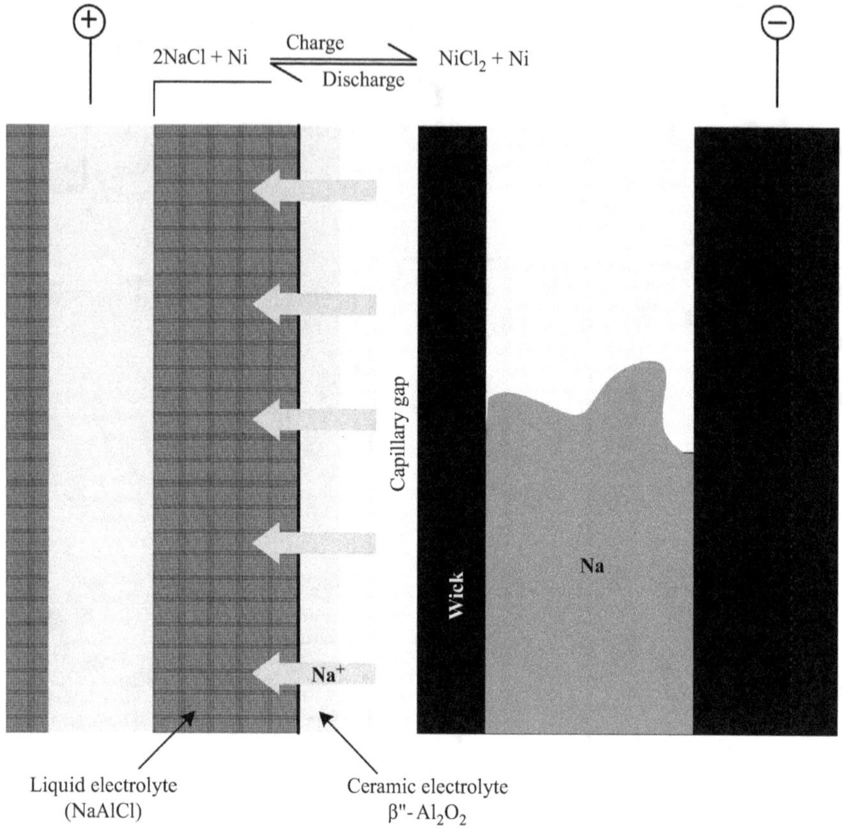

Figure 10.5 Main components of NaNiCl battery

This beta-alumina ceramic acts as an electrolyte and enables the conduction of sodium ions between the anode and the cathode of the cells. The battery temperature is maintained between 270 and 350 °C to keep the electrodes in a molten state; that is, independent heaters are part of the battery system.

The key performance data can be summarized as power range of several MW, energy range from 4 kWh up to MWh, discharge time from 2 h to several hours, cycle life of 4,500 cycles, life duration less than 15 years, reaction time of some milliseconds, efficiency in the range of 85% to 95%, energy density about 100–120 Wh/kg, etc.

10.2 Flow batteries

Unlike the conventional secondary batteries where the energy is charged/discharged in the active materials of the electrodes, in the FB, which is also a rechargeable battery, the energy is stored in liquid active electrolytes of two separate external tanks. The electrolytes are pumped through the electrochemical cells stack that converts the chemical energy directly into electrical energy and vice versa.

FBs are fully reversible by applying the DC voltage to the electrodes during charging process. They have the following attractive features and characteristics:

- The power rating is defined by the size and design of the electrochemical cells in the stack whereas the energy storage capacity depends on the volume of tanks and electrolyte concentration.
- Operation near to ambient temperature.
- Long discharge times.
- Capability of high overloading for short times.
- Response time in milliseconds.
- Low self-discharge in sealed storage tanks.
- Long charge/discharge life cycles.
- High round trip efficiency.

Therefore, with these features, FBs can be fitted to a wide range of stationary applications. Based on the electroactive components of electrolyte, FBs are classified into redox flow batteries (RFBs) and hybrid flow batteries (HFBs).

10.2.1 Redox flow batteries

The RFB has two liquid electrolyte dissolutions containing dissolved metal ions as active masses, which are pumped to the opposite sides of the electrochemical cell. The electrolytes at the negative and positive electrodes are called anolyte and catholyte, respectively.

One of the most mature FB technologies is the vanadium redox flow battery (VRFB) in which a common electrolyte, for example, vanadium sulphate dissolved in sulphuric acid is used to store the energy in charged ions in different valance states. During charging/discharging process, the H^+ ions will be exchanged between the vanadium redox couples of bivalent/trivalent (V^{2+}/V^{3+}), as oxidizing agent, in the anolyte tank and tetravalent/pentavalent (V^{4+}/V^{5+}), as reducing agent, in the catholyte tank. Anolyte and catholyte flow through porous electrodes, separated by a membrane, which allows the ions (H^+) to pass through it for the electron transfer process as shown in Figure 10.6 [3]. During the exchange of charge, a current will flow over the electrodes that can be used by a battery-powered device. The electrodes are continually supplied with the dissolved active masses from the tanks; once they are converted, the resulting product is removed to the tank. The electrochemical reactions are

$$\text{At anode: } V^{2+} \leftrightarrow V^{3+} + e^-$$
$$\text{At cathode: } V^{4+} \leftrightarrow V^{5} + e^-$$
$$\text{Overall chemical reaction: } V^{2+} + V^{5} \leftrightarrow V^{3+} + V^{4+}$$

VRFBs have the advantage of using ions of the same metal on both sides. They are commercialized in wide range of stationary applications with power ratings

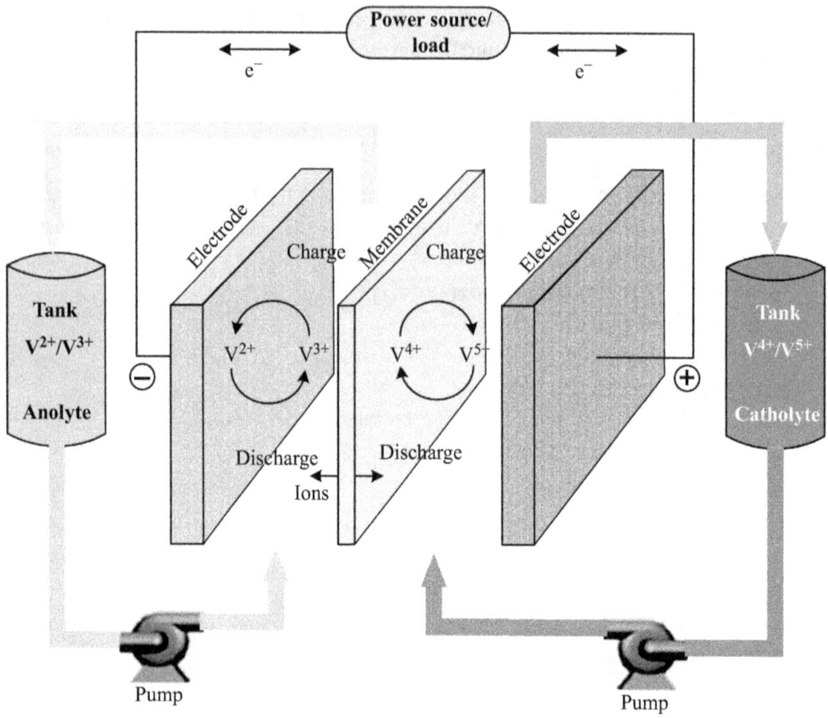

Figure 10.6 Schematic diagram of VRFB structure

from kilowatts to megawatts for power quality improvement, UPS, load levelling and for integration and support of intermittent renewable energy resources. Its standard open-circuit voltage per cell is ~1.26 V.

10.2.2 Hybrid flow batteries

HFEs combine the features of VRFBs and RFBs as one of the active materials is internally stored within the electrochemical cell, whereas the other is liquid electrolyte stored externally in a tank. Therefore, their capacity depends on the size of the electrochemical cell. Zinc bromine (ZnBr) can be used as anolyte yielding an HFB of a cell voltage (1.85 V), energy density (35–75 Wh/L), operating temperature (20–50 °C), life cycles of 10–20 years, large discharge duration (8–10 h) and large *DoD* capability [4].

10.3 Battery modelling

Because of the high non-linearity of batteries operation, accurate models tend to be more complicated mathematical models. The impact of several factors such as charge/discharge rates, battery temperature, battery type and battery age on battery operation should be involved [5]. Therefore, electrochemical models are designed to consider the chemical, thermodynamic and physical qualities of the batteries and are more precise and complex. For instance, estimation of battery state of charge (*SoC*) and *DoD* must be as accurate as possible. It exerts additional control over charging/discharging processes, which can be employed to increase battery life. *SoC* is defined as the ratio of the available capacity of a battery to the battery-rated capacity; that is, how much charge is left in the battery with respect to its nominal capacity. While physically, it is the ratio of the remaining active material to the total active material inside the battery that can be converted into electrical energy from the chemical energy. It can be mathematically estimated at time t by

$$SoC(t) = \frac{Ah_{\text{nom}} - \int_0^t I(t)dt}{Ah_{\text{nom}}} \tag{10.1}$$

where $SoC(t) = 100\%$ at $t = 0$, $I(t)$ is the battery current (positive and negative during discharge and charge, respectively) and Ah_{nom} is the nominal battery capacity in ampere-hour. It is easy to compute the integral discharge/charge current if this current is constant. Actually, the current and thus its integral varies because of the loss variation in battery capacity. In this case, (10.1) will change so that the relation is evaluated in small time intervals (Δt) using (10.2), and the *SoC* at *k*th time interval is obtained by (10.3) [6].

$$\Delta SoC_k = -\frac{I_k \Delta t}{Ah_{\text{nom}}} \tag{10.2}$$

$$SoC_k = SoC_{k-1} + \Delta SoC_k \tag{10.3}$$

To get exactly the value of *DoD*, as a measure of charge drained from a battery in one cycle, it equals to the change in the *SoC* in one discharge cycle, that is,

$$DoD = \Delta SoC \tag{10.4}$$

As an example, a *DoD* of 60% means that the battery is discharged in its first cycle from an *SoC* of 100% until *SoC* of 40% remains. As battery ageing takes place, the *SoC* of a fully charged cell after certain number of cycles decreases to 97%. The next discharge cycle with a *DoD* of 60% will then occur between an *SoC* of 97% and 37% [7].

Modelling of electrochemical batteries in terms of equivalent circuit is another way to model the batteries. It is necessary and easier, especially for simulation of BES in power system applications. It enables the electrical engineers to simulate the battery behaviour in different operating conditions, static and dynamic, as well as to compute useful parameters (e.g. short-circuit currents) for analysing a grid incorporating such batteries and renewable resources. Battery characteristics are predetermined and constant throughout operation in a static battery model, whereas battery voltage, current, charge and temperature are varied and dynamically affecting the battery output in the dynamic model.

The ideal equivalent circuit model is characterized by a voltage source (E_m) only and does not represent internal battery characteristics. Adding an impedance

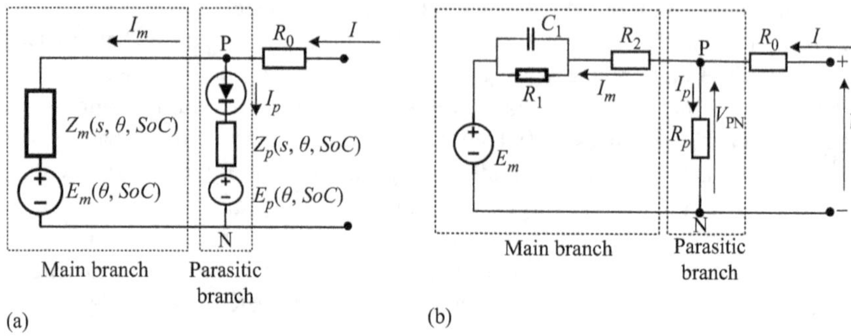

Figure 10.7 Equivalent circuit: (a) battery and (b) LA battery

in series with the voltage source to represent the battery internal impedance forms a simplified model, which can be improved by considering the values of both impedance and voltage source dependent on the battery *SoC*, temperature θ and the Laplace variable s. Further improvement can take place by adding a parasitic branch in parallel as shown in Figure 10.7(a). The parasitic branch represents the irreversible reactions that often present in the battery. The voltage across this branch is separated from the terminal voltage by a resistance, R_0. The current I_p circulates in this branch. It is a part of the total current I and is a lost current that cannot be restored. The charge stored in the battery is only the integral of I_m (the current entering the battery in the main reversible reaction branch).

The equivalent circuit in Figure 10.7(a) can be adapted to model a specific type of batteries, for example, lead acid battery as shown in Figure 10.7(b), where all elements are functions of *SoC*, charging/discharging currents and electrolyte temperature θ. The model includes two main parts: the main branch, which approximates the battery dynamics, and the parasitic branch, which accounts for the battery behaviour at the end of a charge process in addition to a terminal resistance, R_0. The main branch consists of the following [8,9]:

- Internal electromotive force (EMF) or open-circuit voltage of one cell, E_m in volts. It is assumed to be constant when the battery is fully charged and is given by

$$E_m = E_{m0} - K_E(273 + \theta)(1 - SoC) \tag{10.5}$$

  where E_{m0} is the open-circuit voltage at full charge in volts, K_E is a constant in V/°C and θ is the electrolyte temperature in °C.
- Resistance R_1, which is varied with depth of charge (*DoC*) and is a measure of battery's charge adjusted for the discharge current. It is calculated by

$$R_1 = -R_{10} \ln(DoC) \tag{10.6}$$

  where R_{10} is a constant in ohm.

- Capacitance C_1 to model a voltage delay when battery current changed. Its value is given by

$$C_1 = \frac{\tau_1}{R_1} \tag{10.7}$$

where τ_1 is a main branch time constant in seconds.

- Main branch resistance R_2 in ohm, which affects the battery during charging and becomes relatively insignificant for discharge current. It can be computed by the relation

$$R_2 = R_{20}\frac{\exp[A_{21}(1 - SoC)]}{1 + \exp\left(A_{22}\frac{I_m}{I^*}\right)} \tag{10.8}$$

where R_{20} is a constant in ohm, A_{21} and A_{22} are constants, I_m is the main branch current in ampere and I^* is the nominal battery current in ampere.

The parasitic branch current, I_p, is the loss current when the battery is being charged. It depends on the electrolyte temperature and the voltage across the parasitic branch. Therefore, representing the parasitic branch by linear model as shown in Figure 10.7(a) is unrealistic as the dependence of I_p on the voltage V_{PN} is highly non-linear. However, experiences and experiments yield that Z_p can be assumed to be a resistor that is independent of Laplace variable s and SoC. Thus, it is preferred to express this branch in a way that satisfies the algebraic and non-linear nature as shown in Figure 10.7(b). I_p is very small under most conditions except during charge at high SoC and is given by [10]

$$I_p = V_{\mathrm{PN}}G_{p0}\exp\left[\frac{V_{\mathrm{PN}}/(\tau_p s + 1)}{V_{P0}} + A_p\left(1 - \frac{\theta}{\theta_f}\right)\right] \tag{10.9}$$

where V_{PN} is the voltage across the parasitic branch, G_{p0} is a constant in seconds, τ_p is a parasitic branch time constant in seconds, V_{P0} is a constant in volts, A_p is a constant, θ is the electrolyte temperature in °C and θ_f is the electrolyte freezing temperature in °C.

The terminal resistance R_0 in ohm is assumed constant at all temperatures, and varied only with SoC according to the relation

$$R_0 = R_{00}[1 + A_0(1 - SoC)] \tag{10.10}$$

where R_{00} is the value of R_0 in ohm at $SoC = 1$ and A_0 is a constant.

10.4 Comparison of BES technologies

Selection of battery technology as an energy storage is based on both battery characteristics, investments needed and application requirements. Therefore, technical and economical comparison can help the engineers to select the proper type of battery to achieve the desired performance at as lowest a cost as possible, that is, the cost-effective option.

10.4.1 Technical comparison

EcES systems can be distinguished by the mechanisms used to store energy. Batteries such as rechargeable Li- or Na-based store energy within the electrode structure through charge transfer reactions. Compared to fuel cells, which are not rechargeable, the energy is stored in the reactants fed externally to the cells. In addition, both batteries and fuel cells differ from RFBs, which store energy in the redox species that are continuously circulating through the cells.

Batteries have desirable features (pollution-free, high round-trip efficiency, flexible power and energy characteristics) to meet different grid functions, long cycle life, low maintenance and renewable resources integration. Various types of batteries can be used for the applications of power quality, transmission and distribution (T&D) support, for example, load shifting and bridging power at different discharging duration and rated power as illustrated in Figure 10.8 [11] with power and energy densities as shown in Figure 10.9. Advantages and disadvantages of each EcES technology as well as the field of applications are given in Table 10.1.

10.4.2 Economical comparison

For economical comparison, the cost of different battery technologies should be analysed as it is one of the most important factors in the criterion of EcES selection.

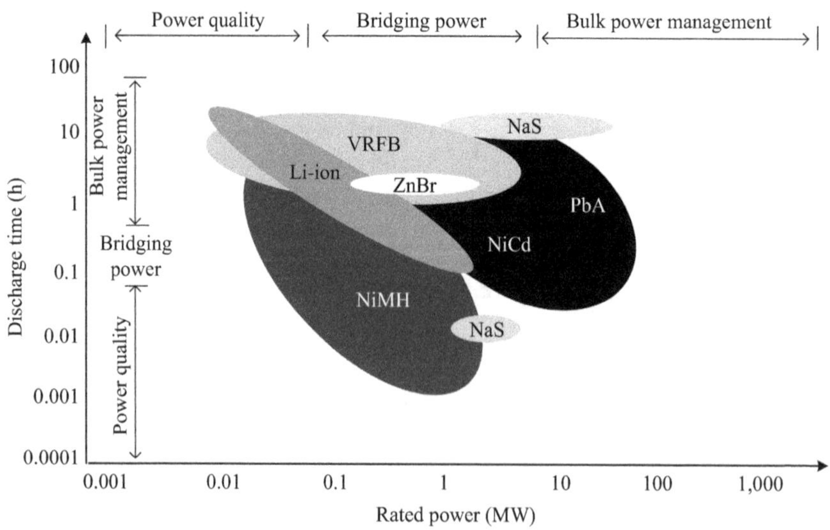

Figure 10.8 Comparison of discharge time and power rating for different BES technologies [11]

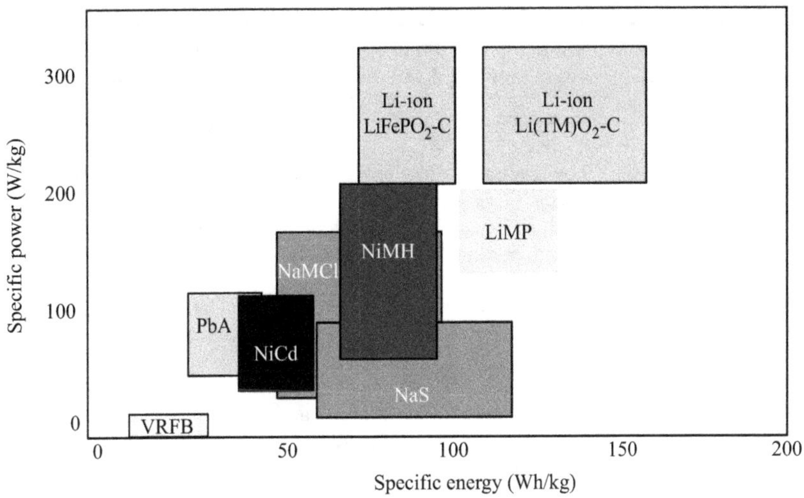

Figure 10.9 Power and energy densities for different rechargeable batteries [12]

In the literature, two methodologies exist to study the cost of each type: total capital cost (TCC) and life cycle cost (LCC) [13–15].

- TCC: It is expressed as cost per kWh or cost per output power rating (kW) and evaluates all expenses of the purchase, installation and delivery of an EcES system including the following:
  - Cost of power conditioning system (PCS) in per unit of power capacity (cost/kW).
  - Energy storage related costs comprise all costs that should be covered for installations of the storage system and expressed in per unit of stored or delivered energy (cost/kWh).
  - Balance of power costs including costs of grid interface and integration facilities, cost of land and accessibility and other services that are not comprised in both C_{PCS} and C_{stor} and expressed in per unit of power (cost/kW) or energy (cost/kWh).
- LCC: It includes TCC in addition to the following:
  - Fixed operation and maintenance (O&M) expenses.
  - Variable O&M expenses.
  - Replacement cost.
  - Disposal and recycling costs.

LCC can be introduced in levellized annual costs (cost/kW year).

A typical economical comparison of various EcES technologies is given in Table 10.2.

Table 10.1 Overview of electrochemical storage technologies

Class	Storage technology	Main advantage	Main disadvantage	Power application	Energy application
HT-batteries	LA	Low capital cost	Limited cycle life	■	●
	aLA-Cee	Low capital cost	Low energy density	■	■
	NaS	High power and energy density	Cost and requirements to run at HTs	■	■
	Li-ion	High power and energy density	Cost and increased control circuit needs	■	▲
	ZEBRA	High power and energy density	Withstand limited overcharge and discharge	■	■
	NiCd-NiMH	Low capital cost	Cost and requirements to run at HTs	▲	■
Flow batteries	ZnBr	Independent power and energy	Medium energy density	▲	■
	VRBr	Independent power and energy	Medium energy density	▲	■

■ Acceptable and fully capable. ▲ Acceptable for this application. ● Feasible but not practical or economical.

Table 10.2 Typical cost of different battery technologies

Battery type	Conventional battery							Flow battery	
	PbA	NiCd	NiMH	Li-ion	NaS	NaNiNl	Zn-air	VRFB	HFB
Power cost	1.75–6	5–15	6–18	12–40	32–40	1.5–3	1–2.5	14–37	18–20
Energy cost	1.5–4	6–24	7.2–28.8	4–25	3–5	2.3–3.45	0.6–1.6	5–8	1–7

Power and energy costs are multiplied by 100 \$/kW and 100 \$/kWh, respectively.

References

[1] Aneke M. and Wang M. 'Energy storage technologies and real-life applications – A state of the art review'. *Applied Energy.* 2016;179:350–77.

[2] Beaudin M., Zareipour H., Schellenberglabe A. and Rosehart W. 'Energy storage for mitigating the variability of renewable electricity sources: An updated review'. *Energy for Sustainable Development.* 2010;14:302–14.

[3] Vetter M. 'Battery system technology'. *Workshop*; Berlin, Germany, April 2011. Fraunhofer ISE; 2011, pp. 1–34.

[4] Park J., Nam G. and Cho J. *Perspective of Zn-air batteries* [online]. 2015. Available from https://www.csm.ornl.gov/BLI8/presentations/J.%20CHO-UNIST-BYI8.pdf [Accessed 8 Feb 2020].

[5] Leao J. F. A., Hartmann L. V., Correa M. B. R. and Lima A. M. N. 'Lead-acid battery modeling and state of charge monitoring'. *Twenty-Fifth Annual IEEE Applied Power Electronics Conference and Exposition (M. B. R. APEC)*; Palm Springs, CA, USA, Feb 2010. IEEE; 2010, pp. 239–43.

[6] Sparacino A. R., Reed G. F., Kerestes R. J., Grainger B. M. and Smith Z. T. 'Survey of battery energy storage systems and modeling techniques'. *PES General Meeting*; San Diego, CA, USA, July 2012. IEEE; 2012, pp. 1–8.

[7] Dogger J. D., Roossien B. and Nieuwenhout F. D. J. 'Charaterization of Li-ion batteries for intelligent management of distributed hybrid-connected storage'. *IEEE Transactions on Energy Conversion.* 2011;26(1):256–63.

[8] Jackey R. A. 'A simple, effective lead-acid batter modeling process for electrical system component selection'. *SAE International Transactions.* 2007;116:219–27.

[9] Moubayed N., Kouta J., El-Ali A., Demayka H. and Outbib R. 'Parameter identification of the lead-acid battery model'. *33rd IEEE Photovoltaic Specialists Conference*; San Diego, CA, USA, May 2008. IEEE; 2008, pp. 1–6.

[10] Ceraolo, M. 'New dynamical models of lead-acid batteries'. *IEEE Transactions on Power Systems.* 2000;15(4):1184–90.

[11] Lucas A. and Chondrogiannis S. 'Smart grid energy storage for frequency regulation and peak shaving using a vanadium redox flow battery'. *Electrical Power and Energy Systems.* 2016;80:26–36.

[12] Dunn B., Kamath H. and Tarascon J. 'Electrical energy storage for the grid: A battery of choices'. *Journal of Science*. 2011;334(6058):928–35.

[13] Zakeri B. and Syri S. 'Electrical energy storage systems: A comparative life cycle cost analysis'. *Renewable and Sustainable Energy Reviews.* 2015;42:569–96.

[14] Chen H., Cong T. N., Yang W., Tan C. and Li Y. 'Progress in electrical energy storage system: A critical review'. *Progress in Natural Science.* 2009;19(3):291–312.

[15] Nadeem F., Hussain S. M., Tiwari P. K., Goswami A. K. and Ustun T. S. 'Comparative review of energy storage systems, their roles and impacts on future power systems'. *IEEE Access.* 2018;7:4555–85.

Chapter 11

Electrostatic, magnetic and thermal energy storage

11.1 Electrostatic energy storage systems

Electrostatic energy storage systems are used for storing electrical energy in the form of charge and then producing electrical power back when required. Supercapacitors (SCs) are devices that can store energy in the form of electrostatic field. SCs are also known as electric double-layer capacitors (DLCs) or ultra-capacitors. Supercapacitor energy storage (SCES) has similarities with both batteries and conventional capacitors. However, unlike batteries, SCs store electrical energy, not chemical energy. Also, unlike conventional capacitors, SCs contain moving ions. SCs take the form of cells in which the energy is stored in the form of electrostatic field between two electrodes separated by a thin layer of insulator or dielectric medium. This is the same principle as the classical capacitors except that the insulating material is replaced by liquid electrolyte ionic conductor between the electrodes, which are made of porous carbon [1].

11.1.1 Working principle of SCES

The principle of SCs is based on the creation of a double electrochemical layer via the accumulation of electrical charge at the interface between an ionic solution (electrolyte) and an electronic conductor (electrode). Unlike batteries, there is no reaction of oxidoreduction. A deposit of activated charcoal on an aluminium film is used to obtain a large surface and, thus, a high specific capacity. The electrodes are submerged in an aqueous or organic electrolyte (e.g. potassium hydroxide or sulphuric acid). The electrical charges are stored at the electrode–electrolyte interface and a porous membrane through which ions transfer between the electrodes takes place to prevent their physical contact as shown in Figure 11.1(a). Thus, a DLC is created between each electrode (A and B) and electrolyte. The corresponding equivalent circuit is depicted in Figure 11.1(b), where it comprises a series combination of two double-layer capacitances, C_A and C_B, with electrode ohmic resistances, R_A and R_B, respectively, in parallel. The two double layers are separated by the ohmic resistance

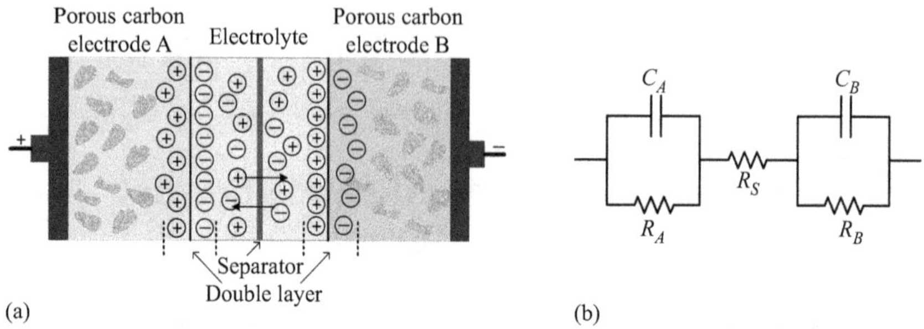

Figure 11.1 SC representation: (a) typical physical representation and (b) the equivalent circuit

of the electrolyte-filled gap, R_S, that is as narrow as possible. The capacitance, C, of a plate capacitor is calculated by the following [2]:

$$C = \varepsilon_0 \varepsilon_r \frac{A}{d} \tag{11.1}$$

where ε_0 is the dielectric permittivity of free space, ε_r is the electrolyte permittivity, A is the anode surface area and d is the thickness of the double layer.

It is to be noted that (i) the use of liquid electrolyte and porous carbon electrodes yields high specific surface area, as well as (ii) the extremely short gap (~1 nm) between them results in the capability of producing high capacitance values (many thousand farads).

11.1.2 Electrical performance

Capacitance of dielectric capacitors is measured in microfarads, nanofarads or sometimes in picofarads. Electrochemical capacitors tap the full range of up to several thousand farads because of their very high energy density. Farad is defined as the ability to store 1 C of electric charge per volt of potential difference between the plates, that is, 1 F = 1 C per volt or 1 A per volt. The term 'capacity' used in battery technology indicates the charge storage Q, in units of coulombs or ampere hours.

Specific energy. Depending on the state of charge (*SoC*), the voltage of a capacitor increases or decreases, respectively, on charging or discharging. The energy stored, W, by a capacitor is calculated in terms of the stored electric charge, $Q = CV$, by the following relation:

$$W = \frac{1}{2} QV = \frac{1}{2} CV^2 \ \& \ C = \frac{Q}{V} \tag{11.2}$$

The capacitance per volume of SCs is 10,000-fold higher than that achievable with the best regular capacitors. Such SCs deliver specific energy of 4–9 Wh/kg (14–32 kJ/kg) and specific peak power in the range of 1–50 kW/kg [2].

Discharge behaviour. Like the most battery chemistries, less than 10% of the stored energy can be used without affecting cycle life. SCs cannot deliver the full charge in a short time as the voltage of an SC drops exponentially with time from rated voltage to zero. However, about 75% of the stored energy can be used at constant current discharge in the range from 100% to 50% of the rated voltage.

Power. SCs are the components dedicated to power storage rather than energy storage. Their specific power density is 10 to 100 times higher than that of conventional batteries (only very high-power lithium batteries can reach the same specific power density). They have significant mass power, theoretically of the order of 10 kW/kg, supporting between 500,000 and 1 million charge/discharge cycles. Several elements, each with a maximum voltage of 2.7–2.85 V, can be connected in series and parallel to achieve the desired voltage value and capacity.

In addition, fast discharge rate of SCs is substantially higher than that of most batteries. Electric power delivered by the capacitor to a load is given by

$$P = VI = I(V_0 - IR_i) \tag{11.3}$$

and the thermal loss is expressed as

$$P_{\text{th}} = I^2 R_i \tag{11.4}$$

where V_0 is the rated voltage, I is the battery current and R_i is the internal resistance.

Efficiency. For batteries and SCs, the electrical efficiency, η_{el}, is given by the voltage efficiency, η_V, multiplied by the Coulombic efficiency, η_C, that is,

$$\eta_{\text{el}} = \eta_V \times \eta_C \tag{11.5}$$

$$\eta_C = Q/Q_0 \tag{11.6}$$

$$\eta_V = V/V_0 \tag{11.7}$$

where Q is the electric charge removed during discharge, Q_0 is the amount of charge that has been supplied during charge (mostly, $\eta_C \approx 99\%$ as there is no role for side reactions during charging), V is the average voltage during discharge and V_0 is the average voltage during charging.

The discharge efficiency, α, and the pulse power of an SC, at constant current discharge in the voltage window between the fully charged and fully discharged conditions at ¾V_0, depend on internal resistance as indicated by the following relations [2]:

$$\alpha = 1 - \frac{IR}{(3/4)V_0} \tag{11.8}$$

and

$$P = (1 - \alpha)\frac{(3/4V_0)^2}{R} = \frac{3}{4}V_0 I \tag{11.9}$$

For a battery,

$$\alpha = \frac{V}{V_0} = 1 - \frac{IR}{V_0} \tag{11.10}$$

and

$$P = \alpha(1 - \alpha)\frac{V_0^2}{R} \tag{11.11}$$

Reversibility and cycle life. The ideal electrostatic charge commonly takes place at the electrode/electrolyte interface, and the material is not subject to any irreversible process and phase changes. Unlike batteries, SCs provide theoretically unlimited cyclability. Cycle lives up to several millions have been observed even for deep discharges (more than 75% of the stored energy), fast rates of voltage change and at low temperatures. Because of the distributed electrolytic and contact resistances within the pore structure of high-area electrode materials in-depth charging and discharging encounter rate limitations. This is opposite to batteries in which slow Faradaic processes involve phase changes of the electroactive materials, which limit cycle life to some thousand charge/discharge acts, depending on the current and temperature.

11.1.3 Lifetime aspect

The lifetime of an SC can reach one million cycles (or 10 years of operation) without any degradation, except for the solvent used in the capacitors as it deteriorates in 5–6 years regardless of the number of cycles. In addition to the lifetime, other terms used in practice are the following:

- *Shelf life*, which is the maximum storage time under open circuit conditions. It depends on self-discharge.
- *Service life or calendar life*, which is the time between initial use and the end of useful life, including periods of rest and standby operation.
- *Operating life*, which denotes the actual time of operation.
- *Float life*, which reflects the lifetime under a certain voltage for long periods of time.
- *Cycle life*, which is determined by the number of useful charge/discharge cycles.

11.1.4 Applications

For low- or high-power applications that require fast charging, SCs are used. Examples of such applications are alarm systems, portable power supplies, automated meter reading equipment, flashlights and so on. Electric propulsion systems in all-electric cars, plug-in hybrids and forklifts require electric short-time storage for load levelling. The capacitor provides peak power density for acceleration or uphill driving, whereas the battery system covers the base load for constant speed. The SC can be advantageously recharged by regenerative braking, where the wheels drive a dynamo which generates electricity.

Furthermore, SCs are also adequate for moderate power demand applications such as rechargeable stand-alone power supplies and motor start-up for large engines (e.g. diesel trucks and railroad locomotives).

In power systems incorporating microgrids, SCES systems offer important benefits to the power system operation such as the following [3]:

- Increasing the penetration of renewable energy resources.
- Decreasing the investment required for new transmission and distribution lines or building new power plants, which in turn lowers the operational cost.
- Providing reserve capacity that can help improve system's stability and reliability in addition to improving the power quality as the start speed is faster than generators.
- Providing an effective mechanism for load management aiming to reduce the cost of power supply at peak hours, and thus lower the electricity price. Moreover, electricity price can be stabilized as the SCES system can avoid energy disruption trading and the loss caused by prediction errors.
- Renewable resources such as wind and solar energy generation are not dispatchable because of their dependency on the weather. With the aid of SCES system, a microgrid that includes such resources becomes dispatchable. This yields more economical energy exchange planning between the microgrid and the utility.

11.1.5 Advantages and disadvantages

SCs offer the following advantages:

- Long lifetime.
- Operation over a wide temperature range.
- Lifetime reaches one million cycles (10 years of operation).
- Durablity.
- High reliability.
- Environmentally friendly.
- Easy to recycle or neutralize.
- The efficiency is around 90%.
- Discharge times are in the range of seconds to hours.
- High specific power density.
- Large number of charge/discharge cycles.

On the other hand, the disadvantages of SCs are the following:

- Low energy density.
- Short storage duration.
- High self-discharge loss.
- High investment costs.

11.2 Superconducting magnetic energy storage

Magnetic coils can store energy in the form of electromagnetic field, analogous to the capacitors that store the energy in the form of electrostatic field. Large flowing currents in the coils are necessary to store a significant amount of energy and consequently the losses, which are proportional to the current squared, will also be high. Thus, the focus on superconducting coils is important as the resistance of the coils becomes zero in the superconductivity state. To achieve this state, the superconducting material must be cooled to a very low temperature, typically below 9.25 K for classical superconducting materials. The cooling process can be achieved with liquid helium (liquefaction point at 4.2 K). However, the transition temperature to get the material into the superconductivity state and cooling liquid depend on the type of material itself, for example, liquid nitrogen for materials with a transition temperature above 77 K.

A schematic design of a superconductive magnetic energy storage (SMES) device is shown in Figure 11.2. The entire SMES unit includes the following subsystems:

- large superconducting coil with the magnet (SCM), which is contained in a Dewar at the desired cold temperature,
- power conditioning system (PCS) to interface the AC utility and SCM by which the power is converted from AC to DC or inversely allowing the SMES system to respond in milliseconds,
- the cryogenic system to cool the SCM below the critical temperature at a desired value, and
- the controller to perform different functions: controlling the modes of charge/discharge/standby by controlling the voltage across the SCM, controlling cryogenic system and controlling the PCS.

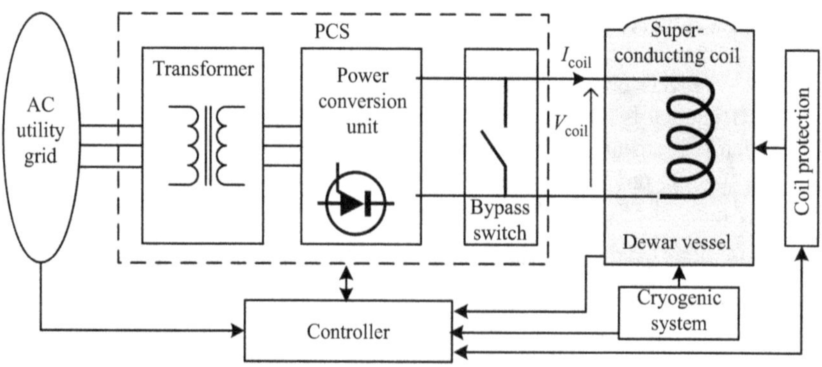

Figure 11.2 Typical SMES unit components

The core element of an SMES system is the SCM of high inductance (L_{coil} in henrys). It stores energy in the magnetic field generated by DC current (I_{coil} in amperes) flowing through the coil. Therefore, the inductively stored energy (E in joules) and the rated power (P in watts) are the common specifications for SMES devices. They are expressed as

$$E = \frac{1}{2} L I_{coil}^2 \tag{11.12}$$

$$P = \frac{dE}{dt} = L I_{coil} \frac{dI_{coil}}{dt} = V_{coil} I_{coil} \tag{11.13}$$

During the charging process, the superconducting coil is fed with direct current from the inverter. The electric current induces a constant magnetic field in which the energy is stored. The discharging process begins with connecting the coil to an external load by a switching system. The energy is then supplied by the magnetic field that drives a current. The magnetic energy and the current are decreased during discharging.

11.2.1 Applications

SMES systems are short-term storage systems for very short periods (power storage). They can be used in three scales of energy storage: (i) as an uninterruptable power supply (UPS) of the order of several kWh [4], (ii) smoothing of production or demand on a local scale (1–1,000 MWh) [5,6] and (iii) stabilization of grids by modulating power transmission, flexible AC transmission, of the order of 1,000 MWh or greater [7–9].

SMES systems offer advantages for these applications compared to flywheels and SCs, especially with respect to efficiency, response time and lifetime as given in Table 11.1 [10]. However, SMES systems are costly and have low storage power density.

11.3 Thermal energy storage

Thermal energy storage (TES) is a technology that stocks thermal energy by heating or cooling a storage medium so that the stored energy can be used later for heating and cooling applications and power generation. TES systems have

Table 11.1 Comparison of alternative storages

Storage system	Efficiency	Power density (kW/m^3)	Response time	Lifetime
Flywheel	0.7–0.9	10–100	Dozens of minutes	100,000 cycles
SMES	>0.9	0.1–5	Milliseconds to a few seconds	20–30 years
SC	0.9–0.95	1–10	Seconds to a minute	500,000 cycles

advantages of very high round-trip efficiency exceeding that of other storage options, better reliability and less pollution of the environment. TES offers high storage efficiency for multiple hours at the modest cost as the investment and running costs are reduced. Additionally, it allows stored power to match peak demand periods, making solar thermal energy technologies competitive with gas turbines as capacity power sources [11].

TES is becoming important for electricity storage in combination with concentrating solar power (CSP) plants, which are explained in Chapter 3, Section 3.3, where solar heat can be stored for electricity production when sunlight is not available. Therefore, it was developed to assist with several applications of energy production and heating, including supplementing the loading of utility scale solar thermal power plants. Based on the method of storage, TES technologies can be classified into three types: sensible heat storage (SHS), latent heat storage (LHS) and heat of reactions as illustrated in the next section.

11.3.1 Classification and characteristics of TES

TES requires the increase or decrease in the internal energy of a substance by heating or cooling. Thermal energy is stored in materials that are classified by one of three methods by which they store energy as heat (Figure 11.3).

- SHS involves a temperature change of the storage material.
- LHS requires an isothermal phase change of the storage material (melting, freezing, vaporization, fusion and crystallization).
- Heat of reaction or thermochemical heat (TCH) is the result of a reversible thermochemical reaction by the storage material.

Typical parameters (capacity, power, efficiency, storage period and cost) that can characterize each type are given in Table 11.2 [12]. In general, to be economically competitive in commercial solar systems, it is desirable for TES to use materials that possess a large change in internal energy per unit volume and/or mass to minimize the space needed to store the desired amount of energy. However, the choice of storage medium depends on the nature of the process. As an example, for water heating energy storage as sensible heat of stored water is logical.

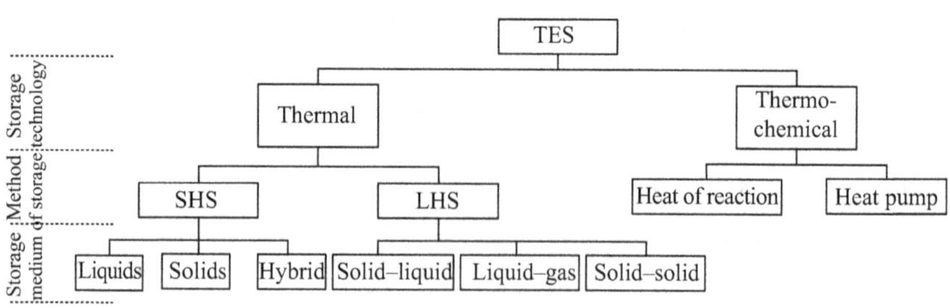

Figure 11.3 Types of TES

Table 11.2 Typical parameters of TES systems

TES system	Capacity (kWh/t)	Power (MW)	Efficiency (%)	Storage period	Cost (€/kWh)
SHS	10–50	0.001–10.0	50–90	Days/months	0.1–10
LHS	50–150	0.001–1.0	75–90	Hours/months	10–50
TCR	120–250	0.01–1.0	75–100	Hours/days	8–100

11.3.2 Sensible heat storage

It is the simplest method of TES as well as it is cheap, and there are no risks associated with the use of toxic materials. It can be hot or cold potential energy with respect to the ambient temperature, stored in a solid, liquid (e.g. water, sand, molten salts, bricks, concrete or rocks), or dual medium that consists of some combination of liquid and solid media. For residential and industrial applications, water is the most popular and commercial heat storage medium.

The amount of stored energy, W_s, in sensible heat depends on mass, specific heat and temperature change of the storage medium during the process of charging/discharging:

$$W_s = \int_{T_1}^{T_2} mC_{\rm p}dt = mC_{\rm p}(T_2 - T_1) \tag{11.14}$$

where W_s represents the heat energy stored in kilojoules, m represents the mass of the material in kilograms, C_p is the specific heat capacity (kJ/kg K) and T_1 and T_2 are the initial and final temperatures of the material, respectively, in kelvin.

SHS materials chosen for TES with sensible heat must be thermally stable and undergo no phase change during temperature extremes. The substance should also have a high heat capacity, high density and an acceptably low vapour pressure. To be economically viable, it must be inexpensive.

11.3.3 Latent heat storage

It is also known as phase-change storage. Latent heat is the amount of energy in the form of heat released or absorbed by a material during a change in physical state or phase transition such as solid to liquid (melting) or liquid to gas (vaporization). The heat is mainly stored in the phase-change process. As heat is added to the material, its temperature increases until a phase change occurs. During the phase change, the solid bonds are broken at a constant temperature as heat is added to the material. This is the fundamental difference between sensible and latent heat. When the phase change is completed (e.g. from solid to liquid), the material again undergoes an internal energy and temperature increase with the application of heat. LHS is attractive for phase-change materials (PCMs) that undergo a significantly high change in internal energy during a phase change. It has the advantages of high-energy storage density and the isothermal nature of the storage process.

The stored capacity, W_s (in joules), with a PCM medium that undergoes a phase change from solid to liquid or liquid to gas or vice versa is given by

$$\begin{aligned} W_{\mathrm{s}} &= \int_{T_{\mathrm{i}}}^{T_{\mathrm{m}}} mC_{\mathrm{ps}}\mathrm{d}t + mf\Delta q + \int_{T_{\mathrm{m}}}^{T_{\mathrm{f}}} mC_{\mathrm{pl}}\mathrm{d}t \\ &= m\left[C_{\mathrm{ps}}(T_{\mathrm{m}} - T_{\mathrm{i}}) + f\Delta q + C_{\mathrm{pl}}(T_{\mathrm{f}} - T_{\mathrm{m}})\right] \end{aligned} \tag{11.15}$$

where T_m is the melting temperature (°C), m is the mass of PCM medium (kg), C_{ps} is the average specific heat of the solid phase between T_i and T_m (J/(kg K)), c_{p1} is the average specific heat of the liquid phase between T_m and T_f (J/(kg K)), f is the melt fraction and Δq is the latent heat of fusion (J/kg). For instance, Glauber's salt has $C_{ps} \approx$ 1,950 J/(kg °C), $c_{p1} \approx 3{,}550$ J/(kg °C) and $\Delta q = 2.43 \times 10^5$ J/kg at 34 °C.

LHS via PCMs. The energy associated with a change of phase for a given material can be used to store energy. The phase change may be one of melting or evaporating, or it may be associated with a structural change, e.g. in lattice form or content of crystal-bound water. When heat is added or removed from a given material, several changes may take place successively, or in some cases simultaneously, involving phase changes as well as energy storage in thermal motion of molecules (i.e. both latent and sensible heat). The total energy change, which can serve as energy storage, is given by the change in enthalpy [11].

The primary advantage of PCMs is their ability to store energy at reduced temperatures and in reduced quantities. Furthermore, their high heat of fusion and other thermal properties allow freezing with little supercooling, no segregation, chemical stability and a sharp melting point. One advantage is that some PCMs have low thermal conductivity in the solid state, making high heat transfer rates necessary during the freezing cycle. They are often flammable, requiring the addition of safety features to the design of storage vessels. In addition, latent heat is much more difficult to transfer than sensible heat [11,13].

11.3.4 TCH energy storage

Thermochemical energy is stored as the bond energy of a chemical compound. During a thermochemical reaction, atomic bonds are broken through a reversible chemical reaction and are catalysed by an increase in temperature, which allows the energy to be stored. After thermochemical separation occurs, the constituents are stored apart until the combination reaction is desired. Recombination of the bonds between atoms releases the stored thermochemical energy.

Advantages of thermochemical storage include high energy density, long-term and low-temperature storage capability. However, the thermochemical process is complex, and the thermochemical materials are often expensive and can be hazardous [13].

Reversible reactions with reactants and products that can be stored easily as liquids and solids are of practical interest for energy storage because of their high density. Reactions that produce two distinct phases such as a solid and a gas are also desirable since the separation of products to prevent back-reaction is facilitated.

The turning temperature, T', is defined as the temperature at which the reactants and products are present in approximately equal quantities. When $T > T'$, the endothermic storage reaction is favoured as it means that heat is necessary for and absorbed during the reaction. Conversely, for $T < T'$, the exothermic reaction dominates, and heat is a product of the reaction.

11.3.5 Storage systems

Two-tank direct storage. In a two-tank direct storage system, the material used to store thermal energy is the same as the heat transfer fluid (HTF) used to collect the thermal energy. The fluid is stored in two tanks: one at a high temperature and the other at a lower temperature as shown in Figure 11.4. The process follows the steps given below:

- fluid from the lower temperature tank flows through a solar collector or receiver;
- where solar energy heats it to the high temperature and it then flows back to the high-temperature tank for storage;
- fluid from the high temperature tank flows through a heat exchanger;
- where it generates steam for electricity production; and
- the fluid exits the heat exchanger at a low temperature and returns to the low-temperature tank.

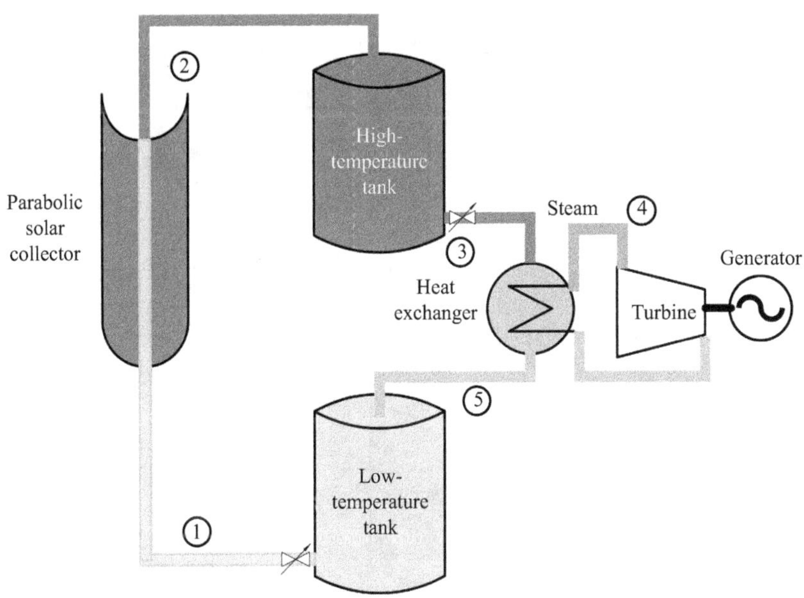

Figure 11.4 Schematic diagram of a two-tank direct storage

It is recommended to use molten salt as an HTF rather than the common fluids, for example, oils. This allows the solar field to operate at higher temperature, which in turn avoids the need for expensive heat exchangers. The use of a cheap heat exchanger and molten salt as an HTF will substantially reduce the total cost of the TES system.

Two-tank indirect storage. The two-tank indirect system operates in the same manner as the two-tank direct system except that different fluids are used for heat transfer and storage. This system is beneficial especially when the HTF is very expensive or not suited for use as the storage fluid. As shown in Figure 11.5, the cycle of the process includes the following: (i) the storage fluid from a low-temperature tank flows through an extra heat exchanger. Thus, it is heated by the high-temperature HTF; (ii) the high-temperature storage fluid then flows back to the high-temperature storage tank; (iii) the fluid exits the heat exchanger at a low temperature and returns to the solar collector or receiver, where it is heated back to the high temperature; and (iv) storage fluid from the high-temperature tank is used to generate steam in the same way as the two-tank direct system. So, the indirect system requires an extra heat exchanger. This adds cost to the system and decreases the overall TES efficiency.

Single-tank thermocline storage. A single-tank thermocline storage system stores thermal energy in a solid medium, usually silica sand, located in a single tank. At any time during operation, a portion of the medium is at high temperature and a portion is at low temperature. The hot and cold temperature regions are

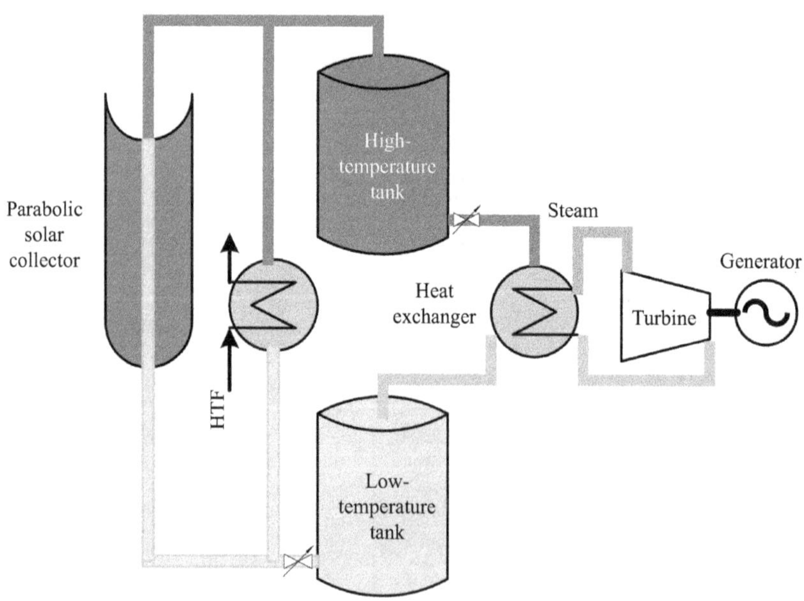

Figure 11.5 Concept of two-tank indirect storage

separated by a temperature gradient or thermocline. High-temperature HTF flows into the top of the thermocline and exits at the bottom at low temperature. This process moves the thermocline downward and adds thermal energy to the system for storage. Reversing the flow moves the thermocline upward and removes thermal energy to generate steam and electricity. Buoyancy effects create thermal stratification of the fluid in the tank, which helps stabilize and maintain the thermocline.

The basic thermocline storage tank concept in which hot and cold materials are stored inside the tank is shown in Figure 11.6 [14]. Thermocline between the two cut-off temperatures, cold and hot, $T_{c,\text{cut-off}}$ and $T_{h,\text{cut-off}}$, respectively, is unusable energy. For $T > T_{h,\text{cut-off}}$, the thermocline cannot be charged because the HTF is very hot to flow into the solar field for heat collection. When $T < T_{h,\text{cut-off}}$, the HTF temperature is not high enough to support thermal power cycle. It is to be noted that the effective capacity of thermocline TES is less than the actual size of the tank.

The thermocline technology has proved advantageous as the reduction of materials used for constructing the tank and storing heat decreases cost and energy input. Using a solid storage medium for only one tank reduces the cost relative to the two-tank systems. The thermocline system can be used at the solar power tower, where steam and mineral oil are used as the HTF and storage fluid, respectively.

Heat pump storage. A true electricity storage system can be obtained via thermal pumping (pumped heat electricity storage). This concept consists of using a heat pump to store energy in thermal form in cheap solid materials with a high expected efficiency. During the load phase, electricity is used to compress the operating fluid and store its heat in a high-temperature container. At discharge, pressurized, high-temperature fluid is used to power a turbo-generator. The efficiency per cycle is estimated at around 70% [12].

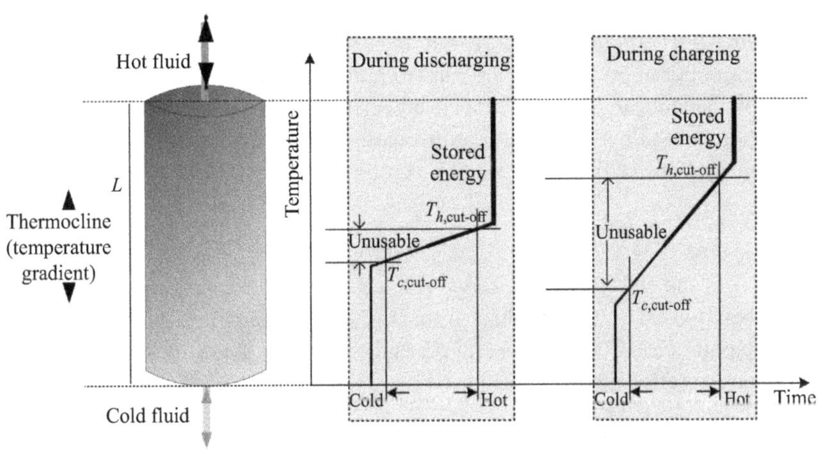

Figure 11.6 Thermocline storage tank and temperature profiles

11.4 TES performance

Three efficiencies can typically be used as a measure of TES performance: storage-tank effectiveness, first-law thermal efficiency and second-law exergy conversion efficiency [14].

Storage-tank effectiveness. It is also called 'discharge efficiency' and defined as a ratio of the amount of heat discharged to the total thermal energy stored when the tank is fully charged; that is, the system capability of providing complete discharge or some residual is necessary. For example, in a two-tank storage system, about 20% of the hot or cold tank storage fluids cannot be used because of the pump need and requirement for submerging the pump in the liquid. In the case of thermocline TES, the tank effectiveness is determined by the usable portion of the stored fluid outside of the mixed-temperature thermocline region. Therefore, the storage-tank effectiveness, η_{eff}, can be written as

$$\eta_{\text{eff}} = \frac{W_{\text{dis}}}{W_{\text{TES}}} \tag{11.16}$$

where W_{dis} is the energy able to be discharged and W_{TES} is the total energy in TES. So, the tank should be oversized to accommodate the unusable residual thermal energy.

First-law efficiency, $\eta_{TES,FL}$: It is defined as the ratio of the energy extracted from the storage to the energy stored in it and can be written in a form as follows:

$$\eta_{TES,FL} = \frac{\textit{energy extracted from the storage}}{\textit{energy stored in the storage}} = \frac{mC_{\text{p}}(T - T_c)}{mC_{\text{p}}(T_h - T_c)} \tag{11.17}$$

where mC_p is the total heat capacity of the storage medium. T and T_c are the hot and cold salt temperature, respectively, of the storage during discharging, T_h is the maximum temperature at the end of charging period.

Second-law efficiency, $\eta_{TES,SL}$: It measures the system ability to generate the same amount of power from TES as from the original energy used to charge the system, W. This means that the second-law efficiency measures the energy quality degradation caused by both charging/discharging temperatures and thermal losses. Thus, exergy transfer through a system can be written as

$$\Delta G = \Delta W - T_a \Delta S \tag{11.18}$$

where ΔG is the storage fluid exergy change, ΔW is the change of the available energy being used (charging/discharging), T_a is the ambient temperature as a reference point (25 °C or 298 K) and ΔS is the entropy change of the storage fluid. The round-trip efficiency of TES based on (11.18) can be written as

$$\eta_{\text{TES,SL}} = \left|\frac{\Delta G_d}{\Delta G_c}\right| = \left|\frac{(\Delta W - T_a \Delta S)_d}{(\Delta W - T_a \Delta S)_c}\right| \tag{11.19}$$

where subscripts *c* and *d* denote TES charging and discharging, respectively. More explanation about losses estimation in TES systems can be found in [14].

References

[1] Nadeem F., Hussain S. M., Tiwari P. K., Goswami A. K. and Ustun T. S. 'Comparative review of energy storage systems, their roles and impacts on future power systems'. *IEEE Access.* 2018;7:4555–85.

[2] Moseley P. T. and Garche J. *Electrochemical energy storage for renewable sources and grid balancing.* Amsterdam: Elsevier Science; 2015. p. 493.

[3] Wei H., Xin W., Jianhua Z. and Jingyan Y. 'Discussion on application of super capacitor energy storage system in microgrids'. *Sustainable Power Generation and Supply, 2009, International Conference on*; Nanjing, China, April 2009, pp. 1–4.

[4] Kim H. J., Seongm K. C., Cho J. W., *et al.* '3 MJ/750 kVA SMES system for improving power quality'. *IEEE Transactions on Applied Superconductivity.* 2006;16(2):574–7.

[5] Nakayama T., Yagai T., Tsuda M. and Hamajima T. 'Micro power grid system with SMES and superconducting cable modules cooled by liquid hydrogen'. *IEEE Transactions on Applied Superconductivity.* 2009;19(3): 2062–5.

[6] Said S. A., Aly M. M. and Abdel-Akher M. 'Application of superconducting magnetic energy storage (SMES) for voltage sag/swell suppression in distribution system with wind power penetration'. *Harmonics and Quality of Power (ICHQP), 2014 IEEE 16th International Conference on*; Bucharest, May 2014, pp. 92–6.

[7] Irie F., Okada H., Ezaki T., *et al.* 'A field experiment on power line stabilization by a SMES system'. *IEEE Transactions on Magnetism.* 1992;28(1): 426–9.

[8] Li Z. and Weizin M. 'Application of SMES system with a comprehensive control for enhancing transient stability'. *Proceedings of the 5th International Conference on Advances in Power System Control, Operation and Management, APSCOM 2000*; Hong Kong, Oct 2000, pp. 225–9.

[9] Huang X., Zhang G. and Xiao L. 'Optimal location of SMES for improving power system voltage stability'. *IEEE Transactions on Applied Superconductivity.* 2010;3:1316–19.

[10] Robyns B., Francois B., Delille G. and Saudemont C. *Energy storage in electric power grids.* London, UK: ISTE Ltd. and John Wiley & Sons; 2015.

[11] Sorensen B. *Renewable energy conversion, transmission and storage.* 3rd edn. Amsterdam: Elsevier, AP; 2007, pp. 234–60.

[12] Sarbu I. and Sebarchievici C. 'A comprehensive review of thermal energy storage'. *MDPI Journals, Sustainability.* 2018;10(1):Article number 191.

[13] Barnes F. S. and Levine J. G. *Large energy storage systems.* Boca Raton, FL: CRC Press; 2011.

[14] Ma Z., Glatzmaierm G. C., Wagner M. and Neises T. 'General performance metrics and applications to evaluate various thermal energy storage technologies'. *Proceedings of ASME 2012 6th International Conference on Energy Sustainability collocated with 9th International Conference on Fuel Cell Science, Engineering and Technology*; San Diego, CA, USA, July 2012, pp. 345–51.

Chapter 12

Energy storage system application: electric vehicles

12.1 Introduction

These days, the trend towards developing electric vehicles technologies, aiming to enlarge the scale of electric vehicles use, is growing rapidly. The catalyst that makes the engineers and researchers motivated to work on this trend is the great concern about the environment, particularly noise and exhaust emissions, in addition to the continuous progress in *batteries and fuel cells* manufacturing and technology. It is, therefore, essential to understand the principles behind the design of electric vehicles as well as the relevant technological and environmental issues.

Of course, internal combustion (IC) engine is the most popular and predominant type used for vehicles at present. However, electric vehicles are looking as a strong contender for transport in the future. IC engine vehicles have greater success as the specific energy of the petroleum fuel is about 300 times its value for lead acid battery (typically ≈9 kWh/kg for petroleum and 0.03 kWh/kg for battery). Even with accounting for the efficiency, the difference is still large. For instance, the efficiency of IC engines, gearbox and transmission is about 20%. It means that a useful energy of 1.8 kWh/kg at gearbox shaft can be obtained. In contrast, efficiency of electric motors is around 90%, which only results in a useful energy of 0.027 kWh/kg. Therefore, to reach the value of useful energy of IC engines using battery, the mass of battery should be about 66.7 times that of IC engine [1].

Moreover, the charging time of lead acid battery, as an example, is considerably longer than it takes to fill a tank with fossil fuel, and its lifetime is limited (~5 years). This requires further investment periodically to replace the batteries. Thus, battery electric vehicles (BEVs) are more expensive compared to corresponding IC engine vehicles.

As their range is limited, the application of BEVs is, to some extent, limited. They are used as mobility devices for the elderly and physically handicapped, city cars, trolleybus or trams and electric bicycles.

Recently, there have been technical developments in electric vehicles design and improvements to motors, controllers and rechargeable batteries with reduced charging time. Another type of electric vehicle is known as a 'hybrid electric vehicle'. It is likely to become the most common type in the near future [2]. It combines an IC engine with a battery, an electric motor and generator. Generally, these two types of electric vehicles (battery electric and hybrid electric) are together referred to as plug-in electric vehicles (PEVs) and denoted as PBEVs and PHEVs, respectively. In addition to PEVs, there are other types with different features, depending on the primary drive, such as photovoltaic electric vehicles (PVEVs) and fuel cell electric vehicles (FCEVs). The photovoltaic (PV) and fuel cells are used rather than rechargeable batteries in these vehicles.

In addition, electric utility systems are proposing to incorporate PEVs into the power system at the distribution level to help manage the load. PEVs can be dealt with as generation source in the parking-discharging state or as loads in charging state.

12.2 PEV technologies

Batteries are the primary power source that supply PEVs with electric energy. They help reduce the need for fossil fuels, but on the other hand, they need to be plugged in to recharge.

A plug-in battery electric vehicle (PBEV) is powered only by an on-board battery that supplies an electric motor with electric energy through a charge controller electronic device. The electric motor provides the mechanical power required for the car wheel through a transmission gear (Figure 12.1). Initially, it is expected for go up to 160 km (100 miles) on a full charge and then the batteries are required to be recharged by plugging into a power source.

A plug-in hybrid electric vehicle (PHEV) is equipped with both an electric motor and an IC engine to power the car. An on-board computer is installed to

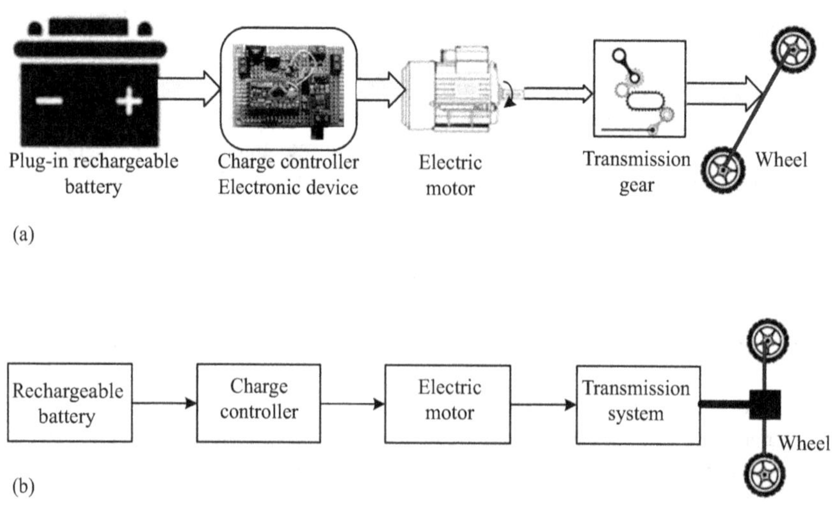

Figure 12.1 PBEV drive train configuration: (a) typical representation and (b) the corresponding block diagram

control the electric and IC engine drives to give the car the power to cope efficiently with the overall distance travelled. The electric motor is supported by an on-board plug-in rechargeable battery. The battery also uses the regenerated energy from the breaking and deceleration to recycle it through the battery to power the car. The car reverts to the standard hybrid operation in case of battery depletion. In PHEVs, a useful variable called the 'degree of hybrid station' (DOH) is used. It is defined as

$$\text{DOH} = \frac{\text{Power of electric motor}}{\text{Power of electric motor} + \text{Power of IC engine}} \quad (12.1)$$

The car uses a smaller IC engine operating close to its optimum efficiency for a greater proportion of the time when the DOH is high.

Three configurations for PHEVs exist: series-hybrid, parallel-hybrid and combined series/parallel hybrid [3].

- Series PHEV: Source of electrical energy is either the battery or a generator driven by an IC engine. The electrical energy is supplied to the electric motor, which is the only power provider to the transmission system (Figure 12.2(a) and (b)). The main disadvantage of this configuration is that all the electrical energy must pass through both the generator and the motor, resulting in a considerable increase in cost.
- Parallel PHEV: Both the electric motor and the IC engine provide power in parallel to the transmission system (Figure 12.3(a) and (b)).
- Combined series/parallel PHEV: This type allows running the car using the electric motor only or both the IC engine and electric motor together depending on the driving conditions that help decide which one of the two options is in an optimal way. It combines the features of series and parallel hybrid. A power split is used to share the output mechanical power from the IC engine to the drive shaft or to the charge controller, as shown by the block diagram in Figure 12.4.

In all configurations, the electric motor works as a generator when used as regenerative brake.

12.2.1 Operation modes

The energy mode to drive PEVs can be determined by what is called state-of-charge index, which is the percentage of the remaining energy capacity in the battery. PEVs have the capability of driving in basically two different modes. First, charge depleting operation mode, which is suitable for urban driving. The vehicle, like all PBEVs, is powered only by the energy stored in the battery. Second, charge sustaining mode, where the battery is recharged over a driving profile to maintain the state-of-charge level. PEVs use the battery pack to a large extent, but they need the IC engine to supplement the power and getting better performance as in PHEVs.

Therefore, the selection criterion of the operation mode is mainly based on (i) the availability of charging facilities from the grid and (ii) the driving distance.

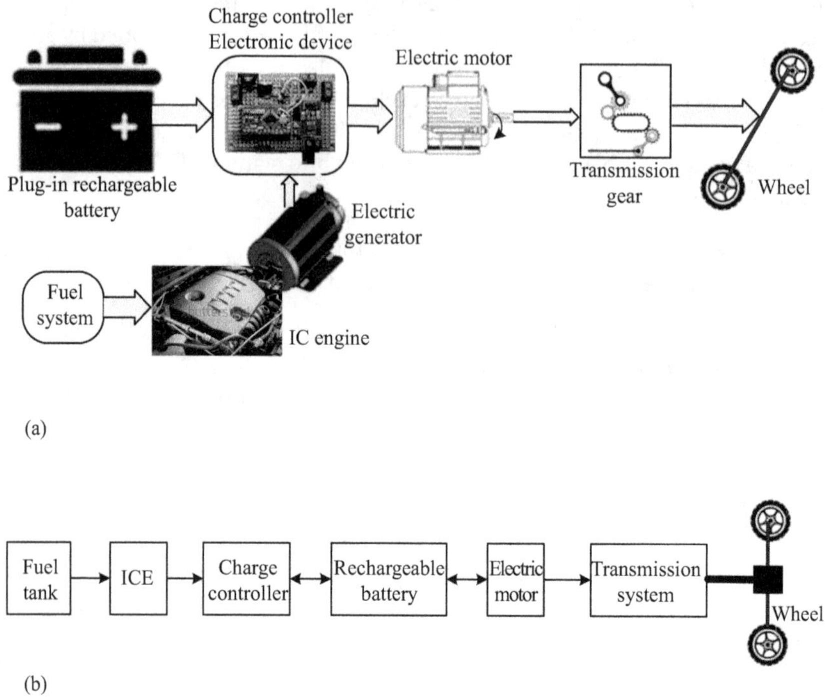

Figure 12.2 Series PHEV drive train configuration: (a) typical representation and (b) the corresponding block diagram

12.2.2 PEV design features

As shown in Figures 12.1–12.3, all types of PEVs use electrical energy as a primary source. They differ from IC engine vehicles, where the fossil fuel is burned by the IC engine to power the vehicle. The most important design features in a PEV drive train are the following.

Large battery packs to store the energy required to power the PEV. Different types of battery packs can be used such as lead acid, NiMH and Li-ion, which is becoming the most dominant pack in PEVs.

Electric motors to convert the electrical energy stored in battery packs to mechanical energy that propels the vehicle forward or backward. The motors may be AC or DC motors.

Electric generators are used to convert the mechanical energy into electrical energy. For instance, the generator converts its input mechanical power coming from the IC engine to the electrical energy being supplied to the electric motor as in the series PHEV.

Inverters: In case of using AC motor, the motor is coupled with the battery pack using an inverter, which converts the battery DC current to AC current.

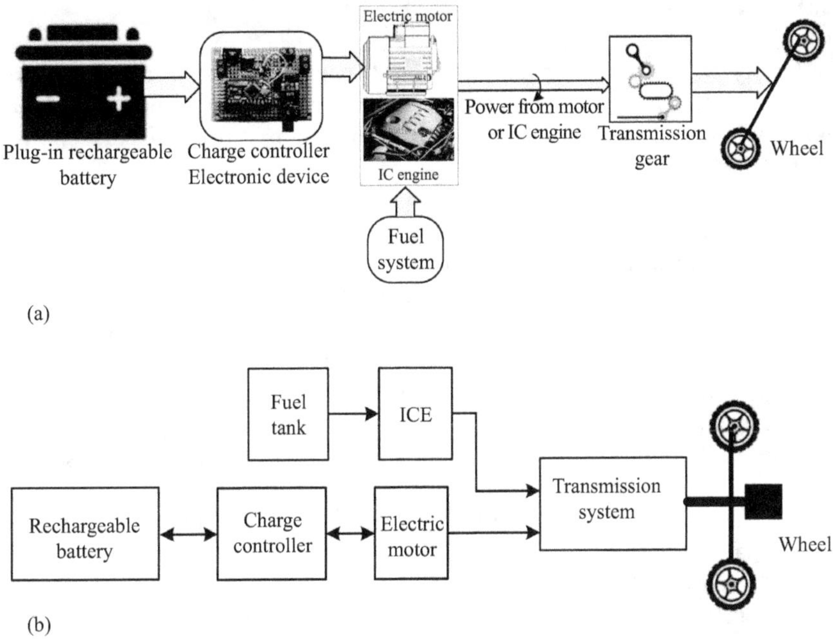

Figure 12.3 Parallel PHEV drive train configuration: (a) typical representation and (b) the corresponding block diagram

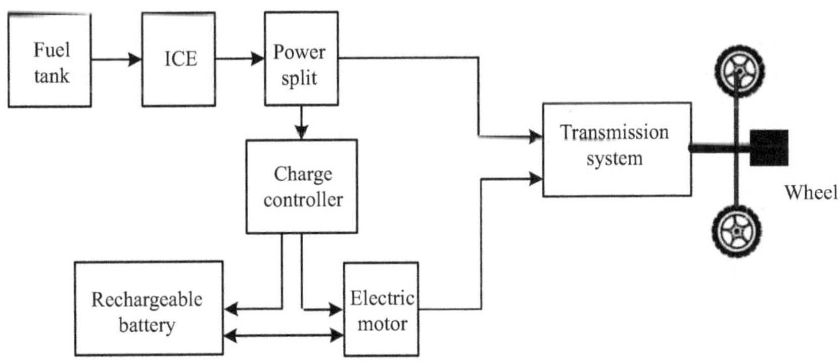

Figure 12.4 Block diagram of combined series/parallel PHEV configuration

Chargers: A charger converts the AC supply into DC and charges the battery pack. The AC supply may be the output of the generator or the utility grid. The chargers are electronic devices, usually, fitted with a controller to optimize the charging process.

12.3 Batteries

The battery is the main source of DC electric supplied to the electric motor in an electric vehicle. The battery performance and cost are of crucial importance in the development of BEVs. The essential performance parameters include power, energy, lifetime, safety and cost.

Power: The desired peak power of battery generally remains constant and depends on the operation mode, mainly the range on charge depleting mode as well as energy management and the total weight of the vehicle.

Energy: An increase in the energy storage capacity of the battery results in a longer distance for electric vehicle to travel, and meanwhile the mass of battery pack may be getting larger. So, the decision of making the energy storage capacity higher depends on the design features including specific energy (energy per kilogram mass) and energy density (energy per volume) of the battery.

Lifetime: The lifetime of the battery is influenced by some factors such as necessity of fast recharge, availability of recycling, desired vehicle mass, compliance with safety standards, withstanding degradation over time, operation mode and thermal management requirements.

Cost of batteries: There are several types of batteries. The cost, generally, differs from one type to another. It includes the cost of: material, manufacturing, transportation, marketing and development. However, high mass production of a specific type of battery can reduce the cost. More details about batteries as energy storage systems are discussed in Chapter 10.

12.4 Electric motors

The electric motor is one of the main components in the drive train of an electric vehicle. It receives either AC or DC electric energy and converts it into mechanical energy to be delivered to the transmission gear of the electric vehicle drive train.

AC induction motors are commonly used in case of supplying the motor with AC electric energy. The battery produces DC energy, and thus an inverter is used for DC/AC conversion to provide the input to the induction motor. More detail about induction machines is provided in Chapter 2, Section 2.4.

DC electric motors, as an alternative to AC motors, are used as well. The battery DC output energy is directly supplied to the DC motor through a controller. Different types of DC electric motors can be used in electric vehicle application. The simplest and classical type that can run with any electronic control is the 'brushed' DC motor. Other advanced types such as brushless DC motors and switched reluctance (SR) motors with sophisticated control are also used in traction applications.

12.4.1 Brushed DC motors

The brushed DC motor is of vital importance and very widely used in several applications including domestic appliances, automated windows and doors, and still used as traction motor.

Operating principles: The motor is equipped with permanent magnets and brushes. North (N) and south (S) poles are arranged alternatively around the circumference of the stator and are fully employed (Figure 12.5). As a result of this arrangement, it, of course, becomes necessary to employ a commutator to get the possibility to group a desired number of conductors in series. The current passing through the coils near the magnet causes a force to be generated in the coil.

To understand the principle of operation, it is necessary to first investigate the constructional feature of the DC motor. As a very basic construction, one conductor of a smoothed-cored armature is assumed, i.e. an armature in which the conductor is not housed in slots, and in which it is, in consequence, acted on directly by the air gap flux. The current carrying armature is connected to the supply end through commutator segments and placed within N and S poles of a permanent or an electromagnet (Figure 12.6(a)). Applying Fleming's left-hand rule, the current carrying conductor is placed in a magnetic field, B, which is perpendicular to the direction of the current, I, and then the conductor experiences a force, F, in the direction perpendicular to both the direction of field and current in the conductor. It is to be noted that, in practice, several conductors in series constituting a coil lie in slots as shown in Figure 12.6(b) to increase the torque. The main flux is shown concentrated into tufts that pass into the tops of the teeth, whereas the armature flux is shown by the dotted fluxes embracing the slots. Supposing that the excitation of the field and the direction of the armature current are as depicted in this figure, the magnetic force on the wires will be set up in an anticlockwise direction and under its influence the armature will rotate. Also, more pair poles may be used to increase the turning force [1,4,5].

The magnitude of the force, F, can be determined by considering the diagram in Figure 12.7. It is known that when an infinitely small charge dq is made to flow at a velocity, v, under the influence of an electric field, E, and a magnetic field, B, then the Lorentz force dF experienced by the charge is given by

$$dF = dq(E + vB) \tag{12.2}$$

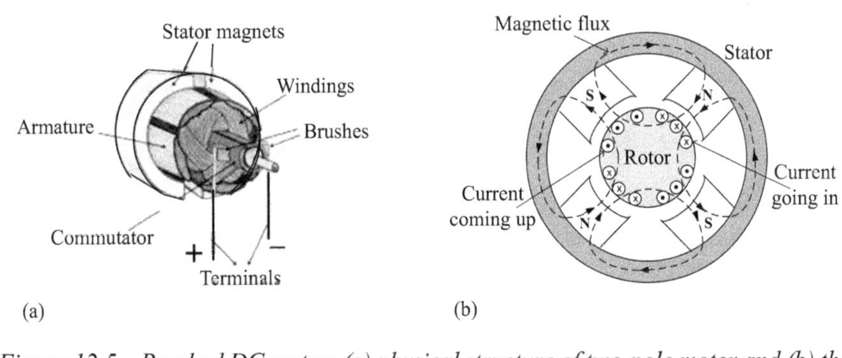

Figure 12.5 Brushed DC motor: (a) physical structure of two-pole motor and (b) the cross section of four-pole motor. The motor torque is anticlockwise

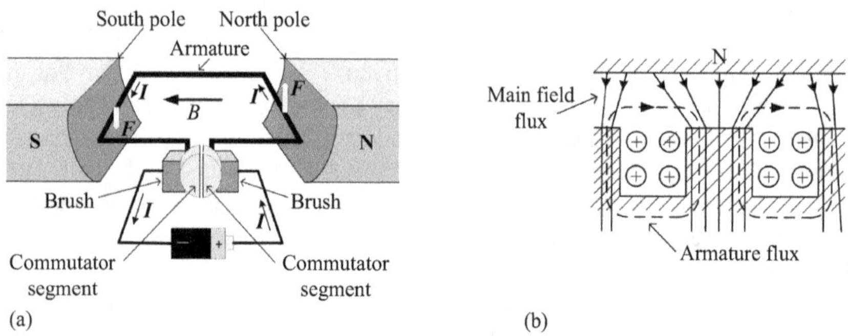

Figure 12.6 (a) Basic construction of a DC motor. (b) Mechanism of torque production

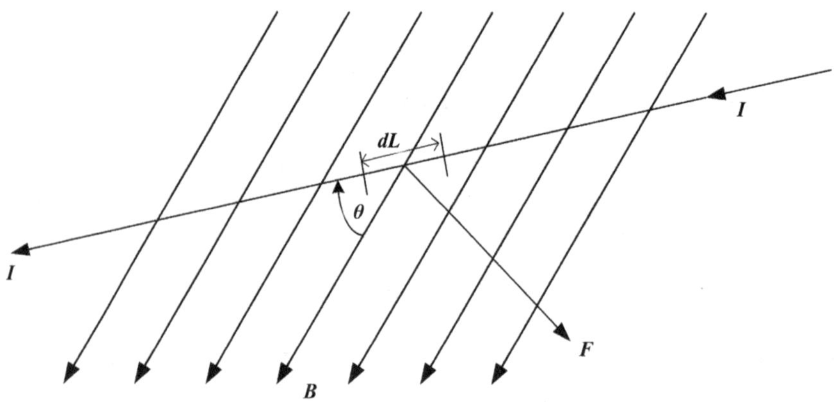

Figure 12.7 Determination of force magnitude

Considering $E = 0$ for the operation of DC motor and substituting $v = dL/dt$, (12.2) becomes

$$dF = dq\frac{dL}{dt} \times B = IdL \times B \tag{12.3}$$

where dL is the length of the conductor (m) carrying charge q and $I = dq/dt$ (A) and B is the magnetic field strength (Wb/m^2).

Integrating both sides of (12.3) and considering the magnetic field component that is perpendicular to the direction of the current, $B \sin\theta$, obtains

$$F = BIL \sin\theta \tag{12.4}$$

It is seen from Figure 12.5(a) that the construction of a DC motor is such that the direction of the current through the armature conductor at all instances is perpendicular to the field (i.e. $\theta = 90°$). Consequently, the force acts on the armature in the direction perpendicular to both the uniform magnetic field and current. Therefore, if the current passing through the right-hand side of the armature conductor is I, the current at its left-hand side will be $-I$ because they are flowing in the opposite direction with respect to each other. Then, the force on the right-hand side and that on the left-hand side are equal in magnitude and in the upward and downward directions, respectively, and separated by armature turn width, w (m). This produces a torque resulting in a rotation of the coil in anticlockwise direction. As given by (12.4), the magnitude of the force on each side of the armature conductor can be expressed as

$$F = BIL \sin 90° = BIL \tag{12.5}$$

Due to the rotation of the armature, the torque expression should be examined when the armature turn is at an angle α with its initial position, that is, an angle between the plane of armature turn and the plane of reference that is taken along the magnetic field direction (Figure 12.8). The torque produced, T, is the multiplication of the force tangential to the direction of armature rotation and the distance w. Hence,

$$T = F \cos \alpha \times w = BILw \cos \alpha \tag{12.6}$$

The torque is not the same at all positions as it varies with the variation of the angle α. The arrangement shown in Figure 12.6(a), where the plane of armature

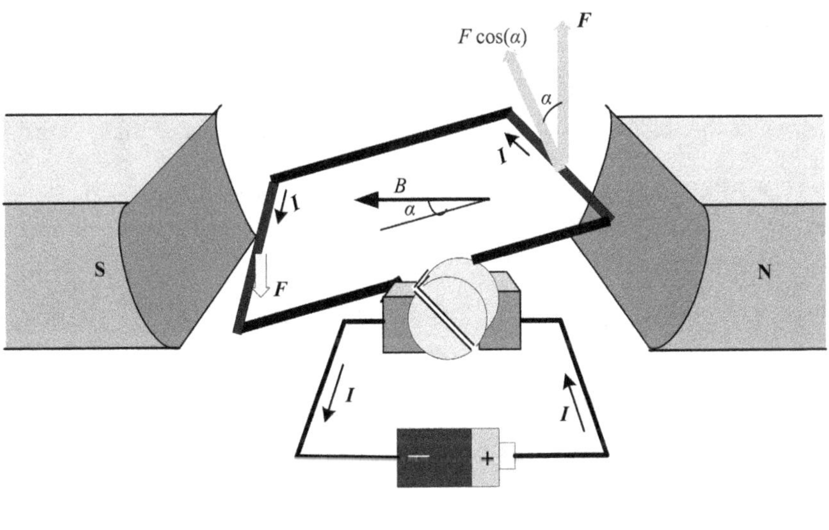

Figure 12.8 Rotation of armature turn with respect to the initial position

turn coincides with the direction of magnetic field B, can be initially considered as a reference position, that is, $\alpha = 0$, and then, the torque is at its maximum value ($T = BIL\omega$) that helps in overcoming the initial momentum of the armature and sets it into rotation. Because of the armature motion, the angle α is increasing in the path of its rotation until it reaches 90° from the initial position. During this period, the torque is decreased and given by (12.6). The instant at which $\alpha = 90°$, the torque is equal to zero, but the armature does not come to a standstill because of its momentum of motion which is designed in such a way that the armature crosses over this null torque point. After that, the angle α decreases and the torque is acting again on the armature.

Supposing that the armature consists of n turns, and φ represents the total flux passing through the coil, i.e. $\varphi = B \times \text{area} = BLw$, (12.6) can be rewritten as

$$T = n\varphi I \cos\alpha \tag{12.7}$$

It is to be noted that the torque does not only depend on the number of turns and their relative position with respect to the magnetic field but also depends on the number of pole pairs and other motor design aspects. So, the average torque can be expressed in terms of the current and magnetic flux by the relation

$$T_{\text{av}} = K_m \varphi I \tag{12.8}$$

where K_m is known as the motor constant and depends on the number of turns in each coil, number of pole pairs and other aspects of motor design.

It is seen from (12.8) that the torque is directly proportional to the armature current I that depends on the source voltage V_s and armature coil resistance R_a. With turning the motor at velocity v, the armature moves in a magnetic field and works as a generator producing a voltage called back electromotive force (BEMF), E_b, which by considering the two sides of the coil is given by

$$E_b = nBLw\omega = K_m \varphi \omega \tag{12.9}$$

where ω is the angular velocity ($v = \omega w/2$) and K_m is the same constant as that in (12.8).

The BEMF acts to reduce the current in the motor as it opposes the source voltage V_s. So, the armature current is produced due to the voltage difference '$V_s - E_b$' and can be obtained by

$$I = \frac{V_s - E_b}{R_a} = \frac{V_s}{R_a} - \frac{K_m \varphi}{R_a} \omega$$

Substituting into (12.8) gets the average torque as

$$T = \frac{K_m \varphi V_s}{R_a} - \frac{(K_m \varphi)^2}{R_a} \omega = \frac{1}{R_a} \left[\mathcal{K} V_s - \mathcal{K}^2 \omega \right] \tag{12.10}$$

where $\mathcal{K} = K_m \varphi$.

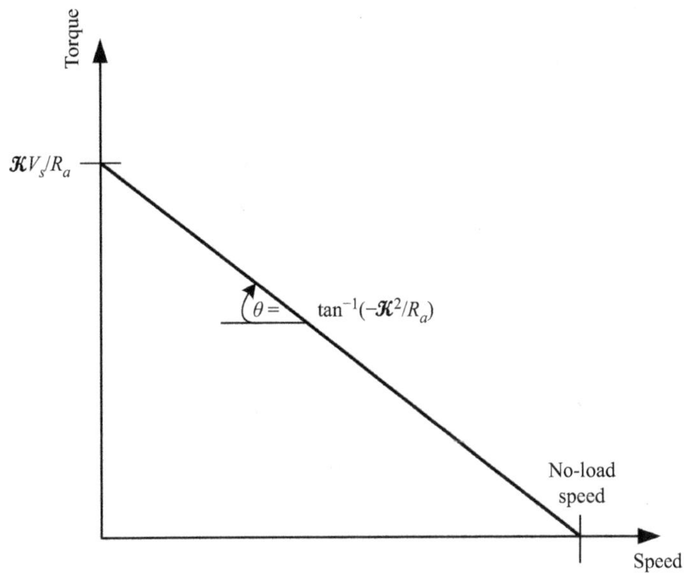

Figure 12.9 Brushed DC motor torque–speed relation

It is noted that in (12.10) the torque needed to overcome the losses due to friction in bearings and at the commutator and windage losses is ignored. However, for this type of constant flux DC motor, the torque–speed relation is a linear relation and the torque decreases as the speed increases with a maximum at zero speed (Figure 12.9).

However, it is seen that the torque is dependent on both the supply voltage, V_s, and the magnetic flux, φ. Two methods can be applied to provide the brushed DC motor with magnetic flux, first, using permanent magnets as shown in Figures 12.5 and 12.6, and second, using electromagnets. For electromagnets, field windings (wdgs) are placed in the stator, and by passing a current through these windings, the magnetic field is created. So, electromagnets give the facility to easily change the flux by varying the current and controlling the torque with great flexibility. Meanwhile, the field windings consume electric current causing additional copper losses. However, brushed DC electric motors with electromagnets are still commonly used as traction motors in electric vehicles application, whereas for smaller motors permanent magnets may be used.

Field windings can be connected either in series, parallel (shunt) or separate with motor armature windings to provide the required magnetic field (Figure 12.10). The separate excited brushed DC motor is more advantageous than the other two types when it is used in this application of electric vehicles. It allows to independently control both the source voltage and the magnetic flux that can be carried out by controlling the voltage across the field windings. So, obtaining the required torque at any speed is ultimately facilitated.

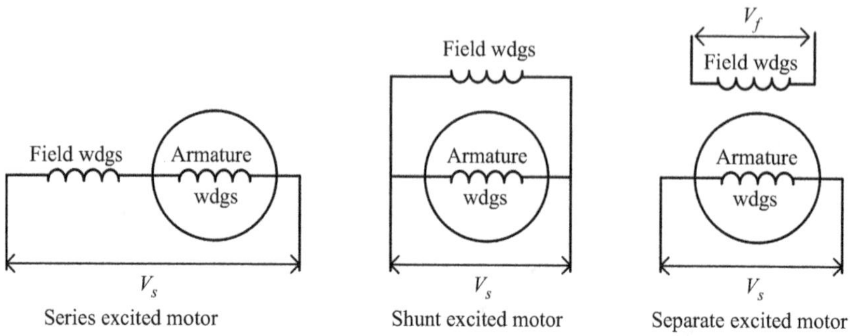

Figure 12.10 Types of excitation for DC motors

In addition, one of the main characteristics that should be determined is the motor losses which in turn affect the efficiency of the motor. Brushed DC motors have the same sources of losses as those of other electric motors. These losses are mainly copper losses, iron losses (hysteresis and eddy currents) and mechanical losses (friction and windage) as well as constant losses representing, for example, the power needed for the electronic control circuits.

Copper losses, P_{cuLoss}, result from the electrical resistance of the windings, R, when the current, I, passes through them and causes heating. Copper losses can be calculated by

$$P_{\text{cuLoss}} = I^2 R \tag{12.11}$$

Incorporating (12.8) and (12.11) gives

$$P_{\text{cuLoss}} = K_{\text{cuLoss}} T^2 \tag{12.12}$$

where K_{cuLoss} has a value depending on R and the magnetic flux φ.

Iron losses, P_{irnLoss}, are produced due to exposing the rotor to changeable magnetic field as the rotor rotates passing, frequently, N pole and then S pole. So, the magnetic field effect on the rotor iron resulting in iron losses which are the sum of what is called 'hysteresis' loss and 'eddy currents' loss. Hysteresis loss is the energy required to magnetize and demagnetize the iron, whereas eddy currents are generated in the iron by rotating the rotor under changeable magnetic field and result in iron heating.

Therefore, iron losses are proportional to the frequency with which the magnetic field changes, and consequently they are proportional to the angular rotor speed, ω. Thus,

$$P_{\text{irnLoss}} = K_{\text{irnLoss}} \omega \tag{12.13}$$

where K_{irnLoss} is a value affected by the magnetic field strength and depends on the method of providing the magnetic field (with permanent magnets, it is nearly a constant value).

Mechanical losses, P_{mecLoss}, consist of (i) friction loss that results from the presence of friction torque, T_{fr}, in bearings and brushes of the motor, it equals to $T_{\mathrm{fr}}\,\omega$, and (ii) windage loss, which is caused by rotor wind resistance (proportional to ω^2), and so the power loss is proportional to the cube of angular speed; that is, it can be expressed as $K_{\mathrm{wLoss}}\omega^3$, where K_{wLoss} is a constant depending on the size and structural design of the motor. Thus,

$$P_{\mathrm{mecLoss}} = T_{\mathrm{fr}}\omega + K_{\mathrm{wLoss}}\omega^3 \tag{12.14}$$

Constant losses, C_{Loss}, are the power needed, for example, to supply the electronic control circuits. These losses independently vary on both speed and torque; that is, they occur even if the motor is stationary.

Therefore, the sum of losses from (12.12) to (12.14) in addition to the constant losses gives the total losses, P_{tLoss} as

$$\begin{aligned} P_{\mathrm{tLoss}} &= P_{\mathrm{cuLoss}} + P_{\mathrm{irnLoss}} + P_{\mathrm{mecLoss}} + C_{\mathrm{Loss}} \\ &= \left(K_{\mathrm{cuLoss}}T^2\right) + \left(K_{\mathrm{irnLoss}}\omega\right) + \left(T_{\mathrm{fr}}\omega + K_{\mathrm{wLoss}}\omega^3\right) + C_{\mathrm{Loss}} \end{aligned} \tag{12.15}$$

By calculating the total losses as in (12.15) and considering the output power, which is the torque T multiplied by the angular speed ω (i.e. $T\omega$), it is easy to determine the motor efficiency, η, by

$$\eta = \frac{T\omega}{T\omega + P_{\mathrm{tLoss}}} \tag{12.16}$$

12.4.2 Dynamic and regenerative braking in DC motors

Dynamic braking means slowing down the motor using a shunt resistance by which the negative torque produced when disconnecting the energy source can be controlled. When a battery of negligible internal resistance is supplying its energy at voltage V_s to a DC motor, the motor starts rotation until reaching its steady state at a speed ω providing a torque T given by (12.10) (Figure 12.11(a)). If the battery is disconnected, the motor continues to rotate, because of its momentum, resulting in a voltage to be generated with a value given by (12.9). Applying this voltage to a resistance R_L, as shown in Figure 12.11(b), causes a current I to flow out of the motor and can be obtained by the following [1]:

$$I = \frac{K\omega}{R_a + R_L} \tag{12.17}$$

The current I causes a negative torque that will slow down the motor. From (12.10) and substituting $V_s = 0$, as there is no battery, the value of the negative torque is

$$T = -\frac{K^2\omega}{R_a + R_L} \tag{12.18}$$

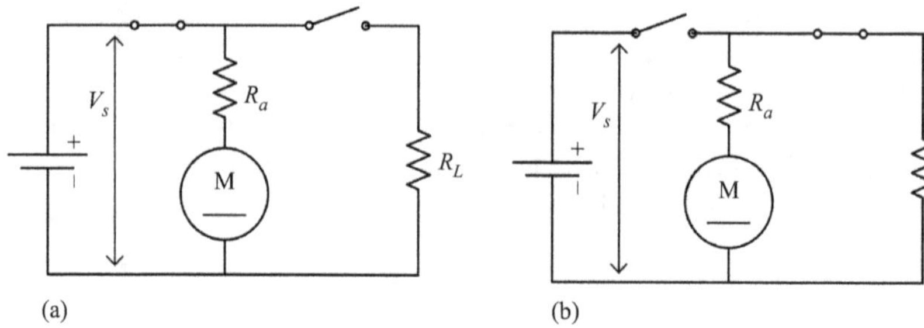

Figure 12.11 Dynamic braking of a DC motor. (a) DC motor at normal operation. (b) DC motor at braking condition

From (12.18), at constant R_L the negative torque falls rapidly as the speed might decrease exponentially to zero. Also, the resistance R_L can be changed aiming to control the negative torque. It is noted that the kinetic energy of the motor is totally converted into heat the same as normal friction brakes. So, it will be worthy to exploit this energy in a useful manner such as storing the electrical energy produced by the motor in a capacitor or battery.

Regenerative braking: A capacitor or a battery is used rather than a resistor to store the braking energy generated by the motor. The problem in this case is that connecting the battery with the motor will cause a current, I_{bat}, depending on the difference between the motor voltage, V_m, and battery voltage, V_{bat}, and is given by

$$I_{bat} = \frac{V_m - V_{bat}}{R_a} = \frac{K\omega - V_{bat}}{R_a} \tag{12.19}$$

By examining (12.19), it is noted the following:

- V_m depends on a varying flux, φ, and varying motor speed, ω, and consequently the braking torque which is proportional to the current I_{bat} will be varying.
- The motor is slowing down (ω is decreased) until reaching an instant at which $V_m = V_{bat}$ resulting in a torque decline and no more braking effect unless the battery voltage is very low.
- For low battery voltage, use of stored energy is difficult.

So, a voltage converter (DC/DC) can be used to solve this problem (Figure 12.12). The input to the converter is the braking power, P_{brak}, delivered by the motor. It equals $V_m I_m$, where V_m and I_m will change with speed and the desired braking torque, respectively. The converter produces an output, P_{conv}, of the same value of the input power, assuming its 100% efficiency, but at an increased voltage and reduced current to be matched with the storing battery (or capacitor). In practice, the efficiency is not 100% as some of the braking electrical power will be lost.

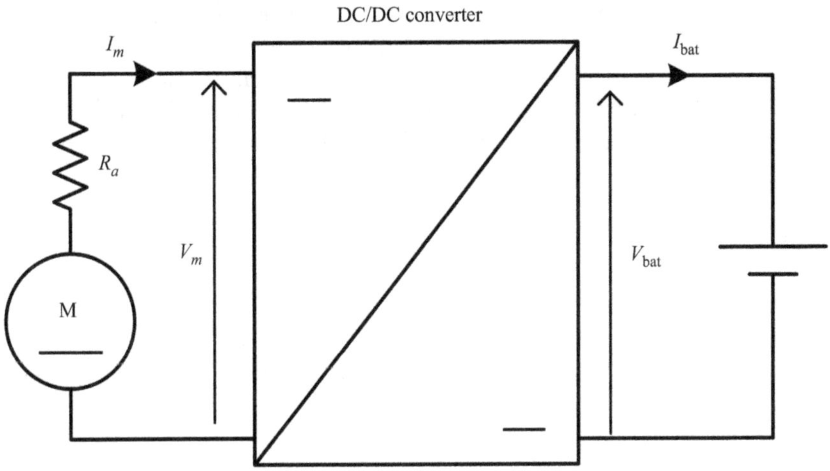

Figure 12.12 DC motor regenerative braking

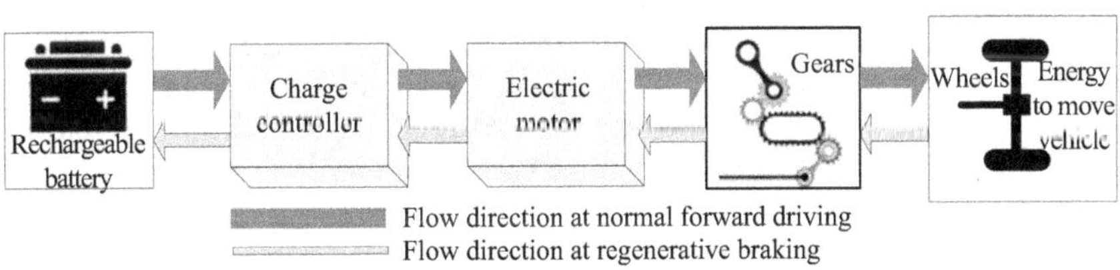

Figure 12.13 Energy flows in a PBEV at normal and braking conditions

However, the function of the converter can be imagined as a transformer transmitting a power at different turns ratios and a relation in terms of converter efficiency, η_{conv}, can be written as

$$P_{\text{conv}} = \eta_{\text{conv}} P_{\text{brak}}$$

Hence,

$$V_{\text{bat}} I_{\text{bat}} = \eta_{\text{conv}} V_m I_m \tag{12.20}$$

The battery might be the same that provides the electrical energy to the motor. So, the energy flows, for example, in a classical PBEV, are shown in Figure 12.13 at the normal forward driving and regenerative braking.

The power produced by the brushed DC motor can be increased by increasing the supply voltage. This increase is limited not only by the insulation of copper wires but also by the heat generated because of the losses above certain power and occurs at the middle of the rotor. So, the heat generated is much more difficult to remove, and then the heat removal is the main limiting factor of power increase. For this reason, in addition to the problem of gradual wearing of both brushes and commutator due to the friction between them, other types of DC motors may be used. Brushless DC and SR motors are commonly used in electric vehicle application where the great majority of the heat generated by losses is in the outer stator, which in turn leads to the heat being removed easily, and thus allows smaller motors to be used.

12.4.3 Brushless DC motors

Brushless DC motors (BLDC motors) are AC motors as the current passing through the motor alternates. The name BLDC comes from the facts that (i) the AC current must be variable frequency and so derived from a DC supply and (ii) the torque–speed characteristic is very similar to that of a brushed DC motor [6–9].

12.4.3.1 Operational theory

The basic principle of BLDC motor operation can be explained by considering that the rotor consists of a permanent magnet whereas the stator has a coil wound either clockwise or anticlockwise along each stator arm. Switches are used to direct the current from a DC source through the coil. In Figure 12.14(a), the current is flowing in the direction of magnetizing the stator to turn the rotor clockwise. The rotor passes between the stator poles until being alignment with stator magnetic axis and the stator current is switched off (Figure 12.14(b)). Momentum carries the rotor on and the stator coil is re-energized, but the current and its resulting magnetic field are reversed as shown in Figure 12.14(c). Thus, the rotor is turning clockwise, and the process continues with switching the current ON–OFF–ON (reversed). So, this switching must be synchronized with the position of the rotor. To do that, the BLDC motor accomplishes commutation electronically using rotor position feedback to determine when to switch the current. Feedback usually entails an attached Hall sensor or a rotary encoder.

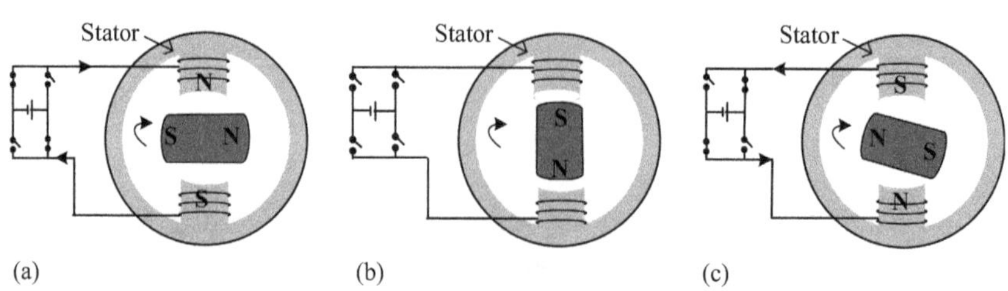

Figure 12.14 Basic principles of BLDC motor operation. The current is switched (a) ON, (b) OFF and (c) ON reversed

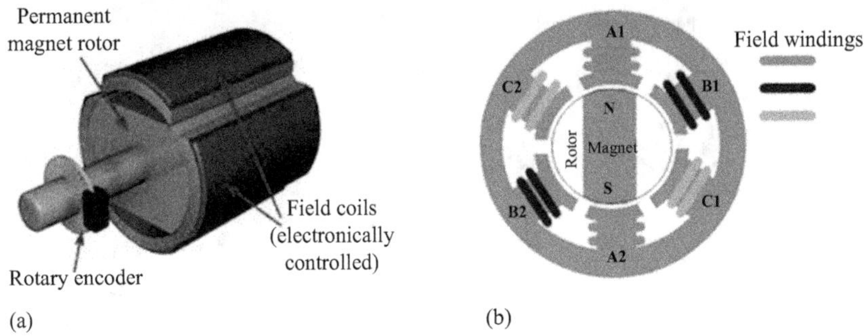

Figure 12.15 Typical examples of BLDC motors; (a) single-phase motor and (b) cross section of three-phase motor

In practice, the rotor may consist of a shaft and a hub with permanent magnets arranged to form between two to eight pole pairs that alternate between N and S poles. The stator has single-phase, two-phase or three-phase windings. Single- and three-phase windings are the most widely used. A BLDC motor with single-phase stator winding forming two pole pairs with electronically controlled field coils is typically shown in Figure 12.15(a). A simplified cross section of three-phase stator windings forming three pole pairs is depicted in Figure 12.15(b). A rotating magnetic field is created by the sequential excitation of the pole pairs by a DC pulse.

Two types of stator windings may be used, trapezoidal and sinusoidal, which refers to the shape of BEMF signal. The shape of this signal is determined by the coil interconnections and the air gap distance. The phase current also follows a trapezoidal and sinusoidal shape.

Motor operation is based on the attraction or repulsion between magnetic poles. Using the three-phase motor shown in Figure 12.16, the process starts when current flows through one of the three stator windings and generates a magnetic pole that attracts the closest permanent magnet of the opposite pole. The rotor will move if the current shifts to an adjacent winding. Sequentially charging each winding will cause the rotor to follow in a rotating field. The torque in this example depends on the current amplitude and the number of turns on the stator windings, the strength and the size of the permanent magnets, the air gap between the rotor and the windings, and the length of the rotating arm.

Therefore, the stator windings work in conjunction with permanent magnets on the rotor to generate a nearly uniform flux density in the air gap. This permits the stator coils to be driven by a constant DC voltage (hence the name brushless DC), which simply switches from one stator coil to the next to generate an AC voltage waveform with a trapezoidal or sinusoidal shape.

12.4.3.2 Switching process

Electric switches are used in BLDC motors to realize current commutation, and thus continuously rotate the motor. These switches are usually connected in an H-bridge structure for a single-phase BLDC motor, and a three-phase bridge

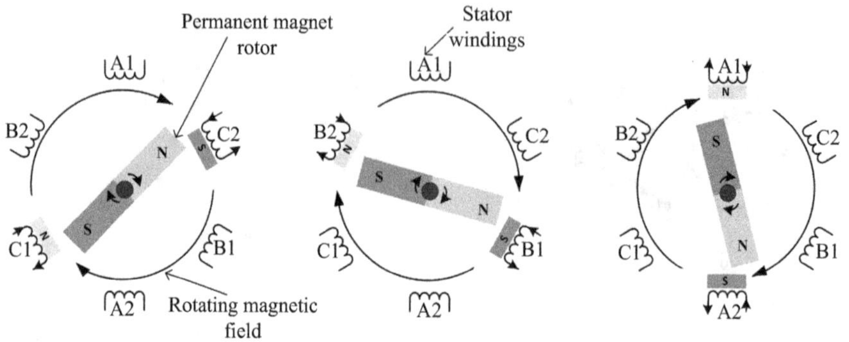

Figure 12.16 Basic principle of three-phase BLDC motor operation

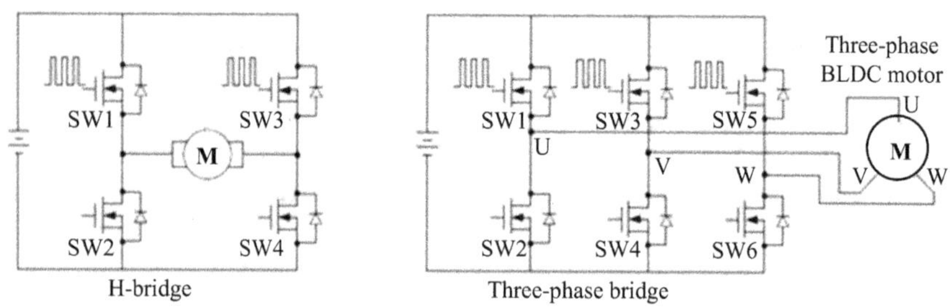

Figure 12.17 Electric driver circuit

structure for a three-phase BLDC motor as shown in Figure 12.17. High-side switches usually are controlled using pulse-width modulation (PWM), which converts a DC voltage into a modulated voltage to easily and efficiently limit the start-up current and control speed and torque. Generally, PWM losses are increased when raising the switching frequency, and on the other hand, lowering the switching frequency limits the system's bandwidth and can raise the ripple current pulses to the point where they become destructive or shut down the BLDC motor driver.

12.4.3.3 Equivalent circuit and general equations

The equivalent circuit in the dynamic state is shown in Figure 12.18(a) as a per-phase diagram. λ_m is the flux linkage of stator winding per phase due to the permanent magnet. For steady-state conditions, both the voltage, v, and electromotive force, e, are assumed sinusoidal at frequency, ω. The equivalent circuit becomes as in Figure 12.18(b), where $X = \omega L$, and V, I, E and λ_m are phasors with rms amplitudes. The steady-state circuit equation can be written as

$$V = E + (R + j\omega L)I \tag{12.21}$$

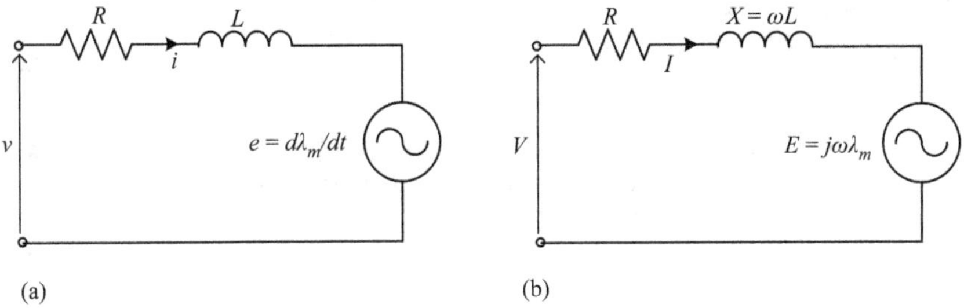

Figure 12.18 Equivalent circuits of BLDC motors: (a) dynamic equivalent circuit and (b) steady-state equivalent circuit

For maximum mechanical power at a given speed, I and E are in phase. This also gives maximum torque/ampere (minimum current/N m). A brushless DC motor has position feedback from the rotor via Hall devices, optical devices, encoder, etc., to keep an angle between V and E, as E is in phase with rotor position and V is determined by the inverter supply to the motor. If $\omega L \ll R$, when I is in phase with E, V will also be in phase with E. Thus, the circuit can be analysed using magnitudes of E, V and I as if it were a DC circuit.

It is noted that when E and I are in phase, the motor mechanical power output (before friction, windage and iron losses), i.e. the electromagnetic output power is

$$P_{\text{em}} = m|E||I| = m\omega|\lambda_m||I| \tag{12.22}$$

where m is the number of phases, $|E|$, $|I|$ and $|\lambda_m|$ are the amplitudes of phasor E, I and λ_m, and the electromagnetic torque is

$$T_{\text{em}} = \frac{P_{\text{em}}}{\omega_r} = \frac{m\omega|\lambda_m||I|}{\omega_r} \tag{12.23}$$

where $\omega_r = 2\ \omega/p$ is the rotor speed in rad/s, and p the number of poles.

Hence,

$$T_{\text{em}} = \frac{mp}{2}|\lambda_m||I| \tag{12.24}$$

The load torque is the actual shaft output torque derived by subtracting the losses in the torque resulting from friction, windage and iron losses. That is,

$$T_{\text{load}} = T_{\text{em}} - T_{\text{losses}} \tag{12.25}$$

Dropping the modulus signs in (12.24) gives

$$T_{\text{em}} = \frac{mp}{2}\lambda_m I \tag{12.26}$$

The voltage, E, in terms of rotor speed can be written as

$$E = \frac{p}{2}\omega_r\lambda_m \tag{12.27}$$

Assuming V and E are kept in phase by the position feedback and $\omega L \ll R$, (12.21) can be simplified and rewritten in an algebraic form as

$$V = E + RI \tag{12.28}$$

Incorporating (12.26), (12.27) and (12.28) gives

$$V = \frac{p}{2}\omega_r\lambda_m + \frac{2R}{mp\lambda_m}T_{\mathrm{em}}$$

Thus,

$$T_{\mathrm{em}} = \frac{m(p\lambda_m/2)}{R}V - \frac{m(p\lambda_m/2)^2}{R}\omega_r = \frac{m}{R}\left(\mathbb{K}V - \mathbb{K}^2\omega_r\right) \tag{12.29}$$

where $\mathbb{K} = p\lambda_m/2$.

It is seen from (12.29) that the torque speed curve is very similar to that of brushed DC motors given in (12.10).

The efficiency is calculated as the ratio of output power, P_{out}, and input power, P_{in}. Hence,

$$\eta = \frac{P_{\mathrm{out}}}{P_{\mathrm{in}}} \tag{12.30}$$

where $P_{\mathrm{in}} = mVI$ and $P_{\mathrm{out}} = T_{\mathrm{load}}\omega_r$.

The input power can be obtained by adding the power losses to the output power. The loss calculations are the same as those of brushed DC motors, Section 12.4.1. A comparison of features between the brushed DC motor and BLDC motor is given in Table 12.1.

12.4.4 SR motors

The basic operation of the SR motor is based on magnetizing the iron stator and rotor by a current through the coil on the stator. As shown in Figure 12.19(a), the rotor is out of line with the magnetic field, and so a torque will be produced to reduce the reluctant in the path of the magnetic field by minimizing the air gap and making the magnetic field symmetrical as well. When the rotor is aligned with the stator, the switch will be OFF and no current is supplied (Figure 12.19(b)). Because of the rotor momentum, the rotor continues turning to the position shown in Figure 12.19(c), where the switch at that instant is in ON position to re-apply the magnetic field in the same direction. Consequently, the field exerts a torque to minimize the air gap and make the field symmetrical. The rotor is pulled forward

Table 12.1 Comparison between brushed DC motor and BLDC motor

Feature	Brushed DC motor	BLDC motor
Life	Shorter	Longer
Maintenance	Periodically required	Less required due to the absence of brushes
Commutation	Brushed commutation	Electronic commutation
Efficiency	Moderate	High (no voltage drop across brushes)
Output power	Moderate as large heat is produced in the middle of rotor and not easy to remove	High as the heat is produced in the outer stator and easy to remove
Torque–speed curve	Moderately flat as brush friction increases at higher speeds	Flat and enables operation at all speeds with rated load
Speed range	Lower because of mechanical limitations by brushes	Higher because of absence of mechanical limitations by brushes and commutator
Rotor inertia	Higher leading to limiting the dynamic characteristics	Low as it has permanent magnets on the rotor and the dynamic response is improved
Electric noise	Higher and results from the arcs in the brushes	Low
Control	Simple and inexpensive	Complex and expensive
Control necessity	It is essential to keep the motor running	It is not needed for fixed speed. Only required if variable speed is desired

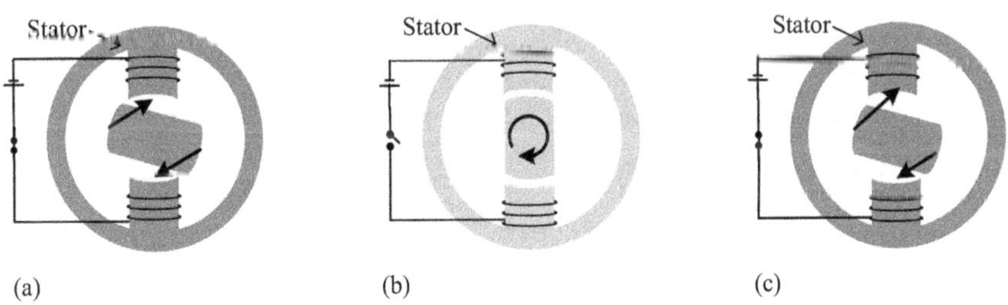

Figure 12.19 Basic principle of SR motors: (a) and (c) power supply is ON, stator and rotor are magnetized and the magnetic field turns rotor to minimize air gap; (b) power supply is OFF and momentum keeps rotor turning

until it is aligned with the stator where the current is switched off and so on. Thus, the SR motor is quite simple and has the following advantages [10,11]:

- The rotor is a piece of magnetically soft iron.
- No need to alternate the current in the coil. So, the cost is potentially reduced.

- Speed control can be carried out by changing the time interval between two successive switching (ON and OFF). The controller should be designed carefully and based on accurate information about the rotor position that is usually provided by sensors or recently inferred from voltage and current patterns. A simplified schematic diagram illustrating the feedback signals from both current detector and position detector to the controller is shown in Figure 12.20.
- No BEMF is generated as in case of BLDC motor because the rotor of SR motor is not a permanent magnet, i.e. enabling higher speeds.
- Higher power density is possible particularly when using a core of high magnetic permeability that leads to higher torque produced within a given volume.

The torque produced by the magnetic field can be made much smoother by multiplying the stator coils as well as using a rotor made of laminated iron and has salient poles (Figure 12.21(a)). In this figure, the SR motor consists of four salient pole rotor and six stator coils. Coil A is magnetized and exerting a clockwise force on the rotor. When the salient poles are aligned with coil A, the current in A is switched off. At this instant, two other salient poles are nearby the line of coil B, which is energized keeping the rotor smoothly rotating. The number of salient poles will mostly be two less than the number of coils. Clearly, correct current switching ON and OFF is needed. A typical physical structure of an SR motor with eight salient pole rotor and ten stator coils is shown in Figure 12.21(b).

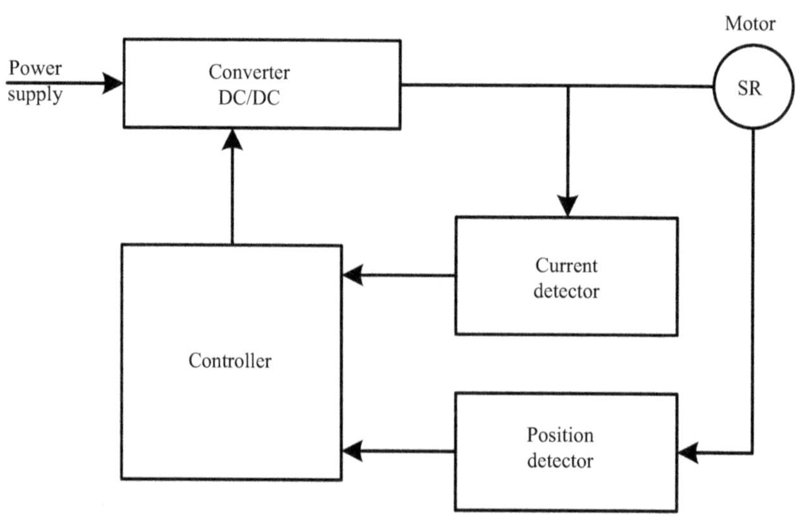

Figure 12.20 Simplified block diagram of SR motor controller

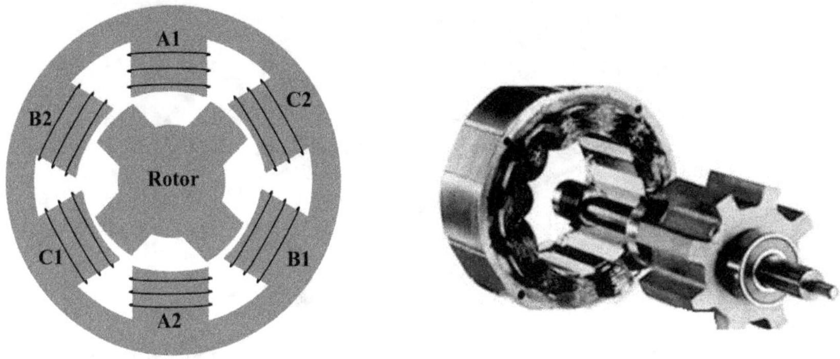

Figure 12.21 SR motor structure: (a) cross section of four salient pole rotor and six stator windings and (b) typical physical structure of eight salient pole rotor and ten stator coils

12.5 Electric vehicles performance

Derivation of equations that express the performance of electric vehicles necessitates the determination of the forces acting on the vehicle during its motion [12,13]. As shown in Figure 12.22, a vehicle of mass m is moving at speed v to climb a hill with a slope of angle Θ, as a general case. Obviously, if the vehicle is moving on a horizontal plane the angle will be zero. The vehicle is exposed to forces as follows:

- A force propelling the vehicle drive train forward, F_{train}.
- A force in opposite to the direction of movement due to the friction of tire on the road, friction in bearings and the gearing system. This force is called 'rolling resistance force', F_{roll}, and is proportional to vehicle weight, Thus,

$$F_{roll} = \mu_{roll} mg \tag{12.31}$$

  where g is the gravitational acceleration and μ_{roll} is the rolling resistance coefficient that depends on the type of tire and the tire pressure, with typical values of 0.015 for radial ply tire or about 0.005 for tires especially developed for electric vehicles.
- The drag force, F_{drg}, due to the friction of the vehicle body moving through the air. It acts in a direction opposite to the direction of velocity and it is a function of frontal area, body shape, duct and air passages, spoilers, etc. The formula expressing the value of this force is

$$F_{drg} = ½\rho A C_{drg} v^2 \tag{12.32}$$

  where F_{drg} is the force in Newtons, ρ is the air density of reasonable value of 1.25 kg/m^3, A is the frontal area in m^2, C_{drg} is the drag coefficient of a typical value around 0.19 depending on the vehicle design and v is the velocity (m/s).

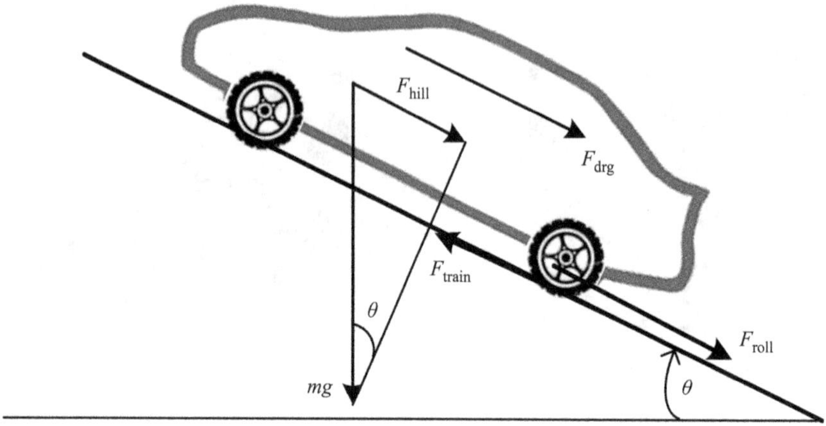

Figure 12.22 Forces affecting the vehicle motion

- Hill climbing force, F_{hill}, which is the force needed to overcome the component of vehicle's weight acting down the slope. It is given by

$$F_{hill} = mg \sin\theta \tag{12.33}$$

and has a negative sign if the vehicle is going downhill.

- A force to accelerate the vehicle is needed and is added to the other forces because the velocity of the vehicle is changing. The acceleration is not only a linear acceleration α, provided by a force, $F_{lin\text{-}acc}$, that is given by

$$F_{lin\text{-}acc} = m\alpha \tag{12.34}$$

but also an angular acceleration that needs a force, $F_{ang\text{-}acc}$, to turn the rotating parts faster. It can be derived as given below.

- The axle torque $= F_{train}r$, where r is the radius of the tire. As the motor torque is transmitted to the tire through a transmission gear with gear ratio n_{gear}, the motor torque T can be obtained by

$$T = \frac{F_{train}r}{n_{gear}}$$

that is,

$$F_{train} = \frac{n_{gear}}{r}T \tag{12.35}$$

The axle angular speed $= (n_{gear}/r)$ rad/s and accordingly the motor angular speed ω and its angular acceleration $\dot{\omega}$ are given by

$$\omega = \frac{v}{r}n_{gear} \quad \text{and} \quad \dot{\omega} = \frac{\alpha}{r}n_{gear} \tag{12.36}$$

Therefore, this angular acceleration needs to provide a torque of a value as

$$T = \frac{a}{r} I n_{\text{gear}} \tag{12.37}$$

where I is the moment of inertia of the motor rotor.

Incorporating (12.35) and (12.37) gives the corresponding force, $F_{\text{ang-acc}}$, at the wheels, as

$$F_{\text{ang-acc}} = \left(\frac{n_{\text{gear}}}{r}\right)^2 I a \tag{12.38}$$

T accounts for the gear system efficiency, η_{gear}, and (12.38) should be written in a more accurate form as

$$F_{\text{ang-acc}} = \frac{1}{\eta_{\text{gear}}} \left(\frac{n_{\text{gear}}}{r}\right)^2 I a \tag{12.39}$$

It is to be noted that both $F_{\text{lin-acc}}$ and $F_{\text{ang-acc}}$ will be negative if the vehicle is slowing down.

Therefore, the force provided the vehicle drive train must be the sum of all other forces, that is,

$$F_{\text{train}} = F_{\text{roll}} + F_{\text{drg}} + F_{\text{hill}} + F_{\text{lin-acc}} + F_{\text{ang-acc}} \tag{12.40}$$

Substituting the various terms given in (12.31) to (12.35) and (12.39) in (12.40) gives

$$\frac{n_{\text{gear}}}{r} T = \mu_{\text{roll}} m g + 0.625 A C_{\text{drg}} v^2 + m g \sin\theta + \left[m + \left(\frac{I}{\eta_{\text{gear}}}\right) \left(\frac{n_{\text{gear}}}{r}\right)^2 \right] \frac{dv}{dt} \tag{12.41}$$

where $a = dv/dt$.

Equation (12.41) can be solved as a first-order differential equation to obtain the value of vehicle acceleration which is a key performance indicator.

12.6 Battery charging infrastructure

Range anxiety problems associated with electric vehicles have been obviated by introducing PHEVs and development of higher density batteries capable of storing more energy in the same space. These days charging capacity is of more concern than the range anxiety as it helps in increasing the popularity of electric vehicles. So, suitable chargers and the charging infrastructure necessary to support the growth in electric vehicles population must be provided. The current-carrying capacity of chargers and charging time must be suitable for the application. However, high charging rate may accelerate battery ageing, which is the reason for the manufacturers to recommend the charging time being between 2 and 3 h. In addition, the charger can be placed on-board of the vehicle (Figure 12.23), or off-board, for example, charging stations [14].

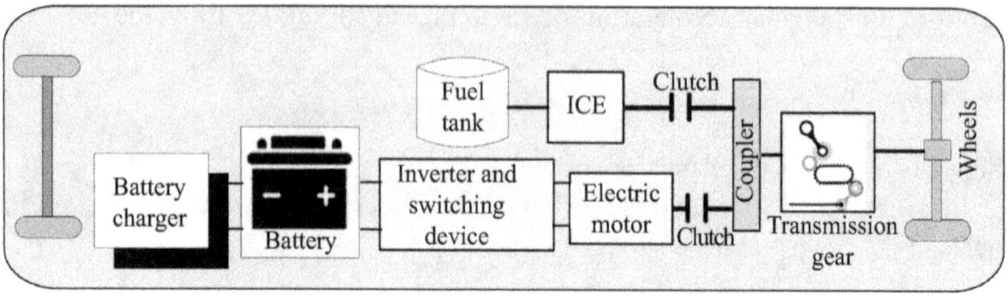

Figure 12.23 On-board battery charger for PHEV

Chargers provide a DC charging voltage from an AC source whether from a common socket outlet or more recently from a purpose-built DC charging station. Most important are the methods of controlling the charge and protecting the battery from over-voltage, over-current and over-temperature. These charger functions are integrated with and unique to the battery.

Chargers for electric bikes are usually low-cost, separate units. To save weight, they are not usually mounted on the bike and charging takes place at home. Their power handling capacity is only enough for charging the relatively low-power bike batteries and entirely unsuitable for passenger car applications.

Chargers for passenger cars are normally mounted inside the car. This is because the vehicle may be used a long way from home, further than the range possible from a single-battery charge. For this reason, they must carry the charger with them on board the vehicle. Charging can be carried out at home from a standard domestic electricity socket outlet, but the available power is very low, and charging takes a long time, possibly 10 h or more depending on the size of the battery. As charging is usually carried out overnight this is not necessarily a problem, but it could be if the car is away from its home base. Such low-power charging is normally used in an emergency and most cars are fitted with a higher-power charging option which can be used in commercial locations or with a higher-power domestic installation. In many countries, this higher-power facility is implemented by means of a three-phase electricity supply.

Commercial electric vehicles need bigger batteries which need higher power charging stations to achieve reasonable charging times, but they also have extra options. Many of them follow prescribed delivery routes within a limited range from base and return to base in the evening. In these cases, off-board charging is possible saving weight and space on the vehicle. Such applications can also be adapted to battery swap options. Each vehicle may have two batteries with one being charged while the other is in use. When used in long-distance shuttle applications, this can double the effective range of the vehicle. The vehicle depletes the battery during each journey and picks up a fully charged battery at the terminus leaving the discharged battery to be recharged ready for the next trip. This shuttle option, however, needs three batteries per vehicle.

12.6.1 Charging stations

Charging stations merely deliver the energy to the vehicle, usually in the form of a high-voltage AC or DC supply. They do not normally have the functions of the charger which must transform the electrical energy into a form that can be applied directly to the battery.

From the wide range of potential consumers noted above, the electric vehicle (EV) community needs several power supply options. Three different power levels have been outlined.

Level 1 refers to single phase AC using grounded receptacles as used in domestic applications (Figure 12.24). In North America, this typically means 16 A at 120 V delivering 1.9 kW of power. In Europe, it may be 13 or 16 A at 240 V delivering 3 kW of power. The EV may incorporate a standard domestic power cord to connect the vehicle to a domestic socket outlet or a Level 1 charging station.

Level 2 delivers up to 20 kW of power from either single- or three-phase AC source of 208–240 V at up to 80 A. The connector and charging cable used in Level 2 applications must follow the standards. These cables are permanently fixed to Level 2 charging station rather than the vehicle with the male connector being mounted in the vehicle itself. The connector is also commonly called a 'coupler'. Individuals can install a Level 2 charging station at home, whereas businesses and

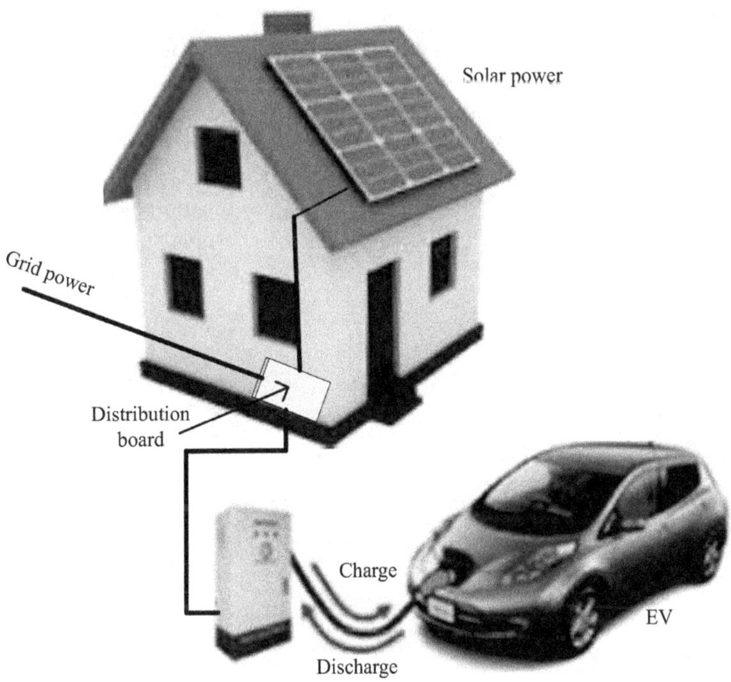

Figure 12.24 Domestic charge–discharge EV

local government can also install Level 2 charging stations in kerbside, car parking and so on (Figure 12.25).

Level 3 refers to DC charging, or 'fast charging'. To achieve very short charging times, Level 3 chargers supply very high currents of up to 400 A at voltages up to 600 V DC delivering a maximum power of 240 kW. Hybrid coupler allows charging from either a conventional 15-A AC wall outlet or a DC connection of up to 90 kW. These systems are being proposed for public fast charging stations (Figure 12.26).

12.6.2 Chargers

The diagram shown in Figure 12.27 illustrates how the key functions of Level 1, 2 and 3 chargers are implemented. All charging systems take AC power from the grid and convert it to DC power at a suitable voltage for charging the battery. In EV applications, except for bicycles, Level 1 and Level 2 chargers are completely contained within the vehicle. In Level 3 charging systems, however, the charging functions are split between the charging station and the vehicle's on-board charger.

12.6.2.1 Level 1 and Level 2 chargers

In low-power, Level 1 and Level 2, applications, the power conditioning which includes the AC to DC conversion, the power control unit which delivers a variable DC voltage to the battery, and various filtering functions are all carried out within the charger and can be implemented at a relatively low cost. The battery management system (BMS) is tightly integrated with the battery. It monitors the key battery operating parameters of voltage, current and temperature and controls the charging rate to provide the required constant current/constant voltage (CC/CV) charging profile and it triggers the protection circuits if the battery's operating limits are exceeded, isolating the battery if needed. The charger will also most likely incorporate controller area network (CAN) bus functionality to integrate with other vehicle systems but not necessarily with the charging station.

(a)

(b)

Figure 12.25 Charging station Level 2: (a) charging station in car park and (b) charging station in kerbside

Figure 12.26 Fast charging station

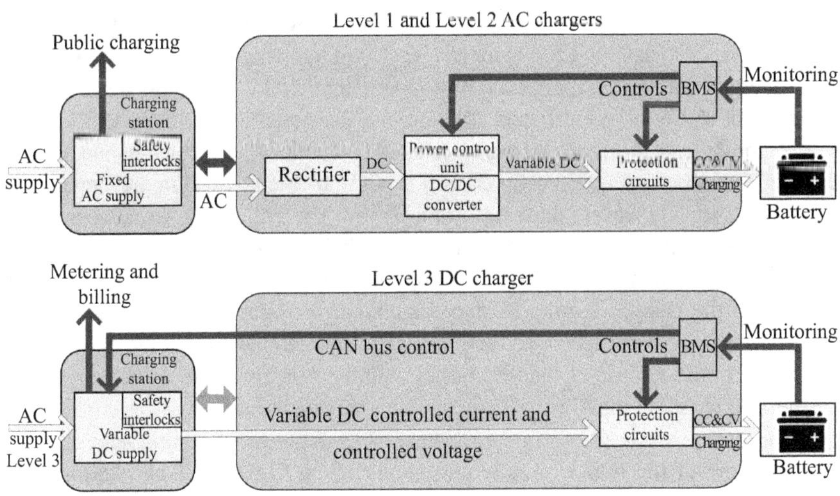

Figure 12.27 Main elements of the charger

Safety measures in the relatively low-power Level 1 charging station are simple and may be limited to a ground fault sensing device and a circuit interrupting device or 'circuit breaker'. However, the charger itself will usually incorporate more comprehensive safety measures in addition to the standard BMS functions including safety interlocks and isolators to prevent power being connected if there is a fault in the battery or the charger, as well as measures to prevent misuse, electric shocks and inadvertently driving away with the power cord still plugged in.

Level 1 charging works from a single-phase AC power outlet and is suitable for private, domestic installations and these do not need authentication and billing.

If the charger is designed to work with public charging stations, as many Level 2 installations will be, it will most likely need to incorporate further intelligence to communicate with the charging station to verify that the user is authorized to draw power from that particular source and to allow it to bill the customer for the energy transferred unless charging is installed at home or as a free service in the workplace or shopping mall.

Level 2 charging stations may use either single-phase or three-phase AC power from the grid.

This self-contained system used for Levels 1 and 2 gives the charger the flexibility to connect to different AC charging sources.

12.6.2.2 Level 3 chargers

Level 3 chargers have the same functionality as the Level 1 and 2 chargers but with the very high-power levels used, the AC/DC conversion and the power conditioning and control circuits become very large and very expensive requiring heavy-duty components. It makes sense to carry out these functions in the charging station rather than the charger so that the equipment can be shared by many users. This allows major cost and weight savings in the vehicle's on-board charger and with a bigger budget for the charging station, more efficient designs may be possible. In this case, the BMS must communicate with the charging station to control the voltage and current delivered to the battery since power control is not implemented within the battery. The charging station, however, does not control how the batteries on the vehicle are charged. That is the function of the charger on the vehicle itself, and its BMS. The charger provides this control by means of the CAN bus which communicates its demands to the charging station. Safety functions for isolating the battery and protecting the vehicle's occupants are also managed by the charger and the BMS.

While putting some of the traditional charger functions into the charging station permits cost savings in the vehicle, the charging station will be correspondingly more expensive. On top of this, there is the cost of providing access to the grid. Because of the high-power requirements of up to 240 kW per station, they cannot just be connected to the grid anywhere. The electricity generating utility must provide a dedicated supply line capable of delivering the high currents demanded. One drawback of this system is that the vehicle can only be charged at compatible public charging stations.

12.6.3 Charging times

Charging rates for very large batteries may be limited by the capacity of the power supply resulting in excessive charging times, and quick chargers have been developed to address this problem. However, just because a fast charging station is available it does not mean that the charging time can be reduced by connecting to it. First, fast charging systems use high-power DC charging systems and, unless the vehicle has a compatible charger on-board, the charging station will be of no use to the user. But more importantly, no matter what voltage or current the charger can deliver, the charging time is ultimately controlled or limited by the battery, and the cells from which it is made and how much current they can accept. Try powering a light bulb with a higher voltage than it was designed for and see how quickly it expires.

Renewable energy sources, such as PV solar panels, wind turbines and fuel cells, can be used to charge the vehicle's battery in addition to the power supplied by utility grid. The controllers of each source can be connected through a wireless network for system management (Figure 12.28) [15]. PV solar arrays can be also

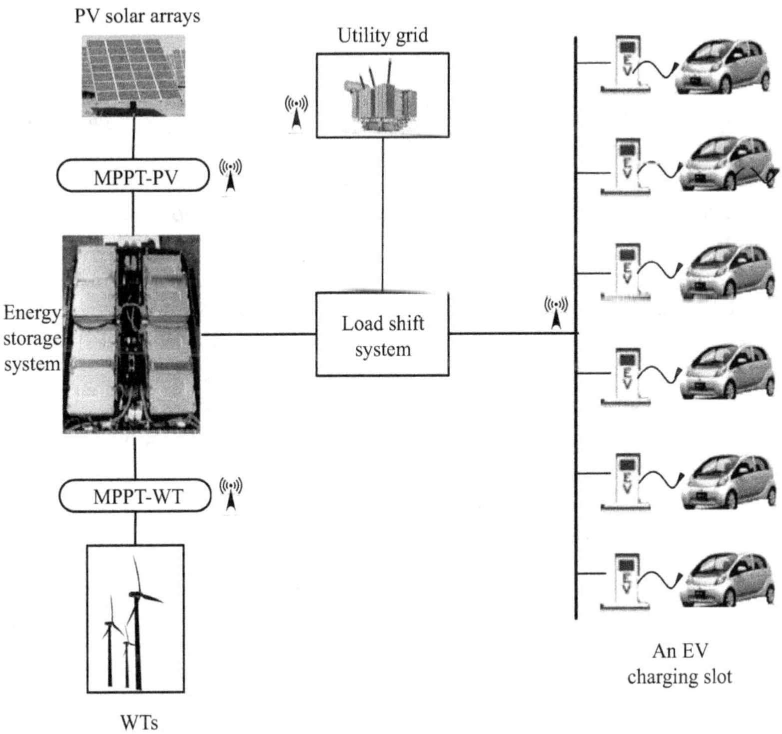

Figure 12.28 Charging station supplied from utility grid, PV panels and wind turbines equipped with maximum power point tracking (MPPT) system

(a)

(b)

Figure 12.29 Installation of PV arrays: (a) on-board PV panels and (b) PV arrays on the roof of a station

installed on-board of the vehicle or on the roof of the charging station or homes (Figure 12.29(a) and (b)).

References

[1] Larminie J. and Lowry J. *Electric vehicle technology explained*. 2nd edn. West Sussex, England: John Wiley & Sons Ltd.; 2012.

[2] Singer M. *Consumer views on plug-in electric vehicles – National benchmark report*. NREL/TP, Technical Report, 2016.

[3] Chakraaborty S., Simoes M. G. and Kramer W. E. (eds.). *Power electronics for renewable and distributed energy sources*. London: Springer; 2013.

[4] Clayton A. E. and Hancock N. N. *The performance and design of direct current machines*. 3rd edn. London: Sir Isaac Pitman & Sons; 1961.

[5] Cotton H. *Electrical technology*. 4th edn. London: Sir Isaac Pitman & Sons; 1961.

[6] Brown W. *Brushless DC motor control made easy*. Chicago, USA: Microship Technology Inc.; 2002.

[7] Yedamale P. *Brushless DC (BLDC) motor fundamentals*. Chicago, USA: Microship Technology Inc.; 2003.

[8] Zhao J. and Yu Y. *Brushless DC motor fundamentals application note*. MPS/AN047 Rev. 1.0 Report, 2014.

[9] Singh B. and Singh S. 'State of the art on permanent magnet brushless DC motor drives'. *Journal of Power Electronics*. 2009;9(1):1–17.

[10] DiRenzo M. T. *Switched reluctance motor control-basic operation and example using the TMS320F240*. Application Report, Texas Instruments Inc., 2000.

[11] Ahn J. W. *Switched reluctance motor*. Rijeka, Croatia, IntechOpen; 2011, pp. 201–52.

[12] Abdel-Kader F. E. S., Elsherif M. Z., Abdel-Rahim N. M. B. and Fathy M. M. 'Control methods of the switched reluctance motor in electric vehicle during acceleration'. *Leonardo Electronic Journal of Practices and Technologies*. 2012;20:127–46.

[13] Abdel-Kader F. E. S., Elsherif M. Z., Abdel-Rahim N. M. B. and Fathy M. M. 'Performance characteristics of the switched reluctance motor in electric vehicle during acceleration at variable turn on and turn off angle'. *International Journal of Science and Emerging Technologies* 2012;4(2):143–53.

[14] Rajakaruna S., Shahnia F. and Ghosh A. (eds.). *Plug in electric vehicles in smart grids: Integration techniques.* London: Springer; 2015.

[15] Yan Y., Qian Y., Sharif H. and Tipper D. 'A survey on smart grid communication infrastructures: Motivations, requirements and challenges'. *IEEE Communications Surveys & Tutorials*. 2012;15(1):5–20.

Part III

Renewables integration

Chapter 13
Power conditioning systems

13.1 General

Power conditioning systems (PCSs) are power electronics devices/circuits that act as electrical interface between the utility power grid or demand and renewable sources or energy storage systems [1]. A PCS is a dedicated device for power processing to output a voltage or current in a form adequate for the end user. The processing is based on one or a combination of some of the following functions [2–8]:

- AC–DC conversion (e.g. rectifiers and converters),
- DC–DC conversion for changing the voltage level (e.g. DC–DC converter),
- DC–AC conversion (e.g. inverters),
- AC–AC conversion (e.g. matrix inverter),
- elimination of voltage wave ripples by filters and
- monitoring and control.

For instance, a PCS for interfacing an energy storage system and utility power grid has three functions: (i) AC–DC conversion for charging the storage device, (ii) DC–AC conversion during storage discharge, and (iii) controlling the power during conversion process to avoid any damage occurring to the storage facility or end user's electrical system [9]. To perform its function, a PCS has a controller integrated with a power electronic device (Figure 13.1).

Each type of storage technology has its own operation methodology during charging/discharging process. Accordingly, a PCS must be properly designed for each type to keep its desired lifespan.

13.2 AC–DC conversion

Rectification is the process of conversion from AC to DC, and the conversion device is called 'rectifier'. For an ideal rectifier, the output voltage should be a pure DC signal without any ripple, and the current in phase with the voltage with no harmonics. Depending on the power electronic device adopted for rectification, rectifiers can be divided into categories, such as (i) uncontrolled rectifiers with diodes, (ii) phase-controlled rectifiers with thyristors and (iii) pulse-width modulation (PWM)-controlled rectifiers with insulated gate bipolar transistors (IGBTs) or metal oxide semiconductor field-effect transistors (MOSFETs).

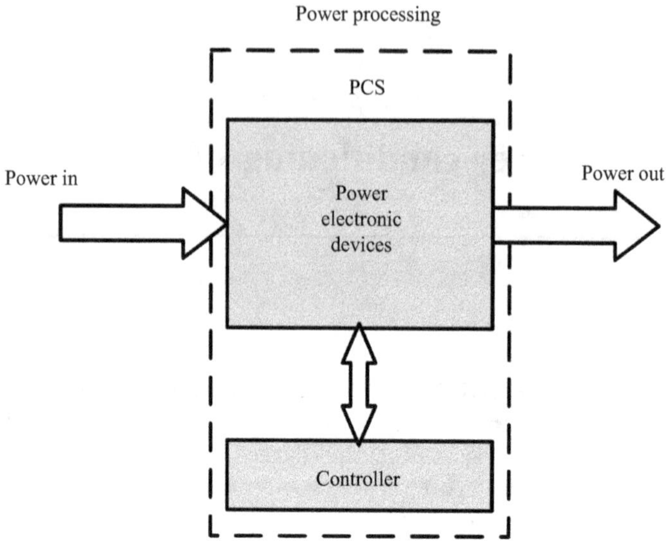

Figure 13.1 Power processing for interfacing

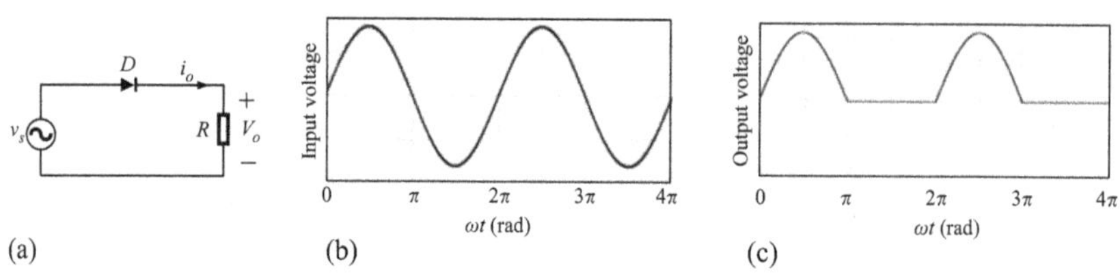

Figure 13.2 Uncontrolled rectifier with diode: (a) single-phase bridge, (b) input voltage waveform and (c) output voltage waveform

13.2.1 Uncontrolled rectifiers

The basic form of rectifiers comprises a diode (Figure 13.2(a)). The diode allows only the positive half cycle of the input sinusoidal AC voltage, Figure 13.2(b), to pass through to provide a DC output voltage in a form shown in Figure 13.2(c). The output voltage encompasses a significant number of ripples and, consequently, the source current waveform is not a pure sinusoid due to the presence of harmonic components. To improve the waveform of both output voltage and source current, a single-phase bridge rectifier is employed by connecting four diodes in a bridge form (Figure 13.3). Compared to the half-wave rectifier with one diode, both half cycles of the AC source voltage are passed to form the DC output voltage that is applied to the load.

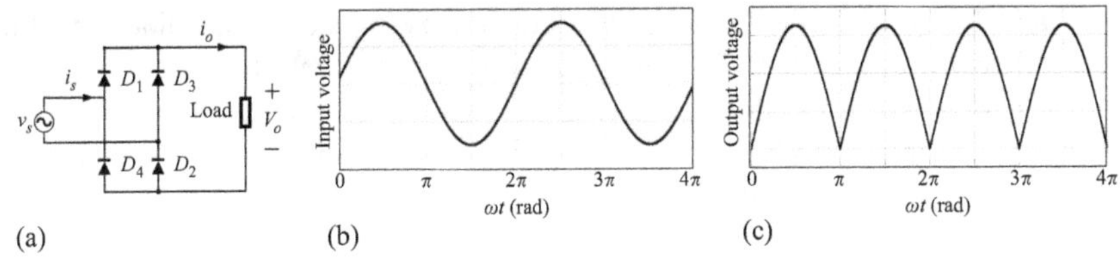

Figure 13.3 Uncontrolled full-wave rectifier: (a) full-wave bridge, (b) input voltage waveform and (c) output voltage waveform

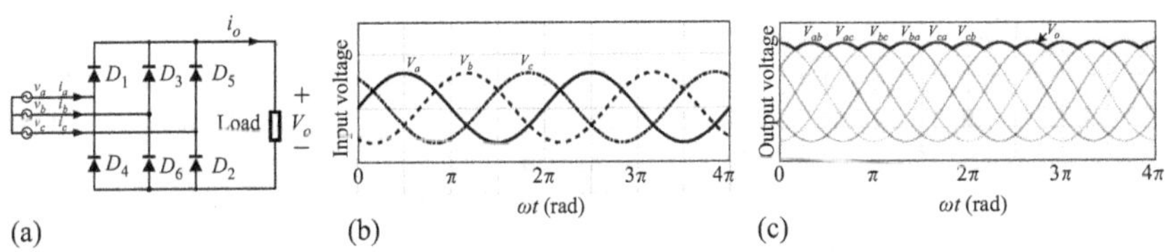

Figure 13.4 Uncontrolled three-phase full-wave rectifier: (a) six-diode bridge, (b) three-phase input voltage and (c) output voltage waveform

This symmetry in the two halves of the cycle reduces both the ripples in the output voltage and harmonic components in the input current. The DC output voltage, V_o, of this full-wave rectifier is given by

$$V_o = \frac{1}{\pi}\int_0^{\pi} \sqrt{2}V_s \sin\omega t \, d\omega t \tag{13.1}$$

where V_s is the RMS value of the source voltage.

For three-phase supply applications, the full-wave bridge converter comprises six diodes (Figure 13.4(a)). The pair of diodes with the highest applied instantaneous line voltage conducts periodically in the order D1, D2 → D2, D3 → D3, D4 → D4, D5 → D5, D6 → D6, D1 → D1, D2 for 120° electrical. Thus, the DC output voltage is the envelope of the line voltages with six ripples each completed cycle (Figure 13.4(b) and (c)), and the DC output voltage performance is enhanced. The DC output voltage is given by

$$V_o = \frac{1}{\pi/6}\int_0^{\pi/6} \sqrt{2}\times\sqrt{3}V_s \cos(\omega t)\, d(\omega t) = \frac{3\sqrt{6}}{\pi}V_s = 2.34V_s \tag{13.2}$$

where V_s is the RMS value of the AC source phase voltage.

13.2.2 Phase-controlled rectifiers

Uncontrolled diode rectifiers can only supply load with constant output DC voltage. The average output DC voltage, V_o, is fixed for a given load and given source (mains voltage 'v_s'). If an autotransformer is used, adjustable DC output voltage can be obtained by varying the source AC voltage. However, this involves a costly and bulky autotransformer. Alternatively, by replacing the diodes in the circuits shown in Figures 13.2–13.4 by thyristors, the output voltage, V_o, can be controlled by varying the thyristor switching angle, and an adjustable output DC voltage can be obtained. Phase-controlled rectifiers with an efficiency greater than 90% are used in many industrial implementations, especially in variable-speed drives.

A phase-controlled single-phase full-bridge rectifier is shown in Figure 13.5(a). The principle of operation is that during the positive half cycle of the source voltage, thyristors T_1 and T_2 are forward biased. If the gate signal is applied to thyristors T_1 and T_2 at $\omega t = \alpha$, the thyristors are turned on. Hence, the load current flows through T_1 and T_2 and the output/input voltage waveforms are as shown in Figure 13.5(b). In case of a large inductive load, thyristors T_1 and T_2 remain conducting even when the AC source voltage becomes negative. Likewise, during the negative half cycle of the AC source voltage, thyristors T_3 and T_4 are forward biased and thyristors T_1 and T_2 become backward biased. When the gate signal is applied to thyristors T_3 and T_4, the thyristors T_3 and T_4 are turned on whereas thyristors T_1 and T_2 are turned off. The load current is provided through T_3 and T_4 instead of T_1 and T_2.

For positive half cycle of AC source voltage v_s during the period $\omega t(\alpha \leq \omega t \leq \pi)$, the source current i_s and source voltage v_s are positive and thus the energy flows from the source to the load. Nevertheless, during the period $(\pi \leq \omega t \leq \pi + \alpha)$, i_s is positive and v_s is negative. Hence, the energy (stored in the large inductor) flows backwards. The line current waveform is depicted in Figure 13.5(c). The DC output voltage is calculated by

$$V_o = \frac{1}{\pi} \int_{\alpha}^{\alpha+\pi} \sqrt{2} V_s \sin(\omega t)\, d(\omega t) = \frac{2\sqrt{2}}{\pi} V_s \cos \alpha \tag{13.3}$$

It can be varied from 0 to $\frac{2\sqrt{2}}{\pi} V_s$ by varying the firing angle α from $\pi/2$ to 0. If the energy stored in the inductor is not enough to keep the load current flowing continuously, the thyristors turn off as the load current becomes discontinuous. It is to be noted that the converter can operate in the inversion mode and feed energy into the grid if a negative source voltage is applied to the DC bus. By varying α

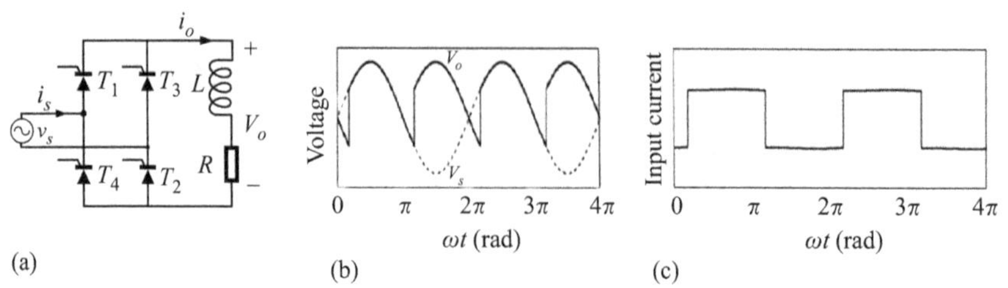

Figure 13.5 Phase-controlled rectifier with a large inductive load when $\alpha = \pi/3$: (a) single-phase full bridge, (b) input and output voltage waveform and (c) line current

between $\pi/2$ and π, energy is fed back to the grid. The corresponding waveforms when $\alpha = 2\pi/3$ are exhibited in Figure 13.6.

For high-power industrial applications, three-phase controlled bridge rectifiers with thyristors as exhibited in Figure 13.7(a) are often employed. The thyristors are triggered at the firing angle α at intervals of $\pi/3$ rad. On applying the triggering pulse to the thyristors that are forward biased, the corresponding applied line-to-line voltage is passed to the load. The waveforms of output voltage for $\alpha = \pi/6$ and $\alpha = \pi/2$ are presented in Figure 13.7(b) and (c), respectively. The DC output voltage is given by

$$V_o = \frac{1}{\pi/3}\int_{\alpha+\frac{\pi}{6}}^{\alpha+\frac{\pi}{2}} \sqrt{2} \times \sqrt{3}V_s \sin\left(\omega t + \frac{\pi}{6}\right) d(\omega t) = \frac{3\sqrt{6}}{\pi} V_s \cos\alpha \tag{13.4}$$

The output voltage can be varied from 0 to $\frac{3\sqrt{6}}{\pi} V_s$ by changing the firing angle α from $\pi/2$ to 0. In a similar manner, the converter operates in the inversion mode when negative supply voltage is applied to the DC bus to deliver energy to the grid (Figure 13.8).

13.2.3 Diode rectifiers cascaded with a boost converter

A significant portion of the source currents of both the uncontrolled diode and phase-controlled rectifiers is comprised of harmonic components, which results in a

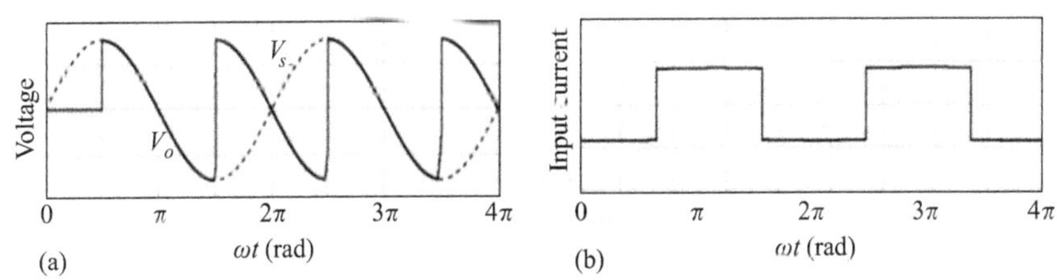

Figure 13.6 Phase-controlled rectifier operated in the inversion mode when $\alpha = 2\pi/3$ with a negative DC bus voltage present: (a) input and output voltage waveform and (b) line current

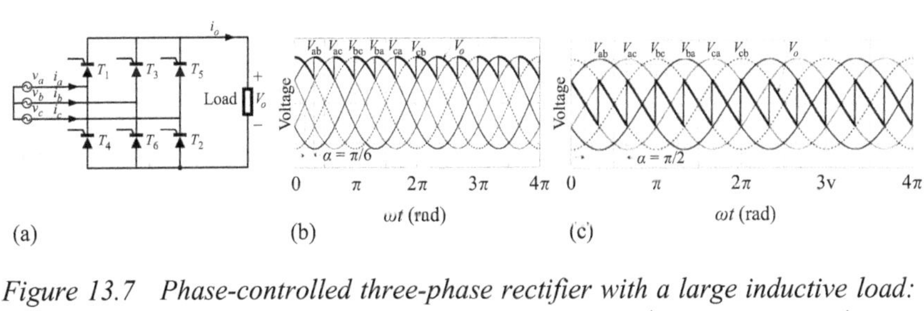

Figure 13.7 Phase-controlled three-phase rectifier with a large inductive load: (a) three-phase bridge rectifier, (b) $\alpha = \pi/6$ and (c) $\alpha = \pi/2$

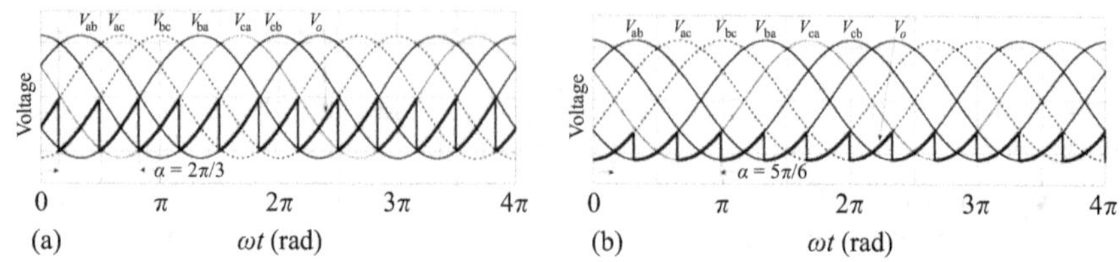

Figure 13.8 Phase-controlled three-phase rectifier operated in the inversion mode when a negative DC bus voltage is present: (a) $\alpha = 2\pi/3$ and (b) $\alpha = 5\pi/6$

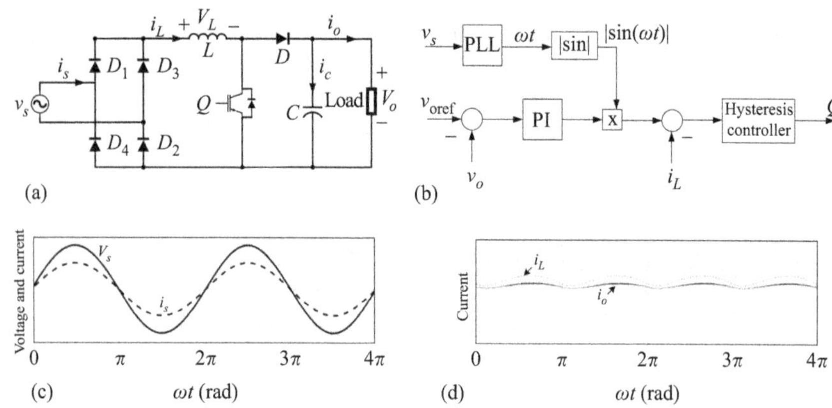

Figure 13.9 Diode rectifier cascaded with a boost converter: (a) diode bridge rectifier, (b) basic hysteresis control (PLL = phase-locked loop), (c) input voltage and input current and (d) i_L and i_o

low power factor. Several current control techniques have been developed to obtain a variable output voltage besides high-quality source currents at the same time [10,11].

One solution is to apply a boost converter in cascade with the DC bus of the uncontrolled bridge rectifier (Figure 13.9(a)). By controlling the boost converter in combination with the basic hysteresis current control technique, the source current can be brought in phase with the source voltage as well as regulate the output voltage (Figure 13.9(b)). The source current becomes sinusoidal while the load current is mainly DC, and the inductor current is full-wave rectified. It is to be noted that the average inductor current is higher than the load current because of the boosted DC output voltage. PWM pulses generated by the hysteresis current controller are used to turn ON/OFF the switch Q according to the difference between the inductor current i_L and its estimated reference, which is in phase with the required output voltage of the uncontrolled rectifier and hence the source voltage.

Thus, the actual inductor current i_L continuously follows the estimated reference current within a hysteresis band (HB) and the source current is in phase with the source voltage [2]. The corresponding circuit waveforms are exhibited in Figure 13.9(c) and (d).

13.2.4 PWM-controlled rectifiers

A bridge rectifier in cascade with a boost converter can enhance the quality of the source current but the power flow is unidirectional from the supply to the load. This problem can be solved by utilizing a bi-directional converter with fully PWM-controlled power semiconductor switches.

A PWM-controlled converter can be used as either a rectifier or an inverter, and a bi-directional flow of power can be achieved if energy is available on the load side. A bi-directional PWM-controlled single-phase H-bridge converter is shown in Figure 13.10(a). As it is mainly operated as a boost converter, an inductor is connected to the supply voltage side. The aim of the control technique applied, Figure 13.10(b), is to make the source current in phase with the source voltage. The difference between the actual output voltage, v_o, and the reference output voltage, v_{oref}, is fed into a PI controller to produce the required amount of the current being absorbed from the supply. It is then multiplied by sin (ωt) of the per-unit source voltage value as a synchronization signal to produce the reference source current.

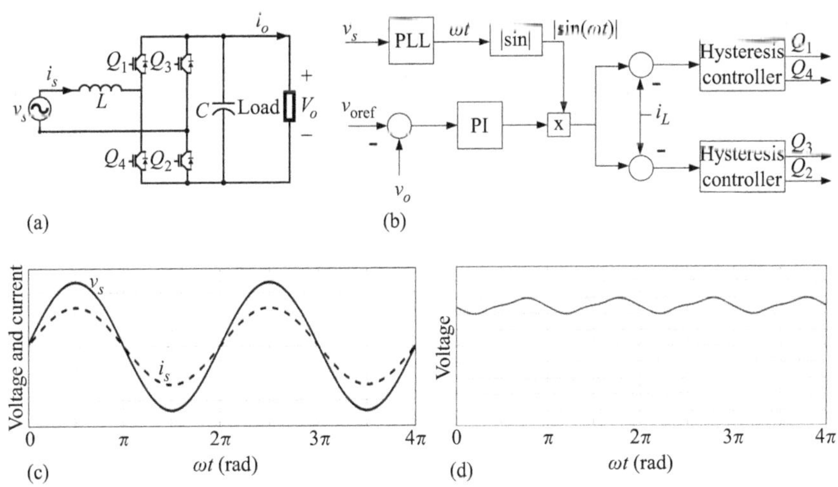

Figure 13.10 PWM-controlled single-phase full-bridge rectifier: (a) PWM-controlled single-phase H-bridge rectifier, (b) basic control strategy, (c) input current and voltage waveform and (d) output voltage waveform

The source current of the converter is controlled through a hysteresis current controller to track the estimated reference source current.

The principle of the hysteresis current control, Figure 13.11, is that when the actual current is below the lower margin of an HB around the estimated reference current, the upper and lower switches are turned ON and OFF, respectively, causing the actual current to increase. When the actual current exceeds the upper margin of the HB, the upper and lower switches again are turned ON and OFF, respectively, causing the actual current to decrease. Consequently, the PWM pulses for the upper switch Q_1 are generated as follows:

$$Q_1 = \begin{cases} \text{ON if } i_{act} < i_{ref} - \dfrac{1}{2}HB \\ \text{OFF if } i_{act} > i_{ref} - \dfrac{1}{2}HB \end{cases} \tag{13.5}$$

The PWM pulses for Q_4 are complementary, and the PWM pulses for Q_2 and Q_3 can be generated accordingly. The waveforms of both source current, input voltage and output voltage of the converter are shown in Figure 13.10(c) and (d). The source current and source voltage are in phase. The output voltage is maintained satisfactory despite the existence of some ripple, which can be handled by other techniques. It is worth noting that the capability of changing the source power factor to any desired value by controlling the phase of the synchronization signal. Three-phase PWM-controlled converter, Figure 13.12(a), can be achieved with the same principle and a slight change in the control strategy, Figure 13.12(b), to accommodate the other two phases. The source and output voltage waveforms are shown in Figure 13.12(c) and (d). The three-phase source currents are controlled to be sinusoidal and in phase with the corresponding phase voltages. In a three-phase system, the power input is constant, and the ripples in the output voltage waveform, Figure 13.12(d), of the three-phase converter are much lower than those in a single-phase converter (Figure 13.10(d)).

13.3 DC–DC conversion

The DC–DC converter is used to convert the output voltage level of a fixed DC source. According to the ratio between the output and source voltages, the

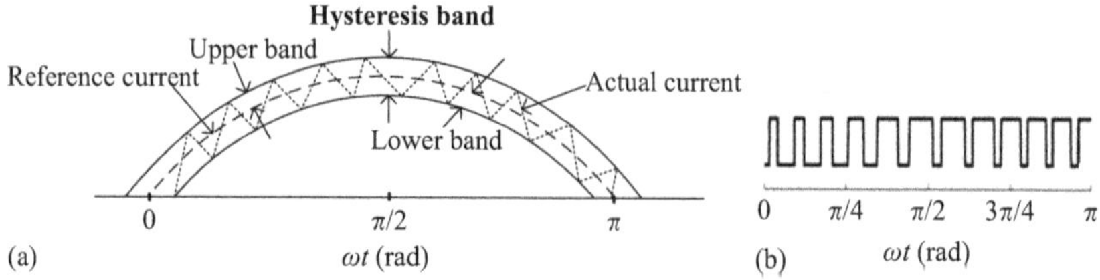

Figure 13.11 Principle of hysteresis control: (a) HB, actual and reference currents; (b) gate signal

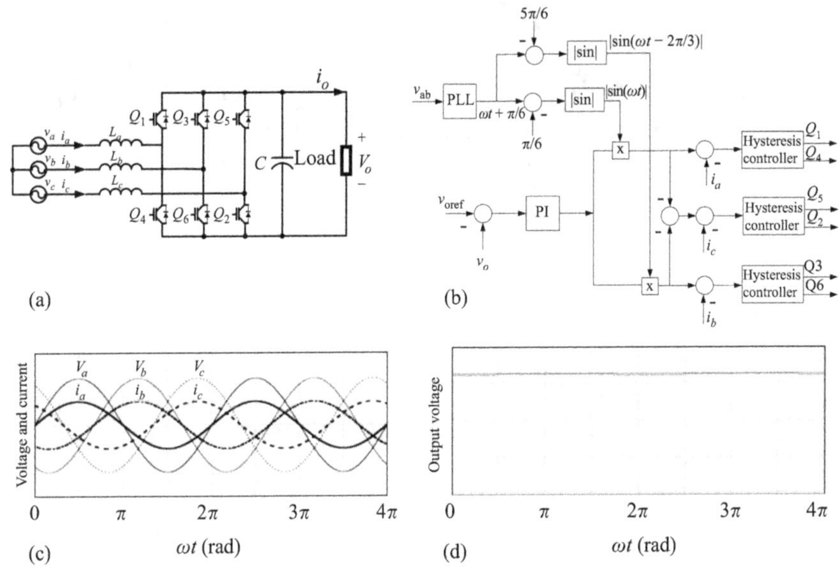

Figure 13.12 PWM-controlled three-phase full-bridge rectifier: (a) three-phase PWM-controlled rectifier, (b) basic control strategy, (c) input voltage and current and (d) output voltage

converter can be designed to step up or step down the output voltage, a function similar to an AC transformer. In this case, the ratio between the source and output voltage is called the conversion ratio α. For $\alpha < 1$, the converter is called a 'buck converter', whereas it is called a 'boost converter' for $\alpha > 1$. If the converter can increase/decrease the voltage level, that is, can operate for both $\alpha < 1$ and $\alpha > 1$, the converter is called a 'buck-boost converter' [4,5].

13.3.1 Buck converter

A buck converter is a step-down DC to DC converter. A typical buck converter, which comprises two switches (semiconductor switch and a diode), a capacitor and an inductor, is shown in Figure 13.13(a). The capacitor and inductor act as passive filters to enhance both the quality of the load current and the output voltage. The switch Q is turned ON and OFF periodically.

If the switching period and the duty cycle are T and k, respectively, then the ON time of one period is kT and the OFF time is $(1 - k)T$. The circuit can operate in two operation modes based on the state of the switch Q: mode #1 and mode #2 when turning ON and OFF, respectively. The corresponding equivalent circuits are shown in Figure 13.13(b) and (c). For mode #1, the inductor current i_L increases linearly as

$$L\frac{di_l}{dt} = v_s - v_o \tag{13.6}$$

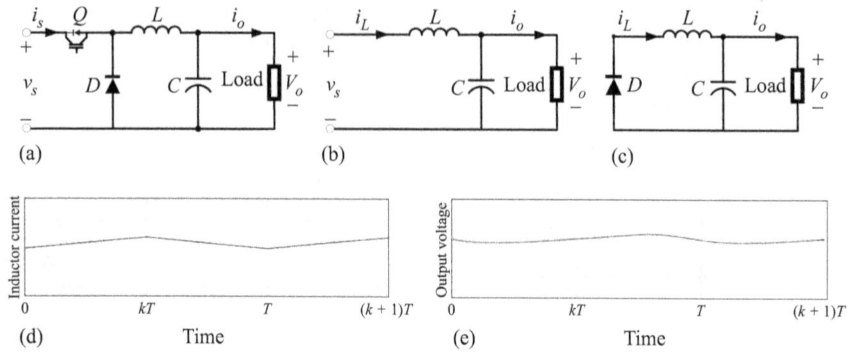

Figure 13.13 Typical buck converter: (a) buck converter topology, (b) mode #1 (Q: ON), (c) mode #2 (Q: OFF), (d) inductor current and (e) capacitor voltage

where v_s and v_o are approximately constant and positive. For mode #2, the inductor current freewheels through the diode and decreases linearly as

$$L\frac{di_l}{dt} = 0 - v_o \tag{13.7}$$

where v_o is approximately constant and negative. The related inductor current waveform is exhibited in Figure 13.13(d).

Ideally, the ripple current should pass through the capacitor and the related voltage waveform is presented in Figure 13.13(e). In the steady state, the net energy exchange in the inductor during one period is almost zero; that is, the increase in current in mode #1 and the decrease in current in mode #2 should be equal. Thus,

$$\frac{kT}{L}(v_s - v_o) = \frac{(1-k)T}{L}v_o \tag{13.8}$$

and the output voltage is

$$v_o = kv_s \tag{13.9}$$

So, it is a buck converter as $\alpha = k$.

13.3.2 Boost converter

The configuration of a typical boost converter (also known as a step-up converter) is shown in Figure 13.14(a). The output voltage is higher than the source voltage, and because of the power balance the output current is lower than the source current. Similar to the buck converter, the boost converter also has two operation

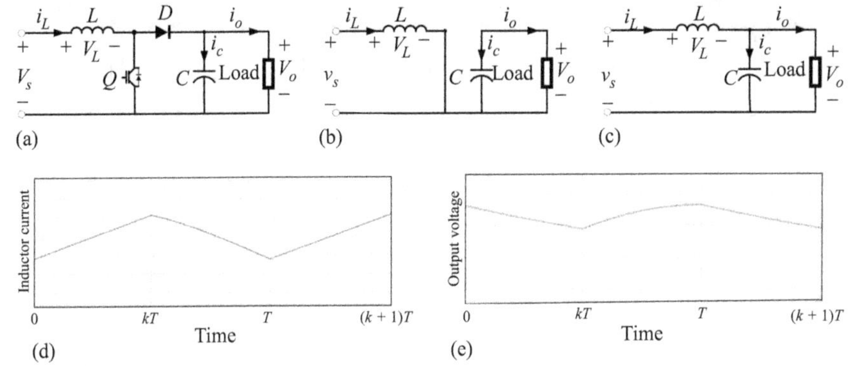

Figure 13.14 Typical boost converter: (a) boost converter topology, (b) mode #1 (Q: ON), (c) mode #2 (Q: OFF), (d) inductor current and (e) capacitor voltage

modes when the switch Q is turned ON and OFF. The equivalent circuits of the two modes are shown in Figure 13.14(b) and (c), respectively. In mode #1, the inductor current increases linearly as

$$L\frac{di_l}{dt} = v_s - 0 \tag{13.10}$$

and the energy from the power source is stored in the inductor, whereas the capacitor discharges to supply the load. In mode #2, the energy stored in the inductor and the energy supplied by the power source are both transferred to the capacitor and the load. The inductor current decreases linearly as

$$L\frac{di_l}{dt} = v_s - v_o \tag{13.11}$$

As in the buck converter, the net exchange of energy in the inductor during one period in the steady state should be zero; that is, the value of both current increase in mode #1 and current decrease in mode #2 should be equal. Thus,

$$\frac{kT}{L}(v_s) = \frac{(1-k)T}{L}(v_o - v_s) \tag{13.12}$$

from which the output voltage can be expressed as

$$v_o = \frac{1}{1-k}v_s \tag{13.13}$$

Hence, it is a boost converter as $\alpha = \frac{1}{1-k} > 1$ for $(0 \leq k \leq 1)$. The inductor current and the capacitor voltage waveforms are exhibited in Figure 13.14(d) and (e), respectively.

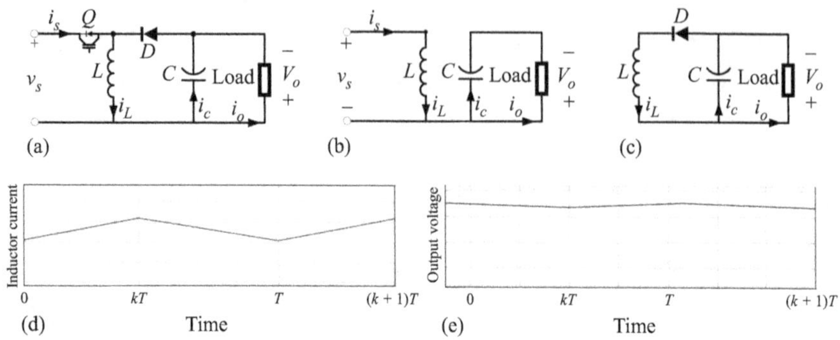

Figure 13.15 Typical buck-boost converter: (a) converter topology, (b) mode #1 (Q: ON), (c) mode #2 (Q: OFF), (d) inductor current and (e) capacitor voltage

13.3.3 Buck-boost converters

A typical configuration of buck-boost converter is depicted in Figure 13.15(a). The output voltage polarity is opposite to that of the source voltage. Like the buck and boost converters, this converter has two operation modes based on the state of switch Q. The equivalent circuits are shown in Figure 13.15(b) and (c) and the corresponding waveforms of the inductor current and capacitor voltage are shown in Figure 13.15(d) and (e).

The output voltage is

$$v_o = \frac{k}{1-k} v_s \tag{13.14}$$

and the converter operates in the buck mode when $k < 0.5$ and in the boost mode when $k > 0.5$.

13.4 DC–AC conversion

The device that adopts the process of DC–AC conversion is known as an inverter. Depending on the type of the DC supply, there are different types of inverters, such as (i) voltage-source inverter (VSI) if the supply is a voltage source in parallel with a large capacitor across the DC bus and (ii) a current-source inverter (CSI) if the supply is a current source with a large inductor connected in series with the DC supply. Similarly, depending on the type of the inverter output, the inverter is called voltage-controlled if the output is controlled to be a voltage source and current-controlled if the output is controlled to be a current source. Thus, there are voltage-controlled VSIs, current-controlled VSIs, voltage-controlled CSIs and current-controlled CSIs [2]. According to the structure of VSI inverter, it can be classified into two-level VSI and multi-level VSI [12] as illustrated in Figure 13.16. The operation of a three-phase two-level VSI is demonstrated in Section 13.4.3.

The output voltage amplitude and frequency of the inverter can be set to a desired value depending on the application. This can be easily accomplished with PWM techniques [13–18]. The main objective of PWM controller is to modulate a triggering signal with certain amplitude into a chain of pulses with adjustable widths to trigger switches and obtain sinusoidal output voltage or current depending on the control mode.

There are different PWM techniques that can be used with DC–AC inverter, such as sinusoidal PWM (SPWM), space vector PWM (SVPWM), selective harmonic elimination PWM [19] and HB controller (as described in Section 13.4.2). The most widely used technique, SPWM, is highlighted in the next subsection.

13.4.1 Sinusoidal PWM

It is required of most inverters that they produce a pure sinusoidal voltage output with a desirable frequency, which is typically much lower than the switching frequency. The desired pure sinusoidal output voltage, used as a modulating signal, can be compared with a triangular carrier wave at the switching frequency to produce a sequence of pulses, as exhibited in the left column of Figure 13.17. The harmonic components of this signal are generally around the multiples of the switching frequency. The output voltage waveform has the same shape as the generated trigger pulses.

The carrier frequency, that is, the switching frequency, is designed to be high so that the harmonic components can be simply filtered out using a low pass filter (commonly, an LC or LCL filter). It is noted that higher switching frequency causes a higher switching loss; thus, switching frequency must be properly designed to

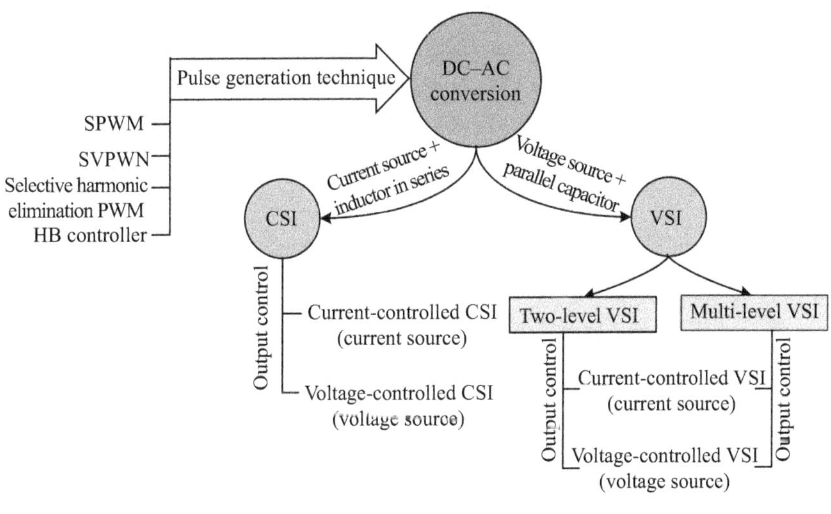

Figure 13.16 Types of DC–AC conversion devices

eliminate the harmonic component without increasing the switching losses. This type of modulation technique is called SPWM. The frequency of the reference signal controls the frequency of the output voltage. Its peak amplitude determines the modulation index and in turn the RMS value of the output voltage. Consequently, the amplitude and frequency of the output voltage can be simply controlled by changing the modulating signal. The SPWM is bipolar, as the carrier alters its sign during the positive or negative half cycle. If the carrier does not alter its sign during the positive or negative half cycle, the resulting SPWM is unipolar, as exhibited in the right column of Figure 13.17.

In both cases, the upper and lower switches on the same leg are operated in a complementary way. Likewise, for three-phase applications, three modulating signals can be compared with the carrier signal to produce the gate-driving signals, as exhibited in Figure 13.18.

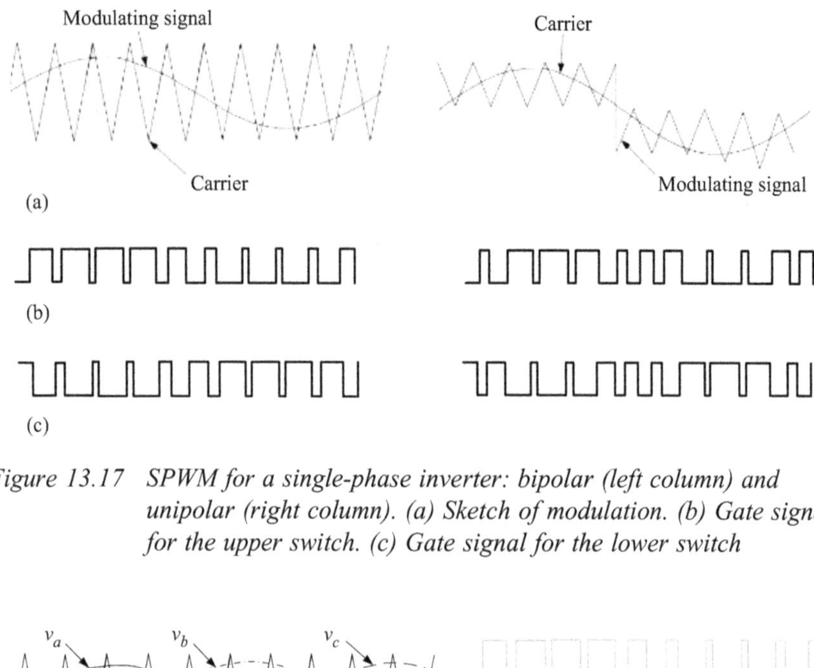

Figure 13.17 SPWM for a single-phase inverter: bipolar (left column) and unipolar (right column). (a) Sketch of modulation. (b) Gate signal for the upper switch. (c) Gate signal for the lower switch

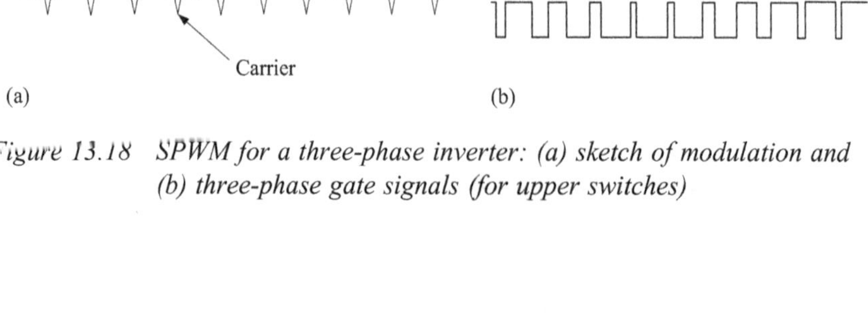

Figure 13.18 SPWM for a three-phase inverter: (a) sketch of modulation and (b) three-phase gate signals (for upper switches)

13.4.2 Operation of single-phase inverters

A single-phase inverter with a DC voltage source is shown in Figure 13.19. To illustrate the principle of operation of the inverter, the DC-bus voltage is split into two halves and the mid-point of the DC bus is considered as a reference point for the two-phase legs.

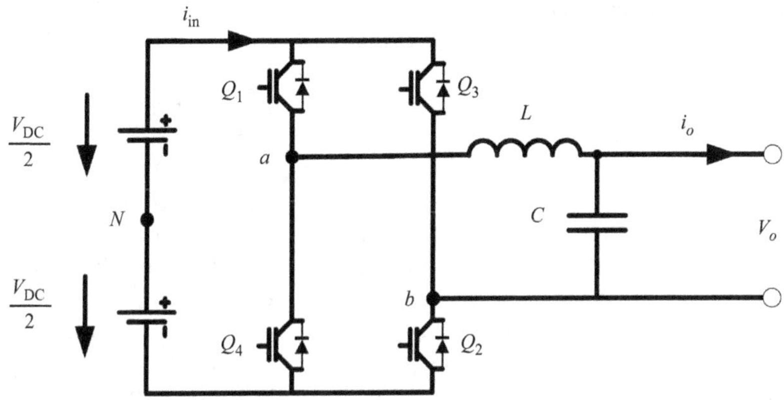

Figure 13.19 Single-phase VSI

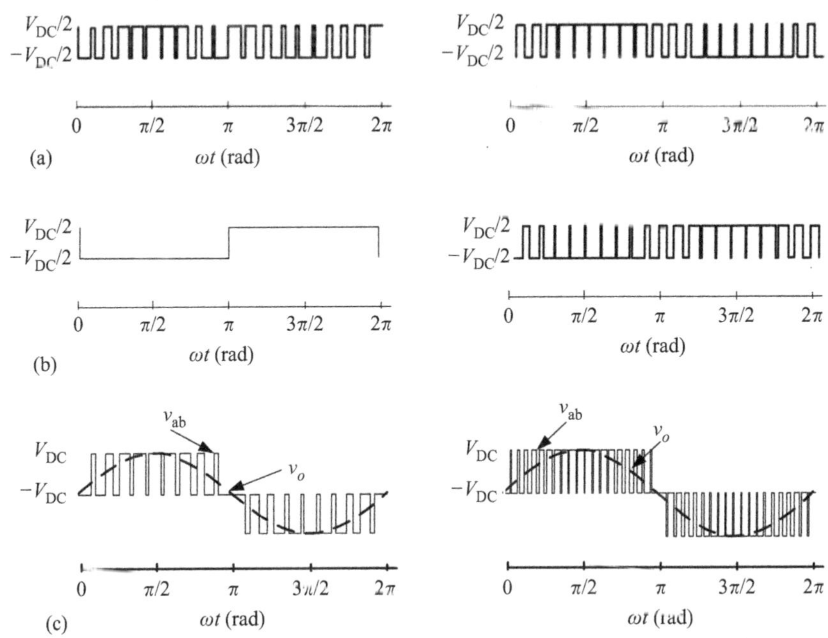

Figure 13.20 Unipolar operation of a single-phase inverter, with only one leg operated at the switching frequency (left column) and both legs operated at the switching frequency (right column): (a) v_{aN}, (b) v_{bN}, (c) v_{ab} and v_o

The inverter can be operated to produce unipolar and bipolar SPWM signals for v_{ab}. In case of operation with the unipolar SPWM signal presented in the right column of Figure 13.17, the voltages v_{aN}, v_{bN}, together with v_{ab} and v_o are exhibited in the left column of Figure 13.20. The phase-leg *a* is operated according to the unipolar SPWM signal and phase-leg *b* is operated according to the polarity of the voltage signal at its frequency. For example, Q_1 and Q_4 are turned ON and OFF when the modulating voltage *v* is positive according to the unipolar SPWM signal, whereas Q_2 is always ON and Q_3 is always OFF when *v* is negative. Q_1 and Q_4 are turned ON and OFF according to the SPWM signal whereas Q_2 is always OFF and Q_3 is always ON. Because the second leg is operated at the frequency of the voltage, depending on the type of the DC supply the switching losses can be decreased.

It is possible to operate the inverter to attain a unipolar SPWM for v_{ab} even if the phase legs are driven by bipolar SPWM signals, as presented in the right column of Figure 13.20. In such a case, the modulating voltage *v* and its opposite are compared with the carrier waveform to produce two sets of bipolar SPWM signals to drive the two-phase legs. Consequently, the voltages v_{aN} and v_{bN} are 180° apart from each other. The voltage v_{ab}, as a difference between the two voltages v_{aN} and v_{bN}, is unipolar at the doubled switching frequency. The switching losses are high as both phase legs are operated at the same high switching frequency. Nevertheless, the output voltage quality is better than the case exhibited in the left column of Figure 13.20 because the resulting v_{ab} has a doubled switching frequency. So, the two-phase legs are operated as two separate phases, which are 180° apart from each other. When both legs of the inverter are operated with the bipolar SPWM as shown in the left column of Figure 13.17, the resulting waveforms are as shown in Figure 13.21. The same SPWM signal is sent to the two-phase legs in a complementary way. That is, (Q_1, Q_2) and (Q_3, Q_4) are two sets of a pair of switches for

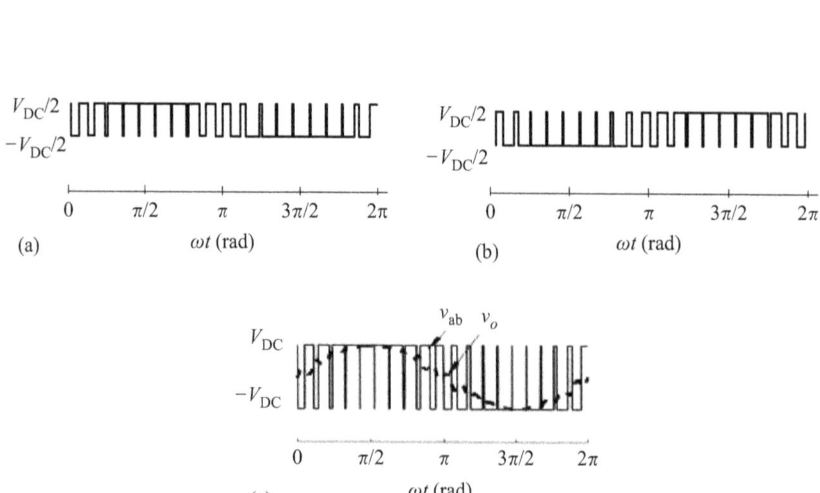

Figure 13.21 Bipolar operation of a single-phase inverter: (a) v_{aN}, (b) v_{bN} and (c) v_{ab} and v_o

each. Switches of each set are operated simultaneously. As a result, $v_{bN} = -v_{aN}$ and $v_{ab} = 2v_{aN}$.

The amplitude of v_{ab} is $\pm V_{DC}$ for the three operation modes and the maximum attainable amplitude is the same as the DC bus voltage.

13.4.3 Operation of three-phase inverters (two-level)

For three-phase inverters, Figure 13.22(a), three sets of bipolar PWM signals are generated by comparing the three-phase voltages with the carrier waveform, Figure 13.18, to drive the three-phase legs individually. The related waveforms are

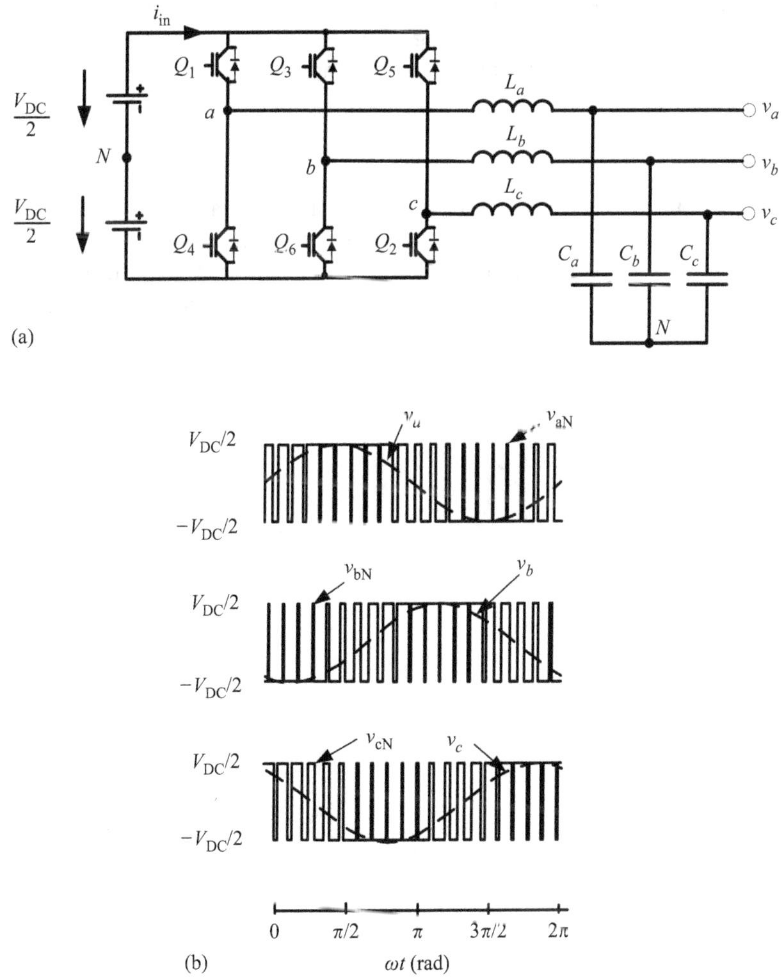

Figure 13.22 Operation of a three-phase inverter: (a) topology and (b) v_{aN}, v_{bN}, v_{cN} and the three-phase voltages v_a, v_b and v_c

shown in Figure 13.22(b). The maximum amplitude of the phase voltages is half of the DC-bus voltage.

13.5 AC–AC conversion

AC–AC conversion can be achieved either indirectly through AC–DC–AC with adding a DC bus in between or directly without using a DC bus. The indirect AC–AC conversion (back-to-back) is mainly a combination of AC–DC conversion and DC–AC conversion, as discussed in Sections 13.2 and 13.4. The direct AC–AC conversion implies bi-directional switches such as thyristors connected in anti-parallel or triacs. One approach to implement direct AC–AC conversion is to operate matrix converters to produce AC outputs with arbitrary amplitude and frequency. The topology of a single-phase AC–AC converter is shown in Figure 13.23(a), where two main control methods can be used: ON–OFF control and phase control.

13.5.1 ON–OFF control

Two thyristors are connected in antiparallel to pass both half cycles of the supply to the load when they are triggered. Triggering signals are provided by an ON–OFF control. The thyristors are turned ON to pass the source voltage to the load and turned OFF when the source voltage is not passed to the load. It can be found that the RMS value of the output voltage is $\sqrt{k}V$ and the source power factor is $\sqrt{k}$,

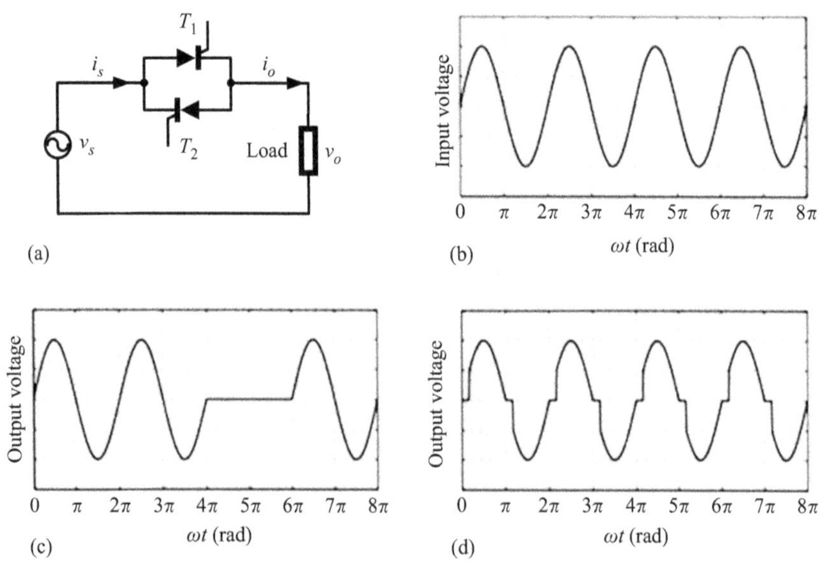

Figure 13.23 Single-phase AC–AC converter: (a) topology, (b) input voltage, (c) output voltage under ON–OFF control and (d) output voltage under phase control

where k is the ratio of the number of ON cycles to the number of total cycles in an operational period and the RMS value of the supply is V. For the source voltage sketched in Figure 13.23(b), the related output voltage is as exhibited in Figure 13.23(c) for one OFF cycle. The thyristors are triggered when the source voltage crosses 0.

13.5.2 Phase control

For the circuit shown in Figure 13.23(a), the voltage phase can be controlled when both half cycles are turned ON through triggering the thyristors at a specific time. It is found that compared to a case with thyristor bridge rectifiers there is no big difference, aside from the fact that both half cycles can be passed to the load. The source voltage as sketched in Figure 13.23(b) produces an output voltage as shown in Figure 13.23(d) for a triggering angle of $\pi/6$ rad. The output voltage encompasses harmonics and the switches are not triggered when the voltage crosses 0.

13.6 Output filters

The output voltage and current waveforms of the inverter often contain harmonics in a multiple of the switching frequency as a result of the effect of PWM pulses. Consequently, a low-pass filter is required to be connected to the output of the inverter to filter out the harmonics of the output voltage. Different configurations of the passive low-pass filter can be used in conjunction with inverters to filter out harmonics, for example, L, LC and LCL filters [20–22].

The L filter is also called choke filter. It consists of an inductor L connected between the inverter output and the load. The inverter output voltage waveform contains AC components in addition to DC components. The inductor provides a high-impedance to the high-frequency components and low-impedance to the low-frequency components. Thus, the high-frequency components of the output voltage are blocked, and only the fundamental frequency component is applied to the load. LC and LCL filters are briefly described in the following subsections.

13.6.1 LC filters

The circuit of an LC filter includes an inductor connected in series with the load and a capacitor connected in parallel across the load. The inductor offers a high-impedance to the high-frequency component and allows fundamental component to flow through to the load, while the capacitor filters out any high-frequency components flowing through the inductor. A passive LC filter can be represented by a circuit shown in Figure 13.24. The equivalent internal resistances of the capacitor and the inductor may be neglected during the design procedure of the filter due to their small values. Practically, the internal resistances can dampen high-frequency components. The capacitance and the inductance should be designed properly to filter out the switching effects considering several factors [23], for example, the cut-off frequency f_c [24,25], the voltage total harmonic distortion (THD), size, the cost function [26–28], resonance damping, efficiency [29], power level and so on. The cut-off frequency f_c of the filter is derived as

$$f_c = \frac{1}{2\pi\sqrt{LC}} \tag{13.15}$$

The most important consideration is the filtering out of switching harmonics. Although it is desirable to keep the cut-off frequency much lower than the switching frequency, it should be selected such that it still provides enough bandwidth for the controller. It is recommended that the cut-off frequency is selected within $\frac{1}{3}$ to $\frac{1}{2}$ of the switching frequency f_{sw} [24] as in the following:

$$\frac{f_{sw}}{3} \leq f_c \leq \frac{f_{sw}}{2} \tag{13.16}$$

It is to be noted that this may cause resonance in the output impedance of the inverter near the cut-off frequency. This amplifies the harmonic current oscillations near the cut-off frequency and might cause high THD in the output voltage. It is, therefore, necessary to not select the cut-off frequency around the band where the major harmonic components of the load current exist. The cost function [26–28] of the LC filter can be expressed as

$$COST = \frac{2Q_L + Q_c}{\sum_{h=1,\text{odd}}^{n} |V_{oh} I_h|} \tag{13.17}$$

and

$$Q_L = \sum_{h=1,\text{odd}}^{n} |I_h|^2 X_{Lh} \tag{13.18}$$

$$Q_C = \sum_{h=1,\text{odd}}^{n} \frac{|V_{Ch}|^2}{X_{Ch}} \tag{13.19}$$

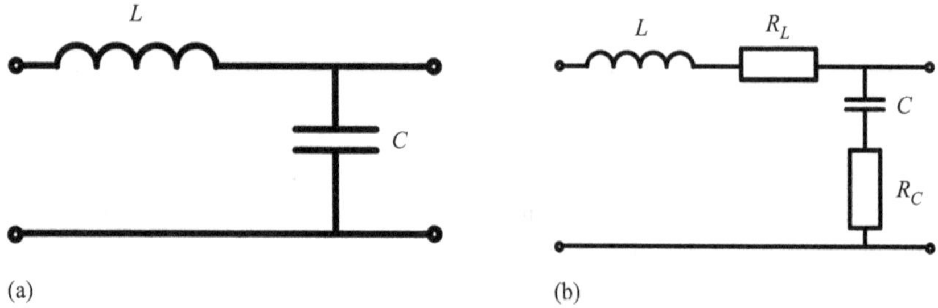

Figure 13.24 Circuit model of a passive LC filter: (a) without considering the internal resistance and (b) with the internal resistance considered

Heuristically, the size of the capacitor C should be small for high-voltage applications and that of the inductor L small for high-current applications maintaining the same cut-off frequency. The values of both capacitance C and inductance L should be small in situations, for example, under low switching frequency where, respectively, large amount of voltage and current harmonics exist. When the internal resistances of the filter, R_C and R_L, are considered or purposely increased, the LC resonance within the cut-off frequency is dampened. Nevertheless, the existence of R_C and/or R_L results in excessive power losses, which is considered as a challenge in the LC filter design for high-power applications. One possible solution is to implement control techniques to add virtual internal resistors to obtain the same results [30,31]. In any case, the ratings, current for the inductor and voltage for the capacitor, must meet the specified requirements of the inverter design.

13.6.2 LCL filters

The passive LCL filter, represented by the circuit shown in Figure 13.25, is commonly adopted in grid-connected inverters. The increase in the order of the LCL filter by 1, because of the additional grid-side inductor L_g, means that better attenuation of the current harmonics can be achieved by an LCL filter than an LC filter. It also provides an additional mechanism for limiting the current harmonics that may be caused by harmonics in the grid voltage.

It is very important to pay attention to selecting the value of the switching frequency in designing the LCL filter. For high switching frequency applications, the LC filter is designed first and then the grid-side inductor is added. For very low switching frequency applications at MW level, it is important that extra attention should be devoted to check the amount of reactive power of the filter capacitor, such that it does not significantly affect the rating of the inverter.

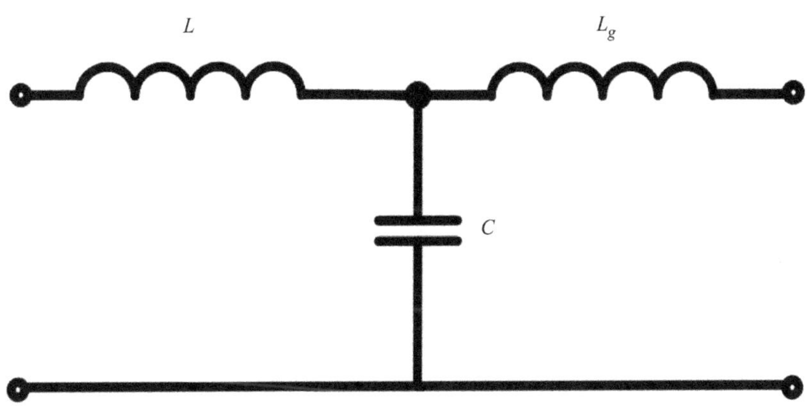

Figure 13.25 Circuit model of a passive LCL filter

13.7 Case studies

The importance and performance of the power electronics-based converters are illustrated through two case studies: Case 1, the application of grid-connected photovoltaic (PV), and Case 2, the application of grid-connected doubly fed induction generator (DFIG) wind turbine (WT).

The two cases are studied using MATLAB®/Simulink® software with various output results.

13.7.1 Case 1: grid-connected PV

A PV module is interfaced to a grid through power electronics two-level converter connected to the secondary of step-up distribution transformer. The PV design parameters are listed in Table 13.1. The converter objective is to obtain the maximum power point of PV and inject the PV-generated power to the grid. The block diagram of the system is shown in Figure 13.26.

The sun irradiance is varied as shown in Figure 13.27 to validate the converter performance in tracking the maximum power supplied by PV. The PV-generated power is injected to the distribution network using two-level converter as shown in Figure 13.28. The DC-link voltage and the waveform of converter output current/voltage are shown in Figures 13.29, 13.30 and 13.31, respectively.

The RMS values for the converter output current and voltages are exhibited in Figures 13.32 and 13.33. The converter is keeping the balance of the RMS values of the output voltages and within the allowable range during the variation of the light irradiance.

13.7.2 Case 2: grid-connected DFIG WT

The WT is connected to a DFIG to generate electrical power using energy extracted from the wind. As the wind speed is varying, the rotor is connected to a back-to-back converter to control the phase and magnitude of rotor voltage and, thereby,

Table 13.1 PV system parameters

System quantity	**Value**	**Unit**
PV module open circuit voltage (v_{oc})	85.3	V
PV module voltage at maximum power point (v_{mp})	72.9	V
PV module short circuit current (i_{sc})	6.09	A
PV module current maximum power point (i_{mp})	5.69	A
PV module max. power (p_{max})	414.8	W
Cells per module	128	Cell
Series modules	11	Module
Shunt modules	17	Module
DC-link voltage	760	V
DC-link capacitance	1,000	μF

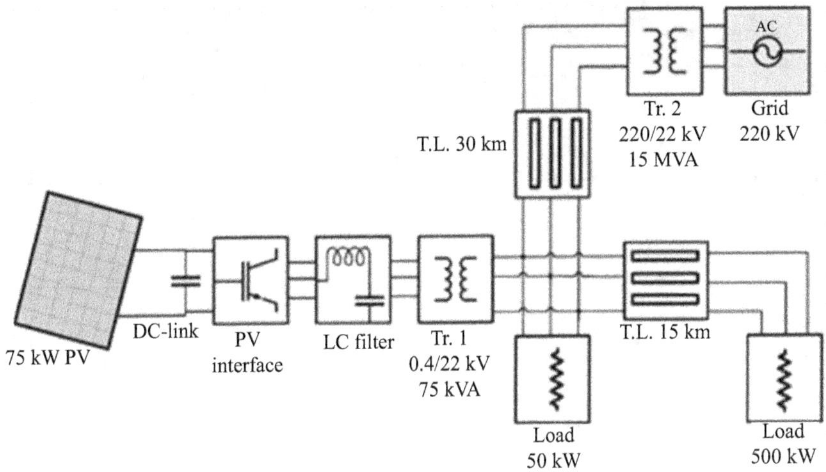

Figure 13.26 Block diagram of a grid-connected PV

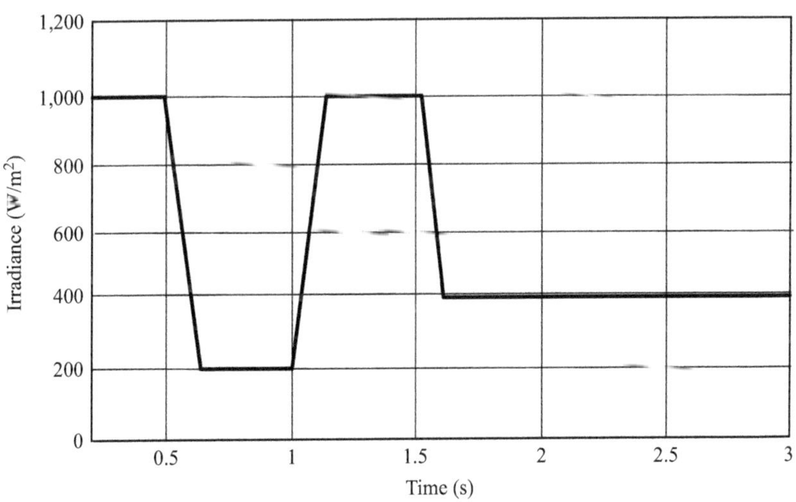

Figure 13.27 Sun irradiance during the study

control the magnetic field frequency or speed. It forms an AC excitation with a variable frequency to match the actual wind speed. This operation gives the DFIG the benefits of a synchronous generator. These benefits include the separate control of reactive and active power, or the control of the WT's power factor. A block diagram of the studied system is shown in Figure 13.34. The system parameters are listed in Table 13.2.

The wind speed is assumed to be varied to assure the converter performance during the different wind speeds (Figure 13.35).

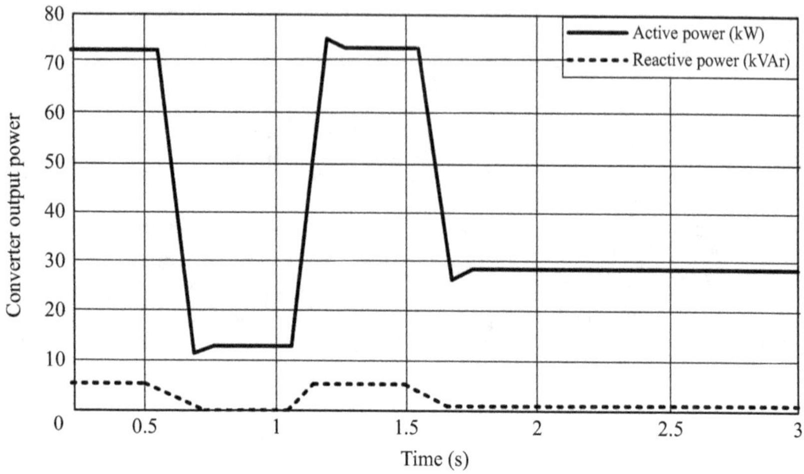

Figure 13.28 PV-injected power

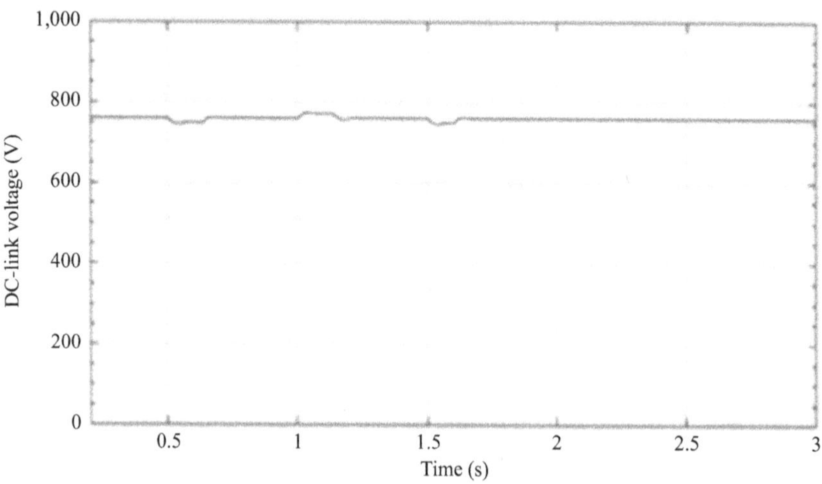

Figure 13.29 DC-link voltage

The active power output generated from the WT during the variation of the wind speed is shown in Figure 13.36. The active power varies according to the wind speed and the converter adjusts the excitation of the DFIG to generate power during low wind speed.

The DC-link voltage of the back-to-back converter during the variation of wind speed remains constant at the pre-specified level as shown in Figure 13.37.

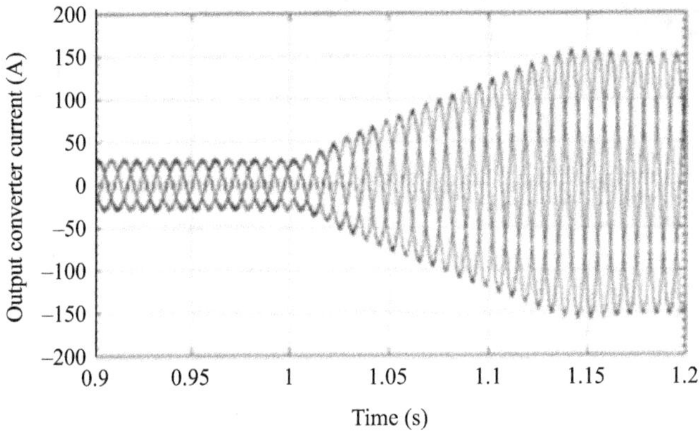

Figure 13.30 Waveforms of the converter output current

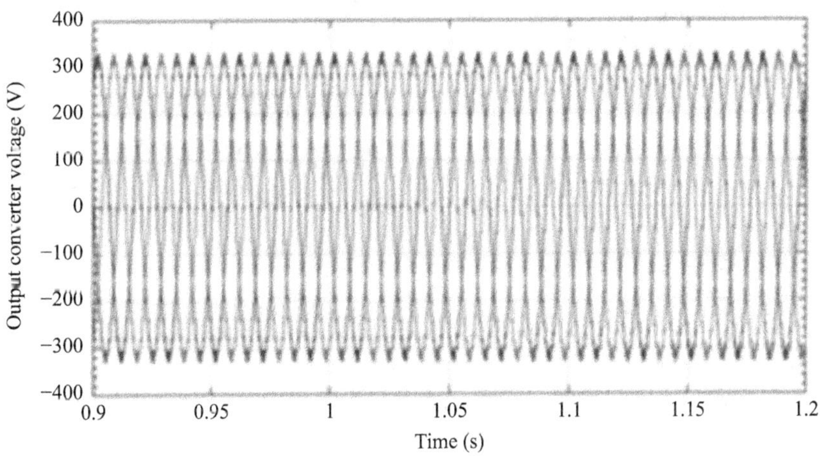

Figure 13.31 Waveforms of the converter output voltage

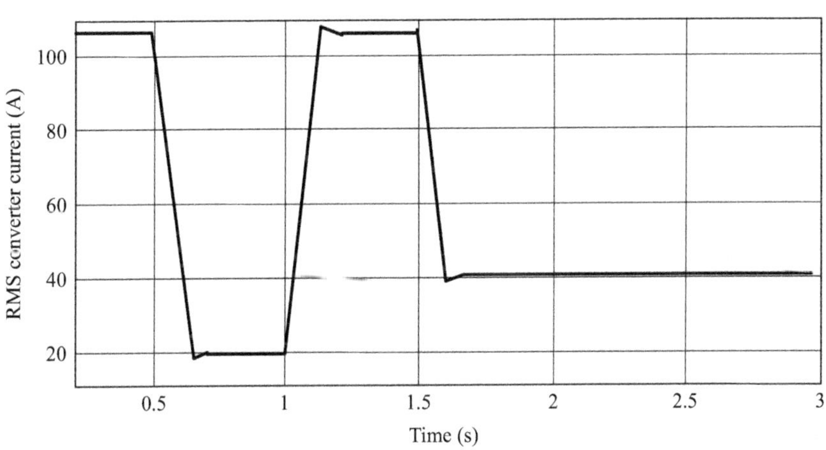

Figure 13.32 RMS values of the converter output current

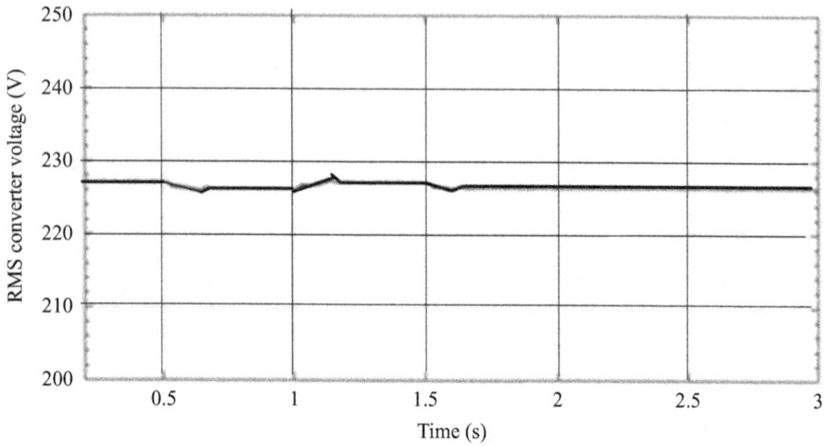

Figure 13.33 RMS value of the converter output voltages

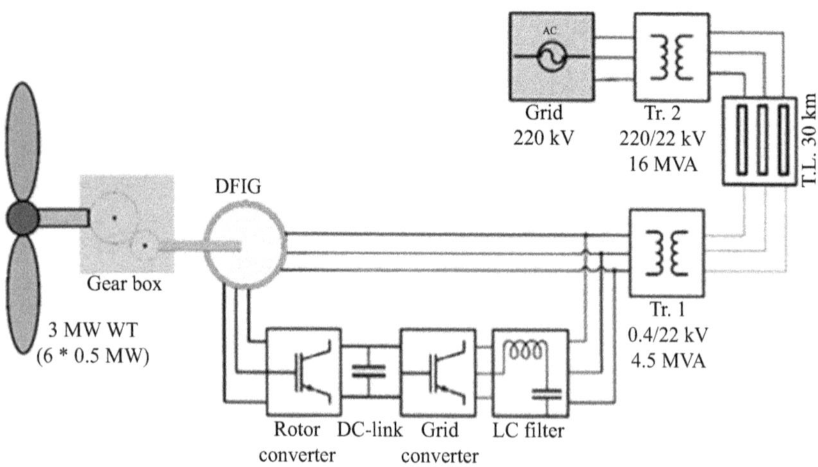

Figure 13.34 Block diagram of a grid-connected DFIG WT

Table 13.2 WT system parameters

System quantity	Value	Unit
DFIG power	500	kW
DFIG stator voltage	400	V
No. of Connected WTs	6	Number
Total connected WT power	3	MW
Wind speed at maximum efficiency	11	m/s
DC-link voltage	1,150	V
DC-link capacitance	5,000	μF

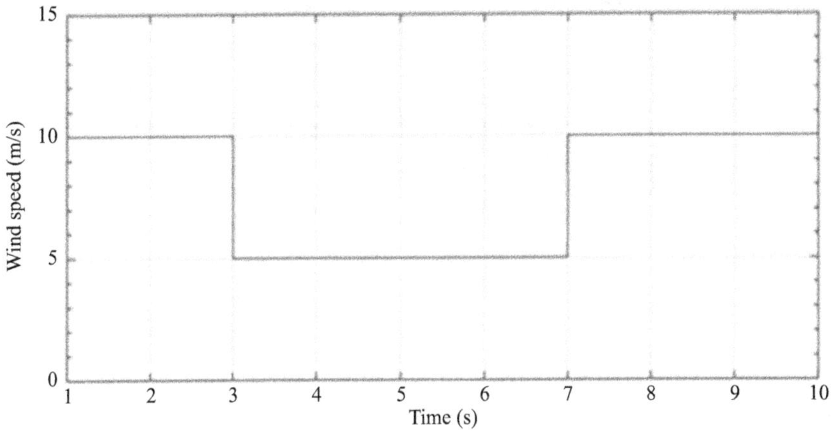

Figure 13.35 Wind speed variation

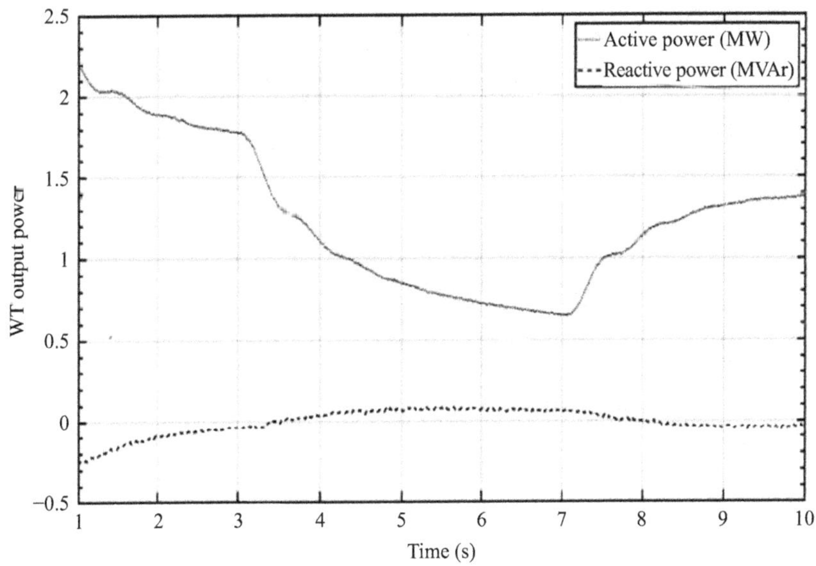

Figure 13.36 WT-generated power

The RMS values of the WT output current during different stages are shown in Figure 13.38. The converter maintains the three-phase RMS current values balanced during the different stages. The three-phase RMS voltage values are kept balanced and within the allowable range as shown in Figure 13.39.

The waveforms of the WT output currents and voltages are maintained balanced and sinusoidal during different generation stages as shown in Figure 13.40(a) and (b).

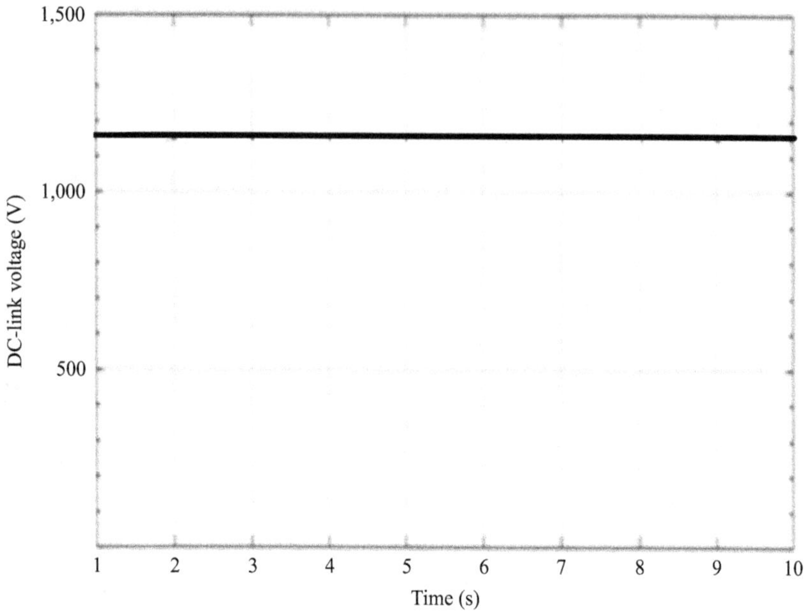

Figure 13.37 Back-to-back converter DC-link voltage

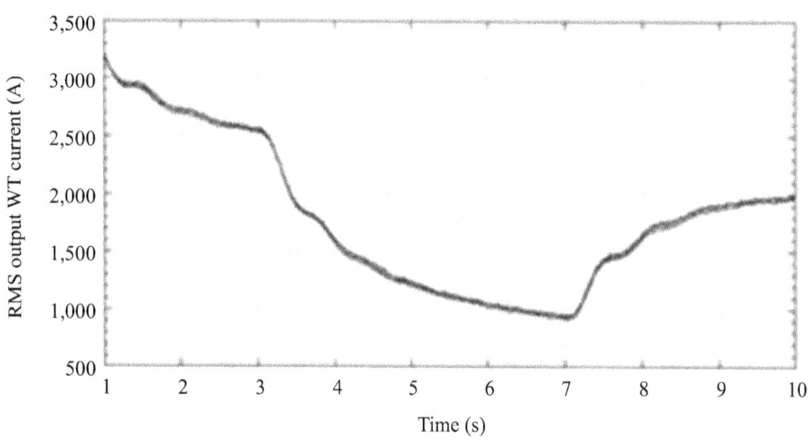

Figure 13.38 RMS values for WT output current

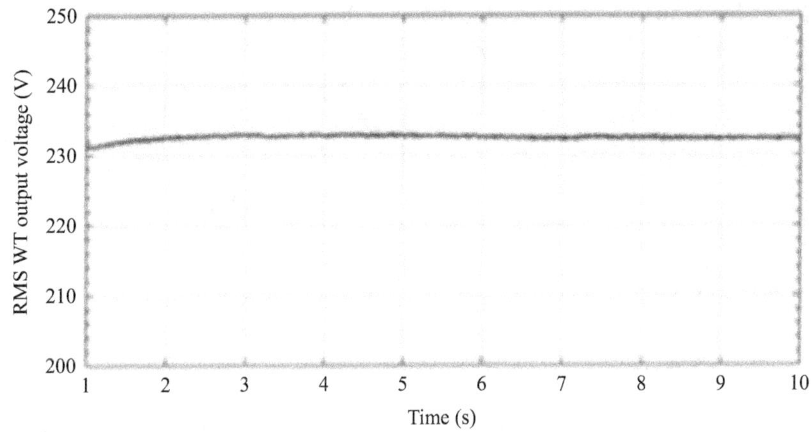

Figure 13.39 RMS values for WT output voltage

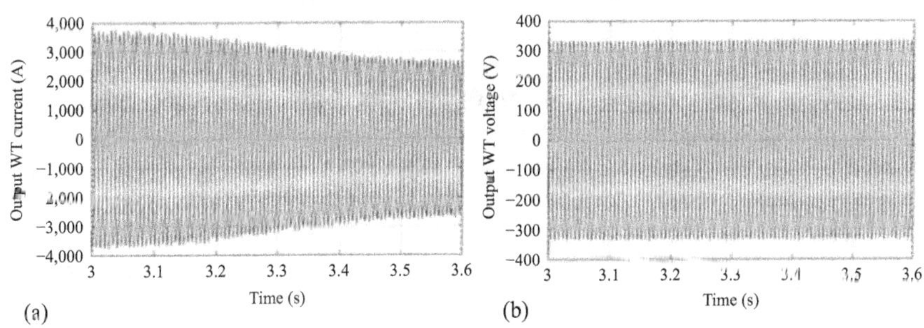

Figure 13.40 (a) Waveforms of WT output current and (b) waveforms of WT output voltage

References

[1] Chakraborty C., Iu H. H. and Lu D. D. 'Power converters, control, and energy management for distributed generation'. *IEEE Transactions on Industrial Electronics.* 2015;62(7):4466–70.

[2] Bimal B. K. *Modern power electronics and AC drives*. Vol. 123. NJ, USA: Prentice Hall; 2002.

[3] Erickson R. W. and Maksimovic D. *Fundamentals of power electronics*. New York: Springer Science+Business Media; 2007.

[4] Mohan N. *A first course on power electronics and drives*. New York: John Wiley & Sons Inc.; 2012.

[5] Rashid M. H. *Power electronics: Circuits, devices and applications*. 4th edn. London: Pearson Press; 2013.

[6] Fisher M. J. *Power electronics*. Boston, USA: PWS Kent Publishing Co; 1991.
[7] Thorborg K. *Power electronics*. New York: Prentice Hall; 1988.
[8] Vithayathil J. *Power electronics: Principles and applications*. New York: McGraw-Hill; 1995.
[9] Baxter R. *Energy storage – A nontechnical guide*. Tulsa, Oklahoma, USA: PennWell Press; 2006 p. 55.
[10] Bollen M. H. J. 'Understanding power quality problems'. *Voltage sags and interruptions*. New York: John Wiley & Sons; 2000.
[11] Sankaran C. *Power quality*. New York: CRC Press; 2001.
[12] Mittal N., Singh B., Singh S. P., Dixit R. and Kumar D. 'Multilevel inverters: A literature survey on topologies and control strategies'. *Proc. Int. Conf. Power Control Embedded Syst.*; Allahabad, India, Dec 2012, pp. 1–11.
[13] Asiminoaei L., Aeloiza E., Enjeti P. N. and Blaadjerg F. 'Shunt active-power-filter topology based on parallel interleaved inverters'. *IEEE Transactions on Industrial Electronics*, 2008;55(3):1175–89.
[14] Holtz J. 'Pulsewidth modulation – A survey'. *IEEE Transactions on Industrial Electronics*. 1992;39(5):410–20.
[15] Lascu C., Asiminoaei L., Boldea I. and Blaabjerg F. 'High performance current controller for selective harmonic compensation in active power filters'. *IEEE Transactions on Power Electronics*. 2007;22(5):1826–35.
[16] Holmes D. G. and Lipo T. A. *Pulse width modulation for power converters: Principles and practice*. Hoboken, NJ, USA: Wiley-IEEE Press; 2003.
[17] Alper Ç. and Ermis M. 'VSC-based D-STATCOM with selective harmonic elimination'. *IEEE Transactions on Industry Applications*. 2009;45(3):1000–15.
[18] Wong P., Xu P., Yang P. and Lee F. C. 'Performance improvements of interleaving VRMs with coupling inductors'. *IEEE Transactions on Power Electronics*. 2001;16(4):499–507.
[19] Shehu G. S., Kunya A. B., Shanono I. H. and Yalcinoz T. 'A review of multilevel inverter topology and control techniques'. *Journal of Automation and Control Engineering*. 2016;4(3):233–41.
[20] Chang G. W., Chu S. Y. and Wang H. 'A new method of passive harmonic filter planning for controlling voltage distortion in a power system'. *IEEE Transactions on Power Delivery*. 2006;21(1):305–12.
[21] Das J. C. 'Passive filters-potentialities and limitations'. *IEEE Transactions on Industry Applications*. 2004;40(1):232–41.
[22] Hamadi A., Rahmani S. and Al-Haddad K. 'A hybrid passive filter configuration for VAR control and harmonic compensation'. *IEEE Transactions on Industrial Electronics*. 2010;57(7):2419–34.
[23] Pasterczyk R. J., Guichon J. M., Schanen J. L. and Atienza E. 'PWM inverter output filter cost-to-losses tradeoff and optimal design'. *IEEE Transactions on Industry Applications*. 2009;45(2):887–97.
[24] Hatua K., Jain A. K., Banerjee D. and Ranganathan V. T. 'Active damping of output LC filter resonance for vector-controlled VSI-fed AC motor drives'. *IEEE Transactions on Industrial Electronics*. 2012;45(2):334–42.

[25] Michels, L., de Camargo R. F., Botteron F., Grudling H. A. and Pinheiro H. 'Generalised design methodology of second-order filters for voltage-source inverters with space-vector modulation'. *IEE Proceedings-Electric Power Applications.* 2006;153(2):219–26.

[26] Dewan S. B. and Ziogas P. D. 'Optimum filter design for a single-phase solid-state UPS system'. *IEEE Transactions on Industry Applications.* 1979;15(6):664–9.

[27] Dewan S. B. 'Optimum input and output filters for a single-phase rectifier power supply'. *IEEE Transactions on Industry Applications.* 1981;17(3): 282–8.

[28] Kim J., Choi J. and Hong H. 'Output LC filter design of voltage source inverter considering the performance of controller'. *PowerCon 2000. Proceedings of 2000 International Conference on Power System Technology;* Perth, WA, Australia, Dec 2000. IEEE; 2000, Vol. 3, pp. 1659–64.

[29] Strom J. P., Korhonen J., Tyster J. and Silventoinen P. 'Active *du/dt* – New output-filtering approach for inverter-fed electric drives'. *IEEE Transactions on Industrial Electronics.* 2011;58(9):3840–7.

[30] Dahono P. A. 'A method to damp oscillations on the input LC filter of current-type AC–DC PWM converters by using a virtual resistor'. *Proceedings of the 25th International Telecommunications Energy Conference, 2003. INTELEC'03*; Yokohama, Japan, Oct 2003. IEEE; 2003, pp. 757–62.

[31] Guo S. and Liu D. 'Analysis and design of output LC filter system for dynamic voltage restorer'. *2011 Twenty-Sixth Annual IEEE Applied Power Electronics Conference and Exposition (APEC)*; Fort Worth, TX, USA, March 2011. IEEE; 2011, pp. 1599–605.

Chapter 14
Integration of distributed energy resources

14.1 Introduction

To solve problems such as high energy costs or low electric power reliability at consumer's facilities, distributed energy resources (DERs) could be the solution that energy managers are looking for.

DERs can deliver the same electricity services provided by centralized resources, including large-scale generators, and transmission and distribution network assets. Many DERs can be deployed at different scales and exhibit economical and technical benefits. For instance, solar PVs can be deployed at the kilowatt scale on residential rooftops, at the scale of several hundred kilowatts to 10 MW on commercial rooftops or ground-mounted arrays, or tens to hundreds of megawatts rating in the utility-scale solar farms. DER technology is developing at a fast rate, and it can provide electrical energy where required. In addition, DER systems can be either connected to the local electric grid or isolated from the grid in stand-alone applications [1].

DER technologies are mainly classified into distributed generation (DG) resources (renewable and non-renewable) and energy storage systems (ESSs) as depicted in Figure 14.1 [2]. It includes the following:

- As explained in Part I, renewable energy sources, such as wind, solar, ocean and biomass energy, (i) can be directly connected to the consumers' load on the distribution systems at the medium- or low-voltage level, (ii) can be dispatched to the grid based on the request of the utility grid operators or (iii) can operate as a stand-alone system to serve a particular load or an area.
- Non-renewable energy sources have the technologies that use fossil fuels (e.g. natural gas, coal and oil) to produce energy for various operations. They can be used for power generation and thermal energy through a co-generation arrangement. Examples of non-renewable technologies are the following:
  - *Microturbines*, which run on natural gas at high speed (~90,000 rpm) to drive an electric generator with power rating of 30–250 kW for small industrial and commercial applications. They have the advantage of very low emissions.
  - *Gas turbines (or combustion turbines)*, which use gas or liquid fuel. They are suited for peaking and load-following applications as well as for base-load operation in large sizes.

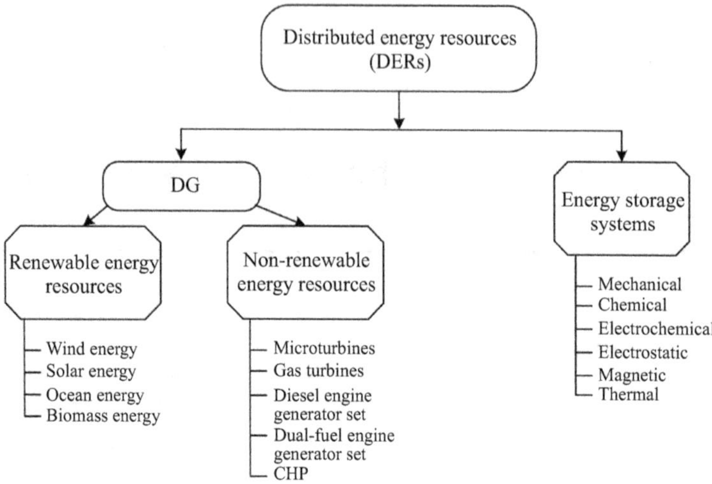

Figure 14.1 Classification of DERs

- *Diesel engine generator sets:* They use diesel as fuel and consist of a compression ignition reciprocating engine as a prime mover coupled to an electrical generator. They operate at a relatively high compression ratio and relatively low speed. Diesel engines are manufactured in a wide range of sizes, from 1 kW to 10 MW, and can be cycled frequently to operate as peak-load or load-following power plants. In stand-alone operation, diesel engines can be run in base-load mode. Major drawbacks are very high level of both emissions and noise, but on the other hand diesel engines generator sets are cost-effective, reliable and a widely used technology. To reduce the level of emissions, the engine is modified using a mixture of diesel (low percentage) and natural gas (as high percentage) fuel. Diesel fuel allows the use of compression ignition whereas the natural gas results in low emissions. The engine generator set with this capability is called 'dual-fuel engine generator set'.
- Cogeneration (combined heat and power (CHP)): Many types of DG can provide useful and valuable thermal energy by capturing excess heat energy produced during electricity generation and using it to heat or cool water or interior space. The excess heat can be also used as energy input to an industrial process, in lieu of burning fuel. This energy capture and reuse process is known as CHP or cogeneration. So, plants including CHP units are designed to produce thermal energy (heating and/or cooling) and electricity, simultaneously, from a single source of energy in one process. CHP units are sized to suit large-scale commercial buildings, institutional facilities and communities with ways to reduce energy costs and total emissions as well as providing more resilient and reliable electric power and thermal energy.

- ESSs are responsible to store the imported electricity from a power grid at off-peak periods in a form that can be converted back into electrical energy when it is required. ESS technologies are classified into mechanical, chemical, electrochemical, electrostatic, magnetic and thermal ESSs as explained in Part II.

It is to be noted that some DER technologies, such as engines and combustion turbines, produce AC power directly and can be connected to the electric grid or AC loads in a straightforward manner. Other technologies, such as photovoltaics (PVs) and fuel cells, produce DC power, which must be passed through a power electronics device called an inverter or power conditioning unit to produce AC output that can then be connected to the AC system. Electric power generated by wind-driven turbines is generally variable frequency and requires rectifier/inverter to convert to constant frequency power.

14.2 Powering DG systems

Different types of fuel based on availability, price and storage can be used to power DG systems as in the following.

14.2.1 Non-renewable resources

Diesel fuel is readily available at a reasonable cost; however, it requires an on-site storage tank and periodic deliveries, and spills or fumes are always possible.

Biodiesel fuel can be an alternative for diesel engine generator sets in areas where it is available. Biodiesel is made by reacting natural oils and fats with an alcohol. It can significantly reduce CO_2, SO_2 and matter emissions, but it may increase NO_x in some cases.

Natural gas is a cleaner alternative to diesel, and comparable in terms of cost. Gas delivery and a storage tank may be required only when it is expensive or difficult to run a gas line. Propane can be also used as an alternative to diesel fuel if the natural gas is not available.

14.2.2 Renewable resources

Biological materials such as agricultural crops, residues from the food, fiber, trees, grasses and forest product industries can be used in place of fossil fuels to generate electricity. In addition, municipal solid waste facilities and sewage treatment plants can also produce fuel to power DG renewable technologies.

PV systems depend on the availability of enough sunlight in a specific area. By paying more attention to the design and placement of the solar panels, and adding ESSs (typically, the deep discharge batteries), the potential of solar installation can be maximized.

Wind energy is highly dependent on location, especially on the average wind speed in a specific locale and the daily and seasonal variations in wind speed. So, adding storage can pay off the overall functionality of a wind system.

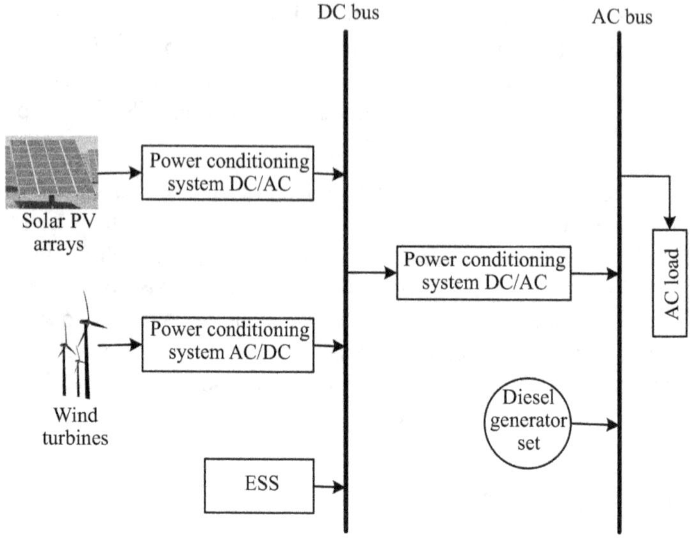

Figure 14.2 PV/wind/ESS/diesel generator hybrid system

Hybrid systems, such as PV and wind in addition to storage and diesel backup generator, Figure 14.2, may be necessary and cost-effective in situations where high reliability and/or stand-alone operations are required. Incorporating both ESS and diesel generator minimizes the risks associated with the uncertainty of renewable output, and consequently, the system reliability and energy services are improved.

14.3 Benefits of DGs

The beneficial effects that result, or are expected to result, from the installation and operation of DG in the electric system may include the following.

Environmental benefits that can be determined by the value of emissions offsets or other environmental benefit such as

- reduction in land-use effects,
- reduction in greenhouse gas (GHG) emission pollutants, environment-friendliness and reduction in health costs with renewable DG.

Technical benefits:

- Generation, transmission and distribution capacity deferral, which can be evaluated by the financial value of deferring or avoiding a capital investment in central generation capacity, transmission system capacity and distribution system capacity, respectively.
- Usage of CHP results in cost savings from exploiting waste heat.

- Ancillary services provided by DG such as spinning reserve, regulation or other ancillary services yield a potential revenue.
- Providing voltage control/var production results in a potential revenue.
- Reduction of system resistive losses and better control of reactive power.
- Reliability enhancement where both capacity release and peak power are reduced as well as the generation diversity is improved.
- Improving the quality of the power at or nearby customer sites.
- Security enhancement for critical loads and power utilities. In addition, security risks to the grid, impacts of cyberattacks and vulnerability of terrorist attacks are reduced.
- Standby generation, which can help provide generation capacity dispatchable by the utility or the customer during emergencies.

Economic benefits:

- Reduction in the cost of installation, auxiliaries' costs, loss-associated costs and operation and maintenance (O&M) costs.
- No fuel cost with renewable DG.
- Deferments of investment in infrastructures.
- Maintaining constant running cost for longer-time period.
- Achieving significant cost savings by utilizing the flexibility of DG, which can start up and ramp quickly.

Furthermore, DG units incorporating ESSs (i.e. DERs) can be installed not only as stand-alone energy resources to serve a specific load or an area but can be integrated with the utility grid. The integration of DERs in a power system is only accepted by utilities if the net benefit exceeds the investment costs of installation. So, some parameters such as the benefit-to-cost ratio, cost of energy, internal rate of return, return on investment and payback period should be estimated and the revenue of the benefits, which result from the integration of DERs, should be evaluated in comparison to the conventional alternatives. However, economics of DER integration benefits can be analysed by considering the following.

Economic benefits of DERs integration in power systems. To evaluate the potential of DERs integration, it is crucial to identify the costs and benefits. DERs compete with conventional generation and network assets to provide electricity services. They can be deployed closer to the point of electricity consumption and in locations inaccessible to more centralized resources. Consequently, DERs can provide value to power systems where and when they can mitigate high marginal network losses, relieve transmission network congestion or defer impending upgrades and enhance reliability. These benefits are highly dependent on the sizing and location of DERs, system configuration and type of DER technology. So, DERs should be at locations in power grids where they are most valuable. If sited at the right locations and operated at the right times, DERs can deliver more locational value. Otherwise, this could result in negative effects such as an increase in system losses, voltage drop, harmonics and low voltage stability.

Locational value is constituted of four major items: (i) the cost of power loss reduction by avoiding the electrical energy to be delivered in areas that experience high marginal losses, (ii) the cost of avoiding or delaying an upgrade when an investment in new network capacity would otherwise be required, (iii) the value that customers place on avoiding service interruptions due to enhancing reliability or resilience by supplying energy to load during network failures [3] and (iv) the recent trend of providing electricity to the end users through low-carbon electricity systems (e.g. renewables), in combination with the base-load conventional plants, means that the provision of flexibility and resilience will increasingly shift towards the flexibility inherent in distributed sources provided by end consumers. Thus, renewable DG will transform passive consumers to active prosumers (professional consumers) who may provide energy and service flexibility to both local and entire systems (Figure 14.3).

Therefore, flexible DG, distributed ESSs as well as demand side management (DSM) will help achieve significant cost savings in comparison to a system in which flexibility and security of supply are delivered by conventional generation [4].

However, economies of size (known as unit scale) should be investigated when using DER technologies. The unit costs of energy technologies typically fall as the technology is installed at larger scales; for example, a 500-MW system of a given technology will typically cost less per megawatt than a 5-MW system of the same type. Fortunately, many technologies of DERs such as solar PVs, fuel cells and electrochemical energy storage can be deployed across a wide range of scales. In addition, the rate at which economies of scale are reduced is much more rapid for DER technologies than that of traditional generation technologies. For instance, economies of two PV systems of scales one to tens of megawatts and tens to hundreds of megawatts have often cost roughly the same per unit installation cost. In contrast, traditional nuclear, hydroelectric, natural gas and coal-fired power plants typically exhaust economies of scale between several hundred megawatts and more than 1,000 MW [3].

Therefore, an optimal trade-off between the value of locational benefits and economies of unit scale should be found.

To select the optimum DGs configuration, application of optimization techniques is necessary. Integration of renewable DG into a power system commonly

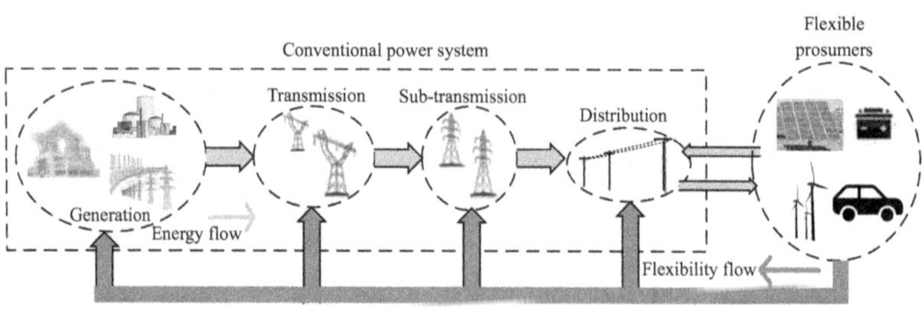

Figure 14.3 Energy and flexibility flow

incorporate an ESS to make it dispatchable and thus more network capacity can be delivered. For example, unit commitment and economic dispatch techniques can be applied to implement economic analysis on different technologies and prioritize them according to a desired criterion.

Emissions reduction (ER) benefits with integration of renewable DG units. One of the major environmental impacts is the GHG emissions. It is widely accepted that integration of renewable DG technologies into a power system can to a large extent reduce the GHG emissions in addition to reduction of other pollutants such as NO_2, SO_2 and CO_2 that can be produced by the traditional fossil fuel power plants.

The percentage of ER for a specific pollutant (ℓ) by incorporating renewable DG can be computed as in the following [2]:

$$ER_\ell = \frac{PE_{\ell/\mathrm{wRE}}}{PE_{\ell/\mathrm{woRE}}} \times 100 \tag{14.1}$$

where $PE_{\ell/\mathrm{wRE}}$ and $PE_{\ell/\mathrm{woRE}}$ are, respectively, the quantity of emissions for the ℓth pollutant with and without application of renewable DG, and

$$PE_{\ell/\mathrm{wRE}} = \sum_{i=1}^{m} x_i y_{\ell i} + \sum_{j=1}^{n} x_j y_{\ell j} \tag{14.2}$$

$$PE_{\ell/\mathrm{woRE}} = \sum_{k=1}^{m} x_k y_{\ell k} \tag{14.3}$$

where

m = number of conventional generating units;

n = number of renewable DG units;

x_i = quantity of electricity produced by the ith conventional power unit with the application of renewable DG (MWh);

$y_{\ell i}$ = quantity of emission of the ℓth pollutant for the ith conventional power unit per MWh of electricity produced;

x_j = quantity of electricity produced by the jth DG unit (MWh);

$y_{\ell j}$ = quantity of emission of the ℓth pollutant for the jth DG unit per MWh of electricity produced;

x_k = quantity of electricity produced by the kth conventional power unit without application of renewable DG (MWh); and

$y_{\ell k}$ = quantity of emission of the ℓth pollutant for the kth conventional power unit per MWh of electricity produced.

If the number of pollutants is $\mathscr{H}$, the total quantity of emissions with and without the application of renewable DG can be given by

$$PE_{\mathrm{total/wRE}} = \sum_{\ell=1}^{\mathscr{H}} PE_{\ell/\mathrm{wRE}} \quad \text{and} \quad PE_{\mathrm{total/woRE}} = \sum_{\ell=1}^{\mathscr{H}} PE_{\ell/\mathrm{woRE}} \tag{14.4}$$

Thus, (14.1) can be rewritten as

$$ER_{\text{total}} = \frac{PE_{\text{total/wRE}}}{PE_{\text{total/woRE}}} \times 100 \tag{14.5}$$

From (14.4), the saving in quantity of pollutant emissions, PE_{saving}, is given by

$$\begin{aligned} PE_{\text{saving}} &= PE_{\text{total/woRE}} - PE_{\text{total/wRE}} \\ &= \sum_{\ell=1}^{\mathcal{H}} PE_{\ell/\text{woRE}} - \sum_{\ell=1}^{\mathcal{H}} PE_{\ell/\text{wRE}} \end{aligned} \tag{14.6}$$

Hence, the cost saving because of pollutant ER, $C_{\text{total/saving}}$, can be computed by

$$C_{\text{total/saving}} = C_{\text{av}} \times PE_{\text{saving}} \tag{14.7}$$

where C_{av} is the average price per unit quantity of pollutant emissions.

14.4 Operation requirements for DERs integration into power systems

DERs, especially inverter-based resources, will need to provide more grid services and functionalities that have been supplied by other means.

14.4.1 Voltage and frequency function

Renewable DG resources (e.g. solar PV and wind) supply a fluctuating power depending on the climatic conditions. To integrate such resources into a power system, the voltage and frequency at the point of coupling in case of normal and abnormal operating conditions must be stabilized. An increase/decrease of resources generation tends to cause over/under-voltage, respectively. In addition, the balance between generation and load is necessary to support the grid frequency.

Renewable DGs are coupled to the grid through power electronic inverters, which are different from the traditional rotating machines. So, inverters should participate in maintaining stable and reliable operation. The most feasible way to do that is by autonomously responding to the AC terminal voltage waveform of the inverter in a way that voltage and frequency are stabilized [5].

Voltage regulation. Inverters have two output parameters, reactive power and active power, available to mitigate voltage deviation through a process called volt-var control. In this process, reactive power is modulated in proportion to voltage deviation, absorbing or producing reactive power for high and low voltages, respectively, with minimal impact on the produced active power. Reducing active power can also bring down the voltage, that is, mitigate over-voltage by a process called volt-watt control, but under this control mode, the produced active power is directly reduced. However, volt-var control is mostly preferred, whereas volt-watt control is only used as an option when the voltage rise is very high.

Frequency regulation. Supporting grid frequency necessitates a balance between generation and load, that is providing more generation or less load when frequency is low and reducing generation or more load when frequency is high. Currently, inverters can reduce power in response to over-frequency events by a process called frequency-watt control. If an inverter has additional active power available, it can increase power in response to under-frequency. A simplified block diagram indicating the inverter associated with volt-var and frequency-watt controls for solar PV resource integration, as an example, is shown in Figure 14.4.

Voltage and frequency ride through functionality. It is the most basic grid-support function required from inverter-based DERs. This function enables DERs to ride-through voltage and frequency disturbances in lieu of tripping offline [6]. Tripping of inverters could exacerbate power system disturbances causing island-wide blackout. IEEE standard-1547 requires DER resources to have ride-through capability according to one of three categories: Categories I, II and III. Each category specifies the DER behaviour within operating regions called continuous operation region, mandatory operation region, permissive operating region and momentary cessation operating region. Each region is defined by upper and lower (U/L) voltage bounds and the corresponding cumulative time tome for ride-through and clearing time for trip as summarized in Table 14.1.

For frequency ride-through, the same methodology is applied and DG behaviour is as given in Table 14.2.

Additional capabilities for inverter ride-through such as high rate of frequency and voltage change, voltage-phase-angle change, phase imbalances and multiple successive disturbances ride-through are required. With such capabilities, inverters can remain connected under abnormal conditions and that is crucial to grid stability.

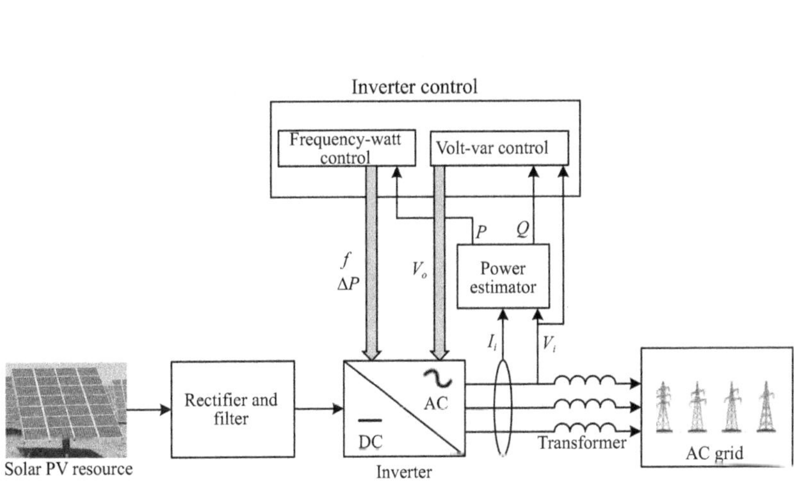

Figure 14.4 Simplified schematic diagram of an inverter control for PV integration

Table 14.1 DER device behaviour for voltage ride-through Categories I, II and III

DER device behaviour (operation regions)	Voltage ride-through categories					
	Category I		Category II		Category III	
	U/L voltage bounds (p.u.)	Clearing time limits (s)	U/L voltage bounds (p.u.)	Clearing time limits (s)	U/L voltage bounds (p.u.)	Clearing time limits (s)
Continuous operation	0.90–1.10	0.01–1,000	0.90–1.10	0.01–1,000	0.90–1.10	0.01–1,000
Mandatory operation	0.70–0.88	0.01–1.20	0.65–0.88	0.01–0.80	0.50–0.88	0.01–0.40
Permissive operation	0.50–0.70	0.01–0.30	0.30–0.50	0.01–0.3	—	—
	1.10–1.20	0.01–1.00	1.10–1.20	0.01–1.00		
Momentary cessation operation	—	—	—	—	0.00–0.50	0.01–1.00
					1.10–1.20	0.01–11.00

For all categories: at regions of U/L 0.65–0.90 p.u. with clearing time 60–1,000 s and U/L 1.20–1.30 p.u. with clearing time 0.01–0.10 s, DG behaviour is 'may ride-through or may trip'. At regions of U/L 0.0–0.5 p.u. with clearing time 1–1,000 s and U/L 1.10–1.30 p.u. with clearing time 10–1,000 s, DG behaviour is 'shall trip'.

Table 14.2 DER device behaviour for frequency ride-through Categories I, II and III

DER device behaviour/ operation regions	Regions for Categories I, II and III	
	Frequency range (Hz)	**Cumulative time for ride-through and clearing time for trip (s)**
Continuous operation	59–61	0.01–1,000
Mandatory operation	57 to <59	0.01–500
	>61–61.7	0.01–500
May ride-through or may trip	56–57 and 61.7–63	0.01–0.2
	56.5 to <57 and 61.7–62.2	0.2–500
Shall trip	56–56.5 and 62.2–63	0.2–500
	56–58.5 and 61–63	500–1,000

The rated frequency is 60 Hz.

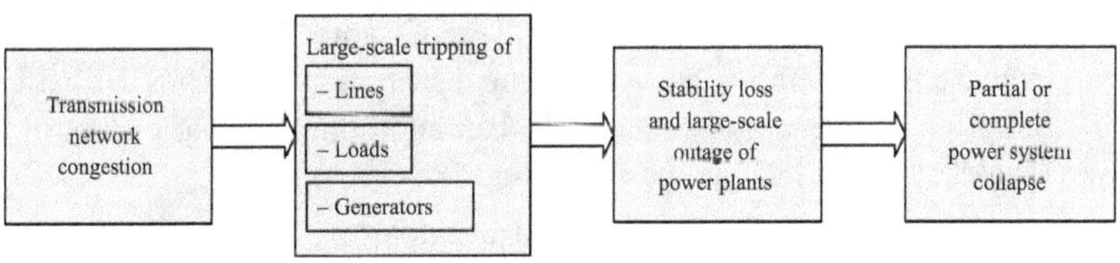

Figure 14.5 Cascade processes of blackout

14.4.2 Power system restoration

As a result of major emergencies (e.g. dangerous congestion or severe faults), blackouts occur in transmission networks due to cascade tripping of transmission lines, loads and generators. These can jeopardize the stable operation regarding voltage and frequency. Thus, loss of stability yields partial or complete collapse of the power system (Figure 14.5) [7]. However, critical infrastructure, such as hospitals and communication infrastructure, has backup power sources to confront power outages, but for a specific period. Therefore, restoring the power system to its stable operation is of utmost importance, especially if the cascade processes develop very fast with a speed exceeding many times the capability of personnel to perceive the situation. Furthermore, the conventional thermal power plants need hours to restart-up.

Restoration procedures can be carried out through three consecutive stages (Figure 14.6) [8]:

1. Collecting all necessary and available status information from the power system control centre, which is securely communicated with local controllers at

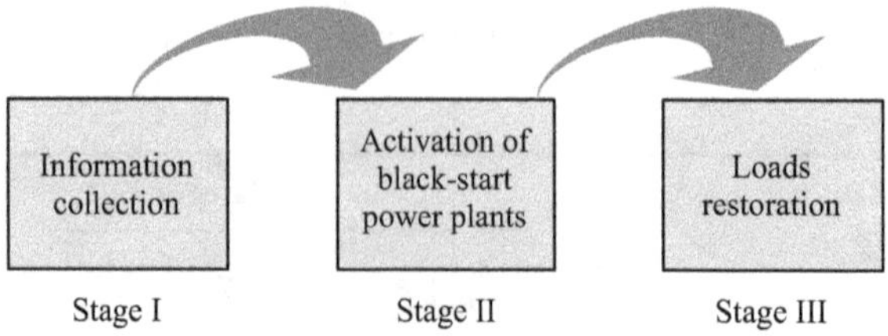

Figure 14.6 Restoration procedures

different desired locations, to be acquainted with the availability of power plants and grid infrastructure.

2. Activating black-start power plants (e.g. hydropower plants) to energize the transmission network assets, lines and transformers, and supply cranking power to non-black-start thermal power plants to enable them to start power generation.
3. Restoring the loads.

Role of DERs. Use of DERs, especially renewable DGs such as wind and solar, to increase the generation capacity of the distribution grid leads to challenges in the restoration process, and meanwhile has additional potential control capabilities. The challenges include the following:

- The active power is dependent on weather conditions and, consequently, may reduce the availability of wind and solar power plants compared to non-renewable DGs (e.g. biomass power plants and CHP units), and that might disturb the restoration process in weak grid situation.
- Large penetration of uncontrollable weather-dependent feed-in from DERs can exceed the power demand, and thus, impact the stable operation of thermal power plants regarding minimum loading requirements and ramp rates.
- Interfacing large-scale DERs through electronic power conditioning circuits can impact the system inertia.
- Communication links might not work in blackout situation leading to unavailability of information that is required in the first stage of restoration procedures.
- Dynamics of frequency-dependent function are defined by the manufacturers, not by most grid codes.
- Limitation of some control parameters, e.g. ramp rate, by manufacturers and grid codes.

On the other hand, DERs are expected to become major contributors to generation-load balance of the future grid. Biomass and hydropower plants are normally interconnected with synchronous generators. Solar and wind power plants use power electronics for grid integration. Control capabilities of synchronous generators include inertia and standard grid-forming modes, whereas the control capacities of inverter-based DGs include very fast control capabilities, possibility of virtual inertia and potential capabilities in power system restoration including the following:

- Starting up with comparably little auxiliary power within seconds or minutes, whereas the thermal power plants need hours.
- Supporting power plants cranking and improving frequency stability.
- Using the reactive power feed-in capabilities to support voltage stability.
- Controlling DERs to maintain generation-load balance.
- Forming grid islands from DERs to reduce blackout duration. Backup generator sets or CHP units are state of the art for critical infrastructure.

14.4.3 Confirming reliability

Reliability is one of the key performance indicators that the planners of utilities use to investigate the power system adequacy and security. Power utilities must strategize different mitigation techniques for power system reliability improvement such as system structure, integration of DER units, installation of protection and switching devices, and so on.

Incorporation of DER units into the electrical distribution system with a goal of supporting the power system reliability entails the following:

- Coordination among all generation, including DER units, in an optimal manner, e.g. specifying the penetration level of DER units under the existing constraints.
- Capability of DER units to maintain a continuous power supply at the load points in the event of a power outage from the utility.
- Serving all necessary distributed system control capabilities by DER units.

However, a substantial improvement in the reliability of the distribution system with penetration of DER units should be achieved. This can be confirmed by calculating the reliability indices of the distribution system as reported in IEEE Std. 1366^{TM}-2003 as in the following.

14.4.3.1 Sustained interruption indices

System Average Interruption Frequency Index (SAIFI): It is the average frequency of sustained interruptions per customer over a predefined area:

$$\text{SAIFI} = \frac{\text{Total number of customers interrupted}}{\text{Total number of customers served}}$$

System Average Interruption Duration Index (SAIDI): It measures the total duration of interruptions. It is cited in units of hours or minutes per year:

$$\text{SAIDI} = \frac{\text{Sum of all customer interupption durations}}{\text{Total number of customers served}}$$

Customer Average Interruption Duration Index (CAIDI): It is the average time needed to restore service to the average customer per sustained interruption:

$$\text{CAIDI} = \frac{\text{SAIDI}}{\text{SAIFI}} = \frac{\text{Sum of all customer interruption durations}}{\text{Total number of customers interrupted}}$$

Customer Total Average Interruption Duration Index (CTAIDI): It represents the total average time in the reporting period that customers who experienced an interruption were without power. This index is a hybrid of CAIDI and is similarly calculated except that those customers with multiple interruptions are counted only once:

$$\text{CTAIDI} = \frac{\text{Sum of customer interruption durations}}{\text{Total number of customers interrupted}}$$

Customer Average Interruption Frequency Index (CAIFI): It gives the average frequency of sustained interruptions for those customers experiencing sustained interruptions. The customer is counted once regardless of the number of times interrupted for this calculation:

$$\text{CAIFI} = \frac{\Sigma N_i}{N_c}$$

where N_i = number of interrupted customers for each sustained interruption event during the reporting period and N_c = total number of customers who have experienced a sustained interruption during the reporting period.

Average System Availability Index (ASAI): It represents the fraction of time (often in percentage) that a customer has received power during the defined reporting period:

$$\text{ASAI} = \frac{\text{Customer hours service availability}}{\text{Customer hours service demand}}$$

Customers Experiencing Multiple Interruptions (CEMI$_n$): It indicates the ratio of individual customers experiencing more than n sustained interruptions to the total number of customers served:

$$\text{CEMI}_n = \frac{N_n}{\text{Total number of customers served}}$$

where N_n = total number of customers that experience more than n sustained interruptions.

14.4.3.2 Momentary indices

Momentary Average Interruption Frequency Index (MAIFI): It indicates the average frequency of momentary interruptions:

$$\text{MAIFI} = \frac{\sum \text{Total number of customer momentary interruptions}}{\text{Total number of customers served}}$$

Momentary Average Interruption Event Frequency Index ($\mathbf{MAIFI_E}$): It has the same definition as MAIFI, but it does not include the events immediately preceding a lockout:

$$\mathrm{MAIFI_E} = \frac{\sum \text{Total number of customer momentary interruption events}}{\text{Total number of customers served}}$$

14.4.3.3 Load-based indices

Average System Interruption Frequency Index (ASIFI): It includes the magnitude of the load unserved during an outage:

$$\mathrm{ASIFI} = \frac{\sum kVA_{\text{sustained}}}{kVA_{\text{served}}}$$

Average System Interruption Duration Frequency Index (ASIDI): It includes the magnitude of the load unserved during an outage.

$$\mathrm{ASIDI} = \frac{\sum kVA_{\text{sustained}} D_{\text{sustained}}}{N_{\text{served}}}$$

14.4.4 Virtual inertia

Virtual inertia is also called 'synthetic inertia'. Some renewable DG resources may still use a thermal cycle for power production like conventional sources, whereas others, such as solar PV systems and most modern wind turbines, are interfaced with utility grids through inverters. This means that the inverter-based generation is decoupled from the grid and its frequency excursions as it does not provide any mechanical inertial response and, hence, compromises frequency stability [9].

Imbalances between active power production and load consumption lead to frequency variations. The frequency drops when the consumption rises rapidly, or a large generating plant shuts down. Three main control actions are used for frequency regulation: (i) primary control by which the governor response takes place within the first few seconds (10–30 s); (ii) secondary control which is the automatic generation control – its action takes place within minutes (10–30 min) and restores the system frequency back to the nominal value; and (iii) tertiary control action is the reserve deployment when actions are taken to get the resources in place to counteract present or future disturbances in the system [10,11].

In the early stage, and before the primary control reacts to changes, the generators cannot respond instantaneously to balance the system. A physical response called 'inertial response' comes from the conventional rotating generators where the kinetic energy stored in the rotating mass is responsible for counteracting the energy imbalance by slowing the rate of change of frequency (RoCoF) until the primary control activates. In case of using renewable DGs, the inertial response decreases, and thus, RoCoF is increased leading to a low-frequency nadir (Figure 14.7).

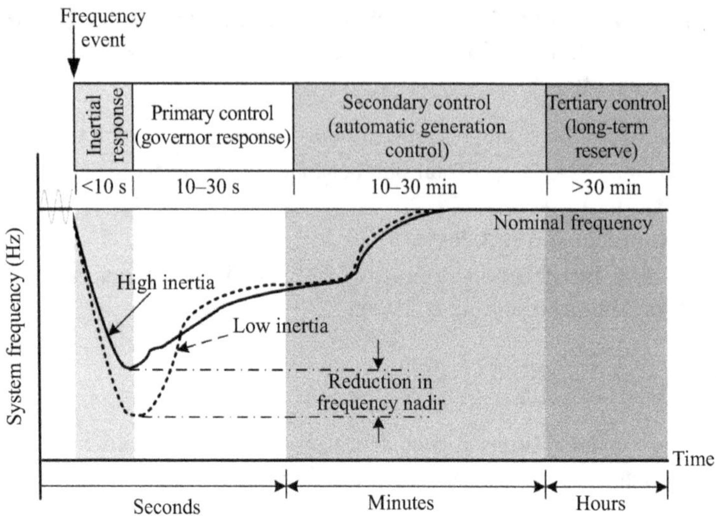

Figure 14.7 Frequency response and control actions

Therefore, in systems with lower inertia, the frequency nadir is lower and RoCoF is getting higher. This can lead to under-frequency load shedding by frequency relay tripping and may yield cascaded outages. Adding virtual inertia to the system in a condition of operation in a very short time interval (<10 s) and in autonomous fashion is the solution to such situations. Furthermore, the system stability is improved and higher level of renewable DGs is permitted when deploying virtual inertia system appropriately.

Various topologies used for virtual inertia implementation are presented in literature [11–13]. They are designed to effectively emulate inertia of synchronous generators through power electronic-based inverters. The concept of virtual inertia topologies can be summarized as in the following.

Using the swing equation, the frequency variation in a power system after a frequency event can be expressed as

$$\frac{2H}{\omega_g}\frac{d\omega_g}{dt} = \frac{\Delta P}{S_g} \tag{14.8}$$

and

$$\Delta P = P_{gen} - P_{load} \tag{14.9}$$

where H, the inertia constant, is expressed as the kinetic energy normalized to apparent power S_g of the connected generators in the system, ω_g is the system angular speed (rad/s), P_{gen} is the generated power and P_{load} is the power demand including losses.

In terms of frequency, f (Hz), (14.8) can be rewritten as

$$\frac{2H}{f}\frac{df}{dt} = \frac{\Delta P}{S_g} \tag{14.10}$$

RoCoF of the system is mathematically given by df/dt, which is inversely proportional to the system inertia. Thus, the system requires additional inertia as more renewable DGs are integrated into the power system. The key idea of the virtual inertia is depicted in Figure 14.8. A typical combination of inverter-based DER units, solar PV system, wind turbines and ESS-like batteries is integrated into a power grid. PV system and ESS have a DC–DC converter, whereas the wind turbines are mostly variable speed wind turbines and have an AC–DC converter. So, all units are interfaced through a DC–AC inverter in the front end and do not contribute to the inertial response. The virtual inertia system receives voltage and current feedback signals from the inverter output to generate gating signals by applying an algorithm. The virtual inertia algorithm is the major part of the virtual inertia emulation system. It presents the different DER units interfaced to the power grid through electronic power conditioning systems as synchronous generators.

As an example, to illustrate the main components and the principle of operation of the virtual inertia system as a controller, one of the topologies called 'Ise lab's topology' is briefly discussed and its schematic diagram is shown in Figure 14.9. This topology is a swing equation-based topology as it solves the swing equation every control cycle to emulate inertia. The principle of operation includes the following steps:

- The controller senses the inverter output current, i, and voltage, v, to compute the grid frequency, ω_g, and the corresponding power, P_{out} by frequency/power

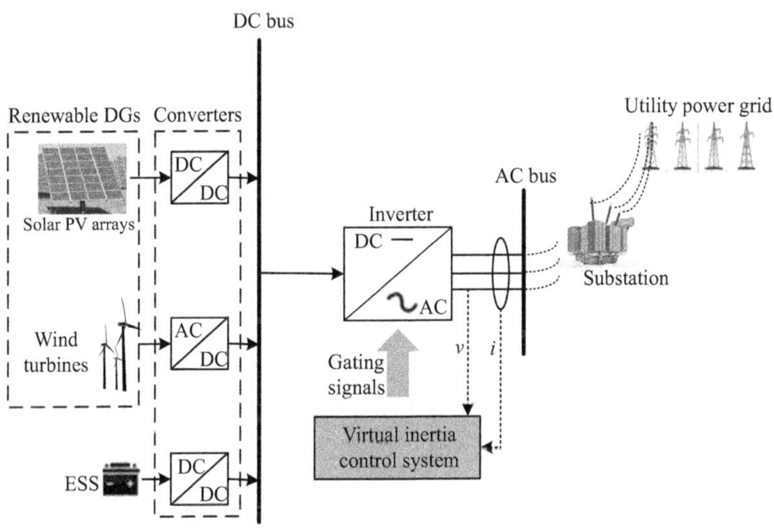

Figure 14.8 Concept of virtual inertia

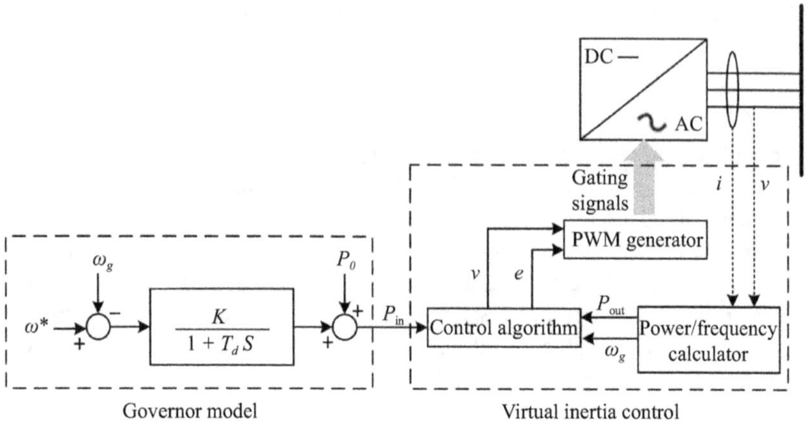

Figure 14.9 Schematic diagram of the key elements of virtual inertia control and governor model

calculation unit. These two parameters along with the prime mover input power, P_{in}, are inputs to the control algorithm unit.

- The control algorithm unit solves the swing equation every control cycle providing the phase command, θ, as well as the voltage reference, e, which is generated through Q-v droop approach. The two parameters, θ and e are inputs to the PWM generator. The typical swing equation of a synchronous generator is

$$\Delta P = P_{\text{in}} - P_{\text{out}} = J\omega_m\left(\frac{d\omega_m}{dt}\right) + D_p\Delta\omega \tag{14.11}$$

and

$$\Delta\omega = \omega_m - \omega_g \tag{14.12}$$

where P_{out} is the inverter output power, ω_m is the virtual angular frequency, ω_g is the grid/reference angular frequency, J is the moment of inertia and D_p is the damping factor. P_{in} is the input power, which is like the prime mover input power in a synchronous generator and can be computed by utilizing a governor model. The governor is modelled, in a simplest way, as a first-order lag element with gain and time constant of K and T_d, respectively. The output of the model, P_{in}, is based on the frequency deviation from a reference frequency ω^* and a continuous power reference, P_0 (Figure 14.9).

- The PWM generator sends the gating signals to the inverter.

14.5 Microgrids

Adding to the benefits of integrating the DER units into the power grid and counteracting the challenges that can be met, the new trend is towards gathering the

DER units with the loads present in their vicinity in a separate gird called 'micro-grid' (MG). Depending on the request of distribution system operator (DSO), the MG can be connected to or disconnected from the utility distribution system through a point of common coupling (PCC). Furthermore, the distribution system may include more than one grid, some within the entire area covered by utility grid and others located remotely and isolated from the utility grid (Figure 14.10).

14.5.1 MG concept

The MG is structured as a group of DERs and loads. It can be considered as a subsystem which can be operated in three possible modes: grid-connected, islanded and isolated. In islanded mode, the MG that is grid-connected disconnects from the main grid, either in a planned fashion or due to fault/disturbance in the main grid. It provides seamless transition between the two modes. The MG in the isolated mode is designed in such a way that it is never connected to the main grid. This is the case of remote sites (e.g. remote communities or remote industrial sites) where an interconnection with the main grid is not feasible due to either technical and/or economic constraints. Therefore, isolated MGs operate permanently in stand-alone mode [14].

A typical MG system is shown in Figure 14.11. It comprises a variety of DERs and different types of end users. Hence, it is considered as a portion of the distribution system, which is located downstream the distribution substation [15].

14.5.2 Operation functions

For each operation mode, MG provides some functions as follows: (i) grid-connected mode, where the MG is connected to the upstream grid (MG can receive energy, totally or partially, from the main grid depending on power sharing. On the other hand, any excess power can be sent to the main grid when the total production exceeds consumption); (ii) island mode, which occurs when the host grid has a failure, or there are some planned actions (e.g., to perform maintenance actions). (MG can smoothly move to islanded operation. Thus, the MG operates

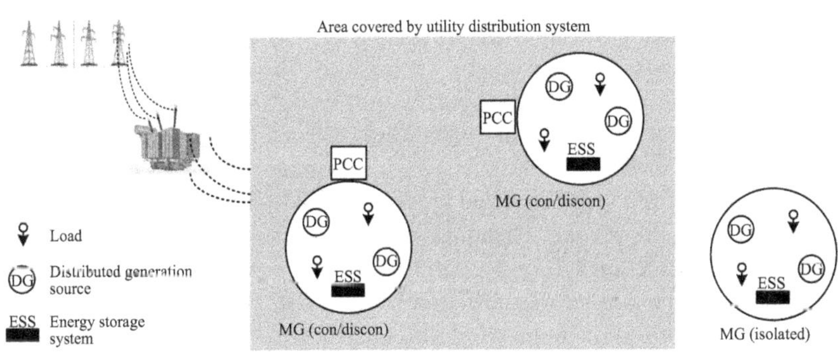

Figure 14.10 Trends towards MGs

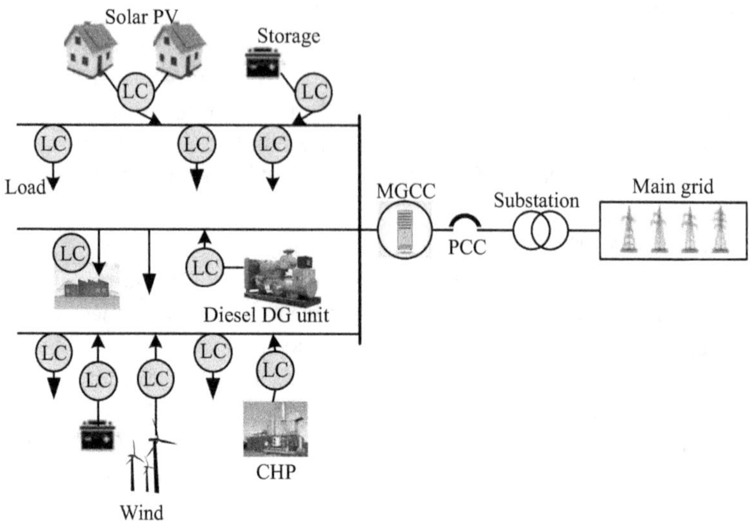

Figure 14.11 Typical MG configuration

autonomously); (iii) ride-through between the two modes [15,16]. The operation of MG whether in grid connected mode or islanded mode requires several reinforcements regarding the control and protection issues.

14.5.3 Challenges facing MG implementation

In this section, the challenges countering the progress of MG deployment are presented. Although MGs have many benefits, expansion of their utilization involves some critical issues, such as technical, regulatory and financial concerns. These concerns are detailed next in this subsection [17,18].

Technical issues. Due to the electrical interface between multiple sources, which may be based on conventional generators (e.g. diesel synchronous generator) or inverter-based renewable energy (wind turbines, PV system, etc.), several interconnection concerns have been raised. Stable operation of MG with voltage quality at certain frequency limits is considered as a serious issue, practically with fluctuating DG power. Moreover, the difficulty of the coordinated control may arise in decoupling the active and reactive power during operation. In addition, generating harmonic currents and voltage between many DGs may lead to conflicting control requirements. One of the most important challenges is the transition ability of the MG from grid connected to islanded mode without losing its stability, practically after fault provoked islanding incidents. This transition gets worse in the presence of dynamic load (e.g. induction motors), especially with intermittent renewable DG units. Further, reconnection with the host grid also poses challenges. Hence, re-synchronization requires carful transition to prevent the voltage disturbance. Other issues have been raised due to designing proper protection scheme of MG. The protection scheme of MG should be able to meet protection requirements such as sensitivity and reliability not only in grid connected mode, but also in the islanded mode. More explanation for MG technical issues are presented in Chapter 19.

Regulatory issues. There are many aspects of legislation that prohibit the integration of MG with utility grid. For example, in Spain, regulations prevent exciting any load in the same circuit between PV system and low-voltage system. Moreover, the integration of generation and storage systems is prevented. Unfortunately, such rules and regulations are applied in Egypt as well, which may lead to constraining the MG concept when connecting MG to the utility grid. From the grid operator point of view, this interconnection may be affected and disturb the safety of the utility grid. Hence, the grid operators create guidelines and codes, differing from one country to other, to manage this interconnection. Unfortunately, most of these guidelines are not directly applied to the MG concept as in case of anti-islanding and faulty regulations, which deal with MG as a potential source of disturbance to the grid. The anti-islanding protection schemes constrain the MG's ability to seamlessly transition to the islanding mode and continue its function locally, since the DGs are forced to disconnect.

Financial issues. Although most of the previous issues have been handled by continuous development and research, financial issue is still the most common obstacle for increasing penetration of MG. Some technologies have already become cost-effective. However, many important technologies such as PV, fuel cells and storage remain expensive and need some sort of financial support and legislation from the interested countries.

14.5.4 Role of ESSs

ESS is considered as a vital part in MG infrastructure because it is responsible for successful operation of the MG, especially with intermittent renewable energy sources. It allows the DG units of the MG to work as dispatched units, with capability of fault ride through during the variation of the primary energy sources (i.e. solar PV and wind). Consequently, it improves the stability, reliability and power quality of the MG. The benefits of storage in the latter application are of interest, because while renewable energy resources are a pillar of the MG, without storage, their generation cannot improve the system reliability and must be duplicated by other means of generation.

A storage unit can provide a functionality like that of the inertia of a synchronous generator by absorbing temporary mismatches between power generation and demand, especially in a low-inertia power electronic-based MG. Therefore, system stabilization can be improved by providing voltage/frequency control in a droop-based scheme [19].

Energy storage has applications in transmission capability improvement, power quality enhancement, MG-islanded operation, active distribution systems and electric vehicle technologies, and may improve dynamic stability, transient stability, voltage support and frequency regulation [19,20]. The power grid can substantially benefit in generation, transmission, distribution and consumption from the availability of stored energy [21]. For example, storage can eliminate or delay expansion of the transmission infrastructure or generation capacity. Storage can be combined with non-dispatchable DER units such as wind and solar energy to turn them into dispatchable units. On the consumers' side, storage can be employed for peak-shaving by storing the locally generated energy until it is needed. An extensive list of applications for energy storage in transmission, distribution and generation is presented in [22].

Despite its benefits, energy storage has not been fully utilized. Among the limiting factors is, besides the cost, the lack of appropriate control and management strategies [23]. Future research is needed to investigate and develop control methodologies for the various forms of energy storage technologies.

14.5.5 Protection strategy

Some technical challenges need to be resolved by power system researchers and engineers. Protection of MGs is one of them that requires more attention. During the grid-connected operation mode, since the main grid provides a large short-circuit current to the fault point, the protection can be performed by the existing protective devices in the distribution networks, but in the islanded mode, fault currents provided by DG units installed in the MG are significantly less than those in the grid-connected mode. Hence, the employment of traditional overcurrent-based protective devices in MG and sub-grids is no longer valid. Most conventional distribution systems operate radially, where the power flows uni-directionally from large power plants to the customers. In such systems, as the magnitude of short-circuit currents is proportional to the fault location, the protection is done by overcurrent-based protective devices [24]. The time-graded coordination between them enables upstream devices to operate as backup for downstream ones [25].

In recent years, the emergence of MG has changed the structure of distribution systems from passive networks into active ones. This change has disturbed the operation of overcurrent-based strategies such that they can no longer protect new structures [26,27]. The fault current contribution of inverter-based DG sources in an MG is limited (only two to three times the maximum load current) due to the low thermal capability of their power electronic devices. Therefore, protective devices in an MG containing inverter-based DG sources would operate very slowly or may not be triggered at all for fault events during islanded mode. In addition, the considerable difference between the magnitude of short-circuit currents in the grid-connected and islanded modes makes single-setting traditional overcurrent relays unable to protect dual-mode operating MG [28,29]. Therefore, overcurrent protective devices cannot be used for protection of MGs and sub-grids including inverter-based DG sources, and some new techniques should be devised.

The traditional overcurrent-based strategies do not have the ability to protect AC MGs and sub-grids due to the drastic difference between the magnitude of fault currents in the grid connected and islanded modes. The main protection strategies are briefly described as follows [30]:

- The DG is automatically taken out of service following a fault on the distribution network, and thus, protection of the distribution network is not affected.

- The capacity of DG is limited, and the distribution network remains unchanged.
- Measures for limiting the fault current are taken; for example, a fault current limiter is used to minimize the effects of DG after a fault occurs, and the distribution network remains unchanged.

An MG should be able to operate in parallel with the grid in islanded mode. It shall meet the following basic requirements:

- In grid-connected operation, if a fault occurs in the MG, the MG protection should operate reliably to clear the fault. For example, when the electrical equipment in the LV distribution network fails, the protection of the distribution network should operate to isolate the equipment to ensure secure and stable operation of the MG.
- When an instantaneous fault occurs in the distribution network, the protection of the distribution network should operate quickly to clear the fault to maintain uninterrupted operation of the MG.
- In case of power loss of the distribution network, the islanding protection of the MG operates to disconnect the MG from the distribution network and the MG switches to islanded operation.
- In islanded operation, when a fault occurs on the MG, the protection should operate reliably to clear the fault to ensure secure and stable operation of the MG.
- Upon recovery of the distribution network, the MG is reconnected to the grid.

Conventional protection schemes face serious challenges when it comes to protecting an islanded MG with inverter interfaced DG (IIDG) units. They need major revision to find new methods based on the limited fault current to detect and isolate the faulty portion. The various possible solutions to cope with the problem can be broadly divided into four categories: use of inverters having high fault current capability, that is, uprating of the inverter [31]; communication between the inverter and protective relays; introduction of energy storage devices that are capable of supplying large current in case of a fault [32]; and in-depth analysis of the fault behaviour of an islanded MG with an IIDG unit to comprehend the behaviour of system voltages and currents [33,34]. This will in turn help in defining alternative fault detection and alternative protection strategies that, in case of a fault, do not rely on a large magnitude of the fault current but rely on other parameters, like change in the voltage of the system [35].

14.5.6 Control of MGs

Generally, there are two levels of MG control system, central control (supervisory control) and local control for each DG or load. Both are discussed in the following [32,36].

Microgrid central control (MGCC). In the grid-connected mode, the MG voltage is influenced by the host utility grid. Thus, the hosted DGs cannot play any appreciable role in determining the MG voltage magnitude or frequency (although they can locally affect the voltage magnitude through their reactive power or, to a lesser extent, real power contributions). Hence, the function of the central control in grid-connected mode is to handle the active- and reactive-power commands for the DG systems. The central control can calculate the commands based on a variety of criteria, such as market signals, economy of the MG, optimal operation, and host grid conditions and requirements. For example, during hours when the grid electricity is cheap, the overall control may decide to reduce the power output of the gas microturbine DG and charge the energy storage devices mainly through the grid power. At the same time, in response to a grid command, it may dispatch the DG systems in such a way that the MG draws reactive power from the host grid to prevent an overvoltage at the PCC.

By contrast, in the islanded mode of operation, the DGs are mainly controlled locally to regulate the MG voltage magnitude and frequency. This process must be fast and reliable and is a function of the DG control. In the absence of a connection to the utility grid, a sustained islanded mode operation also implies that the sum of DG system power outputs equals the aggregate load power. Although, this is strictly and rapidly ensured by the DG control, the central control can specify the steady-state DG system output powers, according to the constraint mentioned above, in such a way that the MG steady-state operation is optimized. Further, the central control can shed loads in the islanded mode, depending on load criticalities, MG energy reserves or other considerations [18,37,38]. For safe reconnection of the MG to the host grid, once the operating mode is to be switched to the grid-connected mode, the central control can also specify the commands for steady-state voltage magnitude and frequency of the DG system if they have drifted away from their nominal values due to the islanded-mode operation.

The central control requires a communications network to exchange information with the DG systems and loads. Even though the communication links may not work fast or be very reliable, their design should be such that their malfunction shall not lead to a system collapse. The malfunction should only lead to the non-optimal operation of the DGs [39].

Local control (LC). The main function of LC is to ensure the stability and robustness of the MG, in transient as well as steady states. Thus, LC ensures that

- DG systems operate in synchronism with the grid in the grid-connected mode of operation. In the islanded mode of operation, the aggregate of the DG system power outputs tracks the aggregated load power. Moreover, some dynamics such as the MG voltage magnitude and frequency are regulated when the DG systems properly share the total load, both in transient and steady-state regimes.
- DG systems operate within their limits and are protected against network faults irrespective of the operating mode.

LC also receives and enforces the commands that are issued by the MGCC. In contrast to MGCC, LC must be fast and very reliable. Otherwise, either the MG voltage magnitude and frequency undergo unacceptable excursions with respect to their nominal values, or the MG collapses completely [39,40]. Therefore, its centralized realization is theoretically feasible. The consensus is that the LC should be implemented based on decentralized methods. The most notable decentralized method is based on droop characteristics. One distinct merit of the droop-based control method is that it grants plug-and-play capability to the DG systems. The DG control may also be applied to the loads, for rapid and continuous power control, or even for load shedding.

14.5.7 Islanding detection

MGs can be viewed as small scale of the classical large centralized electric distribution system and include DERs. They have the capability of switching from one operation mode (grid-connected) to another (islanded). Loads of MG should not be affected when changing the mode of operation.

The interconnection of MGs and/or DERs to the utility grid is controlled by primary (local), secondary (regional) and tertiary (national) controls to be regulated in compliance with the published standards (e.g. IEEE-1547, IEC-62115 and AS-4777). However, it is necessary for each DER (renewable, non-renewable and energy storage) unit to have the ability to detect if it is connected or disconnected from the utility grid, which is normally referred to as islanding detection. Islanding state can be caused by faults, accidental opening of the main switch that connects the MG/DER and the main grid, human errors or natural causes. Islanding should be accurately detected (i.e. the non-detected zone 'NDZ' is as zero as possible) within a time as small as possible (e.g. IEEE-929 std. requires islanding detection within 2 s after the islanding state occurs). Generally, the faster the islanding is detected, the faster the control algorithms of all MG-DERs can be adapted to effectively achieve the desired performance, especially for MG stability and power quality.

14.6 Concluding remarks

- The power system of the future will become more complex as it may comprise different DG units and/or ESSs connected in different locations (integrated into transmission or distribution system) as well as several MGs downstream the distribution system.
- Several technical problems arise because of the integration of DER units, which have different characteristics.
- Cost-benefit analysis should be carried out to decide which type of DG and ESS can be added to the power system.
- Information technology, computer monitoring and control technologies, cyber security techniques, communication networks and advance protective relaying system are essentially needed to get the power system smarter. More details about such grids are explained in Chapter 1, Section 1.4.

References

[1] National Renewable Energy Laboratory, DOE National Laboratory. *Distributed energy resources: A how to guide for federal facility managers* [online]. 2002. Available from https://www.nrel.gov/docs/fy02osti/31570.pdf [Accessed 12 Feb 2020].

[2] Adefarati T. and Bansal R. C. 'Integration of renewable distributed generators into the distribution system: A review'. *IET Renewable Power Generation.* 2016;10(7):873–84.

[3] Burger S. P., Jenkins J. D., Huntington S. C. and Pérez-Arriaga I. J. 'Why distributed? A critical review of the tradeoffs between centralized and decentralized resources'. *IEEE Power & Energy Magazine.* 2019;17(2):16–24.

[4] Strbac G., Pudjianto D., Aunedi M., Papadaskalopoulos D., Djapic P., Ye Y. *et al.* 'Cost-effective decarbonization in a decentralized market'. *IEEE Power & Energy Magazine.* 2019;17(2):25–36.

[5] Hoke A., Giraldez J., Palmintier B., Ifuku E., Asano M., Ueda R., *et al.* 'Setting the smart solar standard'. *IEEE Power & Energy Magazine.* 2018;16(6):18–29.

[6] Vartanian C., Bauer R., Casey L., Loutan C., Narang D. and Patel V. 'Ensuring system reliability'. *IEEE Power & Energy Magazine.* 2018; 16(6):52–63.

[7] Barkans J. and Zalostiba D. 'Blackout prevention and power system self-restoration'. *EUROCON, 2007. The Int. Conf. on 'Computer as a Tool'*; Warsaw, Poland, Sep 2007. IEEE; 2007, pp. 1–8.

[8] Braun M., Brombach J., Hachmann C., Lafferte D., Klingmann A., Heckmann W., *et al.* 'The future of power system restoration'. *IEEE Power & Energy Magazine.* 2018;16(6):30–41.

[9] Taczi I. and Vokony I. 'Possible role and effects of synthetic inertia on the future's power system'. Trivent Publishing, Engineering and Industry Series: *Power Systems, Markets and Renewable Energy Sources in South-Eastern Europe.* 2016;3:32–47.

[10] Eriksson R., Modig N. and Elkington K. 'Synthetic inertia versus fast frequency response: A definition'. *IET Renewable Power Generation, Open Access.* 2018;12(5):507–14.

[11] Tamrakar U., Shrestha D., Maharjan M. and Bhattarai B. P. 'Virtual inertia: Current trends and future directions'. *MDPI, Applied Sciences.* 2017;7(7): 1–29.

[12] Waffenschmidt E. and Hui R. S. Y. 'Virtual inertia with PV inverters using DC-link capacitors'. *IEEE, 18th European Conference on Power Electronics and Applications*; Karisruhe, Germany, Sep 2016, pp. 1–10.

[13] Gonzalez-Longatt F., Chikuni E. and Rashayi E. 'Effects of the inertia from wind power on the total system inertia after frequency disturbance'. *IEEE International Conference on Industrial Technology (ICIT)*; Cape Town, South Africa, Feb 2013, pp. 826–32.

[14] Olivares D. E., Mehrizi-Sani A., Etemadi A. H., *et al.* 'Trends in microgrid control'. *IEEE Transactions on Smart Grid.* 2014;5:1905–19.

[15] Mariam L., Basu M. and Conlon M. F. 'Microgrid: Architecture, policy and future trends'. *Renewable and Sustainable Energy Reviews.* 2016;64:477–89.

[16] Lopes J. P., Hatziargyriou N., Mutale J., Djapic P. and Jenkins N. 'Integrating distributed generation into electric power systems: A review of

drivers, challenges and opportunities'. *Electric Power Systems Research.* 2007;77:1189–203.

[17] Soshinskaya M., Crijns-Graus W. H., Guerrero J. M. and Vasquez J. C. 'Microgrids: Experiences, barriers and success factors'. *Renewable and Sustainable Energy Reviews*. 2014;40:659–72.

[18] Planas E., Gil-de-Muro A., Andreu J., Kortabarria I. and de Alegría I. M. 'General aspects, hierarchical controls and droop methods in microgrids: A review'. *Renewable and Sustainable Energy Reviews.* 2013;17:147–59.

[19] Nourai A. and Kearns D. 'Batteries included: Realizing smart grid goals with intelligent energy storage'. *IEEE Power Energy Magazine.* 2010;8(2):49–54.

[20] Nourai A., Kogan V. I. and Schafer C. M. 'Load leveling reduces T&D line losses'. *IEEE Trans. on Power Delivery*. 2008;23(4):2168–73.

[21] Thomas R. J. 'Putting an action plan in place: The Electricity Advisory Committee's energy storage report'. *IEEE Power & Energy Magazine.* 2009;7(4):26–31.

[22] The Electricity Advisory Committee. *Bottling electricity: Storage as a strategic tool for managing variability and capacity concerns in the modern grid.* EAC report, 2008.

[23] Hajimiragha A. and Zadeh M. R. D. 'Practical aspects of storage modeling in the framework of microgrid real-time optimal control'. *Proceedings of IET Conf. on Renewable Power Generat. (RPG)*; Edinburgh, UK, Sep 2011, pp. 93–8.

[24] Piesciorovsky E. C. and Schulz N. N. 'Fuse relay adaptive overcurrent protection scheme for microgrid with distributed generators'. *IET Generation, Transmission and Distribution* 2017;11(2):540–9.

[25] Liu Z., Hoidalen H. K. and Saha M. M. 'An intelligent coordinated protection and control strategy for distribution network with wind generation integration'. *CSEE Journal of Power Energy Systems.* 2016;2(4):23–30.

[26] Sahoo A. K. 'Protection of microgrid through coordinated directional overcurrent relays'. *Proceedings of the 2014 IEEE Global Humanitarian Technology Conference South Asia Satellite (GHTC-SAS)*; Trivandrum, India, Sep 2014. IEEE; 2014, pp. 129–34.

[27] Choudhary N. K., Mohanty S. R. and Singh R. K. 'A review on microgrid protection'. *Proceedings of the 2014 International Electrical Engineering Congress (IEECON*); Chonburi, Thailand, 2014, pp. 1–4.

[28] Bui D. M., Chen S. L., Lien K. Y. and Jiang J. L. 'Fault protection solutions appropriately used for ungrounded low-voltage AC microgrids'. *Proceedings 2015 IEEE Innov. Smart Grid Technologies – Asia, ISGT ASIA 2015*; Bangkok, Thailand, Nov 2015. IEEE; 2015, pp. 4066–71.

[29] Hussain A., Aslam M. and Arif S. M. 'N-version programming-based protection scheme for microgrids: A multi-agent system based approach'. *Sustainable Energy, Grids and Networks.* 2016;6:35–45.

[30] Fusheng L., Ruisheng L. and Fengquan Z. *Microgrid technology and engineering application*. London: Academic Press; 2016.

[31] Lopes J. P., Moreira C. L. and Madureira A. G. 'Defining control strategies for microgrids Islanded operation'. *IEEE Transactions on Power Systems.* 2006;21(2):916–24.

[32] Jayawarna N., Jenkins N., Barnes M., Lorentzou M., Papthanassiou S., Hatziagyriou N. 'Safety analysis of a microgrid'. *Proceedings of 2005 International Conference on Future Power Systems*; Amsterdam, Netherlands, Nov 2005. IEEE; 2005, pp. 1–7.

[33] Timbus A. V., Rodriguez P., Teodorescu R., Liserre M. and Blaabjerg F. 'Control strategies for distributed power generation systems operating on faulty grid'. *Proceedings of IEEE International Symposium on Industrial Electronics*; Montreal, Canada, July 2006. IEEE; 2006, 2: pp. 1601–7.

[34] Baran M. E. and El-Markaby I. 'Fault analysis on distribution feeders with distributed generators'. *IEEE Transactions on Power Systems.* 2005; 20(4);1757–64.

[35] Tumilty R. M., Brucoli M., Burt G. M. and Green T. C. 'Approaches to network protection for inverter dominated electrical distribution systems'. *Proceedings of 3rd IET International Conference on Power Electronics, Machines and Derives (PEMD 2006)*; Dublin, Ireland, April 2006. IET; 2006, pp. 622–6.

[36] Dimeas A. L. and Hatziargyriou N. D. 'Operation of a multiagent system for microgrid control'. *IEEE Transactions on Power Systems.* 2005; 20(3):1447–55.

[37] Huang W., Lu M. and Zhang L. 'Survey on microgrid control strategies'. *Energy Procedia.* 2011;12:206–12.

[38] Bouzid A. M., Guerrero J. M., Cheriti A., Bouhamida M., Sicard P. and Benghanem M. 'A survey on control of electric power distributed generation systems for microgrid applications'. *Renewable and Sustainable Energy Reviews.* 2015;44:751–66.

[39] Vandoorn T., De Kooning J., Meersman B. and Vandevelde L. 'Review of primary control strategies for islanded microgrids with power-electronic interfaces'. *Renewable and Sustainable Energy Reviews.* 2013;19:613–28.

[40] Green T. and Prodanović M. 'Control of inverter-based micro-grids'. *Electric Power Systems Research.* 2007;77:1204–13.

Chapter 15

Economic dispatch of hybrid thermal and wind plants

15.1 Description of ED problem

Economic dispatch (ED) of power system could be defined as the process of allocating specific generating units and the generation levels to the committed units in the mix with the primary objective of minimizing fuel cost or emissions or both while sustaining system constraints. This means that the total load demand plus system losses are satisfied at the minimum cost while satisfying different technical constraints of the network and power sources.

Prior to the widespread use of alternate sources of energy, the ED problem involved only conventional thermal energy power generators which used exploitable resources such as fossil fuels. It has now become apparent that there is a need for alternatives to thermal energy power generation, and renewable energy sources, such as wind and PV, are now seeing more widespread use. One of the major benefits of renewable energy sources is that, after the initial cost of land, equipment and installation, there is essentially no cost involved in the production of power. Furthermore, environmentally their impact is usually considered to be friendlier than that of thermal energy sources.

The high uncertainty concomitant with the power output of various renewable distributed generations (DGs), for example, wind and solar, is considered as one of the main challenges in the operation of the smart grids. This high uncertainty creates a major task for system operators in keeping the generation–load balance, particularly in regions with high renewable DG penetration [1,2].

Integrating uncertain power sources such as wind power with the conventional thermal units requires specific arrangements, so stochastic ED (SED) is considered. SED is done by characterizing the uncertainty of wind power in the ED problem. The primary problem associated with the SED model is the fact that wind speed, the power source for the wind energy system, forecast at a certain time in future is not exact. So, probability density function (*pdf*) of the actual wind power output should be obtained.

Probability density function is a statistical expression that defines a probability distribution for a continuous random variable as opposed to a discrete random variable (Appendix A). When *pdf* is graphically portrayed, the area under the curve will indicate the interval in which the variable will fall. The total area in this interval of the graph equals the probability of a continuous random variable occurring.

The uncertainty issue has made the researchers assess various aspects of the problem based on stochastic principles [2–5]. The uncertainty of electrical energy cost coefficients and load demand has been considered in the expected objective functions of ED in the literature [6,7]. A proposed stochastic long-term security-constrained unit commitment formulation for representing uncertainties in the availability of generating units and transmission lines, and inaccuracies in load forecasting has been investigated [8]. For a practical problem, like ED, modified intelligent methods should be adequate for solving the ED with more accurate multiple fuel cost functions and constraints.

15.2 Modelling of system comprising thermal units and wind energy sources

ED of a power system comprising thermal and wind energy sources is considered. Precise calculation of ED is based on accurate modelling of power system energy sources, e.g. thermal units and wind. This is a great challenge and more attention must be paid to it because of the stochastic nature of wind energy sources. Therefore, SED is becoming more imperative along with the increased penetration of wind energy sources. Modelling of thermal units and wind energy sources from ED's perspective is described in forthcoming subsections. The modelling is composed of mathematical relations established from the physical behaviour of the system's energy sources. Consequently, the uncertainty associated with wind power is considered in the stochastic dispatch. A *pdf* of the wind power output is determined.

According to probability and statistics, a variable whose value is subject to disparities owing to chance (i.e. randomness, in a mathematical sense) is called a random variable or stochastic variable. As opposed to other mathematical variables, a random variable theoretically does not have a single, fixed value (even if unknown). Somewhat, it can take on a set of possible dissimilar values, each with an associated probability [9]. Then, wind speed is considered a continuous random variable which is affected by a number of factors such as geometric shape, roughness and elevation of the ground surface. Solar radiation (also called clearness index) is similarly considered as a continuous random variable [10].

15.2.1 Wind power modelling using proper pdf

The wind speed distribution primarily determines the performance and the feasibility of wind power systems. Wind energy can be easily defined when the wind speed distribution is identified. Modelling wind speed using a proper *pdf* provides a few key parameters, which can illuminate the characteristics of a wide range of wind speed data. Therefore, it is most desirable to use this modelling in long-term planning problems.

A *pdf* is used to model the wind velocity variation. It provides the probability that an event will occur between two end points. For any two speeds larger than zero, the area under the curve between those two speeds will provide the probability that wind will blow anywhere between those two speeds. It is important to understand that the actual height and shape of a *pdf* curve are determined such that the area under the curve from 0 to infinity is exactly 1. Physically, this means that there is a 100% chance that the wind will blow at some speed in m/s between 0 and infinite.

Many scientific studies in literature concerning wind energy have presented the application of a variety of *pdf* (e.g. normal, lognormal, gamma, Rayleigh and Weibull) to describe wind speed distribution [11,12]. These studies concluded that the Weibull distribution with two parameters might be effectively applied to define the principle wind speed variation. The Weibull probability density given by (15.1) has been found to be a very precise model to define wind velocity variation. This distribution is regularly used in wind energy engineering [13]. Its success originates from the two adaptable parameters, c and k, which can offer great flexibility in fitting the distribution function to the measured values with different performances as shown in Figure 15.1. It is noted that the greater the value of k, the more the curve looks like a reversed bell, though as the value of c increases, the curve spreads out even more.

$$f_v(v) = \left(\frac{k}{c}\right)\left(\frac{v}{c}\right)^{(k-1)} e^{-(v/c)^k} \tag{15.1}$$

This expression is valid for $k > 1$, $v \geq 0$ and $c > 0$. k is the shape factor and c is the scale parameter.

Different methods can be used to calculate the Weibull parameters [13,14]. The shape factor, k, will typically range from 1 to 3. For given average wind speed, a lower shape factor indicates a relatively wide distribution of wind speed around

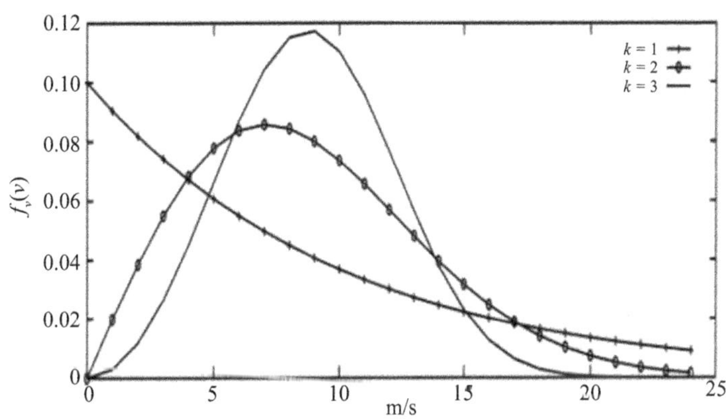

Figure 15.1 Weibull probability distribution function with c = 10 and k = 1, 2 and 3

the average while a higher shape factor specifies a comparatively narrow distribution of wind speed around the average. A lower shape factor will normally lead to higher energy production for a given average wind speed. Figure 15.1 shows the Weibull *pdf* with $c = 10$ and $k = 1$, 2 and 3. Here, the parameter k is calculated, approximately, using the mean wind speed, v_m, and the standard deviation, σ, as follows:

$$k = \left(\frac{\sigma}{v_m}\right)^{-1.086} \tag{15.2}$$

$$c = \frac{v_m}{\Gamma(1 + 1/k)} \tag{15.3}$$

$$v_m = \frac{1}{n}\left(\sum_{i=1}^{n} v_i\right) \tag{15.4}$$

$$\sigma = \left[\frac{1}{n-1}\sum_{i=1}^{n}(v_i - v_m)^2\right]^{0.5} \tag{15.5}$$

where n is the number of non-zero wind speeds.

A Weibull *pdf* with $k = 2$ and $c = 6$, 8, 10 and 12 is shown in Figure 15.2.

At k equal to 2, *pdf* is named Rayleigh *pdf* and is given by (15.6). In this case, *pdf* imitates most wind speed profiles. In case the mean value of the wind speed for a site is identified, the scale index, c, can be considered as in (15.7).

$$f(v) = \left(\frac{v}{c^2}\right)\exp\left[-\left(\frac{v}{2c^2}\right)\right] \tag{15.6}$$

$$c = v_m/1.253 \tag{15.7}$$

Weibull and Rayleigh's functions are generally used to define the random behaviour of the wind speed in a specific location over a specific period.

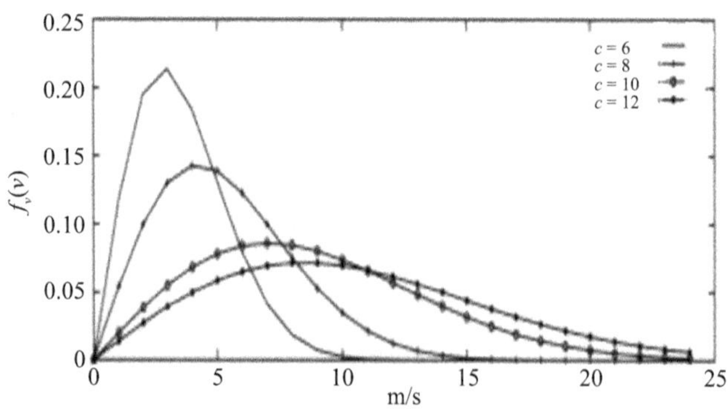

Figure 15.2 Weibull probability distribution function with k = 2 and c = 6, 8, 10 and 12

Furthermore, these two functions can define wind speed distribution for any archetypal hour of the year. The wind power output of wind turbine (WT) generator has a constant zero value lower than the cut-in wind speed and beyond the cut-out wind speed, and it is constant between rated wind speed and cut-out wind speed. The wind power output is a mixed random variable, which is continuous between values of zero and rated power and is discrete at values of zero and rated power output. If it is supposed that the wind speed has a given distribution such as Weibull, it is then essential to change that distribution to a wind power distribution. This transformation may be accomplished in the following manner, where v and w are the wind speed and wind power random variable, respectively, for a linear transformation as follows [15]:

$$f_w(w) = \begin{cases} \left(\frac{klv_i}{c}\right)\left(\frac{(1+\rho l)v_i}{c}\right)^{k-1} \exp\left(-\left(\frac{(1+\rho l)v_i}{c}\right)^k\right) & \text{for } 0 < w < w_r \quad (15.8) \\ 1-\exp\left[-\left(\frac{v_i}{c}\right)^k\right] + \exp\left[-\left(\frac{v_o}{c}\right)^k\right] & \text{for } w = 0 \quad (15.9) \\ \exp\left[-\left(\frac{v_r}{c}\right)^k\right] - \exp\left[-\left(\frac{v_o}{c}\right)^k\right] & \text{for } w = w_r \quad (15.10) \end{cases}$$

where $\rho = \frac{w}{w_r}$, $l = \frac{v_r - v_i}{v_i}$, v_r is rated wind speed, w_r is rated wind power, v_i is cut-in speed and v_o is cut-out speed.

15.3 Formulation of optimization problem

ED provides important functions, like unit commitment, load forecasting, security analysis, scheduling of fuel purchase and such, in modern power systems. A bibliographical survey of ED methods reveals that various numerical optimization techniques have been employed to solve the ED problem. Traditionally, it is solved using mathematical programming based on optimization techniques such as lambda iteration, gradient method and so on [16–18]. ED with piecewise linear cost functions is a highly heuristic, approximate and extremely fast form of ED [19].

Lambda iteration, gradient method [20–22] can solve simple ED calculations. Although they are fast, they are not enough for applications in the deregulated market. Several heuristic methods, among them genetic algorithm, have been applied to solve the real-time economic load dispatch problem [23,24]. An artificial neural network is also used to solve the optimization problem [16]. Other research papers have applied the swarm behaviour to the problem of optimum dispatch as well as the unit commitment problem [25].

Heuristic algorithms are iterative techniques that can search not only locally optimal solutions but also a globally optimal solution depending on problem domain and execution time limit. There are general-purpose searching methods based on principles inspired by the genetic and fruition mechanisms detected in natural systems and populations of living creatures. These approaches have the benefit of searching the solution space more methodically. The chief trouble is their sensitivity to the selection of parameters. Amongst such techniques, particle swarm optimization (PSO) is simple and promising. It necessitates less computation time and memory. It has also standard values for its parameters. A flow chart illustrating the main steps of PSO algorithm is depicted in Figure 15.3.

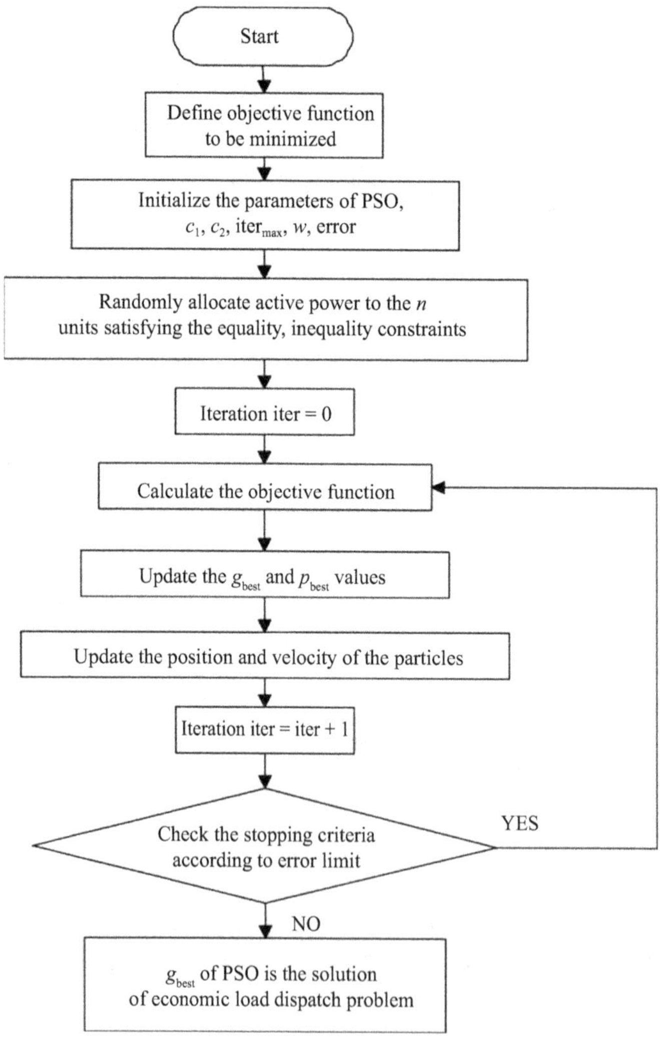

Figure 15.3 Main steps of PSO algorithm

15.4 Single-objective ED model

ED of a power system with thermal and wind units is considered. First, conventional dispatch with only three non-convex thermal units is presented and second, deterministic dispatch incorporating wind power is modelled as described in the next two subsections. The objective is to minimize the combined operating cost without considering imbalance cost due to the mismatch between the actual and scheduled power output of wind and thermal units. Then, the uncertainty associated with the wind power is considered in the stochastic dispatch and the *pdf* of the wind power output is obtained.

15.4.1 Conventional ED with thermal units

In single-objective conventional ED (CED), the objective is to minimize the combined operating cost of the non-convex thermal units. It can be formulated as

$$\text{Minimize} \sum_{i=1}^{N} OC(P_{\text{gi}}) \tag{15.11}$$

Subject to

$$P_{\text{gi(min)}} \leq P_{\text{gi}} \leq P_{\text{gi(max)}} \tag{15.12}$$

$$\sum_{i=1}^{N} P_{\text{gi}} = L \tag{15.13}$$

where

$OC(P_{gi})$: operating cost of non-convex thermal units.
N: number of non-convex thermal units.
P_{gi}: power of the ith generating thermal unit.
L: system load.
$P_{gi(min)}$: minimum generation limit of the non-convex thermal unit.
$P_{gi(max)}$: maximum generation limit of the non-convex thermal unit.

Taking a closer look at the objective function, to effectively solve the ED problem, most algorithms require the incremental cost curves to be of monotonically smooth increasing nature. The generating units with the multi-valve steam turbines exhibit a greater variation by the fuel cost functions, where the valve point results in the rippling form of the heat-rate curve and the cost function contains higher-order non-linearity due to the valve-point effects. The more general fuel cost function of each thermal generator, considering the valve point effect, is expressed as the sum of a quadratic and a sinusoidal function. The total non-smooth fuel cost function in terms of real power output can be expressed as

$$OC(P_{\text{gi}}) = \sum_{i=1}^{N} C_i(P_{\text{gi}}) \tag{15.14}$$

$$C_i(P_{gi}) = a_i + b_iP_{gi} + c_iP_{gi}^2 + \left|e_i \sin\left(f_i \times \left(P_i^{\min} - P_i\right)\right)\right| \tag{15.15}$$

where a_i, b_i and c_i are cost coefficients of ith generator, and e_i, and f_i are the coefficients of that generator reflecting valve-point effects [26].

15.4.2 Deterministic ED incorporating wind power

In a single-objective deterministic ED (DED), the objective is to minimize the combined operating cost of the non-convex thermal unit and wind plant without considering the uncertainty of wind power, i.e. considering the cost coefficient of wind power as constant values. Thus, its formulation can be written as

$$\text{Minimize}\left[\sum_{i=1}^{N} OC(P_{gi}) + \sum_{i=1}^{M} OC_d(W_i)\right] \tag{15.16}$$

Subject to

$$P_{gi(\min)} \le P_{gi} \le P_{gi(\max)} \tag{15.17}$$

$$0 \le W_i \le W_{ri} \tag{15.18}$$

$$\sum_{i=1}^{N} P_{gi} + \sum_{i=1}^{M} W_i = L \tag{15.19}$$

where

$OC_d(W_i)$: deterministic operating cost of wind farms.
W_{ri}: rated wind power from the ith wind-powered generator.
M: number of wind-powered generators.
S: number of solar-powered modules.
W_i: scheduled wind power from the ith wind power generator.

The deterministic operating cost of wind power generators is formulated as

$$OC_d(W_i) = C_{dwi} \times (W_i) \tag{15.20}$$

$$C_{dwi}(W_i) = d_i \tag{15.21}$$

where d_i is the cost coefficient for ith wind power plant.

15.4.3 SED incorporating wind power

Owing to the uncertainty in wind power output and imbalance cost due to the mismatch between the actual and scheduled power outputs of wind units, SED is suggested. The objective of SED is to minimize the stochastic operating cost subject to the system constraints. Weibull *pdf* is utilized to represent the stochastic nature of wind power. Single-objective SED for the proposed system with wind power and thermal units can be formulated as follows [27]:

$$\text{Minimize } \left[OC\left(P_{\text{gi}}, W_i\right)\right]$$

that is,

$$\text{Minimize}\left[\sum_{i=1}^{N} OC\left(P_{\text{gi}}\right) + \sum_{i=1}^{M} OC_s\left(W_i\right)\right] \tag{15.22}$$

Subject to

$$P_{\text{gi(min)}} \leq P_{\text{gi}} \leq P_{\text{gi(max)}} \tag{15.23}$$

$$0 \leq W_i \leq W_{\text{ri}} \tag{15.24}$$

$$\sum_{i=1}^{N} P_{\text{gi}} + \sum_{i=1}^{M} W_i = L \tag{15.25}$$

where

$OC_s(W_i)$: stochastic operating cost of wind farms.

$$OC_s(W_i) = \sum_{i=1}^{M} C_{\text{wi}}(w_i) + \sum_{i=1}^{M} C_{p,\text{wi}} \times \left(W_{i,\text{av}} - w_i\right) + \sum_{i=1}^{M} C_{r,\text{wi}} \times \left(w_i - W_{i,\text{av}}\right) \tag{15.26}$$

On the RHS of (15.26), the first term, $C_{\text{wi}}(w_i)$, is the cost function based on wind speed profile and formulated as in (15.27). The cost coefficient is multiplied by Weibull *pdf* of wind power, $f_w(w)$, as in (15.8)–(15.10),

$$C_{\text{wi}}(w_i) = d_i \times f_w(w) \times W_i \tag{15.27}$$

This cost coefficient is considered in DED as a constant value, whereas in SED it is weighted by Weibull *pdf* of wind speed. The second term in (15.26) is the penalty cost for not using all the available wind power. It is linearly related to the difference between available or actual wind power and scheduled wind power, and can be written as

$$C_{p,\text{wi}}\left(W_{i,\text{av}} - w_i\right) = K_{p,\text{wi}} \int_{\text{wi}}^{\text{wri}} (w - w_i) f_w(w) dw \tag{15.28}$$

where

$K_{p,\text{wi}}$: the penalty cost coefficient of underestimation (over-generation) for the ith wind generator.

$W_{i,\text{av}}$: actual or available wind power from the ith wind-powered generator.

The third term is the reserve requirement cost, which is because the actual or available wind power is less than the scheduled wind power, and can be given by

$$C_{r,\mathrm{wi}}\left(w_i - W_{i,\mathrm{av}}\right) = K_{r,\mathrm{wi}} \int_0^{\mathrm{wi}} (w_i - w) f_w(w) dw \tag{15.29}$$

where $K_{r,\mathrm{wi}}$ is the reserve cost coefficient of overestimation (under generation) for the ith wind-powered generator.

15.5 Economic/emissions dispatch problem

Concentration of greenhouse gases (GHGs) such as CO_2, SO and NO_2 in the atmosphere has been increasing. Due to concerns regarding global warming and air pollution, there has been an international movement promoting renewable technologies for electricity generation and the development of national emissions limits. Several directives controlling emissions are currently in place that have an impact on the electricity industry. For instance, the Kyoto Protocol, which is an agreement under which industrialized countries agreed to reduce their collective emissions of GHGs in 2012 by 5.2% compared with that in 1990. In the Copenhagen Accord, the convention agreed to reduce global emissions to maintain the increase in global temperature below 2 °C. The use of renewable energies can mitigate the GHG effects to meet the strict requirements stated in the Kyoto Protocol [28].

Wind power generation plays an essential role in a grid. Several techniques to handle the environmental ED (EED) problem have been reported [29]. There are three approaches to solve EED problem. (i) The emission is considered as a constraint with a permissible limit [30]. This formulation, however, has severe difficulty in getting the trade-off relations between cost and emission. (ii) Treat the emission as another objective in addition to the usual cost objective [30]. However, the EED problem is converted to a single-objective problem either by a linear combination of both objectives or by considering one objective at a time for optimization. (iii) Handle both fuel cost and emission simultaneously as competing objectives. Stochastic search and fuzzy-based multi-objective optimization techniques have been proposed for the EED problem [31]. However, the algorithms do not provide a systematic framework for directing the search towards Pareto-optimal front and the extension of these techniques to include more objectives is a very involved question. In addition, these techniques are computationally involved and time-consuming.

Unlike genetic algorithm and other heuristic techniques, PSO has a flexible and well-balanced mechanism to enhance and adapt the global and local exploration abilities. It usually results in faster convergence rates than the genetic algorithm [32]. In recent years, PSO has been successfully implemented for different power system optimization problems including the economic power dispatch problem with impressive success. The potential of PSO to handle non-smooth and non-convex economic power dispatch problem has been demonstrated and reported in [33]. However, the problem has been formulated as a conventional dispatch problem with the objective of minimizing only the fuel cost.

15.5.1 Multi-objective stochastic dispatch

Stochastic dispatch can be extended to include multiple objectives with a goal of optimizing two conflicting functions simultaneously, the operating cost of thermal, wind units and emissions caused by the thermal units.

However, some of the approaches that can be used for designing multi-objective PSO (MO-PSO) algorithms consider each objective function separately, whereas others convert the multi-objective functions into a single-objective function using weighted aggregation (WA) method [34]. Each of the objectives is assigned a weight based on their importance level to solve the problem using PSO as a single-objective function. Thus, the objective function is formulated by summing weighted terms, where each term represents an objective.

The key idea of WA method is that systematically changing the weights during evolution will lead the population to the Pareto front. Two possible methods are investigated. One method is to distribute the weights randomly within a population. The other method is to periodically change the weights with the process of evolution. The two methods are briefly investigated in the following [35]:

1. ***Random distribution of weights within a population.*** For the sake of clarity, two objective problems in the current discussion are considered. Their extension to problems with more than two objectives is straightforward. For a conventional aggregation method, the fitness function is the weighted sum of the two different objectives, f_1 and f_2, as

$$\text{Fitness} = w_1 f_1 + w_2 f_2 \tag{15.30}$$

where w_1 and w_2 are two constants determined using prior knowledge about the problem. Using a pair of fixed weights, only one Pareto solution can be obtained.

The algorithm will run many times with a weight combination used for each time. In this way, all Pareto solutions that the Pareto front consists of can be obtained. If the different weight combinations can be distributed among the individuals, the population may be able to approach the Pareto front during the process of evolution. The weight combinations can be distributed uniformly among the individuals in the offspring population. Let,

$$w_1^i(t) = \text{rand}(\Upsilon)/\Upsilon \tag{15.31}$$

$$w_2^i(t) = 1 - w_1^i(t) \tag{15.32}$$

where $i = 1, 2, \ldots, \Upsilon$, and t is the index for generation number. The function rand (Υ) generates a uniformly distributed random number between 0 and Υ. Thus, a uniformly distributed random weight combination (w_1^i, w_2^i) among the individuals can be obtained, where $0 \leq w_1^i \cdot w_2^i \leq 1$ and $w_1^i + w_2^i = 1$.

2. The idea of a uniformly distributed weight combination can straightforwardly be extended to a generation-based approach. However, if a random weight combination is still used, the convergence of the evolutionary algorithm will be

in question. Therefore, a weight combination that is changed gradually and periodically with the process of evolution is used rather than using a distributed weight combination randomly.

In this example, this is realized as follows:

$$w_1(t) = |\sin(2\pi t/F)| \tag{15.33}$$

$$w_2(t) = 1 - w_1(t) \tag{15.34}$$

where t is the number of generations. We can see from (15.33) that $w_1(t)$ changes from 0 to 1 periodically. The change frequency can be adjusted by F. For example, in case of $F = 400$, which means that in every 400 generation, w_1 will change from 0 to 1 and then from 1 to 0 four times. Figure 15.4 shows an example of how the weights change during evolution within 200 generations.

Accordingly, the economic/environmental dispatch for wind and thermal units can, as an optimization problem, be formulated as an objective function subject to the system constraints as follows:

$$\left.\begin{array}{l} \text{Minimize}\left[OC(P_{\text{gi}}, W_i), E(P_{\text{gi}})\right] \\ \text{i.e., Minimize}\left[w_1\left(\sum_{i=1}^{N} OC(P_{\text{gi}}) + \sum_{i=1}^{M} OC(W_i)\right) + w_2 \sum_{i=1}^{N} E(P_{\text{gi}})\right] \end{array}\right\} \tag{15.35}$$

Subject to

$$P_{\text{gi(min)}} \le P_{\text{gi}} \le P_{\text{gi(max)}} \tag{15.36}$$

$$0 \le W_i \le W_{\text{ri}} \tag{15.37}$$

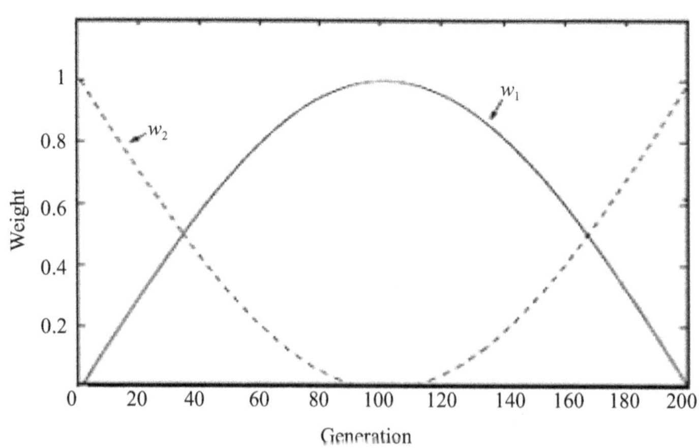

Figure 15.4 Determination of weights: generation-based periodical variation

$$P_i^0 - P_i \leq D_i \tag{15.38}$$

$$P_i - P_i^0 \leq U_i \tag{15.39}$$

$$\sum_{i=1}^{N} P_{gi} + \sum_{i=1}^{M} W_i = L \tag{15.40}$$

where

$E(P_{gi})$: emissions as a function of generator output.

w_1, w_2: non-negative weights, such that $w_1 + w_2 = 1$; w_1 and w_2 are used to make a trade-off between emission and total cost.

D_i, U_i: the down- and up-ramp rate limits of the ith generator, and P_i^0 is the former operating point (hour) of the ith generator, respectively [36].

The first part of the objective function, $OC(P_{gi}, W_i)$, is the expected value of the combined operating cost as given by (15.22)–(15.29) in addition to the ramp rate constraints, (15.38) and (15.39), which limit the generated power of the thermal plants.

It is to be noted that the problem that has been framed as a minimization problem of the fuel cost under load demand constraint and various other constraints at a specific time of interest is known as the static ED problem. It may fail to deal with the large fluctuations of the load demand because of the ramp rate limits of the generators. Thus, due to this large oscillation of the electrical load demand as well as the dynamic nature of the power system, it requires the inquiry of dynamic dispatch (DD) problem. DD problem is a development of the static ED problem to obtain the generation schedule of the committed units to meet the predicted load demand over a certain period at a minimum operating cost under ramp rate constraints and other constraints. Ramp rates are dynamic constraints used to maintain the life of the generators [37].

Although the environmental pollutant gases of fossil-fuel thermal units consist of sulphur oxides SO_x, nitrogen oxides NO_x and dust particles, etc., only the NO_x particles are considered in this study as they are representative of the emissions produced by the fossil-fuel units. Without loss of generality, the total (ton) emissions $E(P_{gi})$ of the NO_x pollutants are given as the sum of the quadratic and exponential function of the thermal units output power [36]:

$$E(P_{gi}) = \sum_{i=1}^{N} \alpha_i + \beta_i P_{gi} + \gamma_i P_{gi}^2 + \xi_i \exp(\lambda_i P_{gi}) \tag{15.41}$$

where $\alpha_i, \beta_i, \gamma_i, \xi_i$ and λ_i are coefficients of the ith generator's emission function. The unit of $E(P_{gi})$ is the mass of nitrogen emitted per unit of energy liberated by the oxidation of the fuel.

15.6 Role of ESS

Renewable energy sources, such as wind and solar, have a massive potential to decrease the dependency on fossil fuels and GHG emissions in the electric area.

Climate change anxieties, state creativities including renewable portfolio standards and consumer efforts are resulting in increased deployments of both technologies. Wind energy has variable and uncertain (sometimes referred to as 'intermittent') output, which is unlike the dispatchable sources used for most of the electricity generation. Though there are clear benefits of using energy storage system (ESS) to enable greater penetration of renewable energy, it is important to consider the potential role of the ESS in relation to the needs of the electric power system.

The variability of these sources has led to concerns regarding the reliability of an electric grid that derives a large fraction of its energy from these sources. Because the wind does not blow all the time at any given location, there has been an increased call for the deployment of ESS as an essential component of future energy systems that use large amounts of variable renewable resources.

To consider wind power as a constant power source, an energy storage device should be coupled with the wind generator. The charging and discharging process of the connected ESS is dependent on the availability of wind power and the connected load of the system. The periods at which the wind power exceeds the connected load, the charging of the ESS is conceivable. However, when the load is in excess of the available wind power, the discharging of the ESS is mandatory. The capacity of the ESS should be selected carefully to be able to cope with the load in the wind energy shortage periods in addition to the possibility to be fully charged in the excess energy periods. A numeric example for further illustration is introduced as in the following.

Example 15.1 A basic system consists of three thermal units (G_1, G_2 and G_3). Two wind plants (W_1 and W_2) and an ESS can be incorporated in different cases of ED application (Figure 15.5). The daily-load curve is shown in Figure 15.6 and the data of the non-convex thermal units is given in Table 15.1. The hourly wind speed is shown in Figure 15.7.

PSO MATLAB® toolbox is used to deal with the problem, and referring to Figure 15.3, its data are [38] the following:

- The initial particles are randomly generated within the feasible range.
- The parameters c_1, c_2 and inertia weight are selected for best convergence characteristic with values of $c_1 = 2.1$ and $c_2 = 2.1$.
- Here the maximum value of w is chosen as 0.9 and the minimum value is chosen as 0.6.
- The velocity limits have a setting of 4 and –4.
- Hundred particles are selected in the population.

ED techniques (CED, DED and SED) are applied to various cases and the results obtained are described as in the following.

Case #1: CED for the basic system

The results obtained, power scheduling and the corresponding cost, in this case, are tabulated in Table 15.2 as well as the total cost is plotted in Figure 15.8.

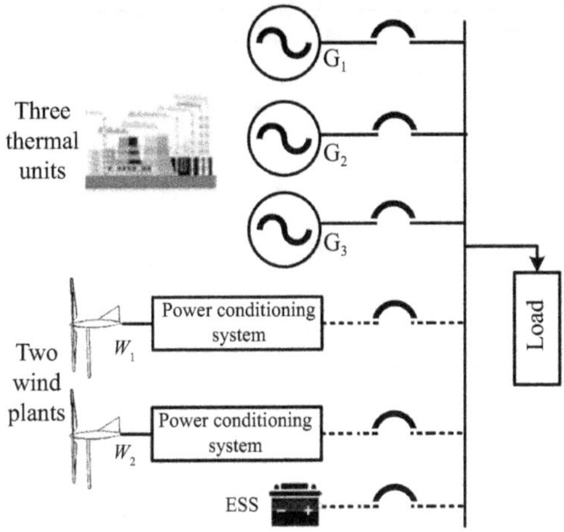

Figure 15.5 The studied system: solid links for the basic system and dotted links for incorporated sources

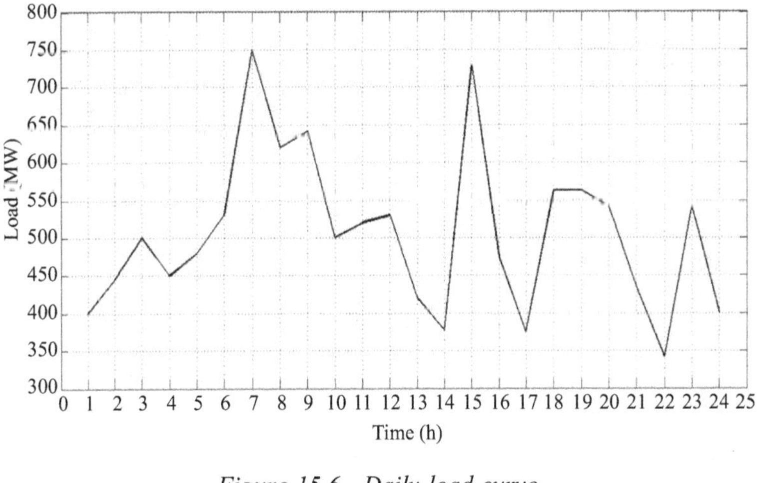

Figure 15.6 Daily-load curve

Table 15.1 Data of the non-convex thermal units [26]

	P_{min}	P_{max}	*a*	*b*	*c*	*e*	*f*
1	100	600	561	7.92	0.00156	300	0.0315
2	100	400	310	7.85	0.00194	200	0.0420
3	50	200	78	7.97	0.00482	150	0.0630

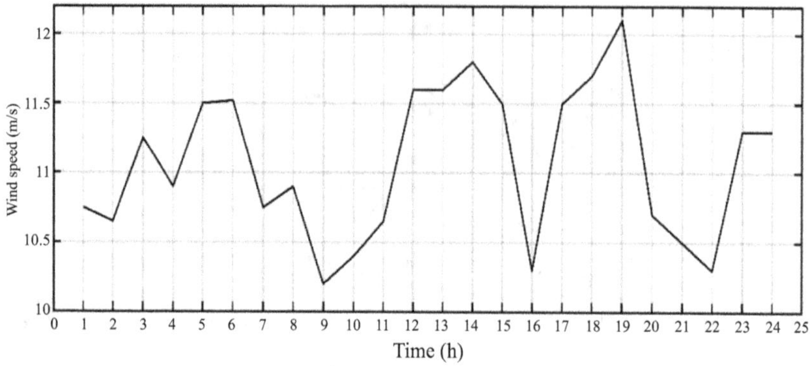

Figure 15.7 Wind speed versus time

Table 15.2 Optimum solution obtained from CED

Hour	Demand (MW)	P_{g1} (MW)	P_{g2} (MW)	P_{g3} (MW)	Min cost × 10^3 (\$/h)
1	400	199.736	100.397	99.8665	4.5482
2	445	100.0000	174.799	170.200	5.0282
3	500	199.733	174.799	125.467	5.4003
4	450	199.9677	174.795	75.2388	5.200
5	480	100	249.599	130.400	5.5188
6	530	199.761	180.505	149.733	6.0227
7	750	399.199	201.067	149.733	8.4963
8	620	199.733	270.533	149.733	6.7143
9	643	299.466	242.667	99.8665	6.8145
10	500	200.1595	249.840	50	5.5257
11	520	100	320.133	99.8665	5.5481
12	530	100	330.131	99.8673	5.0013
13	420	100	170.266	149.733	4.4887
14	377	100.284	176.848	99.8674	4.4017
15	730	399.199	230.934	99.8665	8.3610
16	475	199.733	100.000	175.266	5.0924
17	375	100	175.133	99.8665	4.7377
18	563	299.466	113.800	149.733	6.3902
19	563	100	263.400	199.599	6.0512
20	542	199.733	242.400	99.8665	5.9135
21	433	100	233.133	99.8665	4.9637
22	342	100	174.799	67.2001	4.0779
23	542	299.466	142.667	99.8665	5.6994
24	400	199.756	100.374	99.8700	4.6483

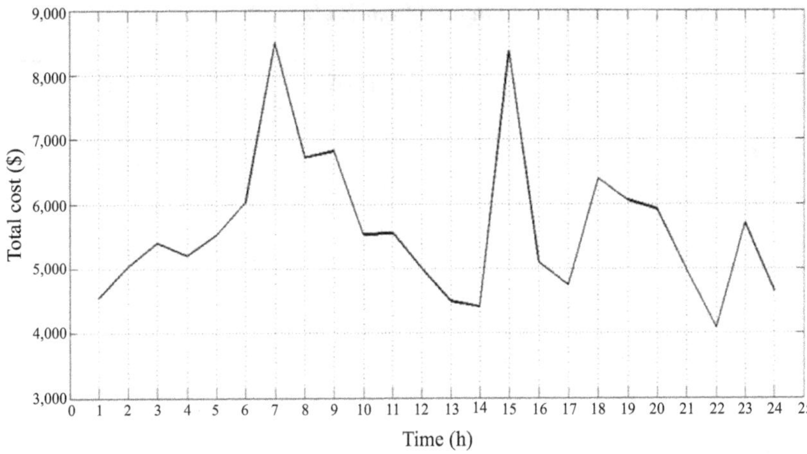

Figure 15.8 Total cost obtained using CED

Case #2: DED for the basic system incorporating wind power plants
Two wind-based DG farms of 45 MW rating are used. Each WT has the following characteristics:

- Cut-in speed = 5 m/s.
- Nominal speed = 15 m/s.
- Cut-off speed = 45 m/s.

The Weibull function parameters of wind are $k = 2$ and $c = 10$ m/s. The results obtained from the DED are given in Table 15.3. The total cost is drawn as in Figure 15.9 and compared to that obtained using CED is shown in Figure 15.10. Figure 15.8, Tables 15.2 and 15.3 and Figure 15.10 show that the total daily cost is reduced when incorporating wind energy sources with the thermal units.

Case #3: SED for the basic system incorporating wind power plants
The penalty and reserve factors are set to $k_{p,\text{wi}} = 2$ and $k_{r,\text{wi}} = 4$ [13]. The deterministic cost coefficients for ith wind power plant are $d_1 = 1$ and $d_2 = 1.25$ [27]. The results are presented in Table 15.4 and the total cost through the day is shown in Figure 15.11.

Case #4: ED for the basic system incorporating wind power plants and ESS
The output power of the two wind plants will be constant at 75 MW as the ESS acts for fixing this power. The total cost through day hours is presented in Figure 15.12. The results of ED application in MW of each unit and the cost in ($/h) are listed in Table 15.5. The charging and discharging state of the ESS through day hours is presented in Figure 15.13.

Case #5: Multi-objective stochastic dispatch
Random distribution of weights within a population method is used to enable PSO to solve environmental/economic MO problem. The diversity of the Pareto-optimal

Table 15.3 Optimum solution obtained from DED

Hour	Load (MW)	P_{g1} (MW)	P_{g2} (MW)	P_{g3} (MW)	W_1 (MW)	W_2 (MW)	Cost ($/h)
1	400	100	139.06	99.86	27.311	33.76	3,955.5
2	445	199.73	174.79	50	4.48	15.9	4,438.7
3	500	199.73	174.79	99.86	6.68	18.91	4,866.8
4	450	100	193.66	99.86	28.23	28.23	4,360.4
5	480	100	249.59	99.86	4.72	25.80	4,680.5
6	530	299.47	109.11	50	32.21	39.19	4,868.8
7	750	299.46	189.45	199.60	27.31	33.76	6,931.7
8	620	199.73	249.53	149.73	0	21	5,958.3
9	642	399.2	174.8	50	0	18	6,201.5
10	500	199.73	174.79	99.86	4.65	20.94	4,865.8
11	520	199.73	174.79	99.86	19.79	25.8	4,883.3
12	530	199.73	249.59	50	4.264	26.40	5,092.4
13	420	100	174.79	99.86	15.84	29.49	4,044.7
14	377	100	174.79	50	24.58	27.61	3,628.0
15	730	399.19	174.79	99.86	24.04	32.08	6,657.1
16	475	100	249.59	99.86	0	25.53	4,675.6
17	375	100.47	110.47	99.87	32.08	32.08	3,615.5
18	563	199.73	174.79	121.62	33.42	33.42	5,240.5
19	563	299.46	100	99.86	27.48	36.17	5,136.9
20	542	299.46	100	99.86	9.24	33.42	5,117.2
21	433	100	100	175.11	25.8	32	4,264.7
22	342	100	136.59	50	24.63	30.77	3,506.5
23	542	100.15	180.7	199.59	30.77	30.77	5,083.2
24	400	100	100	149.73	19.49	30.77	3,870.3

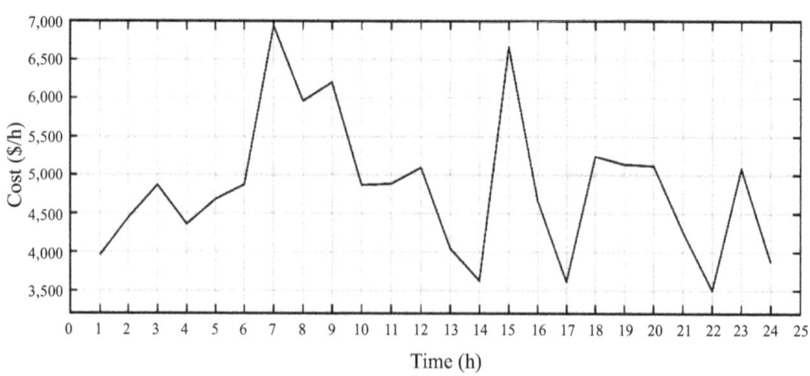

Figure 15.9 Total cost obtained using DED

set over the trade-off is shown in Figure 15.14, which shows that the operating cost of the non-dominated solutions is inversely proportional to their emissions. The data of the generators up and down ramp rates and emissions characteristics are given in Table 15.6 [39] as well as the values of the weights of the two objective functions are shown in Table 15.7. The weights in the proposed MO optimization

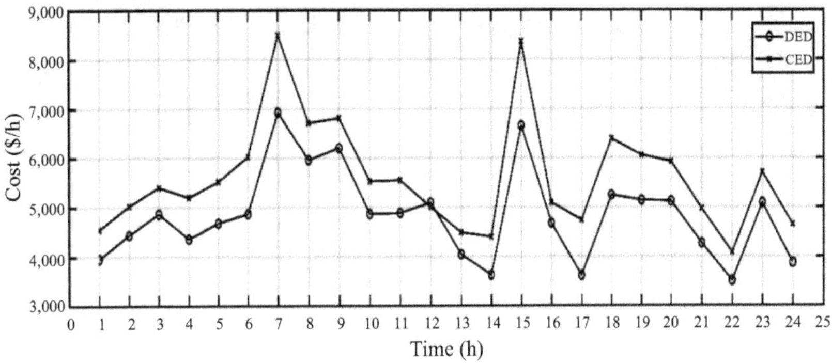

Figure 15.10 Comparison between DED and CED results

Table 15.4 Optimum solution obtained by SED

Hour	Load (MW)	P_{g1} (MW)	P_{g2} (MW)	P_{g3} (MW)	W_1 (MW)	W_2 (MW)	Cost ($/h)
1	400	50	100	181	34.5	34.5	5,262.5
2	445	68.32	108.9	199.99	33.9	33.9	5,758.92
3	500	76.36	217.40	149.57	37.5	37.5	6,399.4
4	450	68.0	111.23	200	35.4	35.4	5,819.2
5	480	77.16	124.8	200	39	39	6,251.1
6	530	86.7	165.0	200	39.12	39.12	6,626.2
7	750	172.30	308.6	200	34.50	34.5	8,444.19
8	620	124.9	224.26	200	35.5	35.4	7,382.7
9	642	133.5	246.1	200	31.2	31.2	7,372.62
10	500	84.8	150.37	199.9	32.39	32.39	6,419.7
11	520	88.28	163.92	199.99	33.89	33.89	6,514.8
12	530	80.2	170.5	200	39.6	39.6	6,561.5
13	420	50	100.0	190.8	39.6	39.6	5,431.6
14	377	50	100	145.4	40.8	40.8	5,008.8
15	730	165.72	286.37	199.9	38.99	38.99	8,326.72
16	475	77.42	125.97	200	37.8	32.7	6,173.65
17	375	50	100.01	152.26	34	39	4,945.9
18	563	100.84	181.76	185	25.2	40.2	6,925
19	563	95.1	182.6	190	42.6	32.6	6,984.4
20	542	97	164.50	173	13.3	40.2	6,771.6
21	433	54.97	100	200	39	39	5,513
22	342	50	100	110.4	43.8	37.8	4,736
23	542	88.94	177.5	200	37.8	37.8	6,660.3
24	400	50	100	174.4	37.8	37.8	5,289.6

are considered as a uniformly distributed random number between 0 and 1. In each run of the program, the weights assigned to each objective function are changed randomly, and after several runs of the program the set of non-dominated solutions can be obtained.

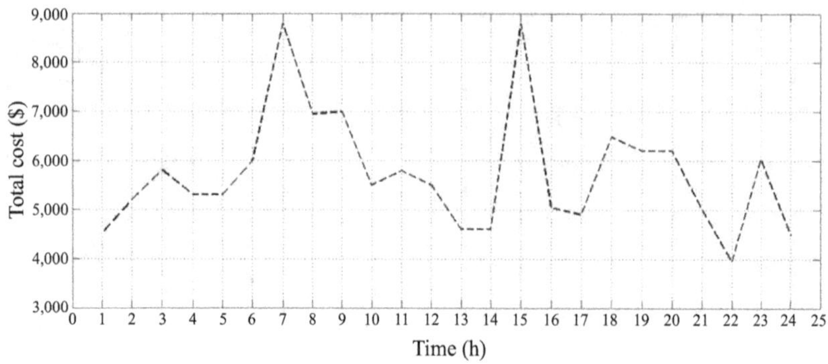

Figure 15.11 Total cost of SED

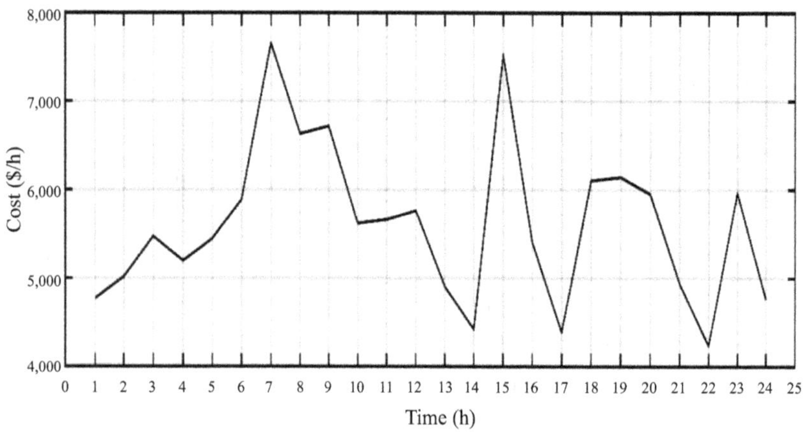

Figure 15.12 Total cost of the power system incorporating wind power and ESS

To better understand the relationship between emissions and generators outputs, Figure 15.15 shows the NO_x emissions versus the power outputs of the three thermal units. Two concepts can be used to solve the EED problem:

1. Deterministic Dynamic Economic Emission Dispatch (DDEED) concept where the imbalance cost due to the mismatch between the actual and scheduled power outputs of wind units is eliminated.
2. Stochastic Dynamic Economic Emission Dispatch (SDEED) concept. Comparing the results of the two concepts as indicated by the Pareto front of each, Figure 15.16, it is found that stochastic dispatch with wind cost and emissions are usually higher than that of deterministic dispatch. This is because the penalties associated with over generation and under generation decrease the

Table 15.5 ED incorporating wind power and ESS

Hour	Load (MW)	P_{g1} (MW)	P_{g2} (MW)	P_{g3} (MW)	$\sum$w (MW)	Cost ($/h)
1	400	100	100	125	75	4,769.4
2	445	100.000	170.13	99.87	75	5,013.3
3	500	100	249.83	75.4	75	5,470.5
4	450	100	175.13	99.86	75	5,194.4
5	480	100	255	50	75	5,443
6	530	100.65	254.46	99.87	75	5,889.3
7	750	100	399.19	175.8	75	7,658.4
8	620	299.98	195.01	50	75	6,629.8
9	642	100	317.27	149.7	75	6,716.6
10	500	199.73	100	125.27	75	5,619.4
11	520	199.73	195.27	50	75	5,660.8
12	530	100.9	254.21	99.87	75	5,763
13	420	100	100	145	75	4,902
14	377	100.0	152.00	50	75	4,423.1
15	730	299.47	174.78	180.73	75	7,527.6
16	475	100	250	50	75	5,406.3
17	375	100	150	50	75	4,389.2
18	563	199.7	100.00	188.27	75	6,103.2
19	563	299.47	100	88.53	75	6,141.2
20	542	100	267.13	99.87	75	5,952.4
21	433	100	174.8	83.2	75	4,935.4
22	342	100	117.0	50	75	4,236.2
23	542	199.73	167.4	99.87	75	5,965.5
24	400	100	100	125	75	4,760

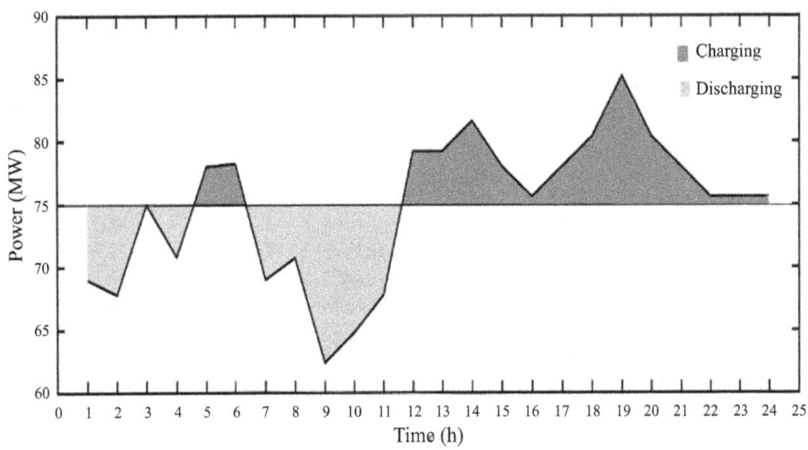

Figure 15.13 Charging and discharging state of EES with wind power

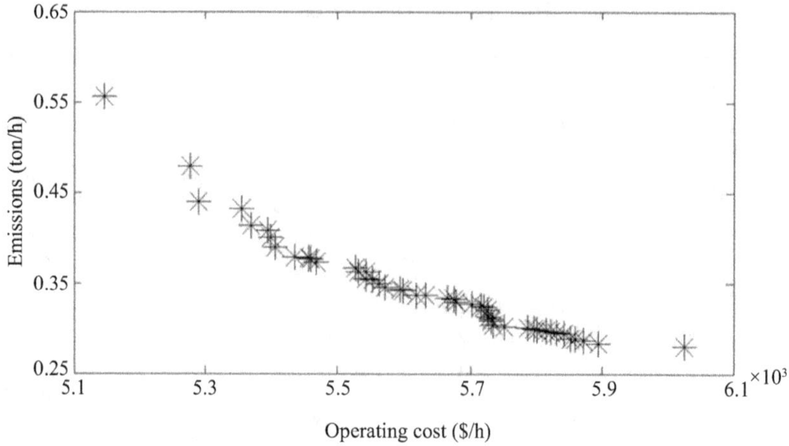

Figure 15.14 Trade-off in cost and emission

Table 15.6 Data of the generators up and down ramp rates and emissions characteristics

	UR	DR	α	β	γ	ξ	λ
1	90	90	0.0041	0.3276	13.8593	0	0
2	90	90	0.0068	−0.5455	40.2669	0	0
3	90	90	0.0046	−0.5111	42.8955	0	0

Table 15.7 Values of weights of the two objective functions through 50 runs

w_1	w_2	w_1	w_2	w_1	w_2	w_1	w_2
0.1190	0.8810	0.4733	0.5267	0.1386	0.8614	0.7792	0.2208
0.4984	0.5016	0.3517	0.6483	0.1493	0.8507	0.9340	0.0660
0.9597	0.0403	0.8308	0.1692	0.2575	0.7425	0.1299	0.8701
0.3404	0.6596	0.5853	0.4147	0.8407	0.1593	0.5688	0.4312
0.5853	0.4147	0.5497	0.4503	0.2543	0.7457	0.4694	0.5306
0.2238	0.7762	0.9172	0.0828	0.8143	0.1857	0.0119	0.9881
0.7513	0.2487	0.2858	0.7142	0.2435	0.7565	0.3371	0.6629
0.2551	0.7449	0.7572	0.2428	0.9293	0.0707	0.1622	0.8378
0.5060	0.4940	0.7537	0.2463	0.3500	0.6500	0.7943	0.2057
0.6991	0.3009	0.3804	0.6196	0.1966	0.8034	0.3112	0.6888
0.8909	0.1091	0.5678	0.4322	0.2511	0.7489	0.5285	0.4715
0.9593	0.0407	0.0759	0.9241	0.6160	0.3840		
0.5472	0.4528	0.0540	0.9460	0.5308	0.4692		

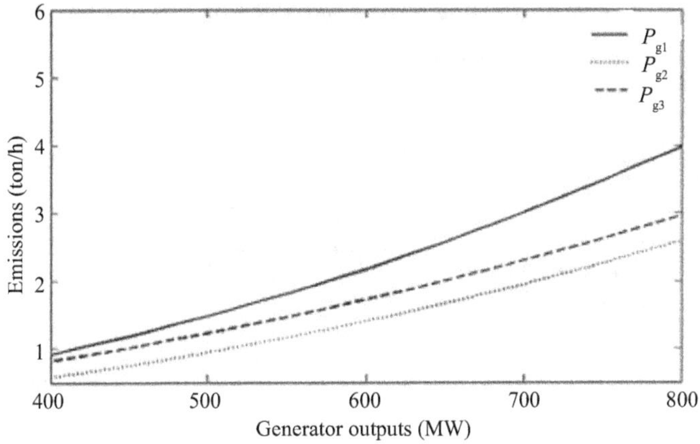

Figure 15.15 Emissions versus the power output of the three thermal units

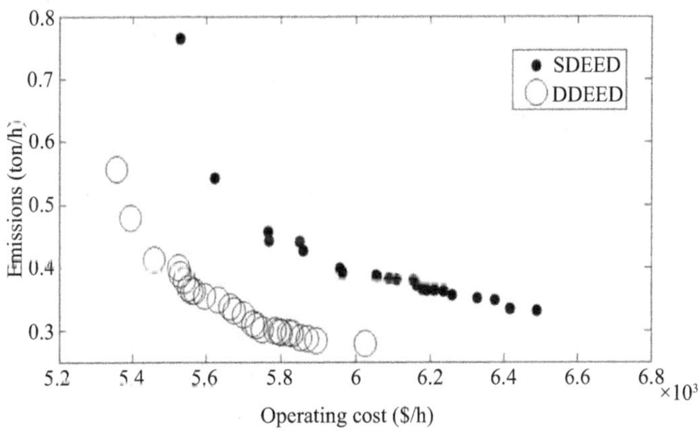

Figure 15.16 Pareto front of DDEED and SDEED concepts

scheduled wind power causing an increase in imbalance charges at a rate higher than the rate of decrease in fuel cost of thermal units. Over day-hours, thermal generation schedule, wind power generation, optimum cost and emissions are obtained as in Table 15.8. It may be noted that, although SDEED has higher cost than that of DDEED, it is preferred from the reliability and security point of view as the uncertainty of wind power is considered and penalized. It is supposed that the total cost and emissions of SDEED could be decreased if wind power forecast is taken into consideration in this model as the mismatch between the actual and scheduled power is decreased.

Owing to the large variation of the customer's load demand and the dynamic nature of the power system, the investigation of SDEED problem is necessary.

Table 15.8 Optimum solution obtained from SDEED

Hour	Load (MW)	Pg_1 (MW)	Pg_2 (MW)	Pg_3 (MW)	W_1 (MW)	W_2 (MW)	Cost × 10^3($/h)	Emissions (ton/h)
1	400	100.0000	100.0000	200.0000	0	0	4.9880	0.3389
2	445	121.3132	127.1361	175.4890	16.4031	4.6586	5.7882	0.3274
3	500	138.7271	150.6407	178.6715	8.7200	23.2408	6.3913	0.3727
4	450	131.5205	129.8715	172.5259	10.2949	5.7873	5.8927	0.3360
5	480	204.1537	143.5667	104.4820	13.6999	14.0977	5.8181	0.3719
6	530	145.0540	213.0200	149.9610	8.5397	13.4253	6.5277	0.4053
7	750	232.843	301.1052	199.7375	13.1569	3.1574	9.0323	0.8522
8	620	150.8029	304.0147	146.3775	14.2972	4.5079	7.6652	0.6022
9	642	139.5146	366.7067	122.9687	6.0960	6.7140	7.9765	0.7611
10	500	120.1499	300.3758	79.4743	9.7881	0	6.0142	0.4854
11	520	136.1991	266.1519	91.3968	15.8314	10.4207	6.6084	0.40134
12	530	166.2866	244.84497	116.842667	0.0018	2.0239	6.8907	0.47538
13	420	111.2687	162.9554	118.4672	12.6375	14.6712	5.0132	0.3251
14	377	133.9896	138.7291	95.6955	1.9287	6.6573	4.9878	0.2651
15	730	236.77743	241.51693	198.48333	25.5000	27.7223	7.9890	0.6853
16	475	132.27633	166.17633	147.47633	12.28	16.63	5.956	0.393
17	375	185.22267	102.62267	63.372667	7.052	16.64	4.875	0.287
18	563	232.9	171.8	125.9	7.475	24.81	6.847	0.496
19	563	238.1	174.7	130.7	12.37	6.864	6.827	0.520
20	542	328.1	136.5	55.13	3.727	18.37	6.913	0.706
21	433	238.7	110.1	61.34	17.52	5.190	5.628	0.391
22	342	161.5	116.3	54.50	7.597	1.913	4.664	0.220
23	542	215.2	199.5	102.7	8.895	15.53	6.490	0.455
24	400	126.9	137.7	104.0	1.643	29.51	5.293	0.240

References

[1] Huang S., Sun Y. and Wu Q. 'Stochastic economic dispatch with wind using versatile probability distribution and L-BFGS-B based dual decomposition'. *IEEE Transactions on Power Systems.* 2018;33(6):6254–63.

[2] Wang L. and Singh C. 'Stochastic economic emission load dispatch through a modified particle swarm optimization algorithm'. *Electric Power Systems Research.* 2008;78(8):1466–76.

[3] Hetzer J., Yu D. C. and Bhattaria K. 'An economic dispatch model incorporating wind power'. *IEEE Transactions on Energy Conversion.* 2008; 23(2):603–11.

[4] Liao G. C. 'A novel evolutionary algorithm for dynamic economic dispatch with energy saving and emission reduction in power system integrated wind power'. *Energy.* 2011;36(2):1018–29.

[5] Dhillon J. S., Parti S. C. and Kothari D. P. 'Stochastic economic emission load dispatch'. *Electric Power Systems Research.* 1993;26(3):179–86.

[6] Liang R. H. and Liao J. H. 'A fuzzy-optimization approach for generation scheduling with wind and solar energy systems'. *IEEE Transactions on Power Systems.* 2007;22(4):1665–74.

[7] Miranda V. and Hang P. S. 'Economic dispatch model with fuzzy wind constraints and attitudes of dispatchers'. *IEEE Transactions on Power Systems.* 2005;20(4):2143–5.

[8] Wu L., Shahidehpour M. and Li T. 'Stochastic security-constraint unit commitment'. *IEEE Transactions on Power Systems.* 2000;22(2):800–11.

[9] Papoulis A. and Pillai U. *Probability, random variables and stochastic processes*. 4th edn. London: McGraw Hill Europe; 2002.

[10] Rice J. A. *Mathematical statistics and data analysis*. 3rd edn. Belmont, CA, USA: Duxbury Press; 2007.

[11] Atwa Y. M., El-Saadany E. F. and Salama M. M. A. 'Adequacy evaluation of distribution system including wind/solar DG during different modes of operation'. *IEEE Transactions on Power Systems.* 2011;26(4):1945–52.

[12] Atwa Y. M. and El-Saadany E. F. 'Reliability evaluation for distribution system with renewable distributed generation during islanded mode of operation'. *IEEE Transactions on Power Systems.* 2009;24(2):572–81.

[13] Jangamshetti S. H. and Rau V. G. 'Site matching of wind turbine generators: A case study'. *Energy Conversion, IEEE Transactions on Energy Conversion.* 1999;14(4):1537–43.

[14] Janbaz G. H., Gholizadeh B. and Soltani B. 'Statistical evaluation of wind speed and energy potential for the construction of a power plant in Baladeh, Nur, Northern Iran'. *International Journal of the Physical Sciences.* 2011; 6(19):4621–8.

[15] Al-Awami A. T., Sortomme E. and El-Sharkawi M. A. 'Optimizing economic environmental dispatch with wind and thermal units'. *IEEE/PES General Meeting*; Calgary, AB, Canada, July 2009. IEEE; 2009, pp. 1–6.

[16] Wood A. J. and Wollenberg B. F. *Power generation, operation, and control.* 2nd edn. New York: John Wiley & Sons; 1996.

[17] Aravindhababu P. and Nayar K. R. 'Economic dispatch based on optimal lambda using radial basis function network'. *International Journal of Electrical Power & Energy Systems.* 2002;24(7):551–6.

[18] IEEE Committee Report. 'Present practices in the economic operation of power systems'. *IEEE Trans. Power Apparatus and Systems.* 1971; PAS-90(4):1768–75.

[19] Chowdhury B. H. and Rahman S. 'A review of recent advances in economic dispatch'. *IEEE Transactions on Power Systems.*1990;5(4):1248–59.

[20] Momoh J. A., Adapa R. and El-Hawary M. E. 'A review of selected optimal power flow literature to 1993, Part I: Nonlinear and quadratic programming approaches'. *IEEE Transactions on Power Systems.* 1999;14 (1):96–104.

[21] Walters D. C. and Sheble G. B. 'Genetic algorithm solution of economic dispatch with valve point loading'. *IEEE Transactions on Power Systems.* 1993;8(3):1325–32.

[22] Tippayachai J., Ongsakul W. and Ngamroo I. 'Parallel micro genetic algorithm for constrained economic dispatch'. *IEEE Transactions on Power Systems.* 2003;17(3):790–7.

[23] Sinha N., Chakrabarti R. and Chattopadhyay P. K. 'Evolutionary programming techniques for economic load dispatch'. *IEEE Transactions on Evolutionary Computation.* 2003;7(1):83–94.

[24] Yang H. T., Yang P. C. and Huang C. L. 'Evolutionary programming based economic dispatch for units with non-smooth fuel cost functions'. *IEEE Transactions on Power Systems.* 1996;11(1):112–18.

[25] Lin W. M., Cheng F. S. and Tsay M. T. 'An improved tabu search for economic dispatch with multiple minima'. *IEEE Transactions on Power Systems.* 2002;17(1):108–12.

[26] Lim S. Y., Montakhab M. and Nouri H. 'Economic dispatch of power system using particle swarm optimization with constriction factor'. *International Journal of Innovations in Energy Systems and Power.* 2009;4(2):29–34.

[27] Hetzer J., Yu D. C. and Bhattaria K. 'An economic dispatch model incorporating wind power'. *IEEE Transactions on Energy Conversion.* 2008; 23(2):603–11.

[28] Hong Y. Y. and Lian R. C. 'Optimal sizing of hybrid wind/PV/diesel generation in a stand-alone power system using Markov-based genetic algorithm'. *IEEE Transactions on Power Delivery.* 2012;27(2):640–7.

[29] Granelli G. P., Montagna M., Pasini G. L. and Marannino P. 'Emission constrained dynamic dispatch'. *Electric Power System Research.* 1002; 24(1):55–64.

[30] Farag A., Al-Baiyat S. and Cheng T. C. 'Economic load dispatch multiobjective optimization procedures using linear programming techniques'. *IEEE Transactions on Power Systems.* 1995;10(2):731–8.

[31] Srinivasan D., Chang C. S. and Liew A. C. 'Multi-objective generation schedule using fuzzy optimal search technique'. *IEE Proceedings of Generation, Transmission and Distribution.* 1995;141(3):233–42.

[32] Poli R. 'Analysis of the publications on the applications of particle swarm optimisation'. *Journal of Artificial Evolution and Applications.* 2008; 2008:1–10.

[33] Park J. B., Lee K. S., Shin J. R. and Lee K. Y. 'A particle swarm optimisation for economic dispatch with non-smooth cost functions'. *IEEE Transactions on Power Systems.* 2005;20(1):34–42.

[34] Berizzi A., Innorta M. and Marannino P. (eds.). 'Multi-objective optimization techniques applied to modern power systems'. *IEEE PES Winter Meeting*; Columbus, OH, USA, Jan/Feb 2001. IEEE; 2001, Vol. III, pp. 1503–8.

[35] Jin Y., Okabe T. and Sendho B. (eds.). 'Adapting weighted aggregation for multi-objective evolution strategies'. *Proceedings of International Conference on Evolutionary Multi-Criterion Optimisation*; Zurich, Switzerland, Mar 2001. Springer, pp. 96–110.

[36] Abido M. A. 'Multi-objective evolutionary algorithms for electric power dispatch problem'. *IEEE Transactions on Evolutionary Computation.* 2006;10(3):315–29.

[37] Xial X. and Elaiw A. M. 'Dynamic economic dispatch: A review'. *The Online Journal on Electronics and Electrical Engineering (OJEEE)*. 2010; 2(2):234–45.

[38] MATLAB/Optimization Toolbox. Available from http://www.mathworks.com/matlabcentral/fileexchange/7506-particle-swarm-optimization-toolbox [Accessed Sep 2018].

[39] Cai J., Ma X., Li Q. and Peng H. 'A multi-objective chaotic particle swarm optimization for environmental/economic dispatch'. *Energy Conversion and Management*. 2009;50(5):1318–25.

Chapter 16

Generation scheduling for power grids with renewables

16.1 Introduction

Electricity demand can exhibit large variations from weekdays to weekends, one day to another, peak to off-peak hours as well as from one season to another. As explained in Chapter 15, optimal loading of each generating unit in a power system encountering conventional thermal units integrated with renewable wind sources can be determined using economic dispatch (ED) techniques. All available generating units can be turned on to meet the demand, but it is usually not economical to continuously run all the units available all the time. Hence, it is desirable to look for an efficient economic short-term generation scheduling that plays an important role in the economic operation of a power system. The principal objective of the short-term generation scheduling is to efficiently determine the commitment of generating units as well as the economic loading of each committed unit so that the forecasted load demand and spinning reserve can be met over a short period ranging from 1 day to 1 week. The resultant schedule should minimize the system production cost during a specified period while simultaneously satisfying system and operational unit constraints. Thus, scheduling power system operation should involve two basic economic decisions:

1. A unit commitment (UC) decision that determines which units should be brought online to meet the expected load demand, power losses and reserve requirements.
2. An embedded ED decision that determines the most economic generation level for each of the committed (synchronized) units.

These two decisions must be taken to minimize the total start-up, shutdown and running costs subject to various constraints such as power balance, spinning reserve and minimum up and downtime constraints [1].

Based on this aspect, the problem of generation scheduling of a conventional power system without and with renewables (wind and solar photovoltaic (PV)) is studied in this chapter. Several critical constraints that govern the operation of synchronized generators and affect the generation scheduling are crucial. The most widely solutions to the generation scheduling are discussed. Emphasis is placed on the advantages and limitations of each method.

Moreover, security-constrained generation scheduling (SCGS) problem for a grid with renewables is formulated. The formulation considers the stochastic nature of both the wind and PV power output, and imbalance charges due to mismatch between the actual and scheduled wind and PV power outputs. A hybrid technique in which the basic elements are a genetic algorithm (GA) with artificial neural network (ANN), and a priority list (PL) can be used to minimize the total operating costs while satisfying all operational constraints considering both conventional and renewable energy generators.

16.2 Thermal generating units

The main generating units in a thermal power system [2] can be categorized under the following three groups:

- Nuclear steam thermal units: These units are characterized by having high capital investment but low online generation costs. Normally, such units are operated as base generation.
- Fossil-fuelled steam thermal units: These units offer the flexibility to follow the changes in load and are normally considered for cycling and peak duties. One characteristic of these units as well as the nuclear units is that a minimum period should elapse before they can be started up (or shut down).
- Gas turbine: These units are another option for dealing with peaks in demand. The reason is that although they have comparatively high online generation costs, they can be started up quickly and shut down after operation for a short period. The start-up costs are comparatively insignificant.
- Combined cycle gas turbine: Modern generating units are increasingly having to cope with complex operation regimes to increase efficiency and reduce costs. A combined cycle unit can combine several subunits such as gas turbine and steam turbine subunits, resulting in several different operating modes. The plant efficiency depends on the number of gas turbines in operation.

16.2.1 Characteristic of thermal units

From an economic point of view, the set of input–output characteristics of a thermal power generation unit is of most interest. The input in this case may be either the heat energy requirements, H, or in terms of total fuel cost, F, measured in MBtu/h or $/h, respectively, whereas the net output is the electrical power output, P, measured in MW, available to the electric utility system. The total operating cost includes the fuel cost, and the cost of labour, supplies and maintenance. However, it is common to assume that the cost of labour, supplies and maintenance is a fixed percentage of the fuel costs.

The input–output characteristic of a steam unit in idealized form is presented as a smooth, convex curve (Figure 16.1(a)). The incremental heat rate characteristic, which is the derivative of the input–output characteristic, for a unit of this type is shown in Figure 16.1(b). The data in this curve are in terms of Btu/kWh

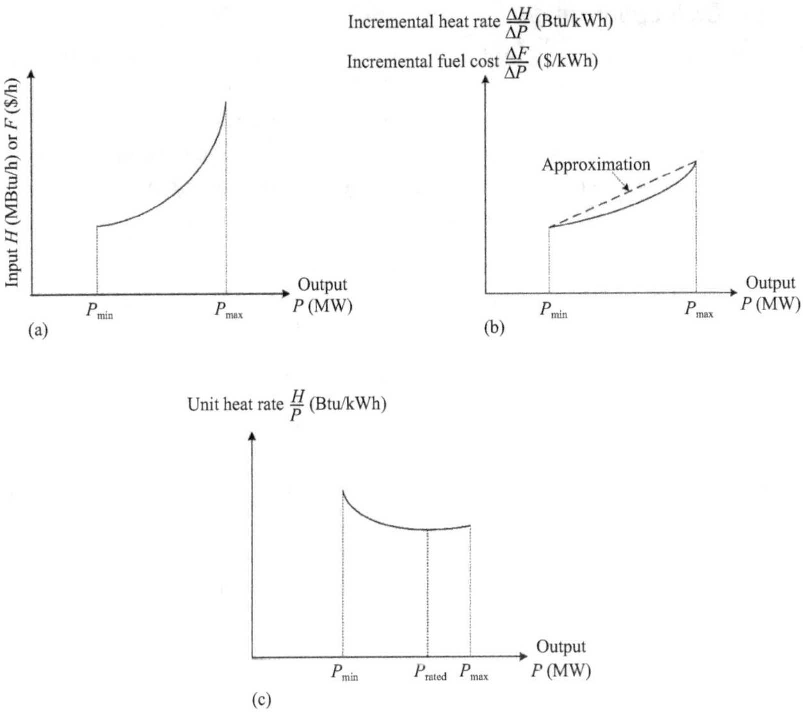

Figure 16.1 Typical steam unit characteristics: (a) input–output characteristic, (b) incremental characteristic and (c) net heat rate characteristic

(or \$/kWh) versus the net power output of the unit in MW. Very often, this characteristic is approximated by a sequence of straight-line segments.

In addition, an important characteristic of a steam unit, which is the unit (net) heat rate characteristic, is represented by H/P versus P as shown in Figure 16.1(c). It is proportional to the reciprocal of the usual efficiency characteristic developed for machinery [3]. However, a series of straight-line segments may be also used to represent the input–output characteristics. The different representations will of course result in different incremental heat rate characteristics.

16.3 Formulation of the thermal scheduling problem

The thermal scheduling problem involves the determination of start-up and shutdown time as well as the power output levels of all system generation units at each time step, over a specified scheduling period T, so that the total start-up, shutdown and running costs are minimized subject to several system and unit constraints. Obtaining an optimal schedule of generation involves the solution of a mixed-integer non-linear optimization problem with many constraints.

16.3.1 Unit commitment

UC represents a challenging problem in the operation of a power system. It belongs to the class of complex, non-linear, mixed-integer programming whose objective function is the system production cost over the entire period. The solution of the UC problem involves non-linear optimization, comprising of both integer and continuous variables, with many equality and inequality constraints.

The UC problem in a thermal system arises because of the following [3]:

- The variation of system load demand during the day. Although some generators can be operated at a fixed output level throughout the day as 'base generation', it is necessary to start up and shut down some generators during the day as 'cycling units', to match the system total generation to the demand.
- The variety in the generating units, in terms of operational characteristics and economic efficiency. Also, the fact that it usually takes some time to start up and shut down a thermal generating unit raises the question as to which combination of generators is the most economical and feasible to meet the system load demand at various times.
- There are several critical constraints that govern the operation of a synchronized generator and affect the UC. It is imperative that these constraints are taken into consideration, to achieve a feasible and realistic solution, while solving the problem of UC.

16.3.2 Problem objective function

The main objective of scheduling in thermal systems is to minimize system operation costs. The total production cost, F_{cost}, for the scheduling period is the sum of the running, start-up and shutdown costs for all the units. Shutdown cost is defined as the amount of the fixed cost for each unit per shutdown and is applied in the hour the unit is taken offline. The total production cost to be minimized is given by

$$F_{\text{cost}} = \sum_{j=1}^{T} \left[\sum_{i \in G} F_i(P_{ij}) + SC_{i,j} \right] \tag{16.1}$$

where

$F_i(P_{ij})$: the production cost of the ith thermal unit in the jth period, and P_{ij} is the power output of ith unit in the period j;
$SC_{i,j}$: unit start-up cost;
T: the time horizon, which is taken as 24 h;
G: the number of generating units.

16.3.3 Fuel costs

Fuel costs of thermal units are usually represented by a quadratic heat rate curve as a function of power output multiplied by the price of the selected fuel. If F_i is the function that relates the generator power output, P_i, to fuel cost, an example of the function often used is

$$F_i = a_i P_i^2 + b_i P_i + c_i \tag{16.2}$$

where a_i, b_i and c_i represent the unit cost coefficients.

Also, a suitable higher-order polynomial may be used to express the unit cost function. In some studies, fuel costs are represented by a fixed linear function of generator output level. However, if more precise and better results are needed, several linear segments (a piecewise approximation) may be implemented.

16.3.4 Transition costs

The start-up cost of a unit is a mixture of fixed and variable downtime-dependent costs. The generator start-up cost, $SC_{i,j}$, depends on the time the unit has been off prior to start-up and can be represented by an exponential cost function [4]:

$$SC_{i,j} = u_{ij}\left[\alpha_i + \beta_i\left[1 - \exp\left(-T_{i,j}^{\text{off}}/\tau_i\right)\right]\right] \tag{16.3}$$

where

α_i: integrates labour start-up cost and equipment maintenance cost of unit i;
β_i: start-up cost of unit i from cold condition;
τ_i: time constant that characterizes unit i cooling speed;
$T_{i,j}^{\text{off}}$: the time a unit has been off prior to start up;
u_{ij}: the status index of ith unit in the jth period (1 for up and 0 for down).

However, there are two approaches to treating a thermal unit during its down period. The first allows the unit's boiler to cool down and then heat, back up to operating temperature in time for a schedule turn on, and in this case, the start-up cost is governed by (16.3). The second is called *banking* and requires that enough energy be input to the boiler to just maintain operating temperature [3].

The objective function F_{cost} is minimized, subject to several system and unit constraints. Each individual power system may specify its own set of constraints depending on certain factors such as network topology, reliability and security load curve characteristic and so on. These constraints are described in the forthcoming sections.

16.3.5 System constraints

System constraints are applied in the power system operating objective function to keep the system operating point within acceptable limits of stability and to meet security requirements. The most common system constraints are discussed in the following subsections.

16.3.5.1 Active power balance

The load demand variation over the scheduling interval provides one of the biggest challenges for the scheduling algorithm. The total power generated by the online units must supply the load demand, P_d, and system losses, P_L, that is,

$$\sum_{i \in G} u_{i,j} P_{ij} = P_{d_j} + P_L \tag{16.4}$$

16.3.5.2 Load forecast error and system reserve requirements

One of the primary objectives of UC is to ensure the security of the system to withstand predicted load uncertainty and generation forced outage. Spinning and supplemented reserves must be allocated in the UC to meet the security criteria set down by the management. Spinning reserve can be defined as the extra generation available on demand from the synchronized generators, while the supplemented reserve specifies an amount of additional reserve that must be available to the system and is composed of hot standby units and rapid start-up units [5].

The hourly spinning reserve requirement, R, that must be met over the whole scheduling period is as follows:

$$\sum_{i \in G} u_{i,j} P_{ij}^{\max} \geq P_{d_j} + P_L + R \tag{16.5}$$

16.3.5.3 Transmission network constraints

The transmission line loading limits must not be violated by the power flow resulting from a given schedule. These constraints are especially crucial for systems that are loaded closer to their thermal limits, that is,

$$-P_m^{\max} \leq P_{mj} \leq P_m^{\max}, \quad m = 1, \ldots, M \tag{16.6}$$

where

$P_m^{\max}$: the maximum transmission capacity of line m;
P_{mj}: the load flow on the line m at period j;
M: the total number of transmission lines.

Transmission capacity constraints present a challenge to researchers in the UC field. Usually, linear DC transmission constraints are considered in UC formulation for system security purposes. However, P_{mj} can be formulated as

$$P_{mj} = \sum_{i \in j} k_{mi} P_{ij} u_{i,j} - \sum_{s=1}^{NB} k_{mj} Pd_{s,j} \tag{16.7}$$

where NB, $Pd_{s,j}$ are the total number of system buses and load at Bus s, respectively. The sensitivity coefficients k_{mi} and k_{mj} can be determined, based on DC or AC power flow corresponding to normal or contingency conditions [6].

16.3.5.4 Emission constraint

With the increased demand for a cleaner environment, the thermal schedules must not violate allowable emission limits, and if possible, the nitrogen dioxide (NO_2), sulphur dioxide (SO_2) and carbon dioxide (CO2) emissions should be minimized [7].

16.3.6 Unit constraints

The generating unit can be subjected to several constraints that must be accounted for in the process; these include the following.

16.3.6.1 Unit minimum up and downtime

Unit minimum up and downtime limit the thermal stresses that the machines would be subjected to if they were started or shut down. Thus, a unit can only be shut down (started) once it has been staying on (off) for a minimum period, called minimum up (down)time. Thus, the limits of minimum up/downtime (MUT/MDT) of units must not be violated, that is,

$$\left(T^{\text{on}}_{i,j-1} - MUT\right)\left(u_{i,j-1} - u_{i,j}\right) \geq 0 \tag{16.8}$$

$$\left(T^{\text{off}}_{i,j-1} - MDT\right)\left(u_{i,j-1} - u_{i,j}\right) \geq 0 \tag{16.9}$$

where $T^{\text{off}}/T^{\text{on}}$ is the unit off/on time.

16.3.6.2 Real power operating limits

It is important that unit rated minimum capacity must not be violated. The maximum and minimum real power limits are generally expressed as

$$P_i^{\min} \leq P_{ij} \leq P_i^{\max}, \quad i \in G, j \in T \tag{16.10}$$

16.3.6.3 Unit ramp rate limits

In the boiler fuel control, there is a limiting factor for increasing the rate of fuel in feed from the nulls and the time lag to convert the chemical energy of coal to heat energy. Similarly, when a unit is in the shutdown process, it will take a while for the turbine to cool down. Before the unit generating capability decreases to its lower limit, the residual energy must be used to meet the load demand [8]. These ramp limits may be expressed as

$$P_{ij} - P_{ij-1} \leq UR_i \tag{16.11}$$

$$P_{ij-1} - P_{ij} \leq DR_i \tag{16.12}$$

where UR_i, and DR_i are the ramp up and ramp down rate limits, respectively.

In addition to the above-mentioned constraints, there are other unit constraints, which include the following:

- The initial unit states at the start of the scheduling period.
- Crew constraints that limit the number of units that can be started at the same time in a specific plant, and unit operating status or mode restrictions.
- Fixed generation units are prescheduled units. Any unit may be prescheduled to 'must be on', 'must be off' or fixed generation for certain intervals of the study period. Specification of such requirements is fed to the UC program as

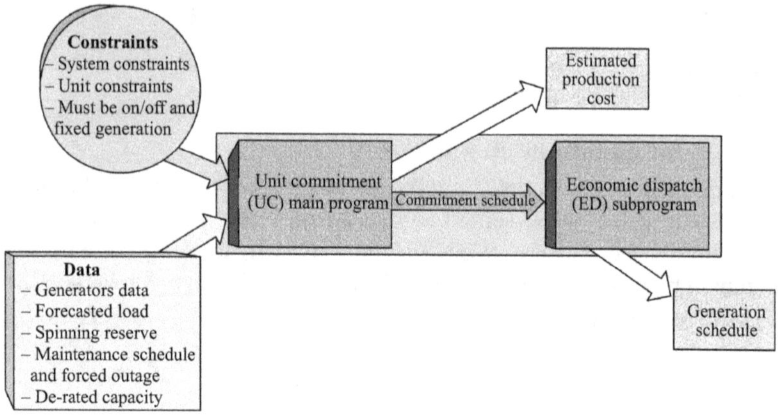

Figure 16.2 UC with ED processes

input data. Units scheduled for maintenance or in forced outage can, therefore, be treated as 'must be off' units.

- De-rated capacity: Partial outages of the units leading to de-rated capacity or changes from de-rated state to full capacity state or to another de-rated state.

A typical block diagram for the input/output of UC with ED program is depicted in Figure 16.2.

16.4 Economic dispatch

Once the generating units in the power system are scheduled and the committed units are identified, it is necessary to determine the optimal allocation of the system power demand among the committed units, while satisfying different constraints. ED problem can be treated as a non-linear optimization problem, which consists of several equality and non-equality constraints. The problem is formulated as follows:

$$\text{Minimize } F_{\text{cost}} = \sum_{i \in G} F_i(P_{ij}) \tag{16.13}$$

subject to (16.4)–(16.6) and (16.10)–(16.12).

The ED problem can be handled from the perspective of a sub-problem of generation schedule.

16.5 Review of thermal scheduling techniques

Generation scheduling in practical power systems poses difficult analytical challenges, as it involves the solution of a large-scale mixed-integer non-linear, optimization problem, with several constraints. The ideal solution to the problem can be obtained by complete enumeration, which cannot be applied to realistic power systems due to its excessive computation time requirement. In the past, a few attempts using both empirical analysis and rigorous mathematical programming techniques have been used for solving the generation scheduling problem [9]. A closer review of some solution methods to the generation scheduling problem is given in the next subsections.

16.5.1 Priority list

The units in this method are committed according to a PL based on full-load average production and transition costs [3]. The PL schemes could be obtained after an exhaustive enumeration of all unit combinations at each load level. The pre-determined order is then used to commit units such that the system load is satisfied.

The merit order dispatch is the simplest dispatch method and it relies on the availability of linear or piecewise linear cost function. The committed generators are indexed in order of increasing incremental cost and initialized at their minimum power output levels. The generators are then considered for loading to their maximum capacities according to their rank in the PL, until the load demand is satisfied. The system incremental cost is then determined by the partly loaded generator in the system.

This method provides reasonably accurate results only if linear cost functions are used. If a more accurate modelling of the unit performance is required, then the merit order scheme fails to provide an acceptable solution. However, its simplicity, speed and ability to guarantee feasible solutions make it one of the most widely used method by electric utilities.

16.5.2 Dynamic programming

Dynamic programming (DP) is one of the techniques that have been most widely researched because of its ability to handle any problem, which can be formulated as a set of separable state transitions [10]. It works by determining the optimal set of state transitions, which will bring the system from an initial state to a final state. It decomposes the problem in time, beginning at the first hour of the scheduling period, committing the units progressively 1 h at a time and storing the unit combinations at each hour and their associated costs. At the end of the schedule, the steps are traced backwards to obtain the optimal combinations that results in the least cost schedule. However, only small systems can be optimized, as the number of states to be stored and evaluated increases exponentially as the number of variables increases. To avoid this curse of dimensionality problem in UC, DP is used to solve the commitment of a single unit, in combination with other techniques such as Lagrangian relaxation (LR). The truncated or variable window DP method is another attempt at reducing the DP storage difficulties [11].

DP method has also been employed for ED [12]. This method, however, suffers from the problem of large dimensionality as it requires exploration of the entire solution space. The ability to find the global optimum solution is dependent on the size of the discrete load step assigned to each generator in the solution process.

16.5.3 Lagrangian relaxation

LR is a mathematical decomposition technique, which is rapidly growing in importance as a solution method for the generation scheduling problem, especially for large power systems. The LR decomposition technique based on duality theory generates a separable problem by integrating some constraints into the objective function through penalty factors, which are functions of the constraint violations [11]. The penalty factors, called Lagrangian multipliers, are determined iteratively and they determine the solution quality. The approach involves decomposition of the problem into a sequence of master problem and easy sub-problems. Each sub-problem consists of determining the commitment schedule for a single unit over the planning horizon and is solved independently using any suitable technique such as dynamic or mixed-integer programming. The only link between these sub-problems is the Lagrangian multipliers, which are adjusted in the master problem by a sub-gradient method or heuristically to ensure that the system constraints are met. An iterative approach usually ensures convergence to a near-optimal solution and generates bounds, which give an indication on the proximity of the solution to the true optimal.

The drawback of this technique is that the quality of the final LR solution depends on the sensitivity of the commitment to Lagrangian multipliers. Slow and unsteady convergence of LR has always been a problem in finding the global optimum solution. The difficulties are often explained by the non-convexity of this type of optimization problem. The augmented Lagrangian method has been applied to overcome the problems of the LR method [12]. In this method, quadratic penalty terms associated with power demand are added to the objective function to improve the convexity of the problem.

16.5.4 Branch and bound

The branch and bound method is a powerful enumeration strategy that helps to reduce the number of combinations of integer variables considered in a mixed-integer non-linear programming problem [13]. It is a technique to solve a discrete variable problem by solving a sequence of simpler problems derived from the original problem. As reported in [14], the UC problem can be formulated as a linear mixed-integer programming and then standard integer programming algorithms are used to solve for the commitment schedule. Several problem approximations and extensions to the basic branch and bound algorithm have been described to make the solution approach tractable for this problem. The advantage of branch and bound technique is that it can provide a sequence of solutions with estimates of their suboptimality. However, it suffers from the curse of dimensionality as the execution time exponentially grows with the size of the UC.

16.5.5 Linear programming

Linear programming [11] is usually combined with other techniques such as DP and mixed-integer programming for solving the continuous ED sub-problem. The advantage of linear programming lies in its fast and reliable solution, whereas its main disadvantage is that all the objective functions and constraints must be linearized, which often leads to suboptimal solutions.

16.5.6 Artificial intelligence

Recently, a new direction in research for reducing the computation time while preserving the system characteristics has pointed out the potentials of using artificial intelligence (AI) techniques. The AI techniques have incorporated the practical operational policies in the mathematical techniques to improve the system models considerably. They, therefore, accommodate more complicated constraints and are claimed to have better solution quality. The AI techniques, unlike strict mathematical methods, have apparent ability to adapt to non-linearities and discontinuities found commonly in physical systems. Although they look promising, they are still evolving.

16.5.6.1 Heuristic and expert systems

Heuristic and expert system methods are based around rules that are derived from the system characteristics and the knowledge of experienced power system operators. The expert system improves the solution of the UC problem by adjusting the program's parameters through interaction with the system operator. Several expert system-based techniques with specified rules for systems have also found widespread usage in utility scheduling [15]. Also, the fuzzy set theory has been applied to solve the UC problem [16]. With this approach, the security constraints, generation costs and load demands have all been expressed in fuzzy set notation. The fuzzy DP has then been performed to reach an optimal commitment schedule [17]. However, the rule based reasoning method has its own drawbacks, primarily because the human knowledge may not be easily expressed and classified by rules, and the solution of a problem based on many rules may be costly and require a long processing time.

16.5.6.2 Simulated annealing

Simulated annealing (SA), a heuristic technique, takes the analogy of the physical annealing of melted particles [18]. Starting from a very high temperature, a melted particle is cooled gradually until it is solidified at a low temperature. This physical annealing process is simulated in the SA technique for the determination of near-global or global optimum solutions for combinatorial optimization problems. Different researchers have used SA method to solve the UC problem. The main drawback of the SA method is the long computation time as well as the inability to guarantee feasible solutions.

An SA-based ED algorithm has been reported in [19]. It has been shown that the algorithm can determine the global or near-global optimum solution for ED. However, the convergence of this algorithm is slow.

16.5.6.3 Artificial neural network

The mechanism of ANN is based on simulating the learning process of the human brain. For implementing ANN, the necessity of the mathematical representation of a physical problem, which on some occasions may create additional problems in achieving a reasonable solution, can be relaxed. The Hopfield neural network mimics the computational capability of biological organisms by utilizing simple computing neurones with high interconnectivity and parallelism. The Hopfield neural network to solve the UC problem has been studied as in [20]. The average execution time was approximately 2.5 times that of the Lagrangian method for 30 generators. In addition, ANN has a great potential in facilitating an economic solution within a reasonable time and a combination of the back-propagation ANN combined with DP have been used to solve the UC problem [21]. The training data set used for back-propagation contains past unit schedules, which are assumed to be optimal. Once trained, the network is presented with a new load profile for which a schedule is desired. Although the ANN can produce fast schedules, its main drawback is the difficulty in providing feasible and near-optimal solutions.

The Hopfield neural network has also been applied to optimal economic/environmental dispatch of thermal generating units [4]. A simulator has been developed and criteria for selecting its parameters have been explored to obtain optimal dispatch in a minimum number of iterations. However, many iterations and oscillation are the major concern in solving the ED problem using the Hopfield neural networks.

16.5.6.4 Tabu search

Tabu search (TS) is an iterative improvement procedure that starts from some initial feasible solution and attempts to determine a better solution in the manner of a greatest-descent algorithm [5]. However, TS is characterized by an ability to escape local optima using a short-term memory of recent solutions. Moreover, TS permits backtracking to previous solutions, which may ultimately lead, via a different direction, to better solutions. An application of the TS method to solve the UC problem is presented in [6]. Numerical results show an improvement in the solution cost, but it suffers from a long execution time.

16.5.6.5 Evolutionary algorithms

All evolutionary algorithm models are based on a learning process using a population of potential solutions to a problem, each representing a search point in the Euclidean space of all the possible solution. The initial population is usually randomly selected and evolves towards the optimal solution according to rules of selection and other operators such as recombination and mutation. Everyone in the population is awarded a fitness measure based on the problem objective function, which determines its ability to propagate its elements in the search process. The recombination allows the mixing of parental information while the mutation operator introduces diversity in the population [22]. There are three main approaches: GA, evolution strategies (ES) and evolutionary programming (EP). These methods differ in terms of representation, operators and selection process. The goal of the EP is to achieve intelligent behaviour through simulated evolution.

GA is one of the most widely studied evolutionary techniques. The GA has been successfully applied to solve the short-term generation scheduling [23].

The results are encouraging, but unrealistic for real-time operation because of long execution time.

GA-based method has also been reported recently for solving the ED problem [8]. On the other hand, extended evolutionary algorithms for solving the optimal reactive power dispatch has been explored [9]. Good and reliable performance has been achieved and validation tests using the standard IEEE57 system have been reported.

However, none of the basic evolutionary computational methods can guarantee an optimal solution to the scheduling problem. They only provide near-optimal solutions, and the quality of each solution provided is affected by either solution time limitations or by premature convergence. Various enhancements and modifications are required to enable the basic evolutionary computational models to be applied successfully in the solution of the scheduling problem.

16.6 Problem representation and solution strategy

The objective is to efficiently determine the effective commitment of the generating units as well as the output level of each committed unit to meet the forecasted demand and spinning reserve at the minimum possible production cost over a specific period, for example, one day. Moreover, this should satisfy system and unit constraints. In practice, short-term generation scheduling often requires a method that is fast to meet system changes and reduce the scheduling error. With the advent of the stochastic global search algorithms, the GA has been successfully applied to solve the short-term generation scheduling problem [23]. The results are promising, but unrealistic for real-time operation because of long execution time. To surmount this problem, a hybrid technique using GA, PL and ANN can be applied accounting for load forecast uncertainty to accelerate and enhance the performance of the short-term generating scheduling solution.

In the UC problem, with ANN computing, most of the time is spent on offline training the network as ANN accumulates knowledge during offline training from the given input/output data pairs. Once the network is completely trained, the online operating time would be very short as compared to that attained for analytical programming techniques [21]. The ANN-supported GA (ANN/GA/PL) is a hybrid of intelligent system and traditional heuristic technique. First, a pre-scheduling of the generating units using the ANN is presented. The GA with PL is then used to search for the best combination of the hesitant units that result from the ANN and satisfying both the system and unit constraints. With its two-stage processing, the algorithm can benefit from the advantage of this technique. The ED sub-problem is then solved for each member of the population satisfying all constraints. The flow chart of the (ANN/GA/PL) hybrid technique is shown in Figure 16.3.

16.6.1 GA implementation

The two main actions that link the GA to the problem it is solving are the means of translating the problem solution to a chromosome representation and an evaluation

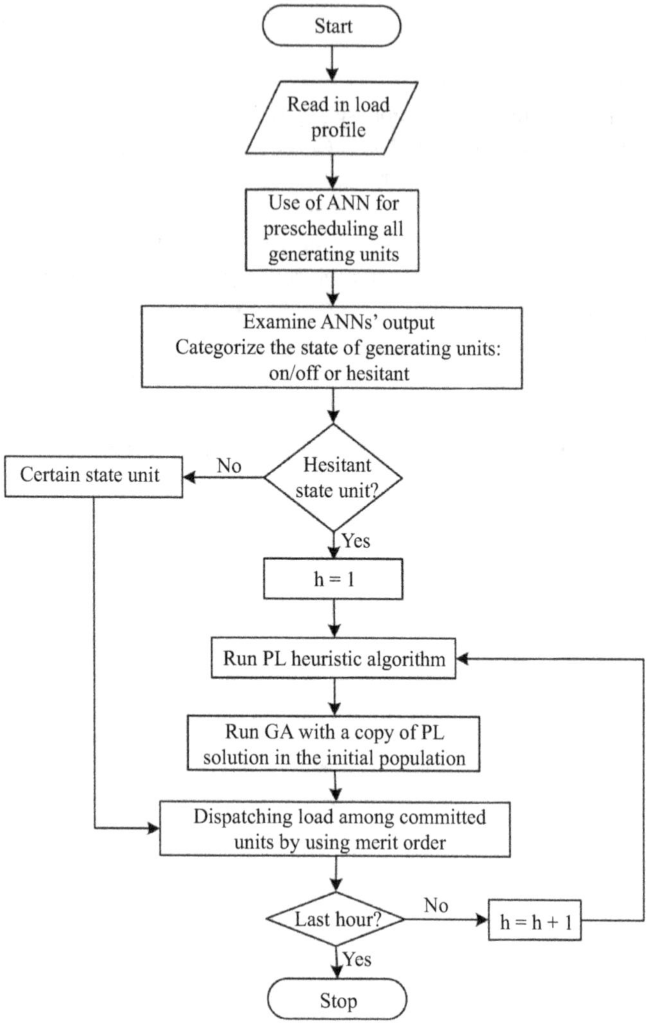

Figure 16.3 Flow chart of (ANN/GA/PL) hybrid technique

function that returns a measurement of worth for any chromosome in the context of the problem. The success of the GA approach is heavily dependent on the problem encoding and the choice of the evaluation function.

The thermal scheduling problem is a mixed-integer non-linear optimization problem that lends itself to a suitable binary representation. For the application of GA to the UC problem, a binary string of ['0', '1'] denoting the off/on state of the unit is randomly generated. The total solution string length is a product of the scheduling period T and the number of generation units G. A typical UC problem representation for scheduling G units over T hours is shown in Figure 16.4.

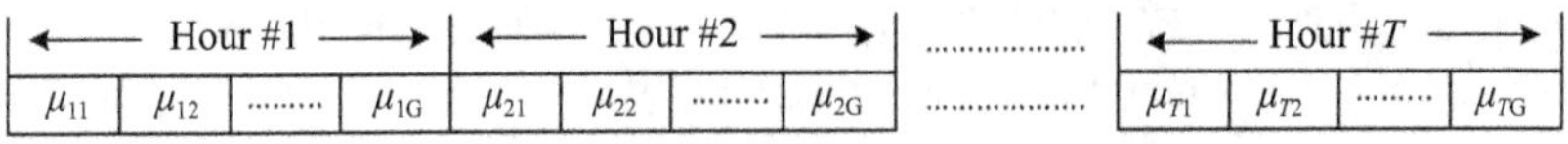

Figure 16.4 An example of binary UC problem encoding ($\mu_{t,i} = 0$ or 1, where t is the time index and i is the unit index)

16.6.2 Decomposition methods

It is expected that as the number of units increases, the string length increases proportionately resulting in a long computational time. To surmount the problem, a sequential GA is used to solve the scheduling problem, and this limits the selection, mutation and crossover of the GA to a single time interval instead of the whole-time interval. However, this process does not strictly partition the problem into a single time span, but involves cumulative time-span partitioning, in that the linking constraints' parameters, such as unit minimum up/downtime, are continuously updated and re-evaluated as the time steps increase. Also, an elitist GA search is used to guarantee that the best solution at each generation is retained and used in the subsequent generation, thereby ensuring that no good solution is being lost during a GA search.

Since a GA works by searching both the feasible and infeasible problem solution search space, it is necessary because of its population-based nature and its selection mechanism to differentiate between the performances of two non-feasible strings. Thus, a string that violates a constraint more than another should receive a higher penalty and expect a corresponding diminished chance of selection. The discrimination between feasible and non-feasible solutions is achieved by making sure that any non-feasible solution is assigned a fitness lower than that of a feasible solution through the application of the penalty algorithm.

The constraints of the problem are handled by including penalty factors within the GA.

The fitness function of the scheduling problem is a linear combination of the total operating cost and the penalty factors and can be represented by (16.14). The minimization cost function is transformed into a maximization problem by the reciprocal of the cost function.

$$F_{\mathrm{GA}} = \sum_{j=1}^{T} OC(P_{ij}) + \alpha_1 FSC(j) + \alpha_2 FDC(j) + \alpha_3 FTR(j) + \alpha_4 FTD(j) + \alpha_5 VTL(j) \tag{16.14}$$

and

$$\sum_{i=1}^{N} OC(P_{ij}) = \left[\sum_{i=1}^{N} FC_i(P_{ij}) + SC_{ij} + SDC_{ij} \right] \tag{16.15}$$

where $FC_i(P_{ij})$, P_{ij}, SC_{ij} and SDC_{ij} are the production cost of ith thermal unit, its power output, start-up and shutdown costs in jth hour. $\alpha_1, \ldots, \alpha_5$ are penalty factors, and *FSC*(*j*), *FDC*(*j*), *FTR*(*j*), *FTD*(*j*) and *VTL*(*j*) are the penalty terms, which are utilized when a unit does not comply with the minimum up/minimum downtime, fails to meet the reserve and the load demand, respectively, and when the transmission line constraints of the system are violated.

Choosing a penalty factor that combines well with the cost function is a difficult task as it affects the performance of the algorithm. Different values of $\alpha_1, \ldots, \alpha_5$ are tried to calculate the effective values that satisfy the violated constraints.

16.6.3 PL implementation

To start a typical run of a GA, an initial population of strings is generated randomly subject to the initial conditions of the studied system. However, an extensive research has indicated the importance of the choice of the initial population [24]. By including the domain knowledge into a hybrid GA technique, the hybrid technique would outperform both the traditional GA and the conventional methods.

The PL scheme is the most popular technique of solving the UC problem, as it is a fast and reliable method. To use a hybrid technique comprising the GA with the PL at each hour, the PL algorithm is run followed by the GA. The latter is run with the inclusion of a copy of the PL solution in its initial population.

16.6.4 ANN implementation

ANN is a high-speed online computational technique, which once trained through an offline algorithm using example patterns, can provide an output corresponding to a new pattern without any iteration in real time. Three-layered feed-forward ANN with a sigmoid function is used in this study to preschedule the generating units and the back-propagation algorithm is used for training the ANN.

Like the human brain, as the dimension of patterns increases, the ANN may find it difficult to remember and recognize each pattern. Thus, each generating unit is prescheduled by a separate ANN as this has proved to be more efficient than training all the units in advance [25]. Hence, the number of ANNs is determined by the number of the generating units. Each ANN has 24 input neurons corresponding to the 24-hourly loads, while the output layer has 24 neurons. The optimum number of hidden neurons is determined by trial and error. In the technique presented here, this is equal to nine.

Load-forecast uncertainty imposes a threat to a firm UC decision. Underestimation or overestimation of load can lead to a failure to provide enough reserve or lead to an unnecessary large amount of spinning reserve, which in turn leads to higher cost [26]. In practice, load demand is very much dependent on weekdays, weekends including holidays and in many countries is dependent on different seasons in a year. An extensive series of studies has shown that to affect a single ANN-based technique to cater for all the foregoing different loads is not very satisfactory. The approach adopted herein is thus based on dividing the load demands into various categories, and in practice, there will thus be several ANN-based algorithms each to deal with a category. Regarding any one category, the practical error margin is a maximum of ±5% [27] and hence, the training and testing data used to illustrate the technique described in this study is realistic in terms of evaluating the performance of the method. After training and testing the ANN using different load curve samples, the ANN produces an output feature value (i.e. the preschedule), which is classified as follows:

- Certain state unit: The output value of this is greater than or equal to 0.9 (with certain 'on' state), or an output feature value of less than 0.1 (with certain 'off' state).
- Hesitant unit: $0.9 >$ the output feature value ≥ 0.1. Here, the GA or the GA/PL is used to solve the problem of scheduling of hesitant units.

16.6.5 ED implementation

In the scheduling process, once a decision is made on whether a unit is to be on or off, a further optimization process must be carried out to optimally allocate the load to the online units, a process which is referred to as ED. The main objective of the ED is to minimize fuel costs whilst allocating the load among the online units. This should also satisfy the physical limitations of units as well as those on the power system. Owing to the ability of the GA to look for the global optimal solution, it is applied to the ED problem. The GA is thus further extended for solving the ED problem concurrently with the UC problem. At each time interval, individual generators are represented by ten bits, where the first bit represents the unit on/off status, whilst the latter nine bits encode the power output level of that unit. These nine bits could easily be decoded and mapped into each unit's own operating range according to the following equation [28]:

$$\text{Mapped value} = \frac{D_{\text{cd}}}{2^L - 1} \times \left(P_i^{\max} - P_i^{\min}\right) + P_i^{\min} \tag{16.16}$$

where

D_{cd}: the decoded value;
L: string length;
$P_i^{\max}, P_i^{\min}$: maximum and minimum capacities of the corresponding generating unit, respectively.

By modifying ten bits for every generator, the GA effectively searches for the optimum solution in the combined UC and ED search space. Also, to a certain extent, its computational time is considerably higher than the techniques currently being used. However, one major advantage of the GA approach to solve the ED problem lies in the fact that it can solve highly non-linear and discontinuous problems without the approximation needed in the classical problem. In addition, with the use of multiprocessors, its computational time can be very significantly reduced and becomes comparable to the time it takes to solve the problem using classical approaches.

One approach to reduce the computational time of the GA is to classify the generating units into three categories: base, cycling and peaking units.

Numerical example 16.1

The hybrid technique based on ANN with GA and PL (ANN/GA/PL) is examined on the 26-unit IEEE reliability system taking into consideration unit and system constraints. The transmission network of the IEEE reliability system consists of 24-bus bars connected by 38 lines and transformers, Appendix B [12,29,30]. A quadratic cost function and exponential start-up cost are considered to meet the system and unit constraints. To show the effect of line flow limits on generation scheduling, flow limits on all lines are reduced to 300 MW.

In both problem formulations, a ±5% load forecast error is aggregated into the ANN training data. Thus, during training the ANN, 11 samples of load curves and its corresponding commitment schedule are used. These samples are the original load curve signified as 'S' p.u. plus five other derived samples above and below the original load curve. These derived samples are denoted by 'S $\pm x$ %', where $x = 1$, 2, 3, 4 and 5. The UC schedules used as training data are obtained by GA.

The basic design of a GA solution method involves making an optimal choice of the various GA parameters, once the appropriate GA model has been chosen. This GA model is composed of the basic sequential processes of fitness evaluation, selection, recombination and mutation. Different GA control variables are tested, and a satisfactory performance is found with the following parameters:

Population size: 100.
Maximum number of generations: 200.
Two-point crossover is used in this study with a probability, $P_{\text{crossover}} = 0.9$.
The mutation probability is not fixed; it begins at 0.04, and then it is changed according to an exponential function as

$$P_{\text{mutation}} = 1 + (\exp(0.5 \times \text{generation}/\text{max\_generation}) + 0.04) \quad (16.17)$$

The GA and PL can be programmed in C language.

Units 17–19 and 24–26 of the IEEE system constitute base-load units, which are responsible for a large portion of system generation; Units 1–9 constitute peaking units; and lastly the other units represent cycling units. The values of penalty factors $\alpha_1, \ldots, \alpha_5$ are taken as 8,000, 8,000, 5,000, 5,000 and 8,000, respectively.

Generally, optimal power flow (OPF) is a tool to regulate the generator active power outputs, voltage, shunt capacitors/reactors, transformer tap-setting and other controllable variables to minimize the fuel cost and the network active power losses. This should also keep the load bus voltage, generator reactive power outputs, network power flows and other state variables in power system within their operational and secure limits. The generators active power is considered as the only control variable, which is used to minimize the total fuel cost while keeping system and units constraints within their limits. However, other control variables such as voltage magnitude and reactive power generation at each bus are ignored for simplification purposes.

The system operational cost with transmission constraints and transmission line losses using AC load flow program is $763,931. By including the ramp rate

Table 16.1 *Unit scheduling generated by the ANN/GA/PL with transmission line, ramp rate constraints and transmission line losses*

Unit	Hour (1–24)
1	1 0 0 0 0 0 0 0 0 0 0 0 1 1 0 0 0 0 1 1 1 1 1 0
2	1 0 0 0 0 0 0 0 0 0 0 0 1 0 0 0 0 0 1 1 1 1 1 0
3	1 0 0 0 0 0 0 0 0 0 0 0 0 0 0 0 0 0 1 1 1 1 0 1
4	1 0 0 0 0 0 0 0 0 0 0 0 0 0 0 0 0 0 1 0 0 1 0 0
5	1 0 0 0 0 0 0 0 0 0 0 0 0 0 0 0 0 0 0 0 0 1 0 0
6	1 0 0 0 0 0 0 0 0 0 0 0 0 0 0 0 0 1 0 0 1 1 0 0
7	0 0 0 0 0 0 0 0 0 0 0 0 0 0 0 0 0 0 0 0 1 1 0 0
8	0 0 0 0 0 0 0 0 0 0 0 0 0 0 0 0 0 0 0 0 0 0 0 0
9	0 0 0 0 0 0 0 0 0 0 0 0 0 0 0 0 0 0 0 0 0 0 0 0
10	1 1 1 1 1 1 1 1 1 1 1 1 1 1 1 1 1 1 1 1 1 1 1 1
11	1 1 1 1 1 1 1 1 1 1 1 1 1 1 1 1 1 1 1 1 1 1 1 1
12	1 1 0 0 0 0 1 1 1 1 1 1 1 1 1 1 1 1 1 1 1 1 1 1
13	1 1 0 0 0 0 1 1 1 1 1 1 1 1 1 1 1 1 1 1 1 1 1 1
14	1 1 1 1 1 1 1 1 1 1 1 1 1 1 1 1 1 1 1 1 1 1 1 1
15	0 0 0 0 0 0 0 1 1 1 1 1 0 0 0 0 1 1 1 1 1 0 0 0
16	0 0 0 0 0 0 0 0 0 0 0 0 0 0 0 0 0 0 1 1 1 1 1 1
17	1 1 1 1 1 1 1 1 1 1 1 1 1 1 1 1 1 1 1 1 1 1 1 1
18	1 1 1 1 1 1 1 1 1 1 1 1 1 1 1 1 1 1 1 1 1 1 1 1
19	1 1 1 1 1 1 1 1 1 1 1 1 1 1 1 1 1 1 1 1 1 1 1 1
20	1 1 1 1 1 1 1 1 1 1 1 1 1 1 1 1 1 1 1 1 1 1 1 1
21	0 0 0 0 0 0 0 1 1 1 1 1 1 1 1 1 1 1 1 1 1 1 1 1
22	0 0 0 0 0 0 0 0 1 1 1 1 1 1 1 1 1 1 1 0 0 0 0 0
23	0 0 0 0 0 0 0 0 0 0 0 0 0 0 0 1 1 1 1 1 1 1 0 0
24	1 1 1 1 1 1 1 1 1 1 1 1 1 1 1 1 1 1 1 1 1 1 1 1
25	1 1 1 1 1 1 1 1 1 1 1 1 1 1 1 1 1 1 1 1 1 1 1 1
26	1 1 1 1 1 1 1 1 1 1 1 1 1 1 1 1 1 1 1 1 1 1 1 1

constraints into the generation scheduling problem, the system operational cost goes up to \$791,332 and the committed schedule is shown in Table 16.1. As the transmission line losses are imposed into the generation scheduling program, a solution which satisfies all required system constraints is reached. In this case, the system operational cost increases even further to \$791,332.

16.7 SCGS for smart grid incorporating wind and solar thermal energy

With increase of the penetration of wind and PV power generation into power systems, more innovative and sophisticated approaches to system operation need to be adopted due to the intermittency and unpredictability of wind and PV power generations.

In the SCGS problem, the main objective is to efficiently determine the commitment of generating units as well as the output level of each committed unit to meet the forecasted demand at the minimum possible production cost. Furthermore, this should satisfy generating units and system constraints in a scheduling horizon over one day. The stochastic nature effect of wind power generation and PV is considered. The objective function is to minimize the total system generation cost of thermal, wind and PV units across the schedule horizon. It can be formulated as

$$\text{Minimize}\left[\sum_{j=1}^{T}\left(\sum_{i=1}^{N} OC\left(P_{ij}\right)+\sum_{i=1}^{M} OC\left(w_{ij}\right)+\sum_{i=1}^{S} OC\left(pv_{ij}\right)\right)\right] \tag{16.18}$$

where T is the time horizon, which is 24 h in this study; $OC\left(P_{ij}\right)$, $OC\left(w_{ij}\right)$ and $OC\left(pv_{ij}\right)$ are the operating cost of the ith thermal unit, wind farm and PV modules in jth hour, respectively; N, M and S are the number of thermal units, wind farm and PV modules, respectively.

It is to be noted that (16.18) comprises three terms to, respectively, represent (i) the operating cost of thermal units, (ii) the combined operating cost of wind farm and (iii) the combined operating cost of PV modules.

Calculation of thermal units operating cost, $\sum_{i=1}^{N} OC\left(P_{ij}\right)$: It can be calculated using (16.15).

Calculation of wind farm operating cost, $\sum_{i=1}^{M} OC\left(w_{ij}\right)$: The wind farm generation cost includes the following:

- A cost based on wind speed profile, $C_{wi}\left(w_{ij}\right)$, which equals the product of a cost coefficient, d_i, and Weibull *pdf* of wind power, $f_w(w)$. Thus,

$$C_{wi}\left(w_{ij}\right) = d_i f_w(w) w_{ij} \tag{16.19}$$

where d_i is the cost coefficient for ith wind farm and w_{ij} is the scheduled output of ith wind farm in jth hour.

- Penalty cost for not using all the available wind power, $C_{p,wi}$. It is assumed to be linearly proportional to the difference between the actual and scheduled wind powers and can be expressed as follows [31]:

$$C_{p,wi} = k_{p,wi}\left(W_{ij,\text{av}} - w_{ij}\right) = k_{p,wi}\int_{w_i}^{w_{ri}} \left(w - w_{ij}\right) f_w(w)dw \tag{16.20}$$

where $k_{p,wi}$ is the penalty cost coefficient for over generation of ith wind farm, $W_{ij,\text{av}}$ is the actual or available wind power from ith wind farm and w_{ri} is the rated output of ith wind farm.

- Penalty reserve requirement cost, which is due to that the actual or available wind power, is less than the scheduled wind power and can be given by

$$C_{r,wi} = k_{r,wi}\left(w_{ij} - W_{ij,\text{av}}\right) = k_{r,wi}\int_{0}^{w_i} \left(w_{ij} - w\right) f_w(w)dw \tag{16.21}$$

where $k_{r,wi}$ is the reserve cost coefficient for under generation of ith wind farm [18].

From (16.19), (16.20) and (16.21), the total wind farm cost can be deduced as

$$\sum_{i=1}^{M} OC(w_{ij}) = \sum_{i=1}^{M} C_{wi}(w_{ij}) + \sum_{i=1}^{M} C_{p,wi}(W_{ij,\mathrm{av}} - w_g) + \sum_{i=1}^{M} C_{r,wi}(w_{ij} - W_{ij,\mathrm{av}}) \tag{16.22}$$

Calculation of total operating cost of PV modules, $\sum_{i=1}^{S} OC(pv_{ij})$: Like to what is followed for wind farm costs, the total operating cost of the PV modules includes the following:

- A weighted cost function, which represents the cost based on solar irradiance profile. The weight is the Weibull *pdf* of PV power and can be expressed as

$$C_{\mathrm{pvi}}(pv_{ij}) = h_i f_{\mathrm{pv}}(pv)(pv_i) \tag{16.23}$$

  where h_i and pv_i are the cost coefficient and the scheduled output from ith PV module in jth hour.

- The penalty cost for not using all the available solar power, which is linearly related to the difference between available or actual solar power and scheduled solar power. It can be formulated as

$$\begin{aligned} C_{p,\mathrm{pv}} &= k_{p,\mathrm{pvi}}(PV_{ij,\mathrm{av}} - pv_{ij}) \\ &= k_{p,\mathrm{pvi}} \int_{pv_i}^{pv(k_{t\max})} (pv - pv_{ij}) f_{\mathrm{pv}}(pv)\,dpv \end{aligned} \tag{16.24}$$

  where $k_{p,\mathrm{pvi}}$ is the penalty cost coefficient for over generation of ith PV module, $PV_{ij,\mathrm{av}}$ is the actual or available PV power from the ith PV module, which is a random variable, and $pv(k_{t\max})$ is the maximum output of ith PV module.

- The penalty reserve requirement cost, which is due to that the actual or available PV power, is less than the scheduled PV power and can be given by

$$\begin{aligned} C_{r,\mathrm{pvi}} &= k_{r,\mathrm{pvi}}(pv_{ij} - PV_{ij,\mathrm{av}}) \\ &= k_{r,\mathrm{pvi}} \int_{0}^{pv_i} (pv_{ij} - pv) f_{\mathrm{pv}}(PV)\,dpv \end{aligned} \tag{16.25}$$

  where $k_{r,\mathrm{pvi}}$ is the reserve cost coefficient for under generation of the ith PV module.

From (16.23) to (16.25), the total operating cost of PV modules can be obtained as follows:

$$\sum_{i=1}^{S} OC(pv_{ij}) = \sum_{i=1}^{S} C_{\text{pvi}}(pv_{ij}) + \sum_{i=1}^{S} C_{p,\text{pv}}(PV_{ij,\text{av}} - pv_{ij}) + \sum_{i=1}^{S} C_{r,\text{pvi}}(pv_{ij} - PV_{ij,\text{av}}) \qquad (16.26)$$

Therefore, the objective function given by (16.18) can be determined in terms of (16.15), (16.22) and (16.26) by which the total operating cost of thermal, wind and PV generation, respectively, are computed.

The following are the system and unit constraints that are considered for solving the optimization problem.

1. Real power balance constraint: The total thermal, wind and PV power must match the total load demand (P_{Dj}), the power losses ($P_{\text{loss}j}$) in transmission lines and spinning reserve (R_j). Thus,

$$\sum_{i=1}^{N} u_{ij}P_{ij} + \sum_{i=1}^{M} u_{ij}w_{ij} + \sum_{i=1}^{S} u_{ij}pv_{ij} = P_{Dj} + P_{\text{loss}j} + R_j \qquad (16.27)$$

   where u_{ij} is the status index of ith unit in the jth period (1 for up and 0 for down). $P_{\text{loss}j}$ can be computed using Newton–Rapson AC power flow.
2. In addition to the constraints given by (16.5), (16.7)–(16.9), (16.11) and (16.12), the real power operating limits of thermal, wind and PV generating units are

$$\left.\begin{array}{c} P_i^{\min} \le P_i \le P_i^{\max} \\ 0 \le w_i \le w_r \\ 0 \le pv_i \le pv(k_{t\max}) \end{array}\right\} \qquad (16.28)$$

Numerical example 16.2

As a supplementary part to Example16.1, two scenarios are considered. In the first, the stochastic nature of both the wind and PV power generators are not considered in the SCGS problem. In this case, the operating cost of the wind and PV power generators are represented by (16.19) and (16.23), respectively. The second scenario considers the stochastic nature of wind and PV power outputs and the imbalance charges due to the mismatch between the actual and scheduled wind and PV power outputs. Thus, the operating cost of wind and PV power generators are calculated by (16.22) and (16.26), respectively.

In both scenarios, 26 thermal units, 2 wind farms and 2 PV power plants are considered. The wind speed parameters v_i, v_r and v_o are 5, 15 and 45 m/s, respectively. The Weibull function parameters of wind are $k = 2$ and $c = 10$ m/s [32,33].

The penalty and reserve factors are set to $k_{p,wi} = 2$ and $k_{r,wi} = 4$. The clearness index parameter $k_{t\max} = 0.755$ and the inclination $\beta = 0$.

The Weibull function parameters of PV are $c' = 0.6251$ and $\lambda = 3.5507$ [34]. The penalty and reserve factors are set to $k_{p,\text{pvi}} = 3$ and $k_{r,\text{pvi}} = 4$. The cost coefficients of the two wind farms and two PV modules are $d_1 = 1.0$, $d_2 = 1.25$ and $h_1 = 1.3$, $h_2 = 1.4$, respectively [32].

To test the effectiveness of both the wind and PV on the SCGS problem, two wind farms and two PV modules are placed at Bus 15, as this bus contains most of peaking units, shown in Figure 16.5. The total installed conventional capacity on the system is approximately 3,225 MW and the assumed peak and minimum loads for the test system are 2,850 and 1,842 MW, respectively. By applying the hybrid technique comprising of ANN/GA/PL and taking into consideration system and unit constraints including the ramp rate and transmission constraints, the total operating costs for both scenarios are $785,660.44 and $784,953.31, respectively. The solution times for both scenarios are 53 and 75 s, respectively. The overall saving for daily power system operation for both scenarios, compared to the cost of the same problem without the consideration of WP and PV power generators, are $37,013.56 and $37,720.69, respectively. This amounts to a net saving over the year of approximately $13.3 million and $13.6 million, respectively. A comparison

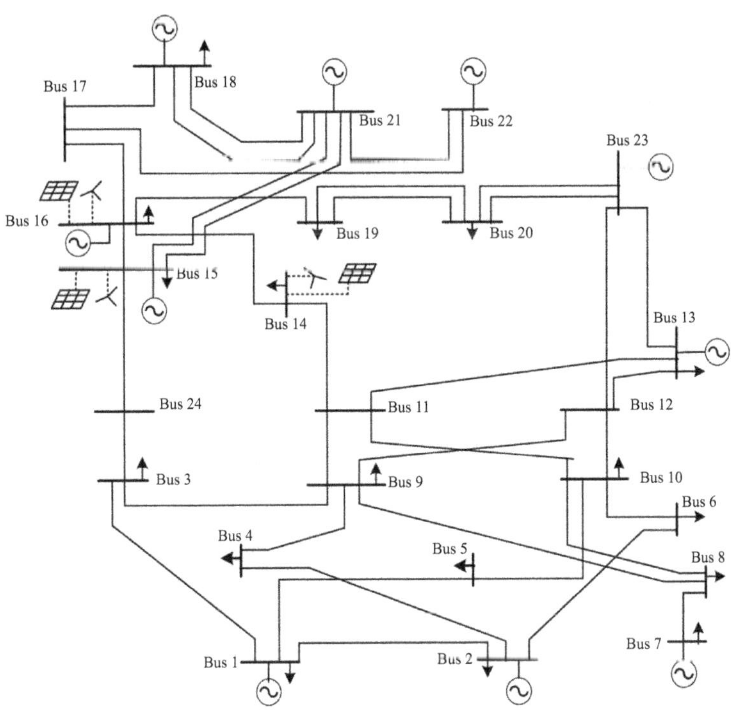

Figure 16.5 Modified IEEE 24-bus system

of flow on the line from Bus 14 to Bus 16, with and without transmission constraints, is given in Table 16.2. A comparison between the hourly cost without embedding wind and PV in the SCGS problem and Scenarios 1 and 2 is demonstrated in Figure 16.6. The second scenario which considers the stochastic nature of both the wind and PV power generators has the smallest cost. It takes more

Table 16.2 Power flows on the line from Bus 14 to Bus 16 (MW)

Hour	Without line limits	With line limits		
		Without wind and PV	With wind and PV first scenario	With wind and PV second scenario
1	314.0	300.0	175.3	12.9
2	335.2	300.0	225.7	117.0
3	323.3	298.0	167.9	271.3
4	322.4	300.0	229.6	112.8
5	300.2	297.8	65.0	240.8
6	328.8	299.4	252.4	248.2
7	323.5	299.8	203.4	215.7
8	339.0	300.0	263.4	254.7
9	307.1	298.8	4.7	112.2
24	317.4	299.1	78.8	5.4

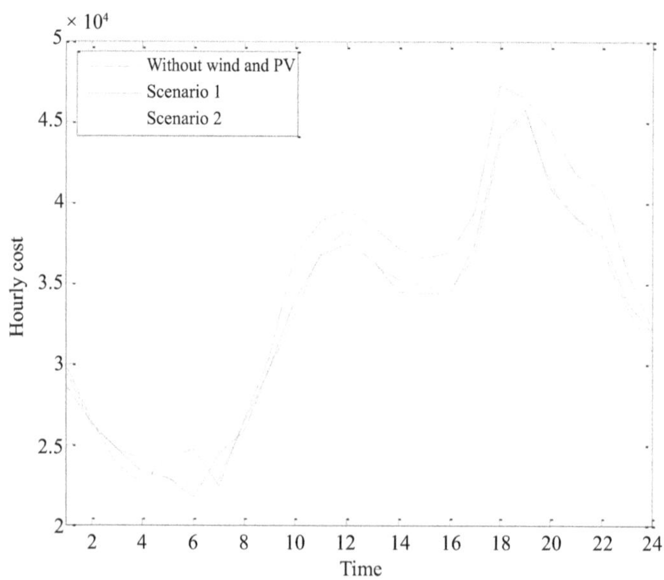

Figure 16.6 Comparison between the hourly cost of the SCGS problem with and without embedding wind and PV generators

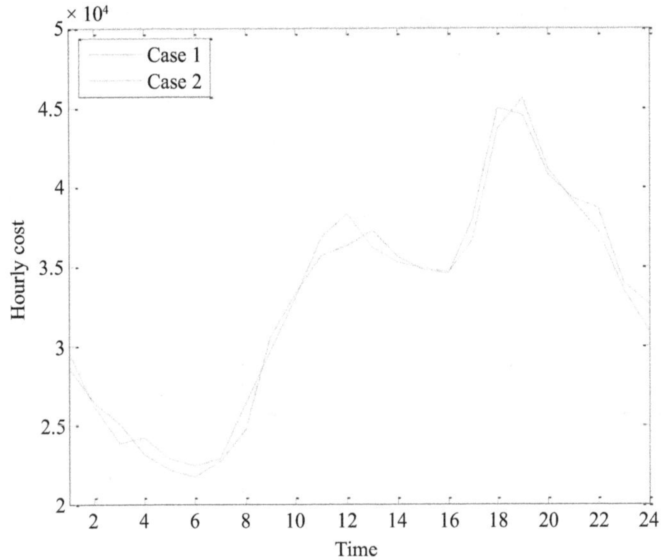

Figure 16.7 Comparison between the hourly cost of the SCGS problem for Cases 1 and 2

Table 16.3 UC with wind and PV (Scenario 2), Case 2

Unit	Hour (1–24)
1	1 0 0 1 1 1 1 1 1 1 0 0 0 1 1 1 0 1 0 1 0 0 0 1
2	1 0 0 0 1 0 1 0 1 0 0 0 0 0 1 0 1 0 0 1 0 0 0 1
3	1 0 0 1 1 1 0 0 0 0 0 0 0 0 1 1 0 0 0 1 0 0 0 1
4	1 0 0 1 1 0 0 1 0 0 0 0 0 0 1 0 0 0 0 0 0 0 0 0
5	1 0 0 1 1 1 0 0 0 0 0 0 0 0 0 0 0 0 0 0 0 0 0 0
6	1 1 1 1 0 0 0 1 1 0 0 1 0 0 0 0 0 0 0 0 0 0 0 0
7	1 1 0 1 1 1 0 1 0 0 0 1 1 0 0 1 1 1 1 1 0 1 1 0
8	0 0 0 0 0 0 0 0 0 0 0 0 0 0 0 0 0 0 0 0 0 0 0 0
9	0 0 0 0 0 0 0 0 0 0 0 0 0 0 0 0 0 0 0 0 0 0 0 0
10	1 1 1 1 1 1 1 1 1 1 1 1 1 1 1 1 1 1 1 1 1 1 1 1
11	1 1 1 1 0 0 1 1 1 1 1 1 1 1 1 1 1 1 1 1 1 1 1 1
12	1 1 0 0 0 0 1 1 1 1 1 1 1 1 1 1 1 1 1 1 1 1 1 1
13	1 0 0 0 0 0 0 0 1 1 0 0 1 1 1 1 1 1 1 1 1 1 1 1
14	1 1 1 1 1 1 0 0 1 1 1 1 1 1 0 0 1 1 1 1 1 1 1 1
15	0 0 0 0 0 0 1 1 1 1 1 1 0 0 0 0 0 1 1 1 1 0 0 0
16	0 0 0 0 0 0 0 0 0 0 0 0 0 0 0 0 0 0 1 1 1 1 1 0
17	1 1 1 1 1 1 1 1 1 1 1 1 1 1 1 1 1 1 1 1 1 1 1 1
18	1 1 1 1 1 1 1 1 1 1 1 1 1 1 1 1 1 1 1 1 1 1 1 1
19	1 1 1 1 1 1 1 1 1 1 1 1 1 1 1 1 1 1 1 1 1 1 1 1

(Continues)

Table 16.3 (Continued)

Unit	Hour (1–24)
20	1 1 1 1 1 1 1 1 1 1 1 1 1 1 1 1 1 1 1 1 1 1 1 1
21	0 0 0 0 0 0 0 0 0 1 1 1 1 1 1 1 1 1 1 1 1 1 1 1
22	0 0 0 0 0 0 0 0 0 0 1 1 1 1 1 1 1 1 1 0 0 0 0 0
23	0 0 0 0 0 0 0 0 0 0 0 0 0 0 0 0 0 1 1 1 1 1 0 0
24	1 1 1 1 1 1 1 1 1 1 1 1 1 1 1 1 1 1 1 1 1 1 1 1
25	1 1 1 1 1 1 1 1 1 1 1 1 1 1 1 1 1 1 1 1 1 1 1 1
26	1 1 1 1 1 1 1 1 1 1 1 1 1 1 1 1 1 1 1 1 1 1 1 1
27	1 1 1 1 1 1 1 1 1 1 1 1 1 1 1 1 1 1 1 1 1 1 1 1
28	1 1 1 1 1 1 1 1 1 1 1 1 1 1 1 1 1 1 1 1 1 1 1 1
29	0 0 0 0 0 0 0 1 1 1 1 1 1 1 1 1 1 0 0 0 0 0 0 0
30	0 0 0 0 0 0 0 1 1 1 1 1 1 1 1 1 1 0 0 0 0 0 0 0

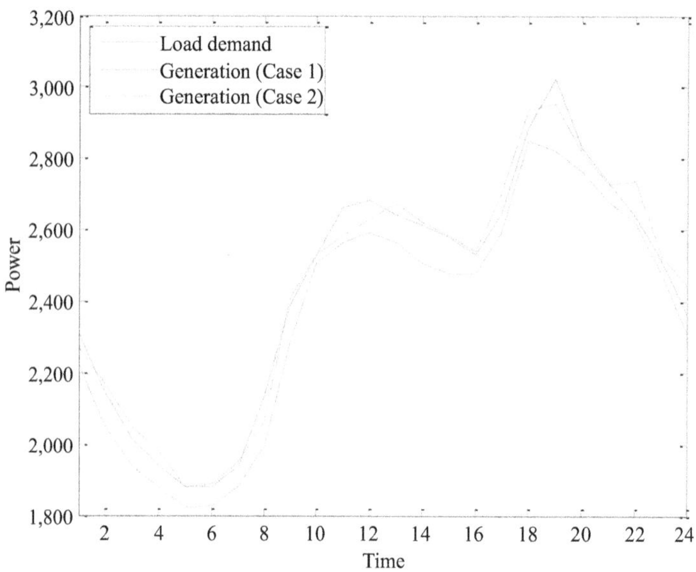

Figure 16.8 Daily load demand and scheduled generation (Cases 1 and 2)

scheduling time than that of the first scenario but still is acceptable for real-time operation.

To ensure that the system security is indeed maintained and to test the feasibility of the hybrid technique, the second scenario is applied to the same system with different locations of wind and PV power generators taking into consideration all the system and unit constraints. First, the same two wind farms and the two PV modules are inserted on Bus 14 (instead of Bus 15) and the SCGS problem is solved using the hybrid technique with the influence of both the ramp rate and transmission constraints. The system operational cost of $784,732.63 with no violation is reported.

Second, after inserting the renewable generators on Bus 16, the system operational cost is $784,076.81, with also no violation reported. The hourly cost of both cases is illustrated in Figure 16.7. The committed schedule of the second case is given in Table 16.3. The numbers (1 or 0) in this table represent on/off states of units at different hours. Units 24–26 constitute base load units, which are responsible for a large portion of system generation, whereas Units 27 and 28 represent the wind farm units and 29 and 30 represent the PV modules generating units. Figure 16.8 shows load demand and generating capability for 24 h, Cases 1 and 2.

References

[1] Marwali M. K. C. and Shahidehpour S. M. 'Coordination between long-term and short-term generation scheduling with network constraints'. *IEEE Transactions on Power Systems.* 2000;15(3):1161–7.

[2] El-Hawary M. E. and Christensen G. S. *Optimal economic operation of electric power systems. Mathematics in Science and Engineering series, no. 142.* New York: Academic Press; 1979.

[3] Wood A. J. and Wollenberg B. F. Power generation operation and control. New York: John Wiley & Sons; 1996.

[4] Yalcinoz T. and Short M. J. 'Large-scale economic dispatch using an improved Hopfield neural network'. *IEE Proceedings of Generation, Transmission and Distribution.* 1997;144(2):181–5.

[5] Glover F. 'Tabu search – Part I'. *ORSA Journal on Computing.* 1989;1(3): 190–206.

[6] Mantawy A. H., Abdel-Magid Y. L. and Selim S. Z. 'Unit commitment by tabu search'. *IEE Proceedings of Generation, Transmission and Distribution.* 1998;145(1):56–64.

[7] IEEE Report. 'Potential impacts of clean air regulations on system operations'. *IEEE Transactions on Power Systems.* 1995;10(2):647–53.

[8] Sheble G. B. and Brittig K. 'Refined genetic algorithm – Economic dispatch example'. *IEEE Transactions on Power Systems.* 1995;10(1):117–24.

[9] Saavedra O. R. and Gomes J. R. 'Optimal reactive power dispatch using evolutionary computation: Extended algorithms'. *IEE Proceedings of Generation, Transmission and Distribution.* 1999;146(6):586–92.

[10] Liang Z. X. and Glover J. D. 'A zoom feature for a dynamic programming solution to economic dispatch including transmission losses'. *IEEE Transactions on Power Systems.* 1992;7(2):544–50.

[11] Tang S. K. and Shahidehpour S. M. 'A combination of Lagrangian relaxation and linear programming approaches for fuel constrained unit commitment problem'. *IEE Proceedings Part C.* 1989;136(3):162–74.

[12] Wang S. J., Shahidehpour S. M., Kirschen D. S., Mokhtari S. and Irisarri G. D. 'Short-term generation scheduling with transmission and environmental constraints using an augmented Lagrangian relaxation'. *IEEE Transactions on Power Systems*. 1995;10(3):1294–301.

[13] Cohen A. I. and Yoshimura M. 'A branch-and-bound algorithm for unit commitment'. *IEEE Transactions on Power Apparatus and Systems*. 1983; PAS-102(2):444–51.

[14] Dillon T. S., Edwin K. W., Kochs H. D. and Taud R. J. 'Integer programming approach to the problem of optimal unit commitment with probabilistic reserve determination'. *IEEE Transactions on Power Apparatus and Systems*. 1978;PAS-97(6):2154–66.

[15] Mokhtari S., Sing J. and Wollenberg B. 'A unit commitment expert system'. *IEEE Transactions on Power Systems*. 1988;3(1):272–7.

[16] Bansal R. C. 'Bibliography on the fuzzy set theory applications in power systems (1994–2001)'. *IEEE Transactions on Power Systems*. 2003; 18(4):1291–9.

[17] Su M. C. and Hs Y. Y. 'Fuzzy dynamic programming: An application to unit commitment'. *Transactions on Power Systems*. 1991;6(3):1231–7.

[18] Mantawy A. H., Abdel-Magid Y. L. and Selim S. Z. 'A simulated annealing algorithm for unit commitment'. *IEEE Transactions on Power Systems*. 1998;13(1):197–204.

[19] Wong K. P. and Fung C. C. 'Simulated annealing based economic dispatch algorithm'. *IEE Proceedings Part C*. 1993;140(6):509–15.

[20] Sasaki H., Watanabe M. and Yokoyama R. 'A solution method of unit commitment by artificial neural networks'. *IEEE Transactions on Power Systems*. 1992;7(3):974–81.

[21] Ouyang Z. and Shahidehpour S. M. 'A hybrid artificial neural network-dynamic programming approach to unit commitment'. *IEEE Transactions on Power Systems*. 1992;7(1):236–42.

[22] Lai L. L. *Intelligent system applications in power engineering evolutionary programming and neural network*. New York: John Wiley & Sons; 1999.

[23] Kazarlis S. A., Bakirtzis A. G. and Petridis V. 'A genetic algorithm solution to the unit commitment problem'. *IEEE Transactions on Power Systems*. 1996:11(1):83–92.

[24] Orero S. O. and Iriving M. R. 'Scheduling of generators with a hybrid genetic algorithm'. *Proceedings of 1st International Conference on Genetic Algorithms in Engineering Systems: Innovations and Applications (GALESIA)*; Sheffield, UK, Sep 1995, IET; 1995, pp. 200–6.

[25] El Desouky A. A., Aggarwal R., Elkateb M. M. and Li F. 'Advanced hybrid genetic algorithm for short-term generation scheduling'. *IEE Proceedings of Generation, Transmission and Distribution*. 2001;148(6):511–17.

[26] El Desouky A. A. and ElKateb M. M. 'Hybrid adaptive techniques for electric-load forecast using ANN and ARIMA'. *IEE Proceedings of Generation, Transmission and Distribution*. 2000;147(4):213–17.

[27] Bunn D. H. and Farmer E. D. *Comparative models for electrical load forecasting*. New York: John Wiley & Sons; 1985.

[28] Li F. and Aggarwal R. K. 'Fast and accurate power dispatch using a relaxed genetic algorithm and a Local gradient technique'. *Expert Systems with Applications*. 2000;19(3):159–65.

[29] Reliability Test System Task Force. 'IEEE reliability test system'. *IEEE Transactions on Power Apparatus and Systems* 1979;PAS-98(6):2047–54.

[30] Tong S. K., Shahidehpour S. M. and Ouyang Z. 'A heuristic short-term unit commitment'. *IEEE Transactions on Power Systems*. 1991;6(3):1210–16.

[31] El Desouky A. A. 'Security and stochastic economic dispatch of power system including wind and solar resources with environmental consideration'. *International Journal of Renewable Energy Research*. 2013;3(4):951–58.

[32] Orgill J. F. and Hollands K. T. G. 'Correlation equation for hourly diffuse radiation on horizontal surface'. *Solar Energy*. 1977;19(4):357–9.

[33] Eladany M. M., El-Desouky A. A., Daoud A. A. and Sallam A. A. 'Optimizing stochastic economic dispatch of wind/solar DG grid'. *International Journal of Distribution Energy Resources*. 2012;8(4):100–12.

[34] Tina G. and Gagliano S. 'Probabilistic analysis of weather data for a hybrid solar/wind energy system'. *International Journal of Energy Research*. 2011;35(3):221–32.

Chapter 17

Load flow analysis for power systems with renewables

17.1 Introduction

The techniques of load flow analysis are of utmost importance to analyse, plan and design the power system. Power flow analysis defines the power system parameters at normal and abnormal operating conditions through the calculation of current or power flow in the lines of the interconnected transmission network, voltages at all nodes of the power system and the power delivered from generating sources.

Several load flow methods, such as Newton–Raphson, fast-decoupled Newton–Raphson, Gauss–Seidel, etc., have been developed for conventional power systems. Based on load flow analysis, protection system and capacity of the power system, assets can be specified, as well as reliability, stability and security can be analysed and enhanced.

Renewable energy sources (RESs) have become an important part of electrical generation and its importance is continuing to increase. But incorporating such sources (e.g. wind farms and photovoltaic (PV) plants) in electrical power system causes some problems because of the fluctuation of power generated by such sources. Forecasting the power generated from renewable sources to predict power output in advance through methods based on numerical weather prediction models and statistical approaches is a recently developed tool for load forecasting, but with a level of accuracy not as great as for the conventional load forecasting techniques. The overall shape of renewable production can be predicted most of the time but with significant errors that can occur in both the level and timing. However, renewable energies forecast for shorter time horizons tend to be more accurate than forecasts for longer time; that is, forecasts for 1–2 h ahead can achieve an accuracy much better than that of one day ahead [1].

Another method, probability distribution, rather than the prediction approach, can be used. For example, the wind speed or solar radiation for PV may be difficult to predict, but the probability of a specific wind speed or specific solar radiation occurring can be estimated. This can be done if the probability distribution is known by assuming it to be a Weibull or Rayleigh distribution. Then, once the wind speed or solar radiation is known, the injected power from the wind farm or PV plant into the grid can be calculated by means of the corresponding power curves [2].

In addition, the renewable sources usually are associated with energy storage systems (ESSs). So, the methods of load flow analysis should be developed in a way that can meet such situations.

17.2 Steady-state load flow calculations

To calculate the load flow in a power system incorporating RESs and ESS and operating in normal conditions, some features must be taken into consideration. These include the following:

- The conventional generating units have a constant power rating. The power delivered from each unit to the power system depends on the system topology, system losses and load demand, up to the rating of the generating unit, P_g.
- The available capacity of RESs is climate dependent, and consequently it is time dependent, $P_{\text{ren}}(t)$. At a prespecified time, the power injected by the RES into the transmission network corresponds to the load demand, system losses and state of ESS, but is restricted by the available capacity at that time.
- The ESS can inject power into the transmission system during its discharging state and receive power from the system during its charge state, P_S, depending on the load demand and the generation over the entire power system.
- The total load is, of course, time variant and its profile is given, $P_L(t)$.

A schematic diagram of the power system configuration as well as the direction of power flow regarding the generating units (conventional and renewable), ESS and loads are illustrated in Figure 17.1.

The charging/discharging power of the ESS can be calculated as the difference, ΔP, between the total power generation (conventional and renewable), P_{Gen}, and the power consumption including the system losses, P_{losses}. Thus,

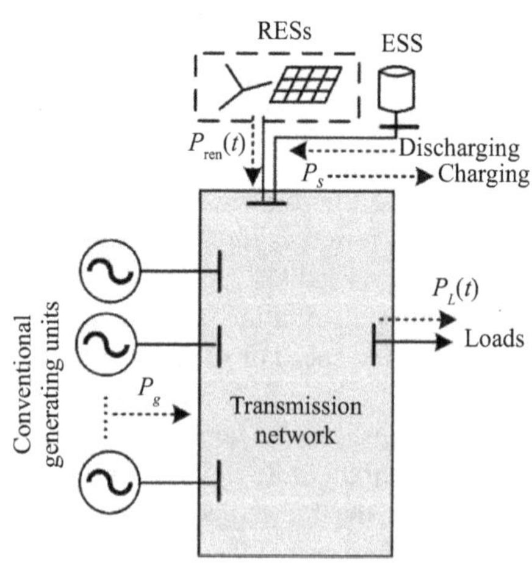

Figure 17.1 Schematic diagram of a power system with RESs and ESS

$$\Delta P = P_{\text{Gen}} - \{P_L(t) + P_{\text{losses}}\} \tag{17.1}$$

and

$$P_{\text{Gen}} = \sum_{g=1}^{N} P_g + \sum_{i=1}^{M} P_{\text{ren},i}(t) \tag{17.2}$$

where N and M are the number of conventional generating units and RESs, respectively, and P_g represents the available capacity of the gth generator.

Hence,

$$\left.\begin{aligned} P_s &\leq \Delta P \text{ for } \Delta P > 0 \text{ for charging state (ESS as a load)} \\ &= \Delta P \text{ for } \Delta P < 0 \text{ for discharging state (ESS as a source)} \\ &= 0 \text{ for } \Delta P = 0 \quad \text{for system balance state} \end{aligned}\right\} \tag{17.3}$$

According to (17.3), the power balance condition can be expressed as follows:

$$P_{\text{Gen}} = P_L(t) + P_{\text{losses}} + P_S \text{ \{for ESS charging state\}} \tag{17.4}$$

and

$$P_{\text{Gen}} + P_S = P_L(t) + P_{\text{losses}} \text{ \{for ESS discharging state\}} \tag{17.5}$$

As the ESS acts as a load or as a source of power depending on its state that is defined by the sign of ΔP, it can be represented by a bus (ESS bus) connected to the power system through a very low reactance not to affect the voltages of different buses in the system. The ESS bus is defined as a load bus in the charging state or PV bus in the discharging state in the load flow studies.

17.2.1 Load flow algorithm

The algorithm of load flow program for a time span of T and starting at time t_0 includes the following steps:

(i) Input the system data incorporating renewables (wind and/or solar PV).
(ii) Specify the time in hour, t_1.
(iii) Read the wind speed and/or solar radiation at the given time.
(iv) Compute the corresponding power generated from wind and/or PV.
(v) Compute the value of ΔP to decide the state of the ESS (negative value for ESS acting as a source, 'discharge state', and positive value for ESS acting as a load, 'charging state', or zero value at which the power is already balanced, i.e. no need to charge/discharge the ESS).
(vi) Proceed with the load flow calculations.
(vii) Output the results: Line flows, bus voltages, power generated from each source, system losses and the role of ESS.
(viii) Advance the time incrementally to $t_1 + \Delta t$ and repeat Steps (iii)–(vii) until reaching the end of time span T.

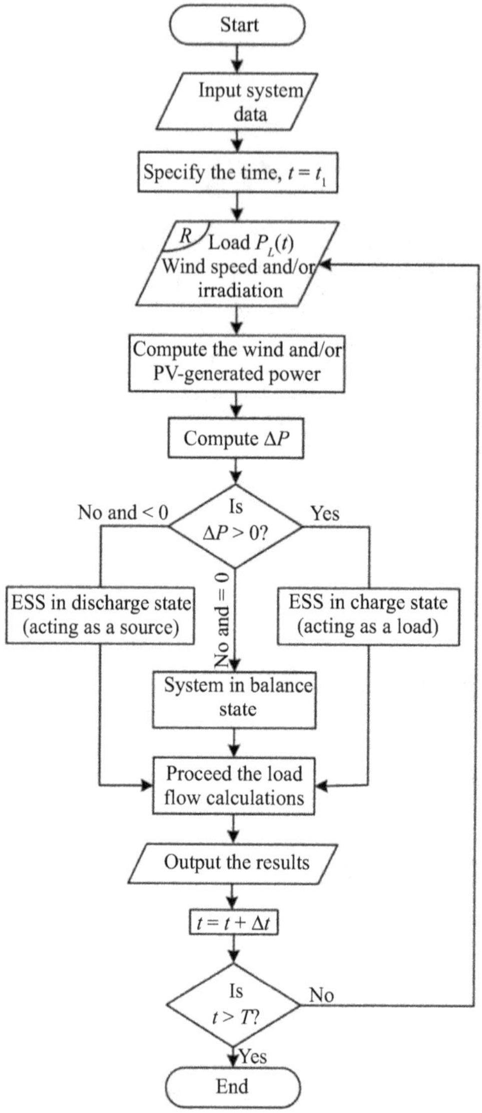

Figure 17.2 Flow chart for load flow calculations

Accuracy of the results increases by decreasing both the incremental time Δt and prediction error of renewables generation [3]. A flow chart indicating the sequence of these steps is shown in Figure 17.2.

Example 17.1 IEEE 14-bus test system (Figure 17.3) and its data are given in Tables 17.1–17.3. It is desired to perform load flow calculations and determine the

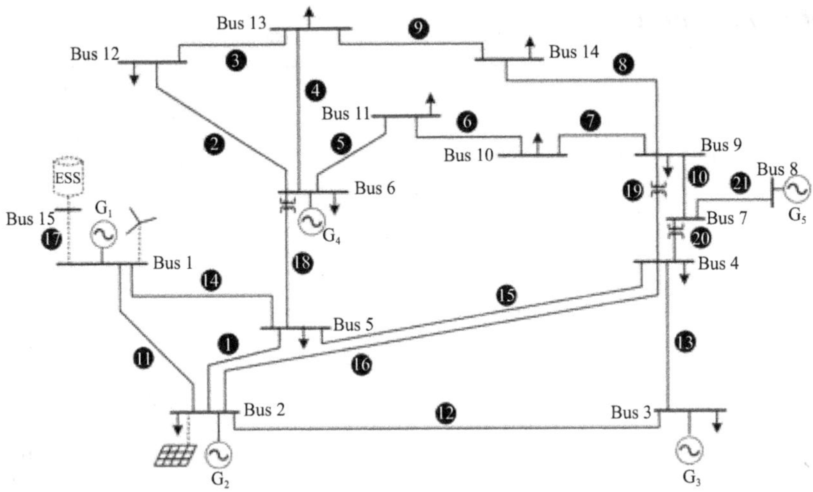

Figure 17.3 Single-line-diagram of IEEE 14-bus system (the dotted lines represent the connections of energy sources that can replace the conventional units)

Table 17.1 Generation data

Generator bus #	1 (G_1)	2 (G_2)	3 (G_3)	6 (G_4)	8 (G_5)
MVA	615	60	60	25	25
PF	0.989 (lead)	0	0.89 (lag)	0	0
	Under-excited	Overexcited	Overexcited	Overexcited	Overexcited
X_L (p.u.)	0.2396	0.00	0.00	0.134	0.134
R_a (p.u.)	0.00	0.0031	0.0031	0.0014	0.0041
X_d (p.u.)	0.8979	1.05	1.05	1.25	1.25
X'_d (p.u.)	0.2995	0.1850	0.1850	0.232	0.232
X''_d (p.u.)	0.23	0.13	0.13	0.12	0.12
T'_{do} (p.u.)	7.4	6.1	6.1	4.75	4.75
T''_{do} (p.u.)	0.03	0.04	0.04	0.06	0.06
X_q (p.u.)	0.646	0.98	0.98	1.22	1.22
X'_q (p.u.)	0.646	0.36	0.36	0.715	0.715
X''_q (p.u.)	0.4	0.13	0.13	0.12	0.12
T'_{qo} (p.u.)	0.00	0.099	0.099	0.21	0.21
T''_{qo} (p.u.)	0.033	0.099	0.099	0.21	0.21
H	5.148	6.54	6.54	5.06	5.06
D	2	2	2	2	2

Bus type: (1) swing bus, (2) generator bus (PV bus) and (3) load bus (PQ bus).

charging/discharging modes of the ESS through 24 h for three cases: (i) the conventional generator at Bus #1 is replaced by a renewable wind farm and ESS; (ii) the conventional generator at Bus #2 is replaced by a solar PV plant and ESS; and (iii) the generator at Bus #1 is replaced by a wind farm and ESS as well as the

Table 17.2 Bus data

Bus #	P_g (p.u.)	Q_g (p.u.)	P_L (p.u.)	Q_L (p.u.)	Bus type	Q_g max. (p.u.)	Q_g min. (p.u.)
1	2.32	0.00	0.00	0.00	1	10.0	−10.0
2	0.4	0.424	0.2170	0.1270	2	0.5	−0.4
3	0.00	0.2737	0.9420	0.1900	2	0.4	0.00
4	0.00	0.00	0.4780	0.00	3	0.00	0.00
5	0.00	0.00	0.0760	0.0160	3	0.00	0.00
6	0.00	0.2251	0.1120	0.0750	2	0.24	−0.06
7	0.00	0.00	0.00	0.00	3	0.00	0.00
8	0.00	0.2516	0.00	0.00	2	0.24	−0.06
9	0.00	0.00	0.2950	0.1660	3	0.00	0.00
10	0.00	0.00	0.0900	0.0580	3	0.00	0.00
11	0.00	0.00	0.0350	0.0180	3	0.00	0.00
12	0.00	0.00	0.0610	0.0160	3	0.00	0.00
13	0.00	0.00	0.1350	0.0580	3	0.00	0.00
14	0.00	0.00	0.1490	0.0500	3	0.00	0.00

Table 17.3 Line data

From bus	To bus	Line #	Resistance (p.u.)	Reactance (p.u.)	Line charging (p.u.)	Tap ratio
1	2	11	0.01938	0.05917	0.0528	1
1	5	14	0.05403	0.22304	0.0492	1
2	3	12	0.04699	0.19797	0.0438	1
2	4	16	0.05811	0.17632	0.0374	1
2	5	1	0.05695	0.17388	0.034	1
3	4	13	0.06701	0.17103	0.0346	1
4	5	15	0.01335	0.04211	0.0128	1
4	7	20	0.00	0.20912	0.00	0.978
4	9	19	0.00	0.55618	0.00	0.969
5	6	18	0.00	0.25202	0.00	0.932
6	11	5	0.09498	0.1989	0.00	1
6	12	2	0.12291	0.25581	0.00	1
6	13	4	0.06615	0.13027	0.00	1
7	8	21	0.00	0.17615	0.00	1
7	9	10	0.00	0.11001	0.00	1
9	10	7	0.03181	0.08450	0.00	1
9	14	8	0.12711	0.27038	0.00	1
10	11	6	0.08205	0.19207	0.00	1
12	13	3	0.22092	0.19988	0.00	1
13	14	9	0.17093	0.34802	0.00	1

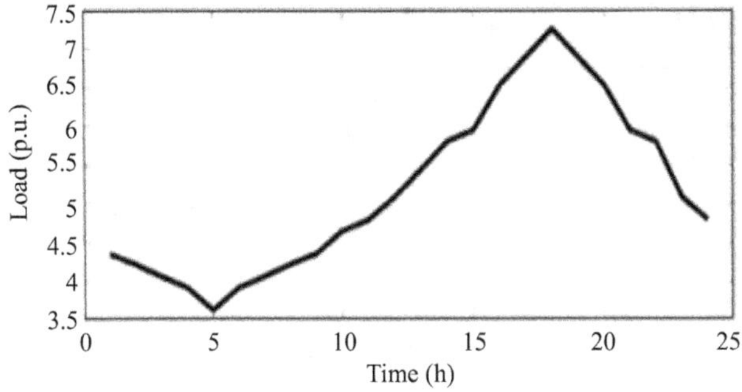

Figure 17.4 Load profile through 24 h

Table 17.4 Load profile data

Time (s)	Load (p.u.) base 100 MVA	Time (s)	Load (p.u.) base 100 MVA
1	4.3513	13	5.4390
2	4.2172	14	5.8016
3	4.0612	15	5.9467
4	3.9159	16	6.5267
5	3.6260	17	6.8894
6	3.9159	18	7.2520
7	4.0612	19	6.8894
8	4.2172	20	6.5267
9	4.3513	21	5.9467
10	4.6413	22	5.8016
11	4.7863	23	5.0764
12	5.0764	24	4.7863

generator at Bus #2 is replaced by a solar PV plant. The load profile is drawn in Figure 17.4 and its data are given in Table 17.4. The time span is 1 day starting at $t = 1$ am, and incrementally increases by 1 h. The base power for per unit values is 100 MVA.

Solution

Case (i): Replacing G_1 by a wind farm and an ESS at Bus #1

Data of wind energy farm are given in Table 17.5. Wind speed and its respective power, computed as explained in Chapter 2, through the 24 h are plotted in Figures 17.5 and 17.6, respectively. The variation of ΔP and P_{losses} through 24 h is computed and depicted in Figures 17.7 and 17.8, respectively.

By using the load flow algorithm, the calculations are implemented 24 times over 24 h, that is, one each hour. As a sample of the results obtained by load flow calculations, the results of line flows at which ΔP is maximum positive and maximum negative, t_6 and t_{18}, respectively, are selected and tabulated in

Table 17.5 Wind farm data

Power, voltage and frequency rating (MVA, kV, Hz)	[610 69 60]
Stator resistance R_s and reactance X_s (p.u., p.u.)	[0.01 0.10]
Rotor resistance R_r and reactance X_r (p.u., p.u.)	[0.01 0.08]
Magnetization reactance X_m (p.u.)	3.00
Number of poles P and gear box ratio	[4 1/89]
Blade length and number (m integer)	[75 3]
Number of wind generators that compose the farm	30
Pitch control gain and time constant K_p, T_p (p.u., s)	[10 3]

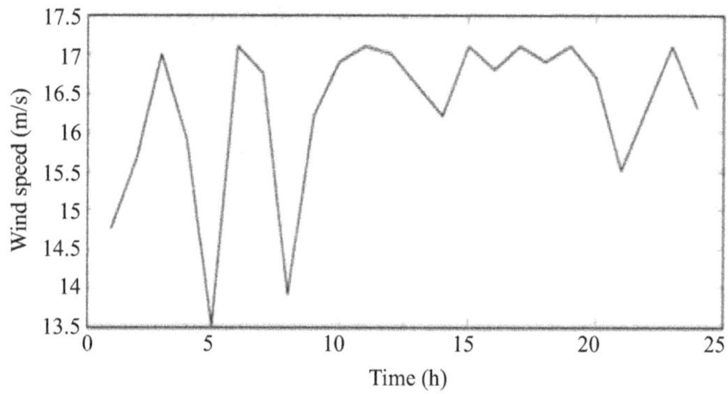

Figure 17.5 Wind speed versus day hours

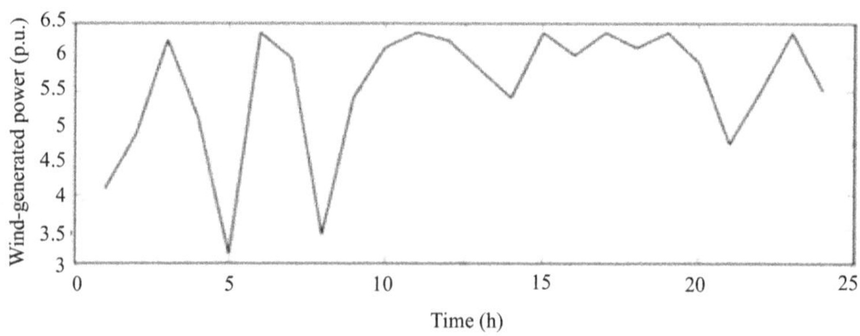

Figure 17.6 Wind-generated power with respect to the speed variation versus hours

Tables 17.6–17.9. The ESS bus is denoted by Bus #15, which is connected to Bus #1 through a reactance of 0.00271 p.u.

Case (ii): Power system incorporating a solar PV plant and an ESS at Bus #2
The procedure of load flow calculations is the same as that used in Case (i) incorporating a wind farm. The solar PV plant is connected to Bus #2, with the following data:

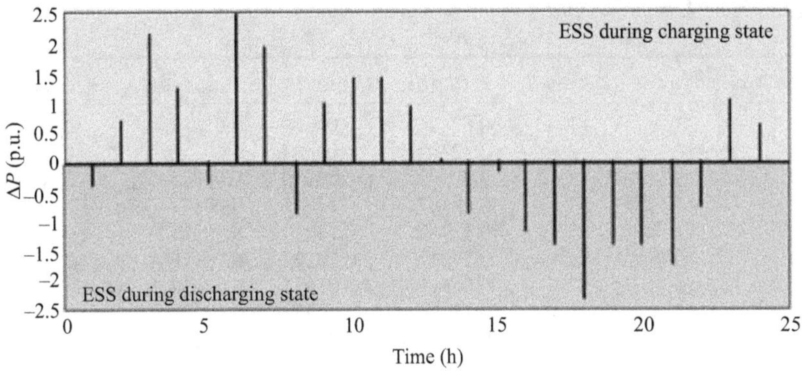

Figure 17.7 ΔP versus day hours

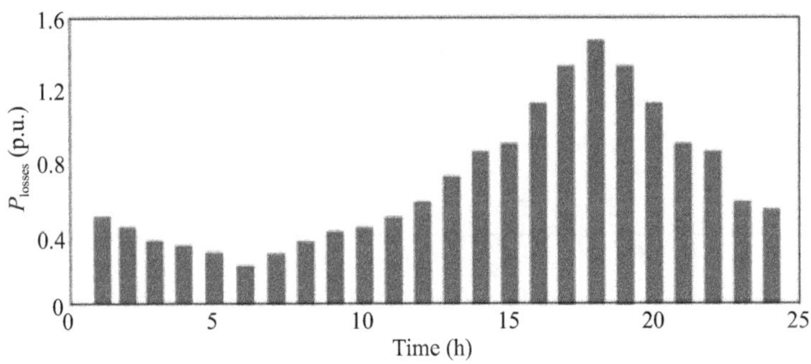

Figure 17.8 Power losses along 24 h

- The output power of each unit in the plant is 10 MW.
- Number of units used six, constituting three models.
- Active power of each model is 20 MW at a rated voltage of 1.045 p.u.
- Daily solar radiation curve is shown in Figure 17.9, and the corresponding generated power is depicted in Figure 17.10.

Changes of ΔP and P_{losses} for 24 h, resulting from load flow calculations, are shown in Figures 17.11 and 17.12, respectively. ΔP is maximum positive and maximum negative at t_5 and t_{18}, respectively, and the corresponding line flows, and bus voltage and power are tabulated in Tables 17.10–17.13.

Case (iii): The power system incorporates a wind farm and an ESS at Bus #1 and a solar PV at Bus #2 without a conventional generating unit at these buses

With the same data as in the first and second cases, load flow calculations result in the changes of ΔP and P_{losses} for 24 has shown in Figures 17.13 and 17.14. ΔP is maximum positive and maximum negative at t_6 and t_{18}, respectively, and the corresponding line flows and bus voltage and power are tabulated in Tables 17.14–17.17.

Table 17.6 Line flows at t_6

From bus	To bus	Line #	P (p.u.)	Q (p.u.)	P_{losses} (p.u.)	Q_{losses} (p.u.)
02	05	1	0.62426	0.08109	0.02084	0.02809
06	12	2	0.12309	0.04667	0.00186	0.00387
12	13	3	0.02903	0.02039	0.00025	0.00023
06	13	4	0.28102	0.14775	0.00582	0.01147
06	11	5	0.12787	0.13218	0.00281	0.00588
11	10	6	0.07217	0.1011	0.00118	0.00277
09	10	7	0.06526	−0.01675	0.00014	0.00038
09	14	8	0.13003	0.00432	0.00211	0.00448
14	13	9	0.09738	−0.07016	0.0025	0.00508
07	09	10	0.40945	0.22868	0	0.02264
01	02	11	2.6641	−0.42397	0.1251	0.32346
03	02	12	−1.0842	0.17793	0.05599	0.18964
03	04	13	−0.3401	0.22551	0.01148	−0.00543
01	05	14	1.2024	0.11556	0.07051	0.23888
05	04	15	0.91916	−0.15627	0.0116	0.02386
02	04	16	0.84653	0.05945	0.03847	0.07783
15	01	17	1e–05	−0.00841	0	0.0083
05	06	18	0.70128	0.06356	0	0.1087
04	09	19	0.23184	0.04332	0	0.02939
04	07	20	0.40945	−0.0669	0	0.03484
08	07	21	0	0.34842	0	0.018

Table 17.7 Bus voltage and power at t_6

Bus #	Bus voltage		P_g (p.u.)	Q_g (p.u.)	P_L (p.u.)	Q_L (p.u.)
	|V| (p.u.)	Phase (rad)				
01	1.06	0	6.03588	−0.27831	2.4922	0
02	1.045	−0.14977	0.4	1.0775	0.3281	0.1778
03	1.01	−0.36367	0	0.67076	1.4243	0.266
04	0.99413	−0.28943	0	0	0.7227	0.056
05	0.99926	−0.24844	0	0	0.1149	0.0224
06	1.07	−0.40311	0	0.4789	0.1693	0.105
07	1.0337	−0.37101	0	0	0	0
08	1.09	−0.37101	0	0.34915	0	0
09	1.0103	−0.41416	0	0	0.446	0.2324
10	1.0097	−0.42008	0	0	0.1361	0.0812
11	1.0342	−0.41475	0	0	0.0529	0.0252
12	1.045	−0.42615	0	0	0.0922	0.0224
13	1.0349	−0.42735	0	0	0.2041	0.0812
14	0.99337	−0.44865	0	0	0.2253	0.07
15	1.045	0	1e–05	−0.000841	0	0

Table 17.8 Line flows at t_{18}

From bus	To bus	Line #	P (p.u.)	Q (p.u.)	P_{losses}	Q_{losses}
02	05	1	1.2068	0.35734	0.08333	0.22101
06	12	2	0.23178	0.06236	0.00618	0.01287
12	13	3	0.05479	0.02709	0.00078	0.0007
06	13	4	0.53615	0.21035	0.01916	0.03774
06	11	5	0.24485	0.20546	0.00848	0.01775
11	10	6	0.13837	0.16251	0.00366	0.00857
09	10	7	0.11793	−0.07102	0.00065	0.00172
09	14	8	0.24064	−0.01361	0.0079	0.01683
14	13	9	−0.18447	−0.10044	0.00852	0.01735
07	09	10	0.76122	0.28364	0	0.07352
01	02	11	5.9947	−0.66673	0.62684	1.8554
03	02	12	−2.0725	0.79697	0.22877	0.91755
03	04	13	−0.56515	0.73375	0.05807	0.11555
01	05	14	2.4045	0.60112	0.29703	1.177
05	04	15	1.6918	−0.40512	0.04622	0.13465
02	04	16	1.6522	0.32518	0.15162	0.4234
15	01	17	2.261	−0.00083	0	0.13347
05	06	18	1.3264	−0.05686	0	0.44205
04	09	19	0.42336	0.04679	0	0.10914
04	07	20	0.76122	−0.12257	0	0.13698
08	07	21	0	0.59583	0	0.05264

Table 17.9 Bus voltage and power at t_{18}

Bus #	Bus voltage		P_{supply} (p.u.)	Q_{supply} (p.u.)	P_{load} (p.u.)	Q_{load} (p.u.)
	|V| (p.u.)	Phase (rad)				
01	1.06	0	6.1382	−0.2248	0	0
02	1.045	−0.33775	0.4	3.503	0.6076	0.1778
03	1.01	−0.77694	5.85e−13	0.67076	2.6376	0.266
04	0.9317	−0.62007	3.54e−12	4.15e−13	1.3384	0.056
05	0.93425	−0.53198	−1.2e−12	4.83e−12	0.2128	0.0224
06	1.07	−0.84891	2.46e−12	0.4789	0.3136	0.105
07	0.99371	−0.78903	7.89e−13	1.42e−13	0	0
08	1.09	−0.78903	−1.4e−15	0.34915	0	0
09	0.96599	−0.87638	8.63e−13	1.75e−13	0.826	0.2324
10	0.9684	−0.88945	1.15e−13	−2.1e−14	0.252	0.0812
11	1.0104	−0.87591	3.67e−14	−1.1e−14	0.098	0.0252
12	1.0296	−0.89579	1.05e−15	1.8e−16	0.1708	0.0224
13	1.0126	−0.90056	−1.2e−15	2.5e−15	0.378	0.0812
14	0.94068	−0.94995	1.17e−13	−3.1e−14	0.4172	0.07
15	1.045	0.05534	2.261	−0.51583	0	0

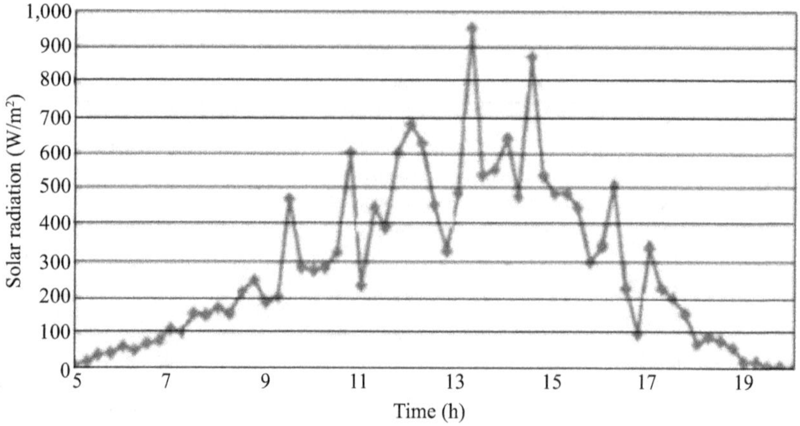

Figure 17.9 Solar radiation through day hours

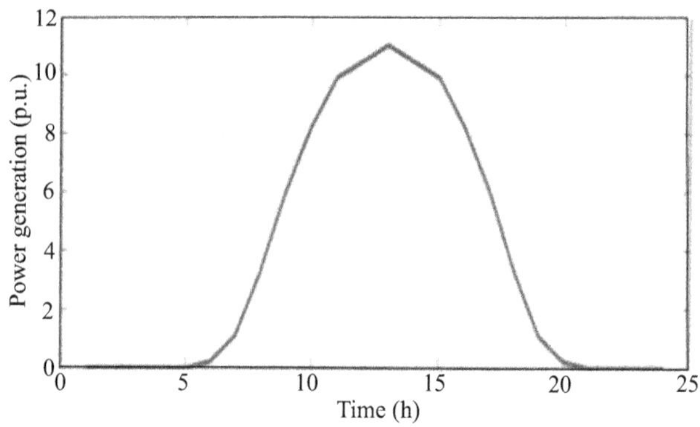

Figure 17.10 PV power versus day hours

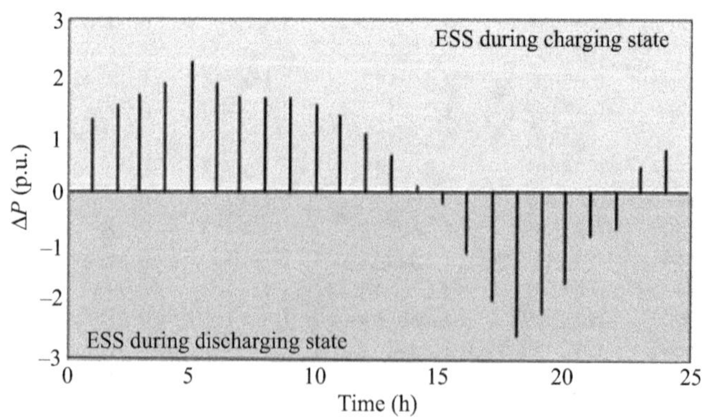

Figure 17.11 ΔP versus day hours

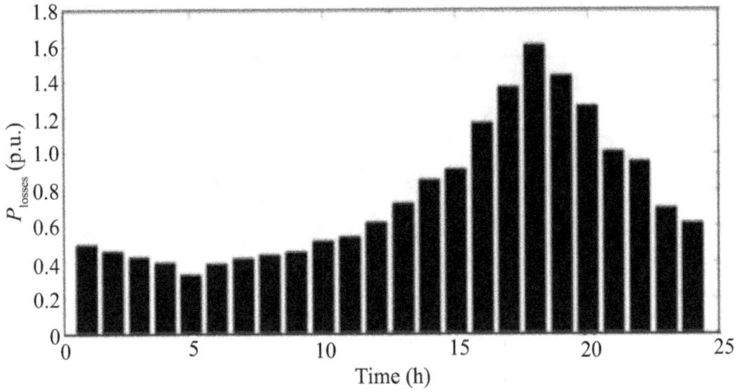

Figure 17.12 Power losses along 24 h

Table 17.10 Line flows at t_5

From bus	To bus	Line #	P (p.u.)	Q (p.u.)	P_{losses} (p.u.)	Q_{losses} (p.u.)
2	5	1	0.37021	0.14556	0.00855	−0.009419
6	12	2	0.11521	0.0457	0.00165	0.00343
12	13	3	0.02816	0.01987	0.00024	0.00022
6	13	4	0.26445	0.14438	0.00525	0.01033
6	11	5	0.12743	0.12796	0.00271	0.005567
11	10	6	0.07573	0.09709	0.00116	0.00272
9	10	7	0.05152	−0.01294	9e–05	0.00023
9	14	8	0.11447	0.00579	0.00163	0.00347
14	13	9	−0.09576	−0.06768	0.00237	0.00482
7	9	10	0.36961	0.22715	0	0.01933
1	2	11	4.9378	−0.65777	0.42735	1.2463
3	2	12	−0.93624	0.10987	0.04118	0.12725
3	4	13	0.38256	0.23647	0.01386	0.00057
1	5	14	1.583	0.13704	0.12181	0.45065
5	4	15	1.0526	−0.23731	0.01554	0.03628
2	4	16	0.62934	0.09407	0.02177	0.02711
15	1	17	−2.2297	0.00618	0	0.01234
5	6	18	0.6639	0.05628	0	0.09739
4	9	19	0.20938	0.04143	0	0.024
4	7	20	0.36961	−0.06814	0	0.0285
8	7	21	0	0.34104	0	0.01724

17.2.2 Role of the ESS

Distribution of ΔP through the 24 h for the three cases collected in one graph as shown in Figure 17.15 as well as the corresponding P_{losses} of the system, which are almost equal, is depicted in Figure 17.16. The time at which the value of ΔP in both charging and discharging state of ESS is maximum is given in Table 17.18.

Table 17.11 Bus voltage and power at t_5

Bus #	Bus voltage		P_g (p.u.)	Q_g (p.u.)	P_L (p.u.)	Q_L (p.u.)
	|V| (p.u.)	Phase (rad)				
1	1.06	−0.27833	6.15208	−0.2248	0	0
2	1.045	0	0	2.345	0.3038	0.1778
3	1.01	−0.27833	0	0.67076	1.3188	0.266
4	0.9957	−0.46084	0	0	0.6692	0.056
5	0.9989	−0.37875	0	0	0.1064	0.0224
6	1.07	−0.47749	0	0.46416	0.1568	0.105
7	1.0349	−0.45218	0	0	0	0
8	1.09	−0.45218	0	0.34104	0	0
9	1.0115	−0.49103	0	0	0.1568	0.2324
10	1.011	−0.49569	0	0	0.126	0.0812
11	1.035	−0.4894	0	0	0.049	0.0252
12	1.0461	−0.4988	0	0	0.0854	0.0224
13	1.0353	−0.49995	0	0	0.189	0.0812
14	0.99602	−0.52103	0	0	0.2086	0.07
15	1.045	−0.28387	1e–05	0.00618	2.2297	0

Table 17.12 Line flows at t_{18}

From bus	To bus	Line #	P (p.u.)	Q (p.u.)	P_{losses} (p.u.)	Q_{losses} (p.u.)
2	5	1	1.375	0.29915	0.10387	0.2835
6	12	2	0.23073	0.06221	0.00613	0.01276
12	13	3	0.05379	0.02705	0.00076	0.00068
6	13	4	0.53221	0.20914	0.01889	0.03721
6	11	5	0.23751	0.20379	0.00813	0.01702
11	10	6	0.13139	0.16157	0.00348	0.00814
9	10	7	0.12479	−0.07038	0.0007	0.00185
9	14	8	0.24522	−0.01302	0.00818	0.0174
14	13	9	−0.18016	−0.10042	0.00819	0.01668
7	9	10	0.76833	0.28316	0	0.07443
1	2	11	3.7014	−0.56731	0.24129	0.67821
3	2	12	−2.1156	0.83171	0.23978	0.96392
3	4	13	−0.52198	0.6845	0.05028	0.09555
1	5	14	2.0061	0.46776	0.20533	0.79818
5	4	15	1.5451	−0.30279	0.0373	0.1064
2	4	16	1.7713	0.30071	0.17244	0.48645
15	2	17	2.4692	0.00758	0	0.01513
5	6	18	1.3141	−0.03438	0	0.42694
4	9	19	0.42767	0.05097	0	0.11069
4	7	20	0.76833	−0.11294	0	0.13783
8	7	21	0	0.58459	0	0.0529

Table 17.13 Bus voltage and power at t_{18}

Bus #	Bus voltage		P_g (p.u.)	Q_g (p.u.)	P_L (p.u.)	Q_L (p.u.)
	|V| (p.u.)	Phase (rad)				
1	1.06	0	5.7076	−0.2248	0	0
2	1.045	−0.20864	0.18	2.163	0.6076	0.1778
3	1.01	−0.65851	0	0.67076	2.6376	0.266
4	0.93552	−0.51372	0	0	1.3384	0.056
5	0.94126	−0.43525	0	0	0.2128	0.0224
6	1.07	−0.74672	0	0.4789	0.3136	0.105
7	0.99553	−0.68326	0	0	0	0
8	1.09	−0.68326	0	0.34915	0	0
9	0.96797	−0.77108	0	0	0.826	0.2324
10	0.9701	−0.7847	0	0	0.252	0.0812
11	1.0114	−0.77249	0	0	0.098	0.0252
12	1.0297	−0.79336	0	0	0.1708	0.0224
13	1.013	−0.79794	0	0	0.378	0.0812
14	0.94202	−0.84568	0	0	0.4172	0.07
15	1.045	−0.20251	2.4692	0.00758	0	0

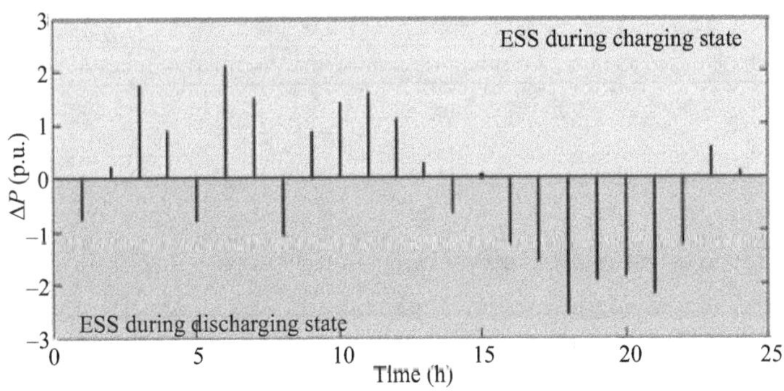

Figure 17.13 ΔP versus day hours

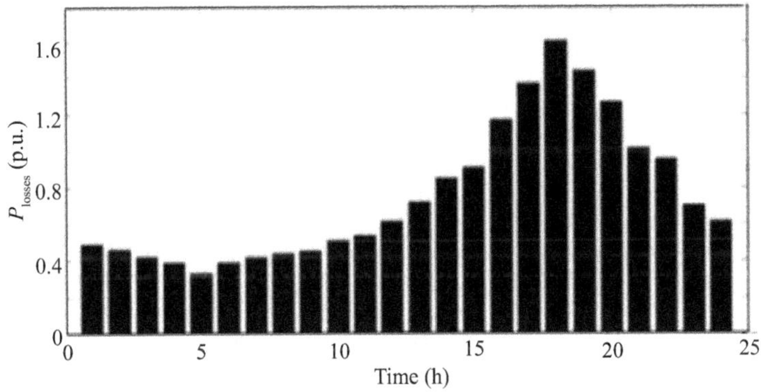

Figure 17.14 Power losses along 24 h

Table 17.14 Line flows at t_6

From bus	To bus	Line #	P (p.u.)	Q (p.u.)	P_{losses} (p.u.)	Q_{losses} (p.u.)
2	5	1	0.59363	0.09061	0.019	0.02249
6	12	2	0.12325	0.04663	0.00186	0.00388
12	13	3	0.02919	0.02035	0.00026	0.00023
6	13	4	0.28166	0.14769	0.00584	0.01151
6	11	5	0.12913	0.13202	0.00283	0.00592
11	10	6	0.0734	0.10089	0.00119	0.0028
9	10	7	0.06403	−0.01654	0.00014	0.00036
9	14	8	0.12925	0.00446	0.00208	0.00443
14	13	9	−0.09813	−0.06997	0.00252	0.00512
7	9	10	0.40818	0.22898	0	0.02256
1	2	11	3.0245	−0.48083	0.16129	0.43394
3	2	12	−1.0745	0.17331	0.05495	0.18523
3	4	13	−0.34981	0.23145	0.01211	−0.00382
1	5	14	1.2712	0.11891	0.07873	0.27283
5	4	15	0.94881	−0.17004	0.01241	0.02642
2	4	16	0.82401	0.06477	0.03652	0.07191
15	1	17	1e–05	−0.00841	0	0.0083
5	6	18	0.70334	0.06183	0	0.10939
4	9	19	0.2311	0.04312	0	0.02922
4	7	20	0.40818	−0.06745	0	0.03466
8	7	21	0	0.34915	0	0.01807

Table 17.15 Bus voltage and power at t_6

Bus #	Bus voltage		P_g (p.u.)	Q_g (p.u.)	P_L (p.u.)	Q_L (p.u.)
	|V| (p.u.)	Phase (rad)				
1	1.06	0	6.0998	−0.2248	2.1042	0
2	1.045	−0.17027	0.012	1.2599	0.3281	0.1778
3	1.01	−0.38211	0	0.67076	1.4243	0.266
4	0.99392	−0.30581	0	0	0.7227	0.056
5	0.99881	−0.26334	0	0	0.1149	0.0224
6	1.07	−0.41855	0	0.4789	0.1693	0.105
7	1.0336	−0.38716	0	0	0	0
8	1.09	−0.38716	0	0.34915	0	0
9	1.0101	−0.43018	0	0	0.446	0.2324
10	1.0095	−0.436	0	0	0.1361	0.0812
11	1.0341	−0.43043	0	0	0.0529	0.0252
12	1.045	−0.44162	0	0	0.0922	0.0224
13	1.0349	−0.44286	0	0	0.2041	0.0812
14	0.99326	−0.46445	0	0	0.2253	0.07
15	1.045	0	1e–05	−0.0005	0	0

Table 17.16 Line flows at t_{18}

From bus	To bus	Line #	P (p.u.)	Q (p.u.)	P_{losses} (p.u.)	Q_{losses} (p.u.)
2	5	1	1.1905	0.36448	0.08157	0.21566
6	12	2	0.23189	0.06239	0.00619	0.01288
12	13	3	0.0549	0.02711	0.00078	0.00071
6	13	4	0.53656	0.21054	0.0192	0.0378
6	11	5	0.24561	0.20575	0.00852	0.01783
11	10	6	0.13909	0.16271	0.00368	0.00862
9	10	7	0.11723	−0.07119	0.00064	0.0017
9	14	8	0.24016	−0.01373	0.00789	0.01678
14	13	9	−0.18492	−0.10051	0.00856	0.01743
.	9	10	0.76049	0.28366	0	0.07343
1	2	11	6.2318	−0.66011	0.67669	2.0076
3	2	12	−2.0685	0.7938	0.22777	0.91334
3	4	13	−0.56914	0.73936	0.05892	0.11774
1	5	14	2.445	0.61791	0.30751	1.2204
5	4	15	1.706	−0.41627	0.04719	0.13771
2	4	16	1.6408	0.32849	0.14973	0.41769
15	1	17	2.481	−0.00305	0	0.15903
5	6	18	1.3277	−0.05975	0	0.44383
4	9	19	0.4229	0.04625	0	0.109
4	7	20	0.76049	−0.12382	0	0.13695
8	7	21	0	0.59733	0	0.0529

Table 17.17 Bus voltage and power at t_{18}

Bus #	Bus voltage		P_g (p.u.)	Q_g (p.u.)	P_L (p.u.)	Q_L (p.u.)
	|V| (p.u.)	Phase (rad)				
1	1.06	0	6.1958	−0.2248	0	0
2	1.045	−0.35108	0.18	2.1658	0.6076	0.1778
3	1.01	−0.78929	0	0.67076	2.6376	0.266
4	0.93118	−0.6312	0	0	1.3384	0.056
5	0.93336	−0.54211	0	0	0.2128	0.0224
6	1.07	−0.85967	0	0.4789	0.3136	0.105
7	0.99347	−0.80013	0	0	0	0
8	1.09	−0.80013	0	0.34915	0	0
9	0.96574	−0.88744	0	0	0.826	0.2324
10	0.96818	−0.90045	0	0	0.252	0.0812
11	1.0103	−0.88679	0	0	0.098	0.0252
12	1.0296	−0.90658	0	0	0.1708	0.0224
13	1.0126	−0.91136	0	0	0.378	0.0812
14	0.94051	−0.96092	0	0	0.4172	0.07
15	1.045	0.06074	2.481	−0.00385	0	0

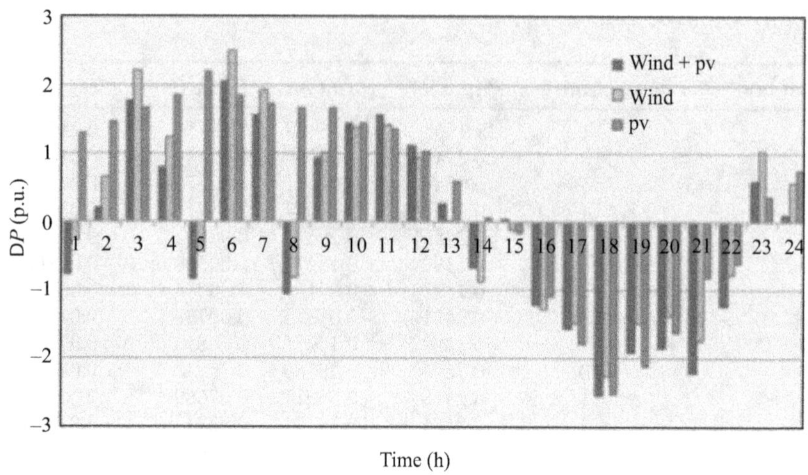

Figure 17.15 ΔP versus day hours

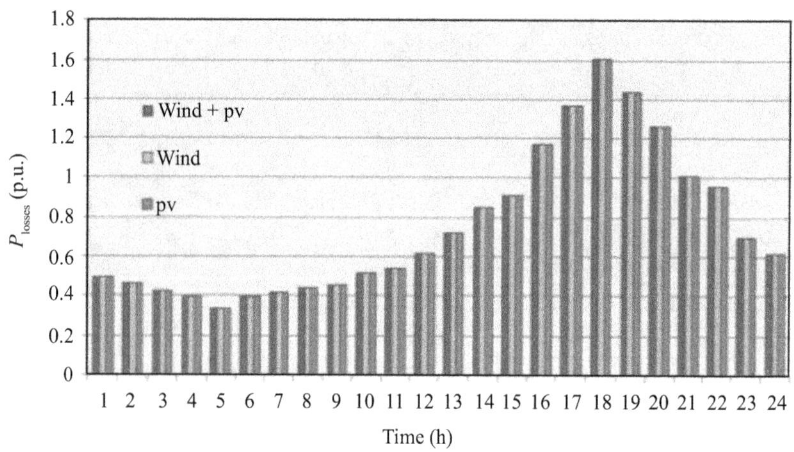

Figure 17.16 Power losses versus day hours

Table 17.18 Maximum ΔP values and time for the different cases

Case #	First case	Second case	Third case
$\Delta P^{+}_{\max}, t_i^{+}$	2.5, t_6	2.25, t_5	2.1, t_6
$\Delta P^{-}_{\max}, t_i^{-}$	2.3, t_{18}	2.6, t_{18}	2.6, t_{18}

$\Delta P^{+}_{\max}$ is the maximum positive value at the ith hour of the day, t_i^{+}.
$\Delta P^{-}_{\max}$ is the maximum negative value at the ith hour of the day, t_i^{-}.

It is to be noted that the positive value of ΔP represents the available power that can be supplied from the system to charge the ESS, whereas the negative value is the power that must be discharged into the system. Therefore, ΔP can be considered as an important parameter by which the size of the ESS in terms of its state of charge and depth of charge can be specified [4–6].

17.3 Load flow calculations for transient stability

It is important to apply the load flow calculations to study the transient stability of power systems with renewables during contingencies such as fault occurrence and system congestion. The following examples illustrate how to identify the rotor angle stability of conventional synchronous generators in the power system that is integrated with RESs, for example, wind and/or solar PV and ESSs.

Example 17.2 As in Example 17.1, with the same data of the IEEE-14 bus system, load profile, wind farm and solar PV plant, investigate the transient stability of the system for the three cases (merging wind, solar PV and combination of wind and PV) when a three-phase fault occurs at Bus #2 at 5 am, and is cleared by isolating line #16.

Solution

The original power system (i.e. without any renewable sources) has a total load of 3.626 p.u. according to the load profile (Table 17.6). Applying the load flow technique to the system by using MATLAB®/Power System Analyse Toolbox (PSAT)®, version 2.1.10 programs [7], it is found that the initial steady-state rotor angle of each generating unit, δ, is as given in Table 17.19. Running the program at different clearing times, it is noted that the critical clearing time (CCT) that keeps the system stability is 290 ms and at clearing time greater than that the system is unstable (Figure 17.17).

Case (i): Replacing G_1 by a wind farm and ESS at Bus #1

Considering a wing generating plant with a doubly fed induction generator (DFIG) operating at the sub-synchronous mode, its equivalent circuit is shown in Figure 17.18(a). The power injected into the grid and equations of motion for the

Table 17.19 Rotor angle of generating units at steady state

Generating unit #	G_1	G_2	G_3	G_4	G_5
δ (rad)	0.53341	0.10487	−0.33367	−0.37167	−0.34132

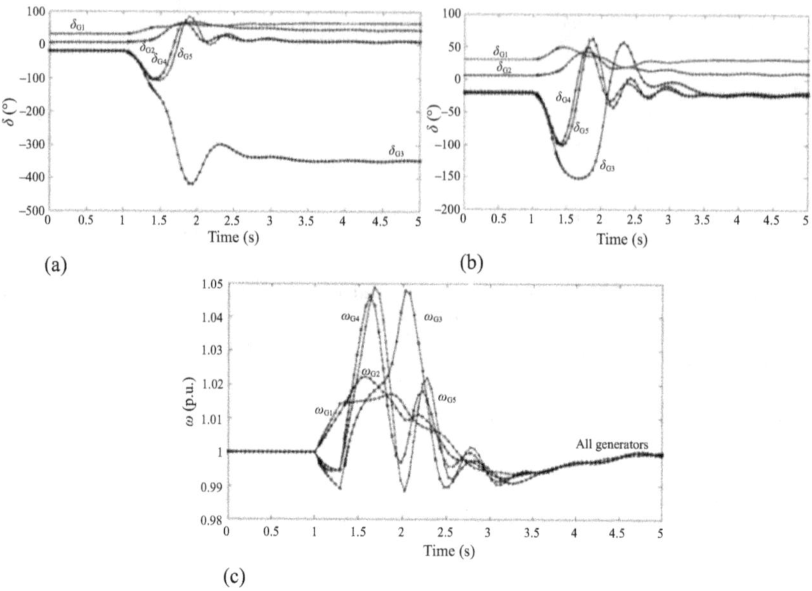

Figure 17.17 Rotor angles response of generating units in the conventional system: (a) angle at clearing time 300 ms, (b) angle at CCT 290 ms and (c) speed variation at CCT 290 ms

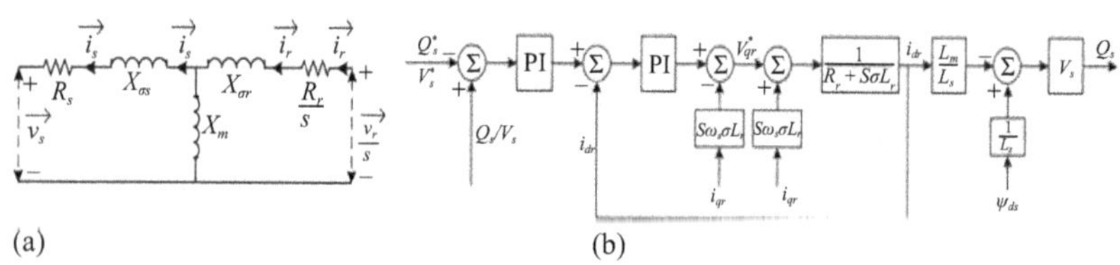

Figure 17.18 DFIG at sub-synchronous speed mode: (a) DFIG equivalent circuit and (b) stator reactive power control via RSC

DFIG dynamic model in *d-q* frame, as explained in Chapter 2, Section 2.4.1.2, can be rewritten as

$$\left.\begin{aligned} P &= v_{\mathrm{ds}}i_{\mathrm{ds}} + v_{\mathrm{qs}}i_{\mathrm{qs}} + v_{\mathrm{dr}}i_{\mathrm{dr}} + v_{\mathrm{qr}}i_{\mathrm{qr}} \\ Q &= -\frac{x_\mu v i_{\mathrm{dr}}}{x_s + x_\mu} - \frac{v^2}{x_\mu} \end{aligned}\right\} \tag{17.6}$$

$$\left.\begin{aligned} \dot{\omega}_m &= \frac{(T_m - T_e)}{2H_m} \\ T_e &= \psi_{\mathrm{ds}}i_{\mathrm{qs}} - \psi_{qs}i_{\mathrm{ds}} \end{aligned}\right\} \tag{17.7}$$

Table 17.20 Wind farm data

Inertia constants H_m (kWs/kVA)	3
Voltage control gain K_v (p.u.)	10
Power control time constant T_e (s)	0.01
P_{max} and P_{min} (p.u., p.u.)	[1 0.00]
Number of wind generators that composes the park	30

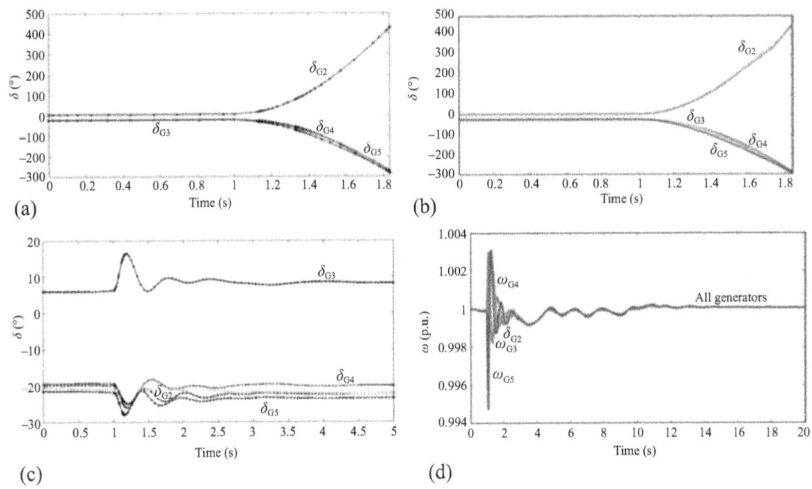

Figure 17.19 Performance of synchronous generators for the system with wind energy source at Bus #1: (a) angle at clearing time 290 ms, (b) angle at clearing time 76.5 ms, (c) angle at CCT 76.4 ms and (d) speed variation at CCT 76.4 ms

The grid-side converter is generally operated at unity power factor mode. The reactive power injected/absorbed via the stator terminals is controlled by the rotor-side converter (RSC) which is considered as a lossless converter as shown in Figure 17.18(b) [7].

The wind farm data given in Table 17.20 in addition to the data tabulated in Table 17.5 are the input parameters to run PSAT program. At the instant of fault occurrence, $P_L = 3.626$, $\Delta P = 2.6297$ p.u., the initial rotor angles, δ, for G_2, G_3, G_4 and G_5 are given in Table 17.19. As ΔP is positive, the ESS is in the charging mode; that is, it acts as a load and consequently Bus #15 is a load bus.

Applying MATLAB/PSAT programs to solve the stability problem, it is found that the CCT is 76.4 ms and for any time more than that the system is unstable (Figure 17.19).

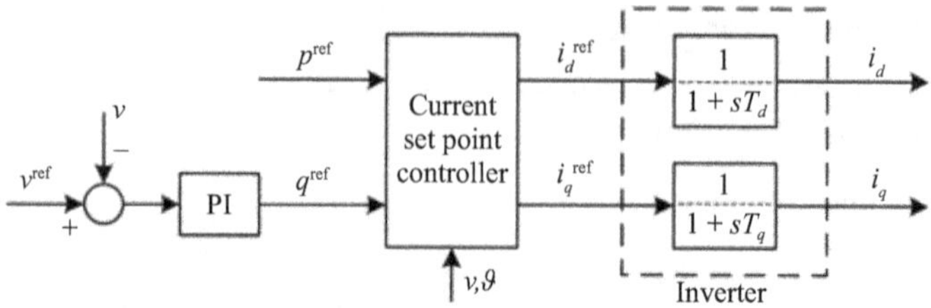

Figure 17.20 Block diagram of PV interface model

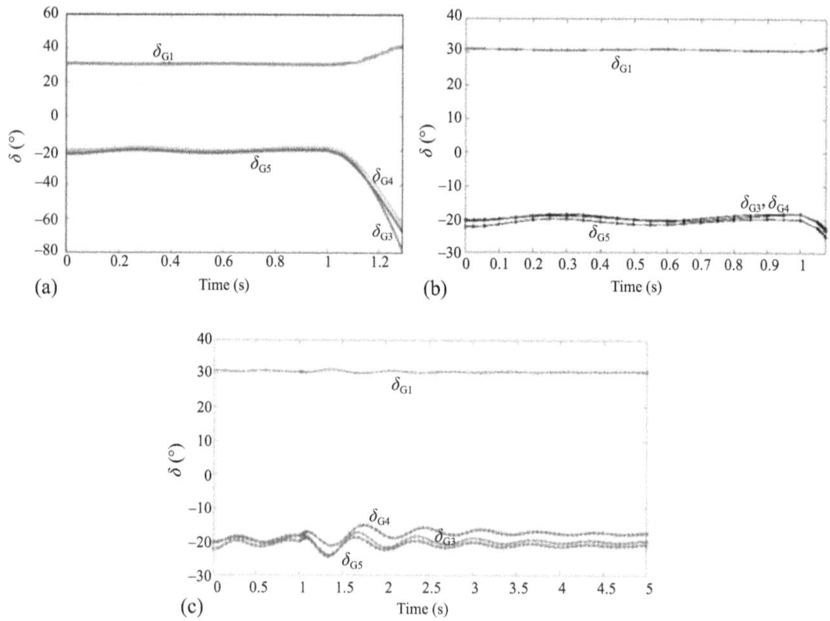

Figure 17.21 Rotor angles of synchronous generators for the system with solar PV energy source at Bus #2: (a) at clearing time 290 ms, (b) at clearing time 76.4 ms and (c) at clearing time 3.24 ms

Case (ii): Replacing the conventional generating unit, G_2 by a solar PV plant and an ESS at Bus #2

At the instant of fault occurrence, $P_L = 3.626$, $\Delta P = 2.2297$ p.u., rated active power = 60 MW, the rotor angles of G_1, G_3, G_4 and G_5 are given in Table 17.19. The solar PV system is interfaced with the grid through an inverter as shown in the block diagram (Figure 17.20). The inverter response times, T_d and T_q, are equal to 0.015 s. The voltage PI controller gains, K_v and K_i, are 0.0868 and 50.9005, respectively, and the voltage reference is 1.045 p.u. [7]. The CCT that keeps the system stability is 3.24 ms, whereas at a clearing time of 290 or 76.4 ms, the system is unstable (Figure 17.21).

The injected active and reactive power to the AC bus can be computed as

$$\left.\begin{aligned} P &= v_d i_d + v_q i_q \\ Q &= v_q i_d + v_d i_q \end{aligned}\right\} \tag{17.8}$$

Case (iii): The power system incorporates a wind farm and an ESS at Bus #1 and a solar PV at Bus #2 rather than a conventional generating unit at these buses

Combining the configuration of Cases (i) and (ii) as a configuration of Case (iii) and applying the load flow algorithm with MATLAB/PSAT programs to study the rotor angle stability of the rotating generators, the variation of rotor angle against time at different clearing times is obtained as shown in Figure 17.22.

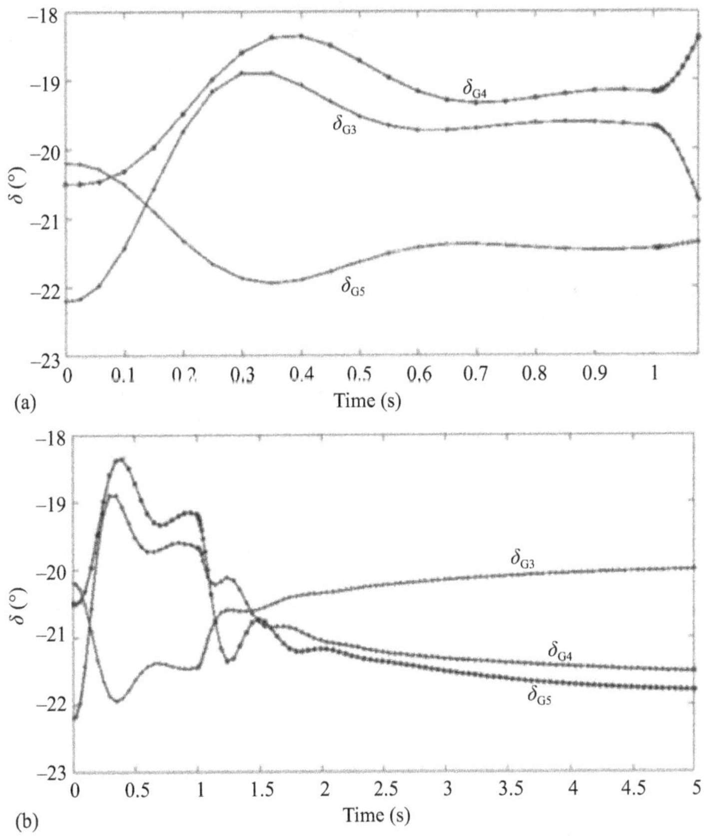

Figure 17.22 Rotor angles of synchronous generators for the system with wind + solar PV energy sources at Buses #1 and #2, respectively: (a) unstable at clearing time 76.4 ms and (b) stable at clearing time 0.324 ms

Table 17.21 Maximum CCT and status of different system configurations

System configuration	**Original**	**Case (i)**	**Case (ii)**	**Case (iii)**
CCT (ms)	290	76.4	3.24	0.324
Status	Stable	Stable	Critically stable, not secured	Critically stable, not secured

The maximum values of CCT that can keep the system stability for the original and three cases of system configurations are summarized in Table 17.21. It is to be noted that the configurations of both Cases (ii) and (iii) are critically stable and may have no stability margin, that is, may not be secure as the value of CCT is too small. In other words, with the addition of renewable energy generation the effective system inertia is reduced significantly and should be enhanced by introducing virtual inertia into the system as explained in Chapter 14, Section 14.4.4, for example, by using hybrid ESS (flywheel + batteries) [8–14].

References

[1] Ackerman T. *Wind power in power systems*. 2nd edn. New York: John Wiley & Sons Ltd.; 2012. pp. 6–20.

[2] Feijóo A. L. and Cidrás J. 'Modelling of wind farms in the load flow analysis'. *IEEE Transactions on Power Systems*. 2000;15(1):110–15.

[3] Su H. I. and El-Gamal A. 'Modelling and analysis of the role of fast-response energy storage in the smart grid'. *Proceedings of Forty-Ninth Annual Allerton Conference*; Allerton House, UIUC, Illinois, USA, Sep 2011, pp. 719–26.

[4] Luba-Rubio R., Trejo-Perea M., Vargas-Vázquez D. and Ríos-Moreno G. J. 'Optimal sizing of renewable energy systems: A review of methodologies'. *Journal of Solar Energy, Elsevier*. 2012;86:1077–88.

[5] Oimm A. J., Cockerill T. T. and Taylor P. G. 'The potential of peak shaving distribution networks using electricity storage'. *Journal of Energy Storage, Elsevier*. 2018;16:231–42.

[6] Reihani E., Sepasi S., Roose L. R. and Matsuura M. 'Energy management at the distribution grid using a battery energy storage system (BESS)'. *Electrical Power and Energy Systems, Elsevier*. 2016;77:337–44.

[7] Federico M., Vanfretti L. and Morataya J. C. 'An open source power system virtual laboratory: The PSAT case and experience'. *IEEE Transactions on Education*. 2008;51(1):17–23.

[8] Taczi I. and Vokony I. 'Possible role and effects of synthetic inertia on the future's power systems'. *Engineering and Industry Series. Volume Power Systems. Energy Markets and Renewable Energy Sources in South-Eastern Europe*. Budapest, Ungaria: Trivent Publishing; 2016, pp. 32–47.

[9] Gonzalez-Longatt F., Chikuni E. and Rashayi E. 'Effects of the synthetic inertia from wind power on the total system inertia after a frequency disturbance'. *IEEE International Conference on Industrial Technology (ICIT)*; Cape Town, South Africa, Feb 2013, pp. 25–8.

[10] Tamrakar U., Shrestha D., Maharjan M., Bhattarai B. P., Hansen T. M. and Tonkoski R. 'Virtual inertia: Current trends and future directions'. *Journal of Applied Sciences*. 2017;7(7):1–19.

[11] Xia S., Zhang Q., Hussain S. T., Hong B. and Zou W. 'Impacts of integration of wind farms on power system transient stability'. *Journal of Applied Sciences*. 2018;8(8):1–16.

[12] Eriksson R., Modig N. and Elkington K. 'Synthetic inertia versus fast frequency response: a definition'. *IET Journal of Renewable Power Generation*. 2018;12(5):507–14.

[13] Waffenschmidt E. and Hui R. S. Y. 'Virtual inertia with PV inverters using DC-link capacitors'. *IEEE 18th European Conference on Power Electronics and Applications (ERE'16 ECCE Europe)*; Karlsruhe, Germany, Sep 2016, pp. 1–10.

[14] Shi L., Dai S., Yao Y., Ni L. and Bazargan M. 'Transient stability of power systems with high penetration of DFIG based wind farm'. *IEEE Power Engineering Society, General Meeting*; Calgary, AB, July 2009, pp. 1–6.

Chapter 18

Integration of electric vehicles with renewables into power grids

18.1 Electrification of transportation

Internal combustion engine vehicles using fossil fuels are at present the dominant energy sources for transportation sector. In the future, some problems may occur because of the fossil fuel availability and the continuous increase in fuel price. In addition, burning the fossil fuels produces greenhouse gases (GHGs) that highly affect the global climate change.

Electrification of transportation sector seems to be one of the feasible solutions to these problems. Electric vehicles (EVs) such as plug-in hybrid electric vehicles (PHEVs) and plug-in electric vehicles (PEVs) will transfer energy demand from fossil fuels to electricity for transportation sector. Thus, reducing the pollution and alleviating issues that may compromise security can be achieved [1].

EVs can be envisioned as distributed battery energy storage. Hence, EVs can help enable the power system to accommodate large-scale intermittent renewable sources (e.g. wind and solar photovoltaics (PVs)) through utilizing the EVs to balance the power generated from these sources.

Moreover, integration of EVs into the power grid in addition to the development of battery technology has a potential to provide grid support by delivering ancillary services (peak power shaving, spinning reserve, voltage and frequency regulations) whenever needed under the concept of vehicle to grid (V2G) presented in Section 18.3.

Penetration of large-scale renewable sources into power system entails large battery energy storage system (BESS) to smoothly support electric grids so that the electrical demand and operating standards are always met [2,3].

For all modes of EVs' operation (delivering ancillary services, supporting the grids and charging the batteries), the vehicles can be clustered in a way that constitute several vehicle fleets based on the similarity of both performance and charging patterns of vehicles in each fleet. Then, the EV batteries of each fleet are aggregated for the network management response as they need to be managed carefully in order to avoid interruption when several EVs are introduced into the system, whatever the state of charge (*SoC*) of the batteries. So, the aggregator clusters EVs in the form of a virtual battery that can be controlled to provide ancillary services such as primary, secondary or tertiary frequency control.

It knows exactly the state in which the EVs are currently in, that is, the vehicles' *SoC* and power capacity. Consequently, the aggregator can determine the virtual storage characteristics. Then, a sophisticated controller is required to regulate multiple EV batteries for each fleet based on signals received from the aggregators. The controller also receives signals indicating the power flow in each vehicle fleet. Consequently, it can decide which fleets are able to afford the power needed by the grid, or the power required from the grid to charge the batteries (Figure 18.1).

Charging/discharging of EVs should be controllable to satisfy both utility and vehicle owners' technical, environmental, economic and social requirements. For instance, technically, emerging large number of EVs to be connected to the grid simultaneously may compromise the quality and stability of the overall power system. Regarding economic and social aspects, it is useful for designing the charger control to understand how people charge their EVs: once per day, charging early or at night, always expecting a full battery or charging on the same day as other neighbours [4]. More details about charging schemes are described in Chapter 12, Section 12.6.2 [5–10].

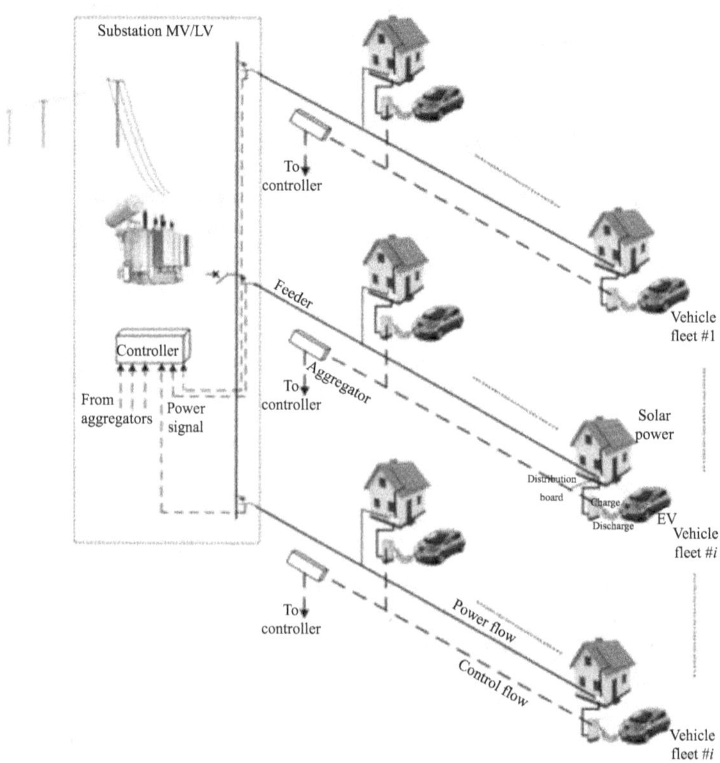

Figure 18.1 Vehicle fleets connected to the substation for energy management

18.2 EVs integration with RESs

Penetration of renewable energy sources (RESs) into the electric power systems (EPSs) causes challenges. The power generated from these sources, especially PV and wind, is fluctuating and unpredictable because of its dependency on the solar radiation and wind speed. This nature of RESs necessitates that it be associated with an energy storage system (ESS) for being dispatchable and enabling energy management. EV batteries can be aggregated and act as dynamic energy storage devices. They can be charged by the excess power generated by the RES or discharged to deliver power to the grid during low power generation. Thus, EVs will act like energy buffer for generation–demand balance and introducing ancillary services. In addition, with adopting EVs capabilities and integrating the RESs specifically, optimal GHG emissions and cost reduction can be achieved by applying multi-objective optimization techniques that maximize customer benefits and minimize disturbances to the grid.

As a perspective of the application of EVs with RESs (PV and wind) integrated into electric distribution grid, an envisioned schematic diagram is depicted in Figure 18.2. There are two configurations of PV solar energy sources: centralized PV solar plant and distributed public solar carport charging stations. EVs are also either in parking lots or in a cluster. EV fleets are integrated to the PV solar plant.

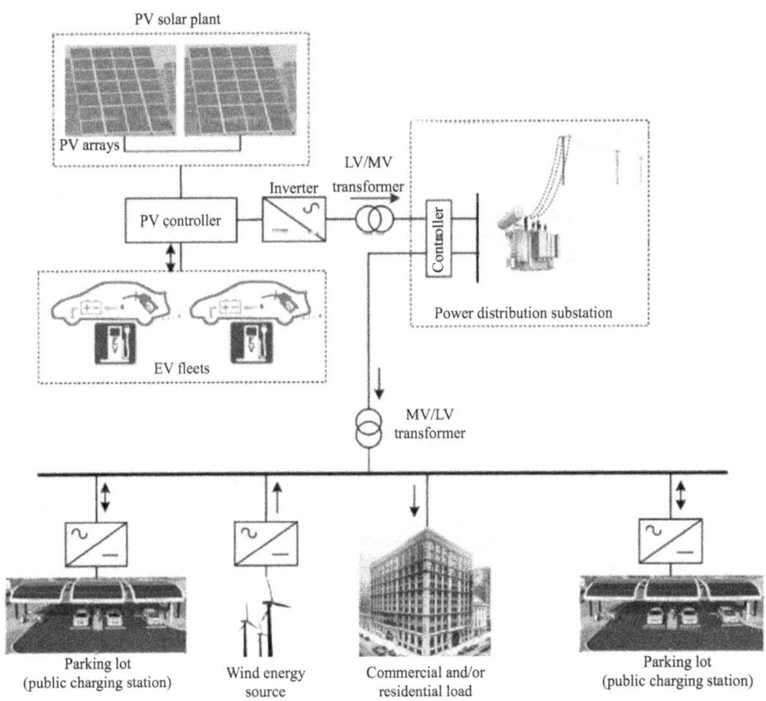

Figure 18.2 EVs with renewable sources integrated into power grid

They act like a load when charging the batteries by surplus energy provided by the PV plant or like a source in discharging state when the PV delivered power is low. The PV plant with EV fleets is integrated into the power grid through PV controller and an inverter to introduce the services required by the grid. However, for a specific area, the residential and/or commercial loads are supplied by power delivered from wind energy source with EV batteries presented in the parking lots as ESS. Meanwhile, these batteries are charged by the solar carport charging stations and utility grid that act as backup source. It is to be noted that the EVs are connected to bi-directional DC/AC power converters depending on the battery state, charging or discharging. The utility power distribution substation is equipped with a controller for energy management [11].

18.3 EVs for ancillary services in power systems

Ancillary service in power systems such as voltage control, frequency regulation, spinning and non-spinning reserve can be provided by aggregated EVs (PEVs and PHEVs). Provision of these services can be local for isolated systems or wide for integrated power systems. To achieve that especially for widespread systems, need of communication infrastructure and information exchange technologies as well as advanced controls and measuring devices is indispensable. The advanced control envisions the EVs as a storage that can be scheduled to either consume power (charging) or feed power to the grid (discharging). This means that the EVs remain connected to the grid and are ready to deliver the energy stored in their batteries when needed. This mode is referred to as V2G.

18.3.1 V2G concept

EVs integrated into power systems can operate as dynamic load drawing power from the grid during charging or dynamic ESS feeding power to the grid during discharging, which is referred to as V2G. Spatial location and low individual storage capacity of EVs entail the aggregation of a large number of vehicles to enable them for V2G service. The aggregation of EVs depends on the control scheme and the service required under V2G concept. The aggregator coordinates all required operational activities like communicating with distribution system operator (DSO), transmission system operator (TSO) and energy system providers, and maintains the link between energy market players and the EV owners. Thus, it helps participation of supporting electric grid regulation and energy management as a single controllable distributed energy source (DES) in a virtual power plant (VPP). The VPP can be understood as a coalition composed of multiple energy producers (e.g. wind and PV solar) and energy storage providers (e.g. EVs) that come together to sell electricity as an aggregate [11,12]. So, within the VPP architecture, the EVs are visible to the DSO, TSO or grid operator through the aggregator and can easily participate in the energy market.

On the other hand, the individual EV owners can play a central role by participating in the energy market through managing the queries from the DSO, TSO and/or energy market players with the aid of the two-way communication and control systems [13].

A schematic diagram to depict the role of VPP control and implementation under the V2G concept is shown in Figure 18.3. The EV aggregator operates as a VPP and the clustered EV fleet at charging station provides status (e.g. *SoC*, available power) to the charging management system (CMS), which communicates with the aggregator control centre (ACC). The aggregated battery power can be dispatched by the VPP control centre to provide ancillary services according to the request of DSO or TSO. The VPP is responsible for communications and power flow management between market players and grid operators. Communication and power lines are illustrated by dotted and solid lines, respectively. The individual EV owners communicate with ACC through EV management system (EVMS) using smart meters (SMs) as measuring devices. It is noted that the operation of EV fleet as demand-side management unit, dynamic load and ESS can be optimized by aiming to minimize the operation cost.

Accommodating RESs and ESSs, or in general DESs, and satisfying the requirements of secure operation make the energy management of the traditional power grid to be more complex and challenging. A grid that is more flexible, efficient, reliable and secure can be configured with a goal of enhancing power generation and distribution. Such a grid encompasses advanced information and communication technologies (ICTs), smart energy metering and advanced control.

18.3.2 EVs for power grid frequency regulation

Electric power grids require that the power generation and loads are in balance to keep the grid frequency constant. In the case of imbalance between generation and

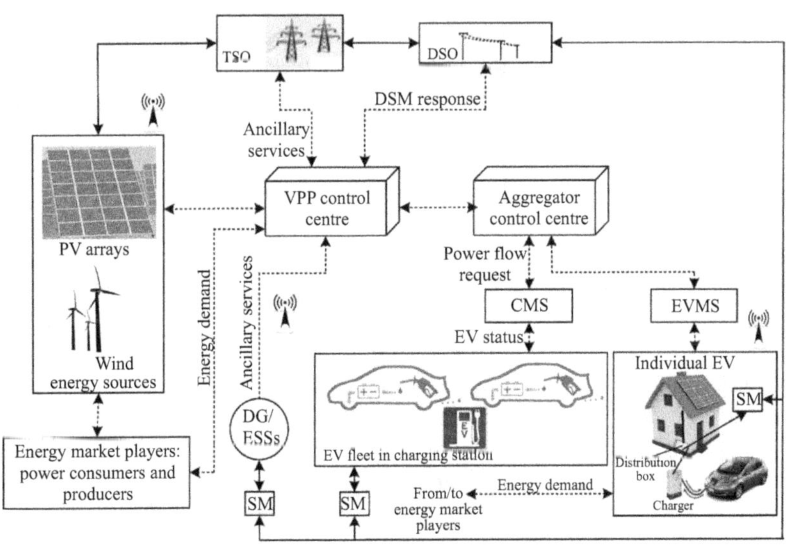

Figure 18.3 Perspective of VPP framework under V2G concept

load, the grid frequency will change. Grid frequency regulation is usually a service performed by the power plants that are controlled to vary their generation output up and down from a nominal value. EVs can provide a similar function by varying their charging rate based on locally measured grid frequency. As EVs reach large-scale adoption, they have the potential for frequency regulation through charge/discharge process.

Concerning the environment, climate change, fossil energy reserve, energy cost and energy security, EVs may have a promising role and are getting great attention of governments and automotive companies in the world. Because of their 'zero emissions' and potential application, the market size of EVs will be enormous. V2G concept has been proposed to make full use of EVs in supporting power grids [14]. It means that EVs can represent a reasonable source of generation and storage capacity for the power system. Several investigations have shown the feasibility of this service [15] and new niche markets [16]. Owing to the fast charging rate of the battery, frequency regulation is regarded as one of the most promising and practical services under V2G concept [17].

By adopting adequate control strategies of different regulation modes (primary/second/tertiary frequency regulation), benefits for the stability of frequency may be achieved. However, frequency regulation with EVs is still in its infancy. A series of adaptive control strategies (effective smart management, coordination between the two parties – 'utility grid and vehicle owners' – and the incentive policy to be in a win-win situation) is needed.

EVs can be used as mobile storage devices, which can provide regulation services [18]. According to the scale, EVs are divided into two types as follows:

- Small-scale EV: The charging points are located in private areas with public or private access, for example, domestic charging. These charging points allow slower charging rate.
- Large-scale EV: The charging points are located in public areas with public access such as charging stations, battery exchanging/swapping stations and fast charging station. Large-scale EV is dedicated to EV fleets and allows for fast charging.

18.3.2.1 Large-scale EV participation in frequency regulation

Nowadays, the development of large-scale RES has brought up new challenges to power system secure operation [19]. With high penetration of renewable energy connected to grid, the proportion of traditional generating units in operation is gradually dropping, and the frequency fluctuation caused by intermittent power generation will exceed the capability of traditional generators for suppressing such fluctuations. Therefore, traditional frequency regulation may be hardly able to meet the balance of supply and demand in the grid integrated with RES.

Meanwhile, management of a 'large-scale EV' requires extra facilities including (i) a smart interface with the grid, (ii) communication with the grid operator and (iii) metering the exchange of power flows information of the grid [20]. These requirements bring more challenges to power grids in the process of frequency regulation such as the coordination between the RES and the EVs, battery charging strategies for different system operating conditions, and technical and economic challenges of adaptive control algorithms in frequency regulation. To accommodate these problems, evaluation of the impact of charging/discharging strategies on power systems, active management solution and control strategies during the batteries' charging periods should be accounted for.

EV fleets more easily accommodate the existing electric market rules [21]. In power markets, the demand response and management are required to achieve the active management target. A scheme for integration into power system can be modelled by energy hubs and demand management [22]. To support large-scale RES and stabilize the grid, the use of EVs under V2G concept to provide storage for large-scale renewable energy generation has been studied [23]. The storage capacity can attenuate the intermittence of RES, and still provide an asymmetrical regulation, such as 'regulation up' or 'regulation down' [24].

When large disturbances such as faults and line trips occur, the power system is subjected to speed and frequency variation, and voltage fluctuations. These fluctuations affect the power quality that customers experience, and if they are strong enough, they could even damage sensitive appliances. Moreover, if the disturbance is not cleared within a specific time (critical clearing time), the systems may loose synchronism and are unable to resume stable operation.

In frequency control, there are three layers of control: primary, secondary and tertiary. EVs can simulate the generator droop characteristic to perform primary frequency regulation and achieve secondary frequency control based on area control error (ACE). Tertiary frequency modulation can be achieved based on economic dispatch [25].

Primary frequency control. In microgrids, a power disturbance would result in a significant oscillation of frequency due to lack of enough generation resources and spinning reserve. EVs connected to the grid can inject power into the grid, which not only benefits the frequency stability, but also increases robustness of operation, especially in isolated system. EVs can be considered as new resources in primary frequency control owing to their fast response to disturbances [26]. Studies such as in [26] make the V2G more noticeable than ever in primary frequency control.

In primary frequency control, there are two loops: first, the droop control that reduces the output of traditional generators, which can cut the fossils consumption, and second, the inertial control that simulates the behaviour of conventional generator. So, these two control schemes autonomously respond to frequency deviation. To alleviate the effect of intermittence of RES, a series of V2G control approaches are adopted in some literature [27]. The effect of EVs as controllable load in primary frequency control and battery *SoC* should be investigated [28] and to satisfy the expected battery *SoC*, an automotive distributed control should be applied [29]. The scheduled charging power can be estimated based on the historic data of frequency deviation, which is dominated by normal distribution [27].

Secondary frequency control. As a centrepiece of secondary frequency control, automatic generation control (AGC) is an important control technology for keeping frequency stability. If EVs participate in dispatching load, AGC would respond to frequency deviation on both generation and load sides simultaneously to offset the insufficient response of traditional generating unit. The feasibility and potential AGC operation with EVs are analysed in [30] and the methodology to make EV participate in AGC in bulk power system is designed [28]. These studies show that EVs can assist AGC. Additionally, some scholars have discussed centralized V2G control method to participate in the load frequency control. In [25], a method for tracking signal is designed. But these studies have not considered how to send signal to EVs.

At present, the study of AGC with EVs is still in initial stages, and the focus is on EV energy storage to reduce the cost of BESS. EV as a distributed controllable load is discussed in [31]. Coordination control method is applied to integrate large-scale wind power and PV power into a large power system [32]. These results show that EVs can effectively reduce the ACE, reduce the tie-line flow deviation and lower the frequency modulation capacity [28].

Tertiary control. EVs participate in tertiary control based on economic dispatch. Stochastic dynamic programming method is used in [33] to optimize EV charging and frequency regulation decisions, taking into consideration the uncertainty of EV and wind power. The economic dispatch model was established in [34]. The results indicate that the EVs plug-in grid can reduce the gap between demand and supply in different periods of time with optimal dispatch schedule. Nevertheless, the economic dispatch with EVs does not include several constraints such as reduction of GHG emissions and loss reduction in power grid. The economic dispatch models containing many uncertain factors and calculating the objective function of the volatility of constraints, using various mathematical programming techniques and intelligent optimization, has received little attention recently.

However, more advanced controllers can be utilized to stabilize the frequency deviation in case of fault in the system. In these cases, fleets of EVs are aggregated into one actuator, which either draws or feeds power from or into the network. The impact of these services on the battery energy content is neglected due to its duration of some seconds only. The dynamic model of batteries is assumed to be homogenous; that is, battery-ageing differences between individual EVs are neglected.

In addition, batteries are equipped with smart chargers that introduce intelligent charging schemes to control the EV charging process. Controllable EV charging surmounts or alleviates the undesired impacts produced by uncontrollable EV charging such as the overloading of the power system facility and increasing power demand resulting in a less efficient electricity supply. SMs can pursue various objectives such as minimization of charging costs in the electricity market, GHG emissions or power system losses. It is crucial to achieve these objectives in an optimal manner through applying optimization techniques and considering the system constraints especially in case of large-scale EV penetration. Several constraints coexist in the real-world implementation as well as their variation depending on the objectives of the deployed EV system.

More description of charge/discharge management and frequency regulation as ancillary services under V2G concept is presented in the following example.

Example 18.1 Scheduling charging/discharging processes of EVs' batteries as well as use of EVs as a source of distributed frequency regulation are investigated through the application to the nine-bus system with the data shown in Figure 18.4 and Tables 18.1 and 18.2. The load curve is given as depicted in Figure 18.5.

Five different car user profiles are included in each fleet:

- Profile #1: People going to work with a possibility to charge their cars at work (100 EVs).
- Profile #2: People going to work with a possibility to charge their cars at work, but with a longer ride (100 EVs).
- Profile #3: People going to work with no possibility to charge their cars at work (100 EVs).
- Profile #4: People are staying home (250 EVs).
- Profile #5: People working on a night shift (250 EVs).

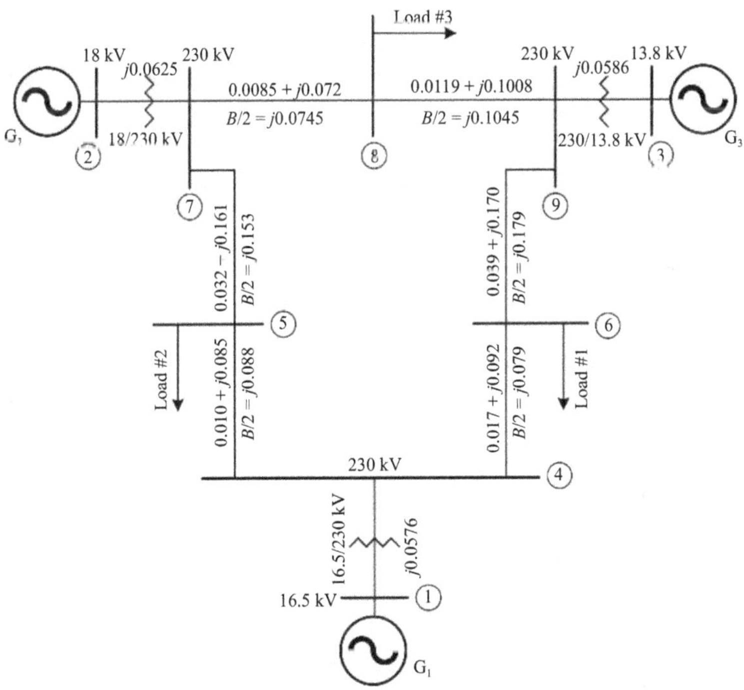

Figure 18.4 Nine-bus system impedance diagram: all impedances are in p.u. on a 100-MVA base

Table 18.1 Data bank of the nine-bus system

Bus no.	Bus type	Generation (p.u.)		Load (p.u.)		Voltage magnitude
		P_G	Q_G	P_L	Q_L	
1	Swing	–	–	0	0	1.04
2	PV	1.63	–	0	0	1.025
3	PV	0.85	–	0	0	1.025
4	PQ	0	0	0	0	–
5	PQ	0	0	1.25	0	–
6	PQ	0	0	0.9	0	–
7	PQ	0	0	0	0	–
8	PQ	0	0	1	0	–
9	PQ	0	0	0	0	–

Table 18.2 Nine-bus generation data

Generator	G_1	G_2	G_3
Rated MVA	247.5	192	128
Power factor	0.85	0.85	0.85
Line–line voltage (kV rms)	16.5	18	13.8
Frequency	60	60	60
x_d	0.8979	1.05	1.305
x'_d	0.2996	0.235	0.252
x_q	0.474	0.646	0.889
x'_q	0.243	0.23	0.22
x_ℓ (leakage)	0.186	0.134	0.134
T'_{d0}	7	601	4.75
T''_{d0}	0.03	0.04	0.068
T''_{q0}	0.0513	0.099	0.21
Stator resistance (R_s)	0.0031	0.0031	0.0031
Inertia coefficient (H)	4.7	5.14	6.54
Friction factor (F)	0	0	0
Pole pairs (P)	1	1	1

18.3.2.2 Control architecture

The multi-agent system used to perform distributed frequency regulation has a hierarchical structure. The vehicle local controllers (LCs) and ACC are working together to determine the regulation needs and control local dispatch. The LC will reside within each EV to monitor and control battery charging and discharging while a vehicle is plugged-in. The objective of the LC is to control charging based on the owner's preference and minimizing the impact on the life of the battery system. Although the LC relies on the ACC to inform its control decisions, it will necessarily make the decisions without regard to what neighbouring LCs are doing. Therefore, the LC acts a decentralized control (Figure 18.6).

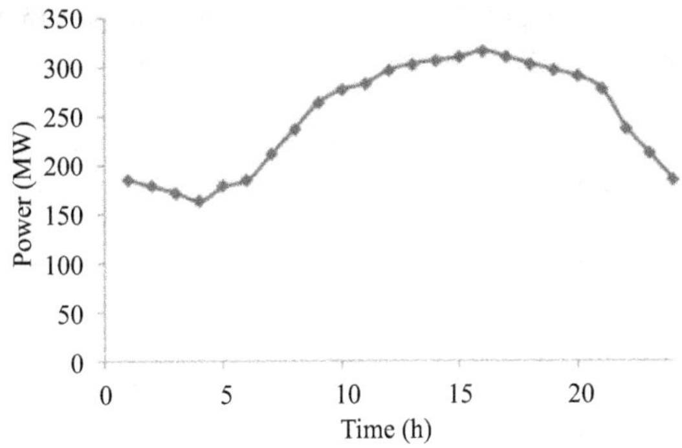

Figure 18.5 Daily load curve

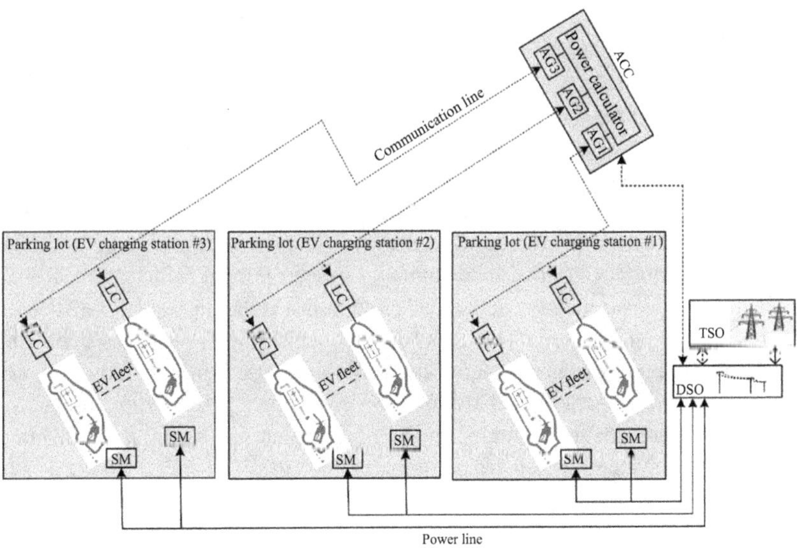

Figure 18.6 Hierarchical structure of control system

- *LC:* It *controls* the battery behaviour and makes the decision as to how much power and energy the battery system can provide for frequency regulation. Determining how much energy a vehicle will supply or absorb for frequency regulation depends on the battery system state and the supply recommendation made by the ACC. As a simplifying assumption in this study, a vehicle will remain plugged in for the full expected charge duration once it returns home, also minimizing the impact on vehicle drivability. In the case that supplying regulation extends the expected charge duration and charging is interrupted, the vehicle could still be driven.

The general approach of the LC is to begin charging at a mid-level rate, e.g. 1 kW, such that there is a margin to increase or decrease charging to supply frequency regulation.

The duration of regulation supply from an EV could be several hours depending on the current battery state and how long the vehicle remains parked. However, as the vehicle may be unplugged or LC overridden at any point, a shorter supply duration may make more sense. Therefore, several supply durations ranging between 5 min and 1 h may be considered to determine the effects on the aggregate supply curve.

The functionality of the LC may also be integrated into a home energy management system (EMS), which monitors household energy consumption in real time and acts as a communications gateway between devices within the home and beyond the point of utility metering. The home EMS may control several loads in the home to smooth or shift the load profile. Here, the EV could aid in managing the energy profile of a home without external system needs. The same LC functionality will be useful in either situation: controlling the charge rate or discharging to supply part of the load over short periods in a smooth manner. The vehicle LC will interact with the ACC, which provides a supply recommendation for frequency regulation, and tracks the aggregate local EV regulation supply based on the control actions taken by the LC.

- *ACC:* Frequency measurements are not reliable at the point of customer service where the EV is connected to the grid. Therefore, the LCs are not able to perform frequency control independently based on local information. Reliable frequency measurements can be made beginning at the high side of distribution substation transformers. The ACC located at a higher voltage level where reliable frequency measurements are available is used in this study to oversee and coordinate local EVs participating in frequency regulation under V2G concept. Here, local EV refers to all vehicles that are below the ACC in the distribution system, which are participating in V2G. The ACC will act as a link between the area in EPS and the local EV. It would also be the natural choice to act as an aggregator to perform market functions for the local EV.

  The aggregator is a central controller in-charge of coordinating all required operational activities, like communicating with the DSO, TSO, grid operator and energy service providers. It also maintains a link between energy market players and the EV owners. The aggregator knows exactly the state in which the EVs are currently in, that is, the vehicles, *SoC* and power capacity. Based on this information, the aggregator can determine the virtual storage characteristics and can directly control their charging/discharging according to TSO demands. It then submits bids to the TSO for ancillary services.

  Because each level of the power system has different operational needs, 'local energy needs' will be defined differently throughout the system. For instance, determining the amount of local energy needed to contribute to improving system frequency performance will depend on the load fluctuations and nearby generation at that point in the system. In addition, the market price for energy and ancillary services will behave differently at different nodes in the transmission system.

The ACC will necessarily interact with local sensing and measurement devices to perform its functions. However, it will also need to interact with the system operator in EPS and of course the LCs.

18.3.2.3 System simulation

In the context of the bulk power system, the basic objective is to make sure that there is enough power being generated to supply the power drawn by the system load, providing customers with a secure and reliable electrical energy supply. There are several types of measures to determine how well the system is operating, including power balance, voltage performance, and financial returns. Each consumer and stakeholder will have a unique set of expectations and desired outcomes with respect to power system performance.

To simulate the use of EV as a source of distributed frequency regulation, a model including power system behaviour, EV loads, and additional controllers is necessary. The ACC acts as an interconnection for energy transfer to and from EV participating in V2G. However, all communications necessary between the EV and the system operator to facilitate distributed frequency regulation are handled by the ACC.

Communication systems are not considered in this study, neither the physical implementation nor the protocols that may be used. Standards for communication systems on the user end are currently being developed and will continue to be refined as the smart grid is implemented. Future studies may consider the communication systems to get a more realistic picture of how well the distributed control works.

The simulator is designed to study the speed response of generating units to changes in system loading, and to facilitate testing of control strategies for balancing load and generation. The ACC tracks the system frequency of the bus and provides the LCs with the regulation control signal, which combines the elements of primary and secondary frequency regulation.

The single-line-diagram and the simulation model of the nine-bus system with EVs are illustrated in Figures 18.7 and 18.8, respectively. The system includes three generators, three load buses with a load curve shown in Figure 18.5 for each. Three EV fleets are embedded into the system at buses 4, 7 and 9, with one fleet in a parking lot (a charging station) for each.

The five load profiles are simulated as shown in Figure 18.9(a). Initialization of *SoC* and the state of cars (plug-in or plug-out) can be identified by the simulation model as depicted in Figure 18.9(b). Once a mileage range is identified, the average distance driven in that range during a specific hour is used to determine the battery's remaining *SoC* assuming that each vehicle consumes 0.2 kWh/mi in its all-electric range with a minimum *SoC* of 20%.

Consideration of a maximum and a minimum *SoC* is necessary for two reasons: first, battery protections often prevent the battery from reaching too extreme values to avoid cell degradations, and second, because the battery cannot recharge/discharge at its maximum rate when approaching extreme *SoC* values. Maximum *SoC* of 0.9 is assumed as for $SoC > 0.9$, the charging rate decreases and a regular

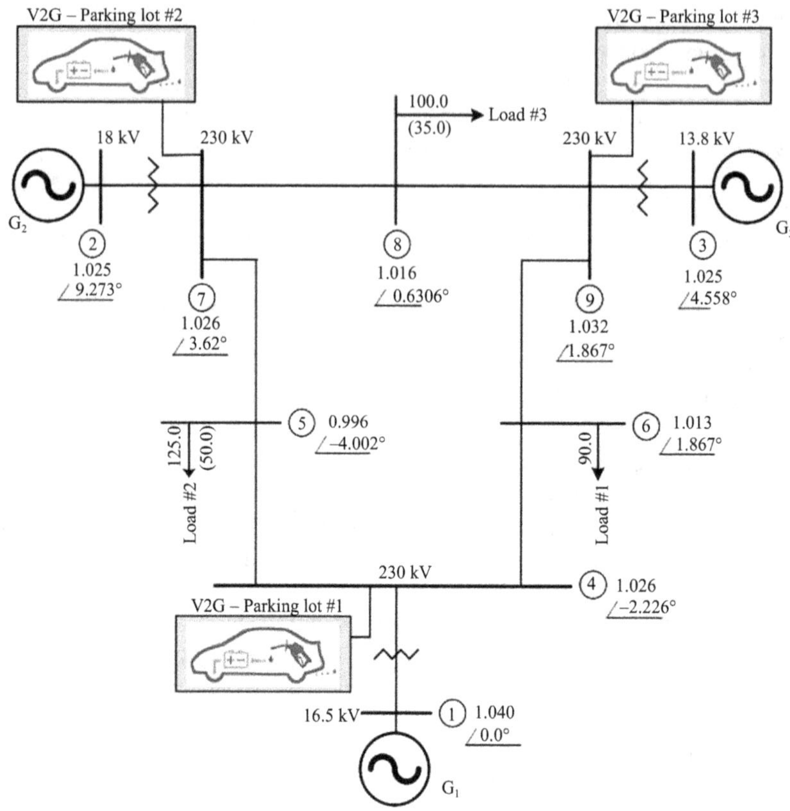

Figure 18.7 Nine-bus system with three parking lots showing the bus voltages (in kV and p.u.)

contribution to frequency regulation cannot be provided. Minimum *SoC* of 0.4 is also taken because of the battery protection tripping at *SoC* below 0.2, and 20% more for home-to-work journeys. When the *SoC* drops below 0.4 because of a journey, the EV will only be allowed to take part in regulation 'down', in a condition that it will be recharged when necessary.

The generator speed (Hz) is monitored and controlled to be kept at a desired value (reference value) by varying the load and thus deciding which of regulation, up/down, will be done? Accordingly, the ACC estimates the number of EVs to participate in V2G and gives decision of charging or discharging based on the state of plugging and regulation up/down. The simulation models of speed control of both the generators and the aggregator are shown in Figure 18.10(a) and (b).

18.3.2.4 Scheduling the charging/discharging processes for EVs fleet

The financial transactions recorded at a parking lot for 24 h are presented in Figure 18.11. Most electricity purchases should be shifted to the off-peak periods as

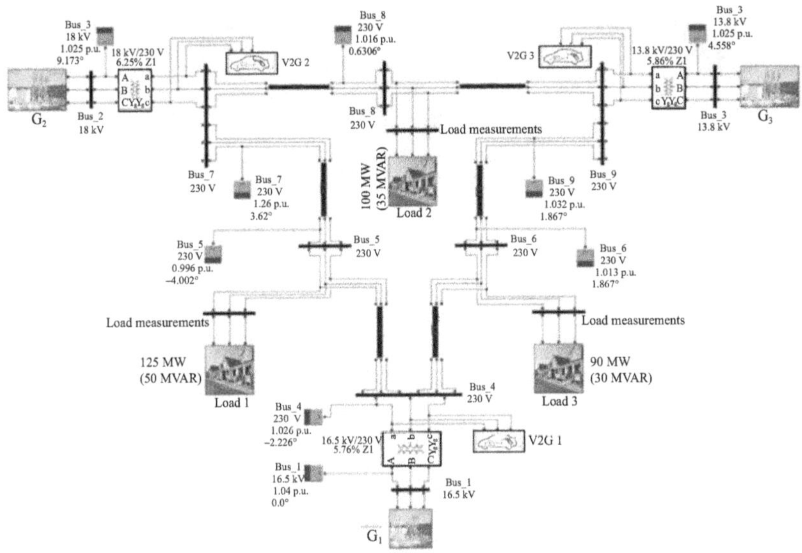

Figure 18.8 Simulation model of nine-bus system with EVs

the price is low, whereas electricity sales are centred around the peak periods as the price is high. Therefore, more connected EVs are scheduled for charging during the periods when electricity tariffs are relatively low. The available power supplied by V2G operation of EVs at these periods is insufficient to meet the huge amount of charging load, and so, the CMS needs to purchase electricity from the grid to substitute the power deficits. For scheduling the periods with most expensive electricity prices, the CMS tends to discharge the energy stored in most connected EVs. Since the EV charging demands at these periods are relatively small, the CMS can sell the excess electricity to the power grid for more revenue [35].

The variations in average *SoC* for existing EV profiles in a parking lot and the rated MW in response to the dynamic electricity tariff are given in Table 18.3. In this table, EV plug-in state indicator is represented by '1' during charging/discharging or standby states and by '0' for EVs in driven or out of service states. The power that can be supplied to/delivered from EVs in charging/discharging state is positive and negative, respectively, whereas in standby state zero power is assumed as the EV is only participating when needed (like spinning reserve), e.g. Profile 4 at hours 21 and 22.

The average *SoC* of EVs starts to grow when the electricity tariff is relatively low (Figure 18.12). When the electricity price reaches its peak, the majority of connected EVs are scheduled by the CMS to discharge their stored energy to the grid, leading to decrement of *SoC* [36,37].

Applying the load flow technique (e.g. Newton–Raphson method) and MATLAB®/Simulink® programs to the nine-bus system without EVs, the values of dispatch power of each generator, system losses and the load over day hours are

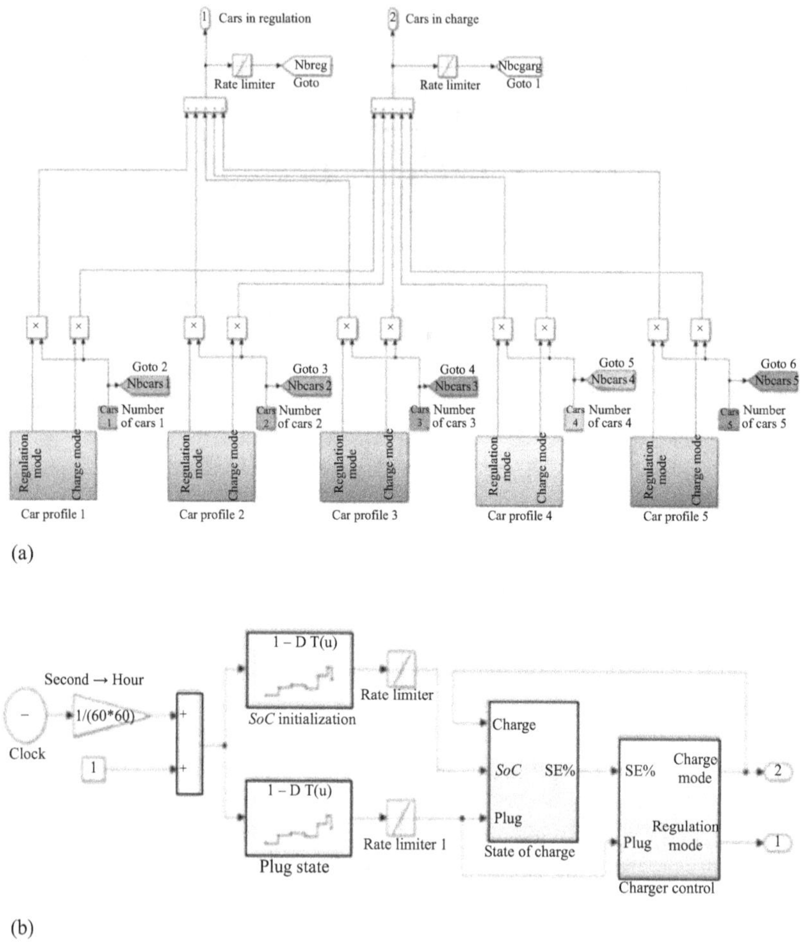

Figure 18.9 Simulation model of (a) car profiles and (b) SoC initialization and plug-in state of each profile

obtained as given in Table 18.4. In the case of the system incorporating EVs, these values in addition to the net power (the difference between the load power and EVs powers) are obtained by using the simulation models of the system, aggregator and load profiles as tabulated in Table 18.5.

The EVs participation in feeding the load and the corresponding power delivered from the conventional synchronous generators (net power) through day hours are drawn in Figure 18.13.

It is to be noted that, in Figure 18.13, the EV acts as a load during out of peak times as it is in charging process, and discharges energy at peak times to operate as a source of electricity. Therefore, there is a reduction in the load during the peak

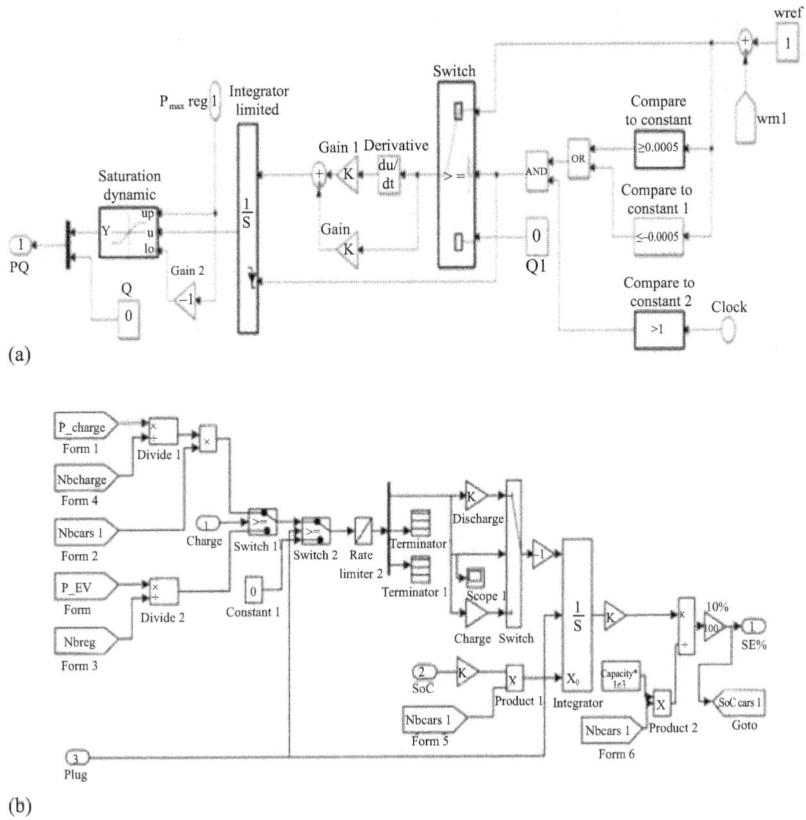

Figure 18.10 MATLAB/Simulink implementation of (a) speed control and (b) aggregator model

times, which yields a reduction of the electricity consumed from the conventional synchronous generators.

18.3.2.5 Application of EVs as a source of frequency regulation

Because of the load/generation imbalance resulting from disturbance occurrence or congestion in the grid, the frequency will change – increasing if load is less than generation and decreasing if load exceeds generation.

AGC acts as an interconnected system to ensure that frequency is brought back to the nominal value and inter-area power flow is regulated. AGC can restore system frequency in several minutes. So, EVs can be considered as new resources in secondary frequency control owing to their fast response to disturbances. When frequency goes downwards, EVs can be easily switched partially or fully from acting as a load to act as power producer to prevent frequency drop. The feasibility of EVs as controllable load to provide effective system secondary frequency response can be approved. Regarding management, the aggregator can alter EVs' mode with the rate of frequency changes.

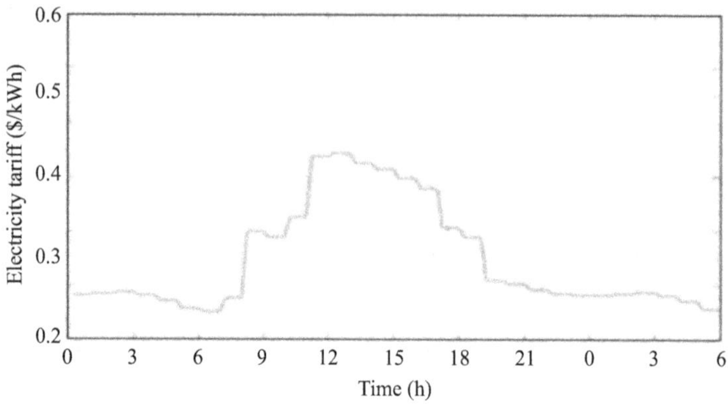

Figure 18.11 Financial transactions versus day hours

PEVs not only improve the load curve on peak time, but also can surmount frequency deviations in a few seconds. Different types of disturbances in IEEE nine-bus system are simulated such as oscillatory transients (OTs) caused by switching processes and faults like single line-to-ground (L-G), double line-to-ground (2L-G) and three-phase short circuit (3-Ph short circuit). The fault is located on the line between Bus 1 and Bus 4 very close to Bus 1. It occurs at 1:00 a.m. and at its clearance the system restores its original topology.

For OTs, EVs help damp the frequency fluctuation fast as well as the range of change is getting less (Figure 18.14(a)). For instance, as in Table 18.6, the time to reach the allowable frequency change (TR-AFC), which is $\pm 0.1\%$ of the rated value, is reduced from 6 to 3.7 s as well as the maximum range of frequency change is also reduced from 0.017 to 0.0094 p.u.

When the fault occurs, the frequency drops and EVs act as an energy source to mitigate the frequency variation. As explained in Chapter 14, Sections 14.4.1 and 14.4.4, the governor response as a primary control is active for the duration of 10–30 s from the instant of fault occurrence and is taken in this simulation as 10 s. So, V2G is important to play a role in the first 10 s to reduce TR-AFC and damp the frequency fluctuation fast by reducing the rate of change of frequency (RoCoF). In the next 10 s, the governor control is in action rather than the EVs. It is to be noted from Figure 18.14(b)–(d) that the change of frequency is instantly increased and then starts to decay as well as the instant of fault occurrence is considered as the zero time for plotting the curves. The numerical results of simulation are given in Table 18.6.

It is concluded that the EVs can improve power grid transient stability when the power grid is subject to large disturbances, including bus faults, generator and branch tripping, and sudden large load changes.

EV behaviour as an aggregate load and source of frequency regulation is better viewed from the bulk system level. This is mainly because the frequency response

Table 18.3 The average SoC and plug-in state of EV profiles versus day hours

Time (h)	Prof. 1			Prof. 2			Prof. 3			Prof. 4			Prof. 5		
	SoC (%)	Plug-in state indicator	MW	*SoC* (%)	Plug-in state indicator	MW	*SoC* (%)	Plug-in state indicator	MW	*SoC* (%)	Plug-in state indicator	MW	*SoC* (%)	Plug-in state indicator	MW
1	40	1	4.8	40	1	4.8	50	1	6	40	1	12	40	0	0
2	50	1	6	50	1	6	60	1	7.2	50	1	15	30	0	0
3	60	1	7.2	60	1	7.2	70	1	8.4	60	1	18	30	0	0
4	70	1	8.4	70	1	8.4	80	1	9.6	70	1	21	30	0	0
5	80	1	9.6	80	1	9.6	90	1	10.8	80	1	24	50	1	6
6	90	1	10.8	90	1	10.8	90	0	0	90	1	27	70	1	12
7	90	0	0	90	0	0	90	0	0	90	1	0	80	1	15
8	80	0	0	90	0	0	80	0	0	90	1	0	90	1	18
9	80	1	0	80	0	0	80	1	−1.2	90	1	0	90	1	0
10	80	1	0	80	0	0	80	1	−1.2	90	1	−3	90	1	−3
11	80	1	0	80	0	0	70	1	−2.4	80	1	−6	80	1	−6
12	80	1	−1.2	70	0	0	70	1	−2.4	80	1	−6	80	1	−6
13	70	1	−2.4	70	0	0	60	1	−3.6	70	1	−9	70	1	−9
14	60	1	−3.6	70	0	0	60	1	−3.6	70	1	−9	70	1	−9
15	50	1	−4.8	60	0	0	50	1	−4.8	60	1	−12	60	1	−12
16	40	1	−6	60	0	0	40	1	−6	60	1	−12	60	1	−12
17	40	0	0	60	0	0	40	0	0	50	1	−15	50	1	−15
18	30	0	0	50	0	0	30	0	0	40	1	−18	40	1	−18
19	30	1	−1.2	50	1	−1.2	30	0	0	30	1	−21	40	1	−18
20	20	1	−2.4	40	1	−2.4	30	1	−1.2	20	1	−24	40	1	−18
21	20	1	0	30	1	−3.6	20	1	−2.4	20	1	0	40	0	0
22	20	1	0	20	1	−4.8	20	1	2.4	20	1	0	40	0	0
23	20	1	2.4	20	1	2.4	30	1	3.6	20	1	6	40	0	0
24	30	1	3.6	30	1	3.6	40	1	4.8	30	1	9	40	0	0

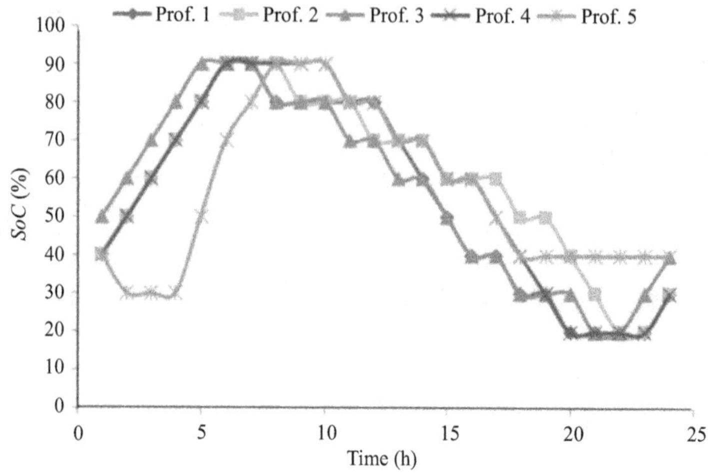

Figure 18.12 The average SoC of EV profiles versus day hours

Table 18.4 IEEE nine-bus simulation results (generators, losses and load) without EVs versus day hours

Time (h)	Gen. 1 (MW)	Gen. 2 (MW)	Gen. 3 (MW)	Load (MW)	Losses (MW)
1	80.872799	62.755885	41.911348	184.011671	1.528361
2	77.948526	60.547738	40.416703	177.449282	1.463686
3	75.078261	58.277691	38.932414	170.886708	1.401659
4	71.643945	55.58818	37.1151	163.015989	1.331236
5	77.988806	60.515676	40.411403	177.452391	1.463494
6	80.876709	62.741207	41.923988	184.01357	1.528334
7	92.479023	71.65935	47.946722	210.267017	1.818078
8	104.010048	80.730547	53.934071	236.51584	2.158825
9	115.732023	89.642538	59.941626	262.765404	2.550784
10	121.42111	94.246493	62.990441	275.889382	2.768661
11	124.321996	96.511318	64.500537	282.451403	2.882448
12	130.22057	100.963012	67.51449	295.578123	3.11995
13	133.069069	103.281795	69.033115	302.13891	3.245068
14	134.523535	104.407009	69.797954	305.419777	3.308721
15	135.982403	105.544905	70.54695	308.700927	3.373332
16	138.913513	107.786959	72.068195	315.263491	3.505177
17	135.983248	105.543958	70.546867	308.700726	3.373346
18	133.095961	103.245999	69.040751	302.137932	3.244779
19	130.20566	100.974822	67.514404	295.574855	3.120031
20	127.244122	98.762194	66.004946	289.011676	2.999586
21	121.460248	94.187855	63.000074	275.880054	2.768123
22	103.986806	80.74285	53.938237	236.50899	2.158902
23	92.42523	71.734781	47.920565	210.262147	1.81843
24	80.856502	62.769389	41.914657	184.012092	1.528457

Table 18.5 IEEE nine-bus simulation results (generators, losses, load and net MW) incorporating EVs versus day hours

Time (h)	Gen. 1 (MW)	Gen. 2 (MW)	Gen. 3 (MW)	EV (MW)	Losses (MW)	Load (MW)	Net (MW)
1	80.8728	62.7559	41.9113	27.6701	1.5284	184.0117	211.6818
2	77.9485	60.5477	40.4167	34.2011	1.4637	177.44923	211.6504
3	75.0783	58.2777	38.9324	40.8005	1.4016	170.8867	211.6872
4	71.6439	55.5882	37.1151	47.4514	1.3312	163.0159	210.4674
5	77.9888	60.5157	40.4114	60.0110	1.4635	177.4524	237.4634
6	80.8767	62.7412	41.924	60.6349	1.5283	184.0136	244.6485
7	92.4790	71.6594	47.9467	15.0120	1.8180	210.2670	225.2790
8	104.0100	80.7305	53.9341	18.0130	2.1588	236.5158	254.5288
9	115.7320	89.6425	59.9416	−1.2670	2.5508	262.7654	261.4984
10	121.4211	94.2465	62.9904	−7.2147	2.7687	275.8894	268.6747
11	124.3220	96.5113	64.5005	−14.4030	2.8824	282.4514	268.0484
12	130.2206	100.9630	67.5145	−15.6120	3.1199	295.5781	279.9661
13	133.0691	103.2818	69.0331	−24.0608	3.2451	302.1389	278.0781
14	134.5235	104.4070	69.798	−25.2106	3.3087	305.4198	280.2092
15	135.9824	105.5449	70.547	−33.6060	3.3733	308.7009	275.0949
16	138.9135	107.7869	72.0682	−36.1321	3.5052	315.2635	279.1314
17	135.9832	105.5439	70.5469	−30.1050	3.3733	308.7007	278.5957
18	133.0960	103.2460	69.0408	−36.1380	3.2448	302.1379	265.9999
19	130.2057	100.9748	67.5144	−41.4720	3.1200	295.5749	254.1029
20	127.2441	98.7622	66.0049	−48.0824	2.9996	289.0117	240.9293
21	121.4602	94.1879	63.0001	−6.0579	2.7681	275.8801	269.8222
22	103.9868	80.7429	53.9382	−2.4140	2.1589	236.5089	234.095
23	92.4252	71.7348	47.9206	14.4792	1.8184	210.2622	224.7413
24	80.8565	62.7604	41.9147	21.1050	1.5285	184.0121	205.1171

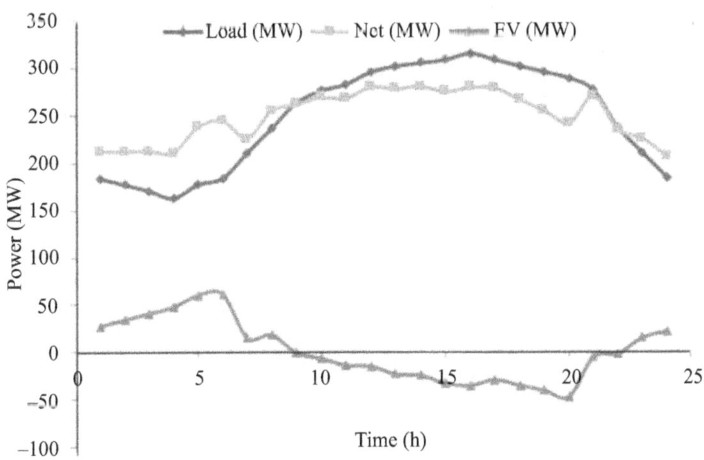

Figure 18.13 Load, EV and net (MW) versus day hours

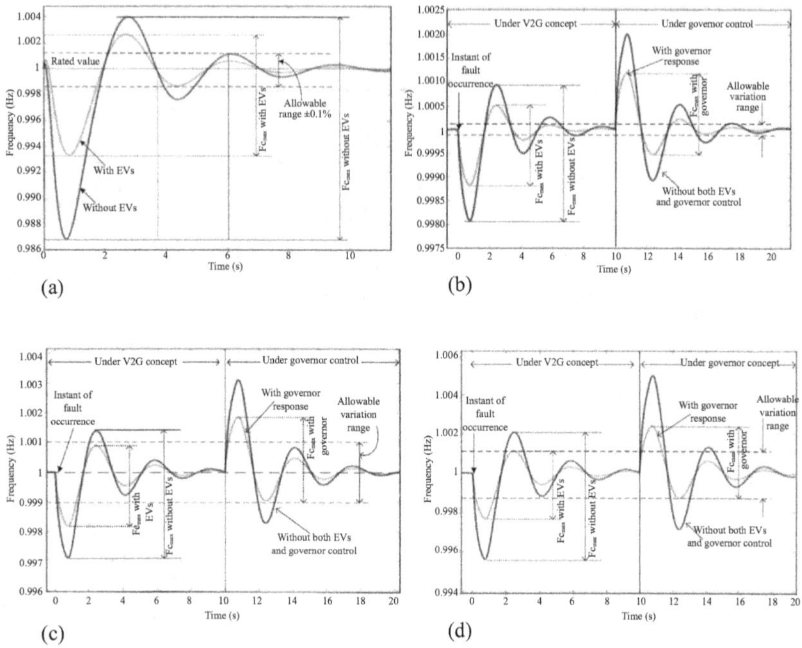

Figure 18.14 Frequency variation for nine-bus system with/without EVs: (a) OT, (b) L-G fault, (c) 2L-G fault and (d) three-phase short circuit

Table 18.6 Frequency variation at different faults in the system with/without EV

Type of fault	Maximum range of frequency change (Hz) without EVs/GR (p.u.)		TR-AFC (s)	Maximum range of frequency change (Hz) with EVs/GR (p.u.)		TR-AFR (s)
	From	**To equal**		**From**	**To equal**	
OT	0.987	1.004 = 0.017	6	0.9933	1.0027 = 0.0094	3.7
L-G V2GGR	0.998	1.001 = 0.003	6.8	0.9988	1.0005 = 0.0017	5
	1.002	0.999 = 0.003	4.7	1.0012	0.9995 = 0.0017	3
2L-G	0.997	1.0015 = 0.0045	3.6	0.9982	1.0009 = 0.0027	1.5
V2GGR	1.0032	0.9982= 0.005	3.5	1.0019	0.999 = 0.0029	2
3-Ph sc	0.9955	1.002 = 0.0065	3.5	0.9978	1.0011 = 0.0033	1.6
V2GGR	1.005	0.997 = 0.008	3.5	1.0024	0.9988 = 0.0036	1.7

of central generation needs to be incorporated into a system simulation model to study how effective EV could be as a source of frequency regulation. As a first step in modelling EV as distributed energy resources, a transmission level model of the power system is used and an aggregate EV load will be tied to each load bus in the system. The power system model is based on a modified version of the power system simulator.

Power calculations. At the ACC, the aggregated delivered power from each parking lot can be calculated as shown below.

Some parameters are defined as follows:

- *Arrival time* (T_{arr}): The time at which the EV arrives the parking lot.
- *Time of request* (T_{req}): The time at which the stored power in EV battery is required to be delivered to the grid.
- SoC_{max}: The maximum value of *SoC* of EV battery.
- SoC_{min}: The minimum value of *SoC* of EV battery.
- *EV plug-in state coefficient* (λ): It equals 0 if the EV is on drive, out of service, $SoC \leq SoC_{\min}$ or in charging, and equals 1 when the EV is in discharge or standby mode.
- *Actual battery capacity* (E_{actual}): It is the present energy stored in the battery.
- *Battery capacity* ($E_{charge/disch}$): The energy that can be provided or absorbed by EV battery. It can be calculated by the following equation:

$$E_{\text{charge/disch}} = E_{\text{actual}}(SoC_{\max} - SoC_{\min}) \tag{18.1}$$

Each EV in the parking lot is identified by ID number, $SoC_{\max}$, $SoC_{\min}$, T_{arr} and the load profile number that the EV belongs to. Hence the total energy that can be supplied to/delivered from EVs' batteries during charging/discharging in the three parking lots can be calculated by the ACC using the following equation:

$$E_{\text{total}} = \sum_{i=1}^{N}\left[\sum_{j=1}^{L}\left(\sum_{k=1}^{M} E_{i,j,k}\right)\right] \tag{18.2}$$
$$i = 1, \ldots, N,\ j = 1, \ldots, L,\ k = 1, \ldots, M$$

where

N = number of parking lots (i.e. no. of fleets);

L = number of load profiles;

M = number of EV/load profile.

The time of stay for the kth EV belonging to both jth load profile and ith fleet, $\Delta T_{i,j,k}$, can be computed by

$$\Delta T_{i,j,k} = \left(T_{\text{req}} - T_{\text{arr}}\right)_{i,j,k} \tag{18.3}$$

The power is the energy per unit time, so (18.2) can be adapted for calculating the total EVs power during discharging state and rewritten as

$$P_{\text{total}} = \sum_{i=1}^{N}\left[\sum_{j=1}^{L}\left(\sum_{k=1}^{M} \frac{\lambda_{i,j,k} E_{i,j,k}}{\Delta T_{i,j,k}}\right)\right] \tag{18.4}$$
$$i = 1, \ldots, N,\ j = 1, \ldots, L,\ k = 1, \ldots, M$$

18.4 V2G challenges

Economic and technical challenges might delay V2G implementation in the real world as described in the following.

Economic challenges:

- V2G should be in competition with the other deployed ESS technologies (e.g. pumped hydroelectric, flywheel, concentrating solar power).
- EV battery technology is not preferable to be operated in a frequent switch between charging/discharging modes. In addition, it is expensive and that negatively impacts on V2G from becoming a reality.
- Minimization of the overall charging cost is based on the time-of-use/real-time price.
- It is not clear yet whether the economic incentives justify V2G from the utility's perspective. Several issues (e.g. complexity of distribution system required, battery technology, lack of support for smart grid technologies) may slow the progress in the deployment of V2G.
- For large-scale EV adoption, batteries have a lack of either low energy density or power capacity. In addition, batteries need to be large to allow the EV for distant driving and that makes them heavy, unreliable and expensive.
- The risk of choosing inadequate design, number and location of charging stations is high.

Technical challenges:

- A reliable two-way communication-enabled infrastructure is greatly needed. Its effort is anticipated to be high and delays could heavily affect the performance of the provided services.
- Two-way power flow control is the key enabling technology that will make V2G become true.
- Bi-directional inductive power interface should be proposed to facilitate V2G technology and identify the impact of communication architecture on the V2G ancillary services in terms of availability, reliability and value of vehicle-provided ancillary services.
- Controlling the vehicle charging rate remotely based on grid conditions and user preference as well as reducing additional load at peak times.
- Ensuring high power quality from the sources to allow easier integration of renewable sources.

References

[1] Su W., Eichi H. R. and Zeng W. 'A survey on the electrification of transportation in a smart grid environment'. *IEEE Transactions on Industrial Informatics*. 2012;8(1):1–10.

[2] Galus M. D., Vayá M. G., Krause T. and Andersson G. 'The role of electric vehicles in smart grids'. *WIREs Energy Environment*. 2013;2(4): 384–400.

[3] Pang C., Dutta P. and Kezunovic M. 'BEVs/PHEVs as dispersed energy storage for V2B uses in the smart grid'. *IEEE Transactions on Smart Grid*. 2012;3(1):473–82.

[4] Quirés-Tortés J., Ochoa L. and Butler T. 'How electric vehicles and the grid work together'. *IEEE Power and Energy Magazine*. 2018;16(6):64–76.

[5] Lu C. F., Liu C. C. and Wu C. J. 'Dynamic modelling of battery energy storage system and application to power system stability'. *IEE Proceedings of Generation, Transmission and Distribution*. 1995;142(4):429–35.

[6] Su W. and Chow M. 'Sensitivity analysis on battery modelling to large-scale PHEV/PEV charging algorithms'. *IECON 2011 - 37th Annual Conference of the IEEE Industrial Electronics Society*; Melbourne, VIC, Australia, Nov 2011, pp. 1–6.

[7] Yao L., Damiran Z. and Lim W. H. 'Optimal charging and discharging scheduling for electric vehicles in a parking station with photovoltaic system and energy storage system'. *Journal of Energies*. 2017;10(4):1–20.

[8] Moses P. S., Deilami S., Masoum A. S. and Masoum M. A. S. 'Power quality of smart grids with plug-in electric vehicles considering battery charging profile'. *IEEE OES Innovative Smart Grid Technologies Conference Europe (ISGT Europe)*; Gothenburg, Sweden, Oct 2010, pp. 1–7.

[9] Shabaan M. F., Eajal A. A. and El-Saadany E. F. 'Coordinated charging of plug-in hybrid electric vehicles in smart hybrid AC/DC distribution systems'. *Renewable Energy, Elsevier*. 2015;82:92–9.

[10] Capasso C., Iannuzzi D. and Veneri O. 'DC charging station for electric and plug-in vehicles'. *The 6th International Conference on Applied Energy (ICAE2014); Energy Procedia, Elsevier*. 2014;61:1126–9.

[11] Vasirani M., Kota R., Cavalcante R. L. G., Ossowski S. and Jennings N. R. 'An agent-based approach to virtual power plants of wind power generators and electric vehicles'. *IEEE Transactions on Smart Grid*. 2013;4(3):1314–22.

[12] Pudjianto D., Ramsay C. and Strbac G. 'Virtual power plant and system integration of distributed energy resources'. *Renewable Power Generation*. 2007;1(1):10–16.

[13] Mwasilu F., Justo J. J., Kim E. K., Do T. D. and Jung J. W. 'Electric vehicles and smart grid interaction: A review on vehicle to grid and renewable energy sources integration'. *Renewable and Sustainable Energy Reviews, Elsevier*. 2014;34:501–16.

[14] Kempton W. and Letendre S. E. 'Electric vehicles as a new power source for electric utilities'. *Transportation Research Part D: Transport and Environment, Elsevier*. 1997;2(3):157–75.

[15] Tomic J. and Kempton W. 'Using fleets of electric-drive vehicles for grid support'. *Journal of Power Sources*. 2007;168(2):459–68.

[16] Kempton W. and Tomic J. 'Vehicle-to-grid power fundamentals: Calculating capacity and net revenue'. *Journal of Power Sources.* 2005;144(1):268–79.

[17] Brooks A. N. *Vehicle-to-Grid demonstration project: grid regulation, ancillary service with battery electric vehicle*. California Air Resources Board and California Environmental Protection Agency report, 2002.

[18] Kempton W. and Tomic J. 'Vehicle-to-grid power implementation: From stabilizing the grid to supporting large-scale renewable energy'. *Journal of Power Sources.* 2005;144(1):280–94.

[19] Shakoor S. A., Black M., Pudjianto D. and Boppc T. 'Impact of wind generation on the operation and development of the UK electricity systems'. *Electric Power Systems Research.* 2007;77(9):1214–27.

[20] Guille C. and Gross G. 'A conceptual framework for the vehicle to grid (V2G) implementation'. *Energy Policy.* 2009;37(11):4379–90.

[21] Ríos A. D. L., Goentzel J., Nordstrom K. E. and Siegert C.W. 'Economic analysis of vehicle-to-grid (V2G)-enabled fleets participating in the regulation service market'. *Proceedings of 2012 IEEE PES Innovative Smart Grid Technologies (ISGT)*; Washington, DC, USA, Jan 2012, pp. 16–24.

[22] Geidl M., Koeppel G., Favre-Perrod P., Kloeckl B., Andersson G. and Froehlich K. 'Energy hubs for the future'. *IEEE Power and Energy Magazine.* 2007;5(1):24–30.

[23] Sekyung H., Soohee H. and Sezaki K. 'Development of an optimal vehicle-to-grid aggregator for frequency regulation'. *IEEE Transactions on Smart Grid.* 2010;1(1):65–72.

[24] Shimizu K., Masuta T., Ota Y. and Yokoyama A. 'Load frequency control in power system using vehicle-to-grid system considering the customer convenience of electric vehicles'. *Proceedings of 2010 International Conference on Power System Technology*; Hangzhou, China, Oct 2010, pp. 1–8.

[25] Tian W. Q., He J. H., Niu L. Y., Zhang W. G., Wang X. J. and Bo Z. Q. 'Simulation of vehicle-to-grid (V2G) on power system frequency control'. *Proceedings of IEEE PEST Innovative Smart Grid Technologies Conference*; Tianjin, China, May 2012, pp. 1–4.

[26] Baboli P. T., Moghaddam M. P. and Fallahi F. 'Utilizing electric vehicles on primary frequency control in smart power grids'. *Proceedings of IPCBEE Petroleum and Sustainable Development Conference*; Singapore, May 2011, pp. 6–10.

[27] Ota Y. 'Autonomous distributed vehicle-to-grid for ubiquitous power grid and its effect as a spinning reserve'. *Journal of International Council on Electric Engineering.* 2011;1(2):214–21.

[28] Almeida P. M. R., Lopes J. A. P., Soares F. J. and Vasconcelos M. H. 'Automatic generation control operation with electric vehicles'. *Proceedings of IREP Symposium*; Rio de Janeiro, Brazil, Apr 2010, pp. 1–7.

[29] Ota Y., Taniguchi H., Nakajima T., Liyanage K. M., Baba J. and Yokoyama A. 'Autonomous distributed V2G (vehicle-to-grid) satisfying scheduled charging'. *IEEE Transactions on Smart Grid.* 2012;3(1):559–64.

[30] Kempton W., Tomic J., Letendre S., Brooks A. and Lipman T. *Vehicle to grid power: Battery, hybrid, and fuel cell vehicles as resources for distributed electric power in California*. California Air Resources Board and California Environmental Protection Agency report, 2001.

[31] Mullen S. K. *Plug-in hybrid electric vehicles as a source of distributed frequency regulation*. PhD dissertation, US, University of Minnesota, 2009.

[32] Pillai J. R. and Jensen B. B. 'Integration of vehicle-to-grid in the Western Danish power system'. *IEEE Transactions on Smart Grid*. 2011;2(1):12–19.

[33] Donadee J. and Llic M. 'Stochastic co-optimization of charging and frequency regulation by electric vehicles'. *Proceedings of North American Power Symposium (NAPS)*; Champaign, IL, USA, Sep 2012. IEEE; 2012, pp. 1–6.

[34] Xie F. X., Huang M., Zhang W. G. and Li J. 'Research on electric vehicle charging station load forecasting'. *Proceedings of International Conference on Advanced Power System Automation and Protection*; Beijing, China, Oct 2011. IEEE; 2012, pp. 2055–60.

[35] Vayá M. G. and Anderson G. 'Combined smart-charging and frequency regulation for fleets of plug-in electric vehicles'. *IEEE PE, General Meeting*; Vancouver, BC, Canada, July 2013. IEEE; 2013, pp. 1–5.

[36] Mu Y., Wu J., Ekanayake J., Jenkins N. and Jia H. 'Primary frequency response from electric vehicles in the Great Britain power system'. *IEEE Transactions on Smart Grid*. 2013;4(2):1142–50.

[37] Wu C., Mohsenian H., Huang J. and Jatskevich J. 'PEV-based combined frequency and voltage regulation for smart grid'. *IEEE-PES Innovative Smart Grid Technologies (ISGT)*; Washington, DC, USA, Jan 2012. IEEE; 2012, pp. 1–6.

Chapter 19

Microgrids: modelling and control

19.1 Introduction

The global warming and other environmental hazards of conventional sources of electrical energy push the energy sector continuously towards distributed energy resources (DERs), especially renewable sources [1]. The fluctuating power produced by renewable sources such as wind and solar photovoltaic (PV) systems may bring challenges in generation/load power balance. With increasing penetration of renewable energy, the function of such distributed generation (DG) is changing from an auxiliary role to a primary role in the energy sector. DERs, renewable or non-renewable, and energy storage systems (ESSs) are integrated into what is known as a microgrid (MG). An MG deals mainly with the issues associated with the integration of such intermittent renewable energy sources (RESs) and ESSs. It gives many advantages not only for power balance but also for environmental challenges, economic benefits and grid reliability requirements [2]. Moreover, MG handles many technical issues in decentralized form [3,4]. There are intensive interests to develop and study the future horizon of MG concept throughout the world [4]. Generally, MG is a small-size, discrete electricity framework consisting of a collection of DG units and loads. It may operate in grid-connected or islanded mode and provide seamless transition between the two modes. MG may also include conventional generators such as diesel-based synchronous generators to mitigate the effects of the intermittency nature of RESs. Furthermore, many RESs act as inverter-based sources as inverters are used to interface them with the host grid or AC loads.

Many issues pertain to the operation of an MG and the integration of DERs in an MG. Critical among these are the following: (i) the stable operation of MG at different modes of operation and loading conditions during normal and after fault provoked islanding incidents; (ii) the impact of different inverter control strategies and schemes on the dynamic performance of MG; and (iii) the power quality (PQ) that should be in compliance with the standards as the interfacing power conditioning systems (PCSs) and non-linear loads may cause negative impact on PQ.

Integration of DERs, constituting an MG, entails a control strategy that can be categorized into three levels: (i) primary control, which is a local control for each source and each load in the MG – it is based on the local measurements of the parameters necessary for managing and controlling the MG; (ii) secondary control that is located at the point of common coupling (PCC) and is responsible for MG operation according to the current mode of operation, grid connected or isolated; and (iii) tertiary control, which is the highest level of control. It coordinates multiple groups of integrated DERs (i.e. multiple MGs) interacting with one another in the system and communicates requirements from the utility grid.

Primary and secondary controls are associated with the MG operation itself, whereas the tertiary control is considered as part of the utility or host grid. Regardless of the kind of operation mode, MG components should be interoperable and with interfaces complying with functional standards defined by the MGs as well as the secondary control should enable the MGs to manage themselves and operate autonomously. One of these standards that must be met is the MG PQ especially when implying integrated RESs. Harmonics distortion and voltage stabilization are listed as major PQ issues in the literature because of load non-linearities and renewable sources' uncertainties. Nowadays, industrial and commercial loads have numerous electronic drives, information and communication devices, and computers as well as PCSs are used for RESs interface. All these devices have a catalyst nature to produce harmonics and cause PQ degradation. Passive, active and hybrid power filters can be used for mitigating harmonic distortion. Meanwhile, multifunctional voltage source inverters or flexible alternating current transmission system (FACTS) devices can be utilized to process these problems that may yield harmonic distortion and/or voltage instability in the MG. In the next sections, the methodology of using PQ enhancement device (PQED) to mitigate the harmonic distortion and keep the voltage at its nominal value at the PCC is described. It can be summarized by the following steps:

- Specifying the MG architecture as an integration of DERs.
- Identifying the mathematical model for each of MG components.
- Constructing the simulation model of the MG.
- Selecting the type of PQED to be embedded at the PCC.
- Examining the PQ aspects of the MG with/without PQED.

19.2 MG architecture

Three DERs, solar PV, wind energy system (WES) and proton-exchange membrane fuel cell (PEMFC), in addition to a battery energy storage system (BESS) with data given in Tables 19.1–19.4 are integrated into an MG as shown in Figure 19.1. Any additional data are provided where needed. As a primary control level, a local control (LC) is connected to each source and each load, which in turn is connected to the MG central controller (MGCC) as a secondary control level. The LC consists of a breaker for switching operation and maximum power point tracking (MPPT) circuit. It is used for achieving reliable and economical operation as well as interacting with the distribution system operator (DSO) through the MGCC to access the data of PQ requirements (e.g. electricity, fuel costs and emission mitigation).

Table 19.1 Parameters of PV module

Open-circuit voltage, V_{oc} (V)	32.9
Short-circuit current, I_{sc} (A)	8.21
Module voltage at the MPP, V_m (V)	26.3
Module current at MPP, I_m (A)	7.61
Maximum power, P_m, of a PV module (W)	200
Reference temperature (°C)	25
Reference solar radiation (W/m^2)	1,000
Number of series PV modules in a string	8
Number of parallel connected PV strings	63

Table 19.2 Wind generation data

Rated power (kW)	V_{ci} (m/s)	V_{co} (m/s)	V_r (m/s)	Hub height (m)	Swept area (m^2)
60	2.5	16	12	12	10.235

Table 19.3 PEMFC parameters

Type	Stake nominal power (kW)	Nernst voltage (E_n) (V)	FC resistance (Ω)
PEM	50	1.1342	0.66404

Table 19.4 BESS parameters

Capacity (Ah)	Efficiency (%)	Minimum charge (%)	Maximum charge (%)	Maximum charging rate (kW/h)	Maximum discharging rate (kW/h)
2,160	85	20	80	−40	20

The MGCC manages the operation of DERs and leads integration with the goal of the following:

- Providing the individual power and voltage set point for each power flow according to DER control strategy.
- Satisfying the electrical loads' requirements.
- Alleviating emissions and network losses.
- Exploiting the operational efficiency of the DERs.

DERs are connected to commercial and residential loads at 380 V, 50 Hz. A PQED such as filter, compensator or multifunctional device (e.g. filter-compensator) can be embedded into the MG at the PCC.

19.3 Modelling and simulation of MG components

The components of the MG shown in Figure 19.1 should be mathematically modelled and then simulated as in the following to study how to improve the PQ of an MG with integrated RESs.

19.3.1 PV system

The equivalent circuit and mathematical model of the PV module and array are explained in Chapter 3, Sections 3.2.5 and 3.2.6. The *I–V* and *P–V* characteristics are also described in Chapter 3, Section 3.2.7. It is to be noted that PVs have non-linear characteristics, where the performance and output power are directly affected by the change of operating conditions. The output power of PV is directly proportional to the amount of solar irradiance falling on it, and inversely proportional to its temperature. As the temperature and solar irradiance change, the point at which the power is maximum will accordingly change. This means that a DC–DC converter (boost converter) must be used to change the array terminal voltage with a goal of MPPT.

Real values of PV module parameters are given for Kyocera KC 200 GT, manufactured by Kyocera [5]. *I–V* and *P–V* characteristics of the PV module at different irradiance levels and constant temperature (25 °C) are shown in Figure 19.2.

The output power of PV arrays is mainly influenced by the irradiance (amount of solar radiation) and temperature. Moreover, for a certain irradiance and

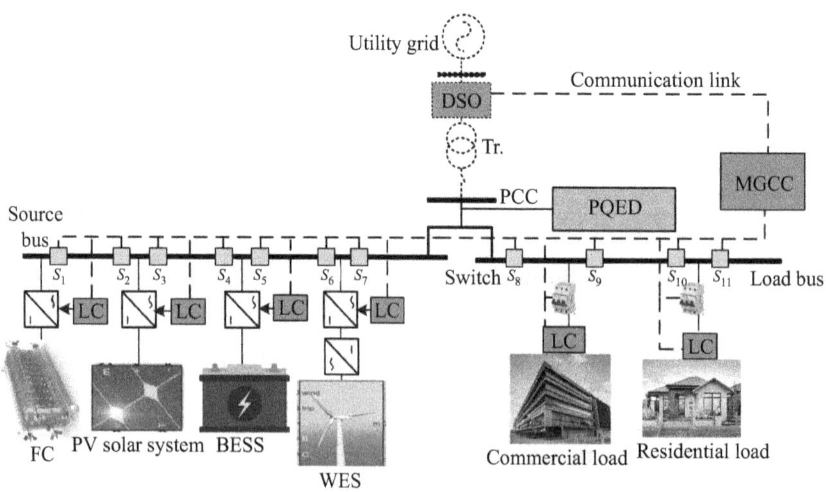

Figure 19.1 Layout of MG structure

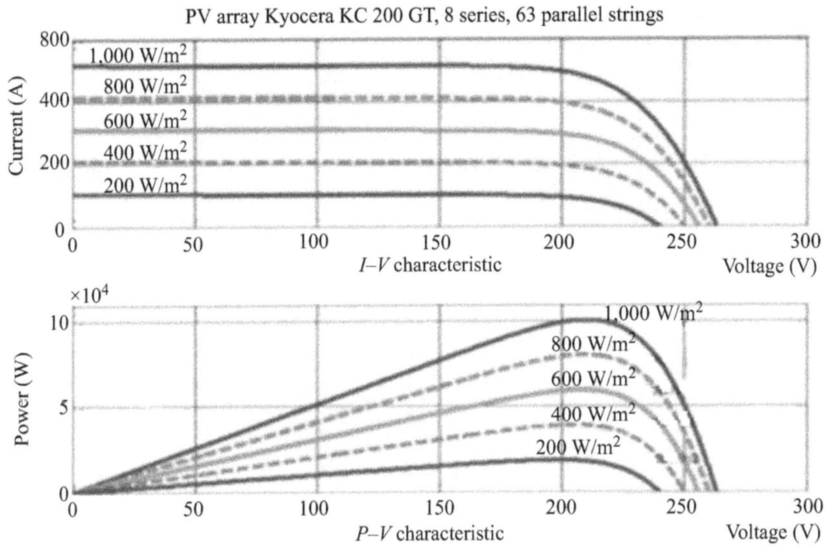

Figure 19.2 PV characteristics

temperature, the output power of the PV array is a function of its terminal voltage and there is only one value for the PV's terminal voltage at which the PV panel is utilized efficiently. The procedure of searching for this voltage is called 'MPPT'. Recently, several algorithms have been developed to achieve the maximum power point (MPP) such as perturb and observe (P&O), incremental conductance, fuzzy or neural-based techniques [6,7]. However, the insulation levels and the cell temperature determine only the limits of the best obtainable matching. The array voltage determines the real matching, which can be improved by using the MPPT controller to locate the local MPP in the *P–V* characteristic of the solar panel [8].

MPPT: The PV panel has a non-linear characteristic and its output power depends mainly on the irradiance (amount of solar radiation) and the temperature. Moreover, at the same temperature and irradiance the output power of a PV panel is a function of its terminal voltage. The procedure of seeking the voltage where the power is maximum can be obtained either in a single stage or in a double stage. In a single stage, a DC–AC converter is employed, whereas in a double stage DC–DC and DC–AC converters are utilized. PV characteristics, Figure 19.2, show that the MPP for the panel lies at values approximately between 75% and 80% of array's open-circuit voltage. Consequently, the MPPT algorithm scans the *P–V* curve at predefined voltage of 75% of the array's open-circuit voltage. The MPPT technique is associated with the DC–DC converter, which interfaces the PV array to the MG through an inverter. This can be achieved by controlling the input voltage of the DC–DC converter (Figure 19.3).

MPPT system tracks the new MPP in the corresponding curve whenever temperature and/or insolation variation occurs. MPPT is used for extracting the

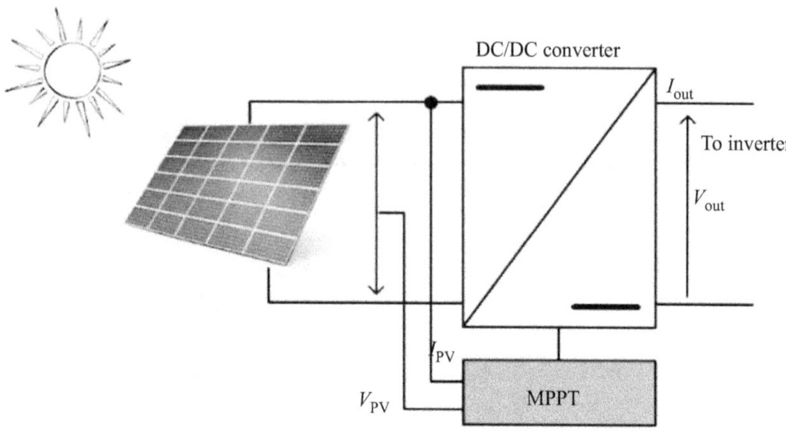

Figure 19.3 Block diagram of MPPT system

maximum power from the solar PV array to be transferred to the MG. A DC–DC (step-up/step-down) converter acts as an interface between the inverter and the array. The MPPT changes the duty cycle to keep the power transfer from the solar PV array to the MG at the maximum value [8,9].

P&O algorithm: It is widely used because of its ease of implementation and its low cost. It is based on the following criterion: in Figure 19.4(a), if the operating voltage is perturbed in a given direction and $dp/dV > 0$, it indicates that the perturbation moves the operating point towards the MPP and the P&O algorithm would then continue to perturb the operating voltage in the same direction [10–12].

Otherwise, if $dp/dV < 0$, the change in operating point moves that point away from the MPP and the P&O algorithm reverses the direction of the perturbation. In other words, the system works by increasing or decreasing the operating voltage and observing its impact on the output power.

P&O has been implemented to extract maximum power at each instant. Its control action is summarized by the steps shown in Figure 19.4(b). The operating voltage is perturbed with every MPPT cycle. As soon as the MPP is reached, it will oscillate around the desired operating voltage and the perturbation is narrowed down until the oscillations are damped.

Boost converter: Based on the description of boost converters in Chapter 13, Section 13.3.2, the simulation model of the boost converter control and the local control action of the PV system are illustrated in Figures 19.5 and 19.6, respectively. Accordingly, the MATLAB®/Simulink® model of the PV system including the elements, PV array, boost converter, MPPT system and the local control are depicted in Figure 19.7.

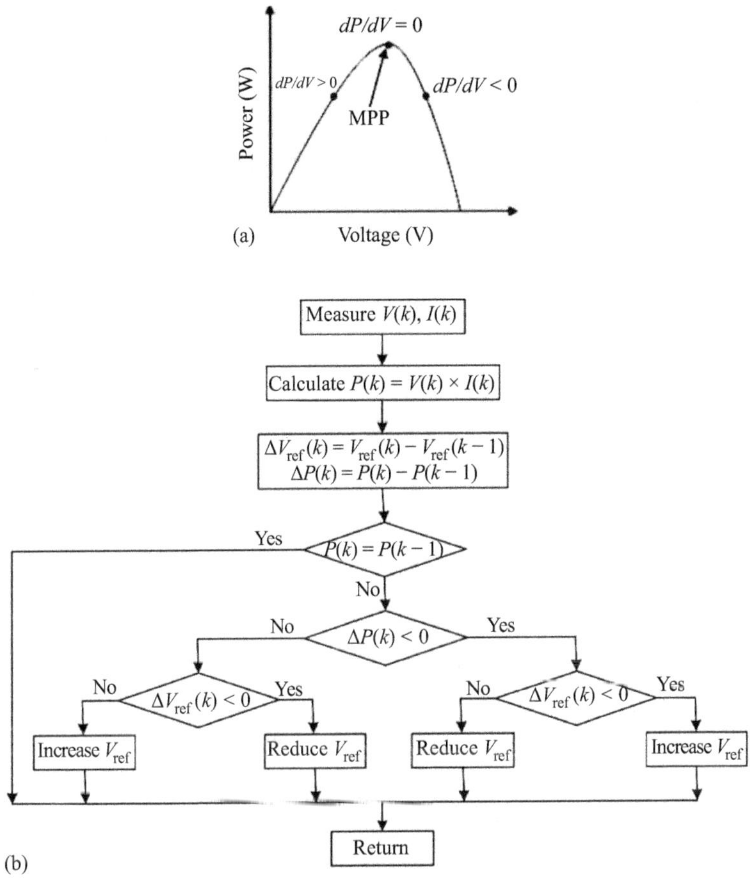

Figure 19.4 P&O application for MPPT: (a) searching mechanism and (b) flow chart of P&O algorithm

19.3.2 Wind energy system

Wind turbine characteristics are discussed in Chapter 2. In particular, the generated wind power is explained in Sections 2.2 and 2.3. In addition, the wind turbine generator, in this study, is a permanent magnet synchronous generator (PMSG), which is also explained in Section 2.4.2.

However, the following relations are rewritten for convenience.

The output power of the wind turbine is given by

$$P_w = \begin{cases} 0 & \text{for} \quad V_w < V_{ci} \\ aV_w^3 - bP_r & \text{for} \quad V_{ci} < V_w < V_r \\ P_r & \text{for} \quad V_r < V_w < V_{co} \\ 0 & \text{for} \quad V_w > V_{co} \end{cases} \tag{19.1}$$

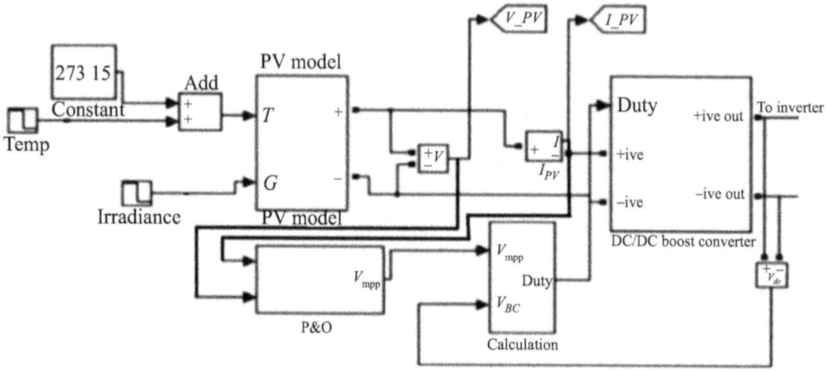

Figure 19.5 Boost converter control model

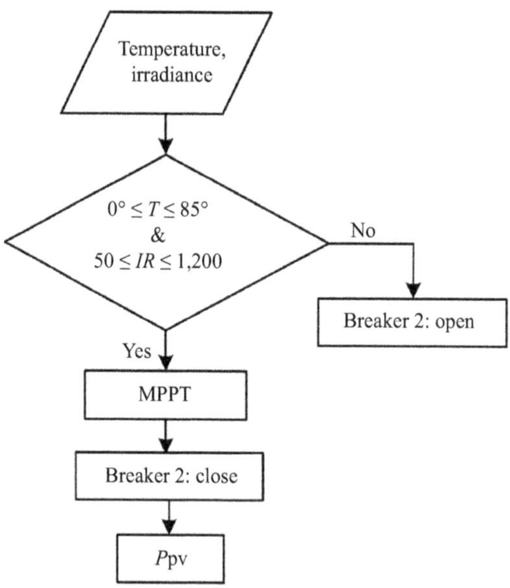

Figure 19.6 Local controller of PV system

where $a = \frac{P_r}{\left(V_r^3 - V_{\text{ci}}^3\right)}$, $b = \frac{V_{\text{ci}}^3}{\left(V_r^3 - V_{\text{ci}}^3\right)}$, P_r is the rated power of the wind generator, and V_{ci}, V_{co} and V_r are the cut-in, cut-out and rated speeds of the wind turbine, respectively. The extracted real power from wind turbine is expressed as

$$P_w = \frac{1}{2}\rho A C_p(\lambda, \beta)\left[\frac{R w_{\text{opt}}}{\lambda_{\text{opt}}}\right]^3 \tag{19.2}$$

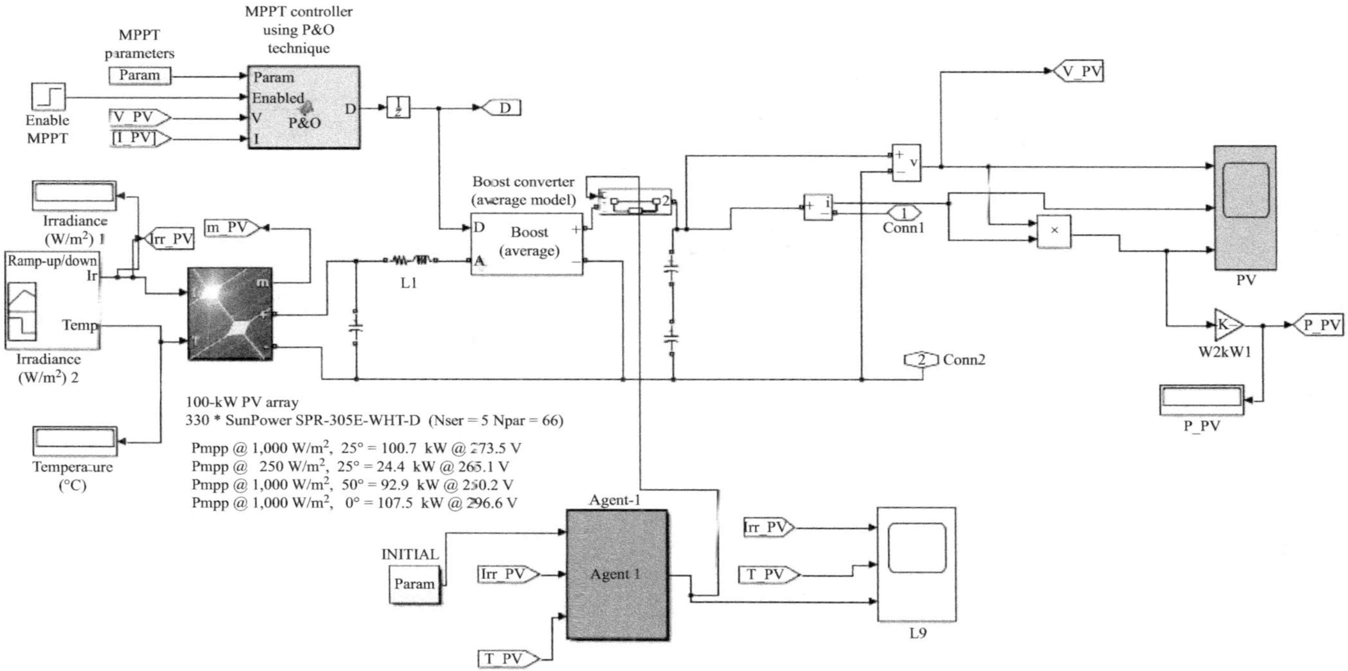

Figure 19.7 Model of PV system in MATLAB/Simulink

where ρ is density of the air ($\rho = 1.2250$ kg/m^3), A is the blade swept area, R is the length of the blade, w_{opt} is the optimum speed of wind turbine and λ_{opt} is the optimum tip-speed ratio. The power coefficient C_p is a function of the pitch angle β and the tip speed ratio λ, which is given by

$$\lambda = \frac{\omega R}{V_w} \tag{19.3}$$

where ω is the rotational speed of the wind turbine and V_w is the wind speed. P&O algorithm is used for MPPT, and real WES module data (E3120) manufactured by Endurance are given in Table 19.2 [13].

The action of the local controller and the corresponding simulation model of the WES including PMSG and P&O technique for MPPT are depicted in Figures 19.8 and 19.9, respectively.

19.3.3 Proton-exchange membrane fuel cell

The equivalent circuit and mathematical modelling of PEMFC are described in Chapter 9, Sections 9.6 and 9.7, respectively. P&O algorithm is applied based on measuring the current and voltage variables, I_{FC} and V_{FC}, at each instant to calculate the corresponding power, $P_{FC}(t)$, that being compared to the power at the preceding instant, $P_{FC}(t-1) = I_{FC}(t-1) \times V_{FC}(t-1)$. The algorithm continues to perturb the system in the same direction of reference voltage increase when the

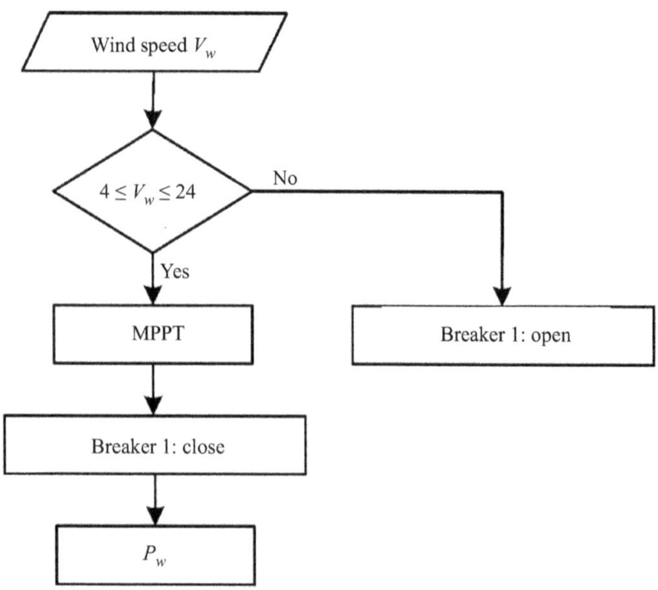

Figure 19.8 Local controller of WES

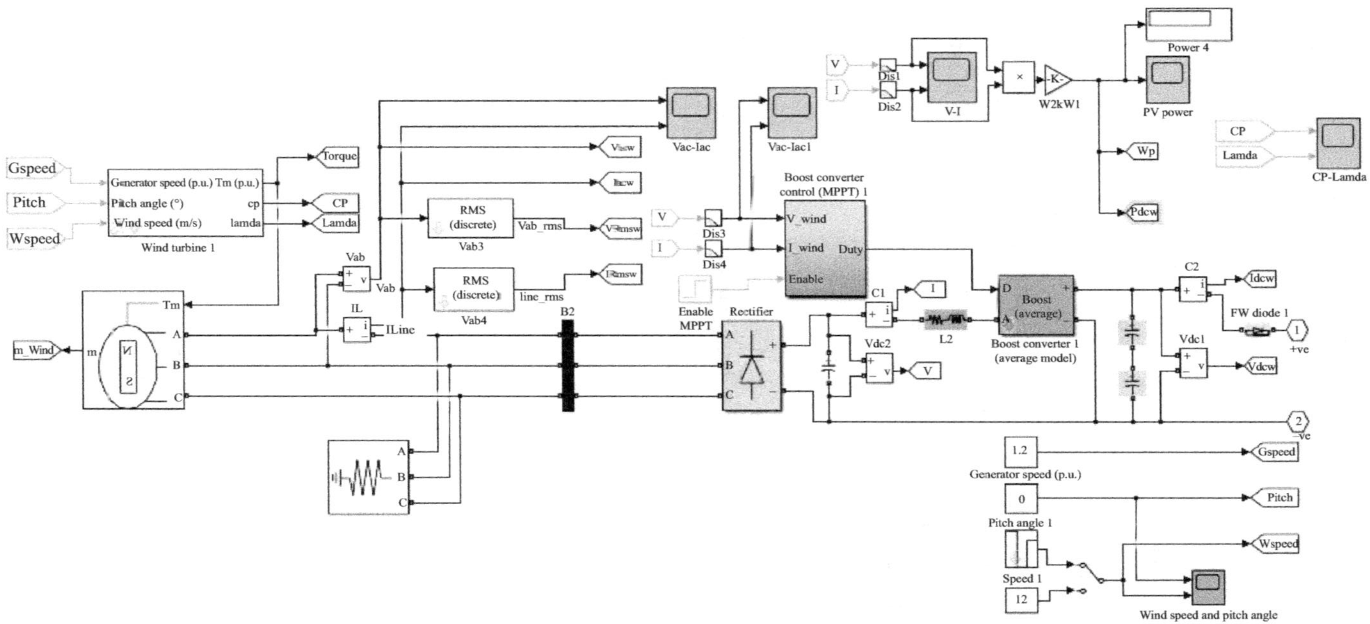

Figure 19.9 MATLAB/Simulink model of PMSG with MPPT using P&O technique

power and voltage variations are positive. Otherwise (when the variations are negative) the direction of perturbation is reversed. The flow chart of P&O algorithm is shown in Figure 19.10 and the corresponding simulation model is illustrated in Figure 19.11.

19.3.4 Battery energy storage system

The battery bank stores the surplus electrical energy in charging state and delivers energy in discharging state when the consumption is higher than the energy produced by the DERs. The most frequent type used is the lead acid battery, which can be modelled as described in Chapter 10, Section 10.3.

The goal of the battery local controller is to control the charging and discharging of the battery according to the available renewable power, state of charge (*SoC*) and load demand. Real parameters of BESS model (S2-2160GEL) manufactured by Rolls battery are given in Table 19.4 [14]. The local control action follows the steps indicated by the flow chart in Figure 19.12. Thus, the simulation model of the BESS with LC can be drawn as depicted in Figure 19.13 as well as the charger controller construction is shown in Figure 19.14.

19.3.5 Load profiles

Real load data of Hurghada city, Egypt, are given. It includes the load curves for active and reactive power of (i) commercial load, Figure 19.15(a) and (b), and (ii) residential load, Figure 19.15(c) and (d).

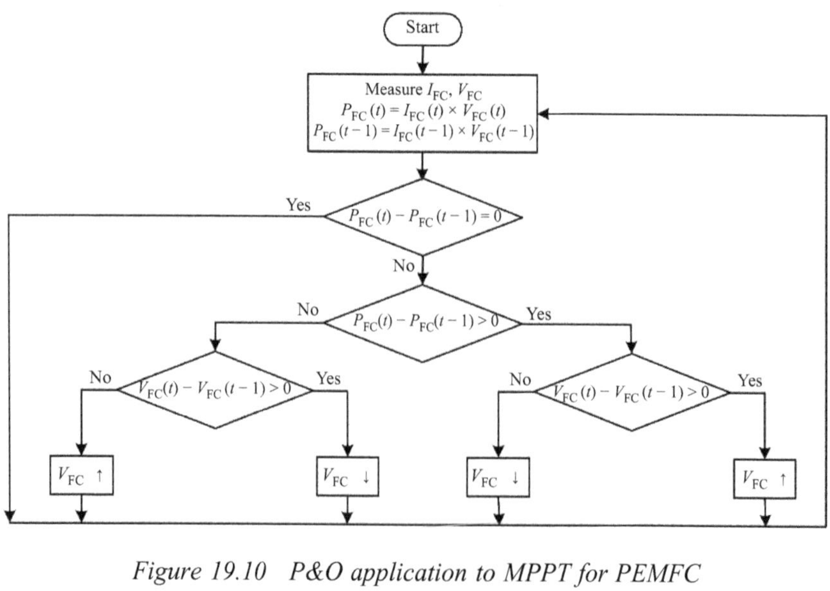

Figure 19.10 P&O application to MPPT for PEMFC

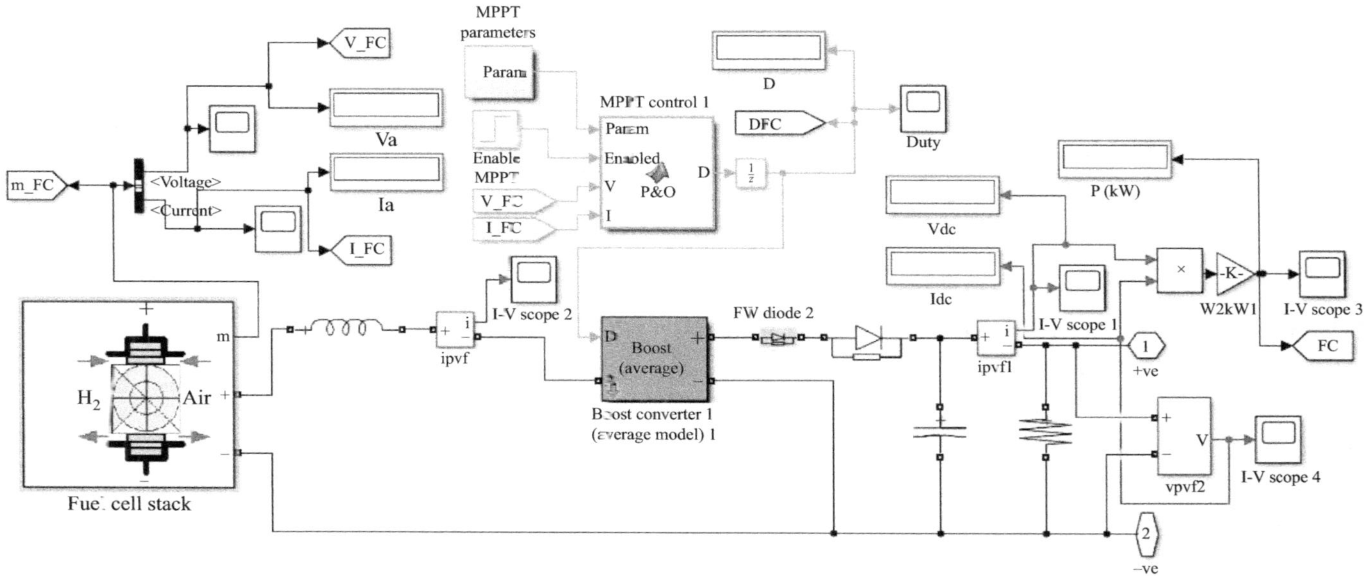

Figure 19.11 PEMFC system simulated in MATLAB/Simulink

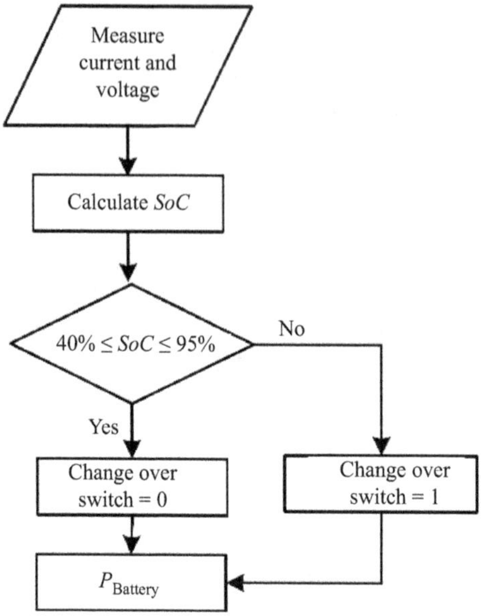

Figure 19.12 Local controller of BESS

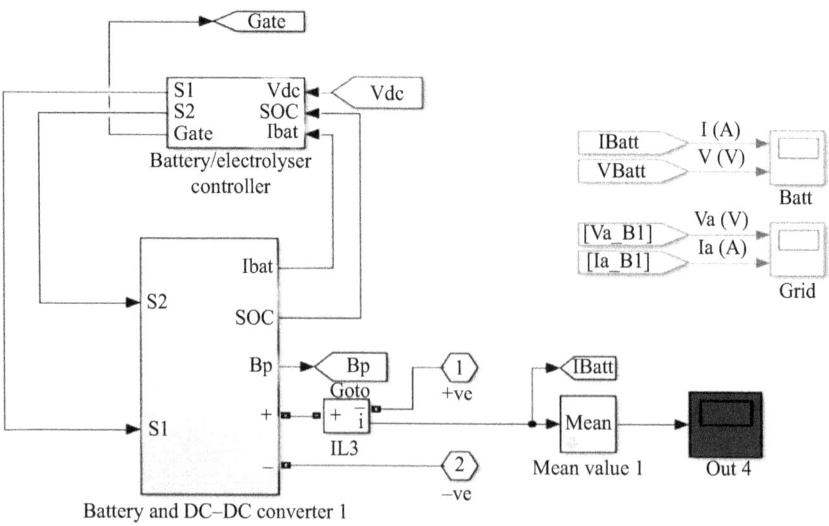

Figure 19.13 MATLAB/Simulink model of BESS including local controller

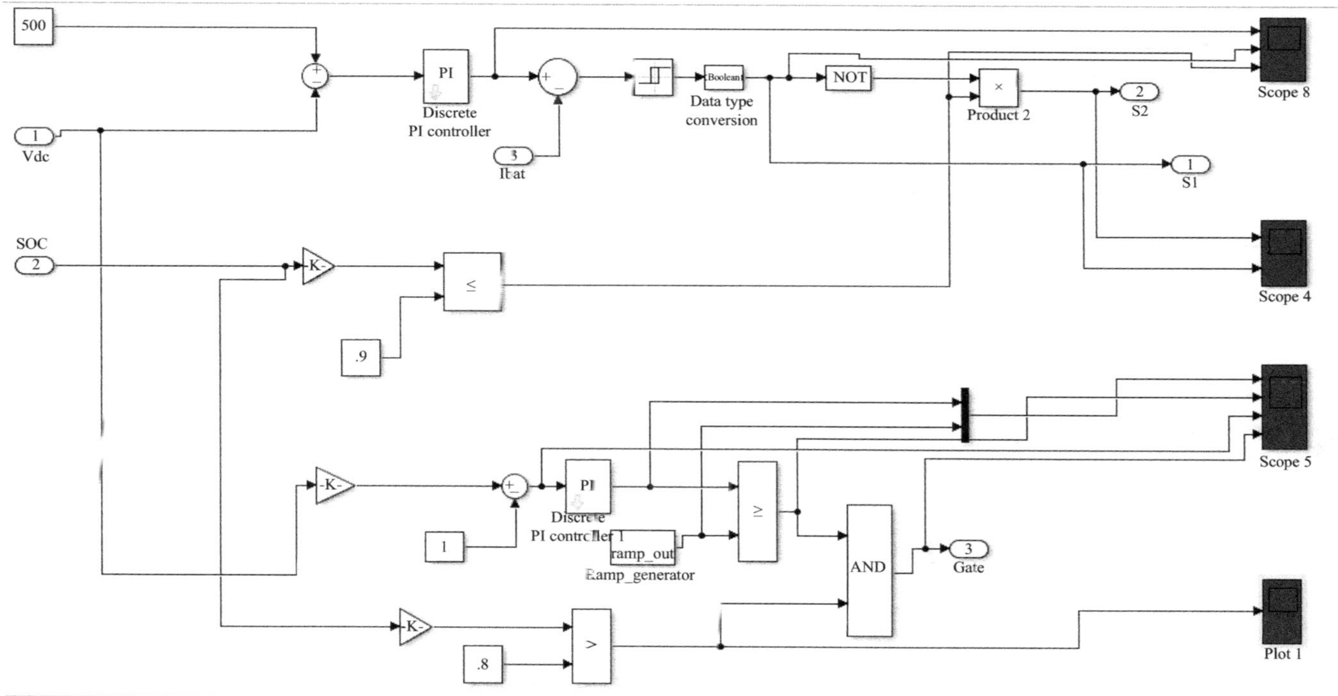

Figure 19.14 Charger controller construction of the BESS in MATLAB/Simulink

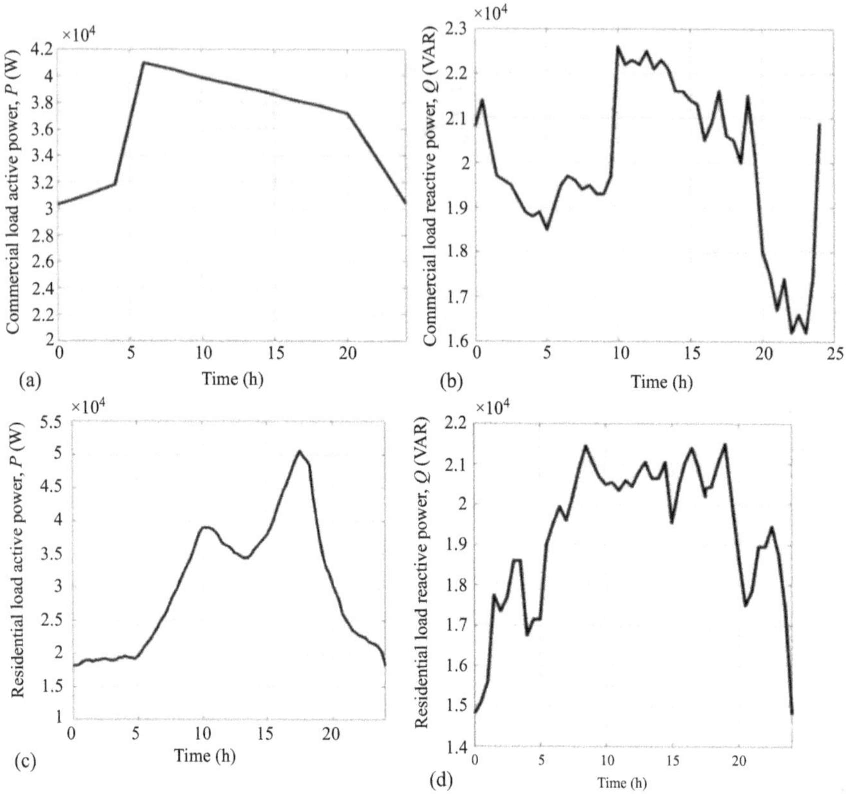

Figure 19.15 Load profile curves of active and reactive power for (a and b) commercial load and (c and d) residential load

Combining the models of individual components of the MG, the integrated model of the MG in MATLAB/Simulink is shown in Figure 19.16. However, for better and clearer illustration, the model has been divided into two parts as shown in Figure 19.17(a) and (b).

Applying the simulation model of the MG with the load profiles shown in Figure 19.15, the frequency spectrum and the voltage variation at the PCC are shown in Figure 19.18(a) and (b). It is to be noted that the total harmonic distortion of the voltage, THD_V, without a PQED is 19.07%, whereas the voltage is approximately constant at 1 p.u.

19.4 Application of D-STATCOM

D-STATCOM is applied as a PQED. It is one of the most popular fast-response D-FACTS topologies that is used to enhance poor PQ issues throughout various

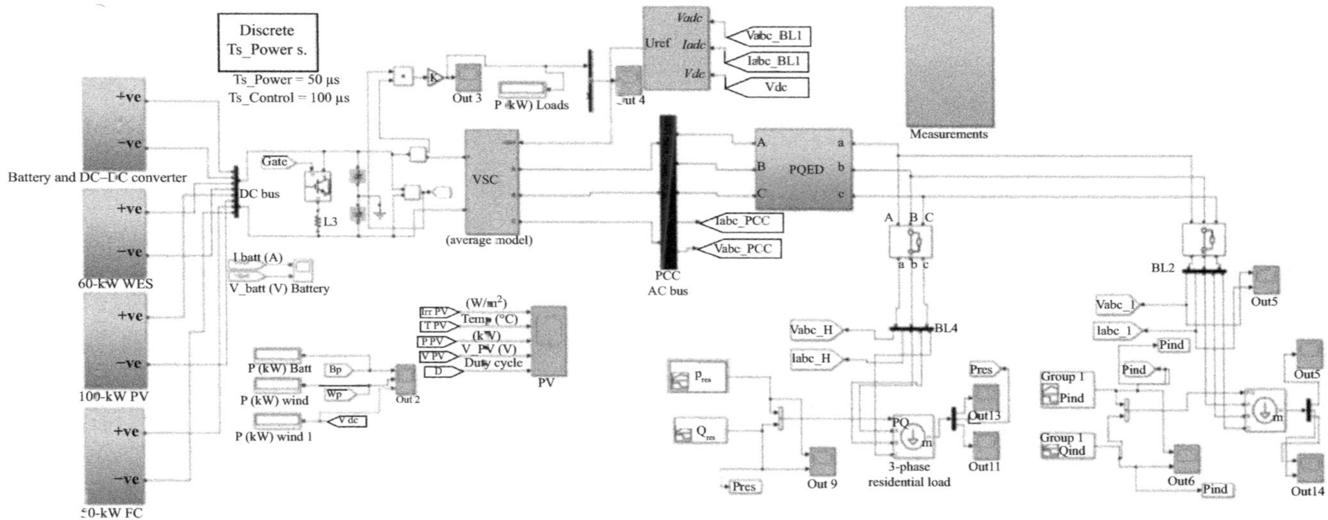

Figure 19.16 Integrated model of MG in MATLAB/Simulink

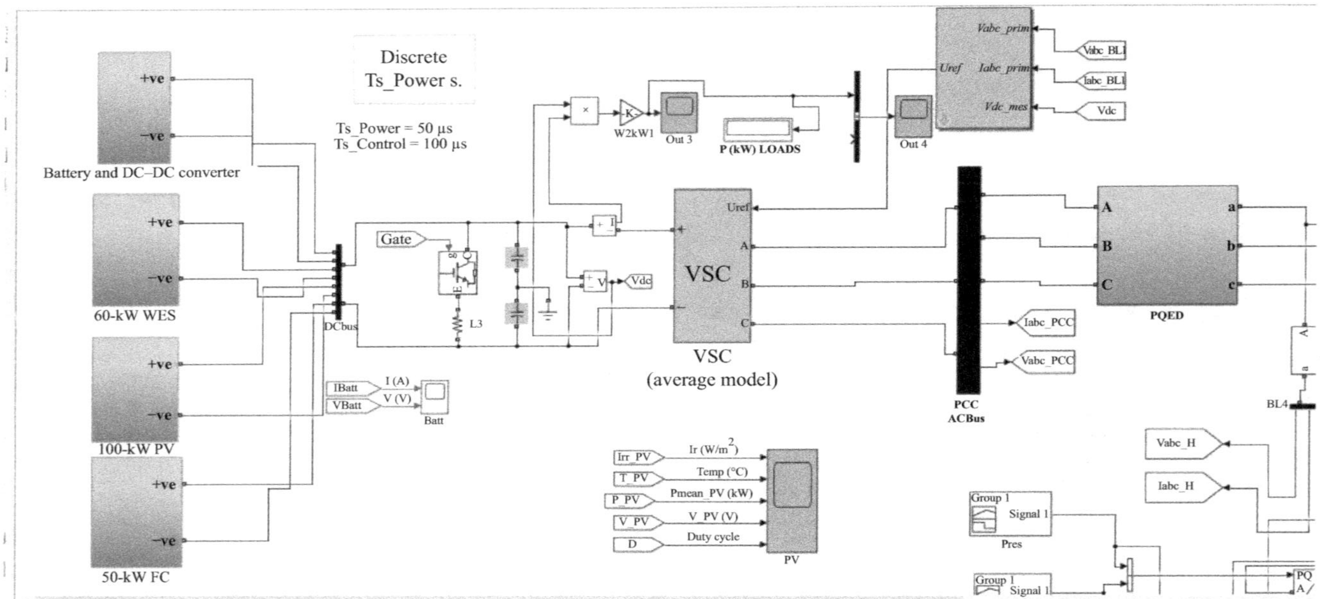

(a)

Figure 19.17 MG simulation model split into two parts: (a) Part I and (b) Part II

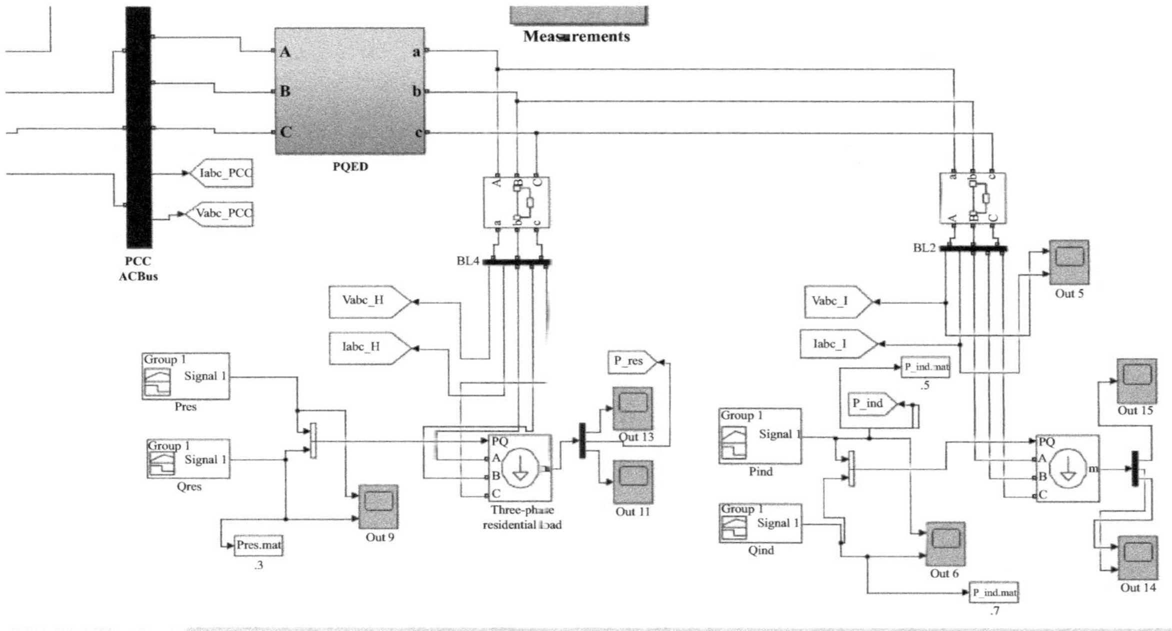

(b)

Figure 19.17 (Continued)

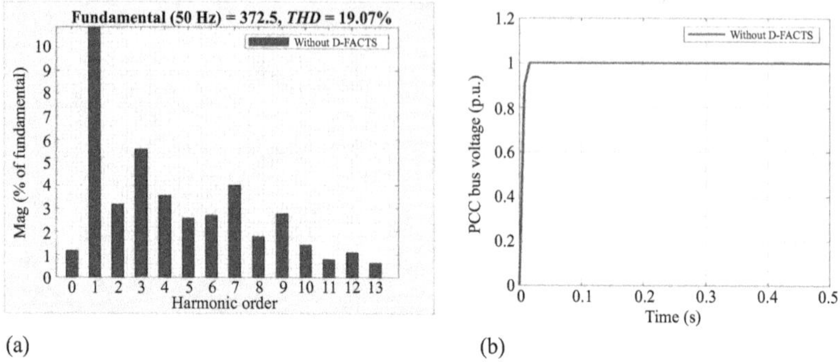

Figure 19.18 Dynamic response: (a) frequency spectrum and (b) voltage variation at PCC without PQED

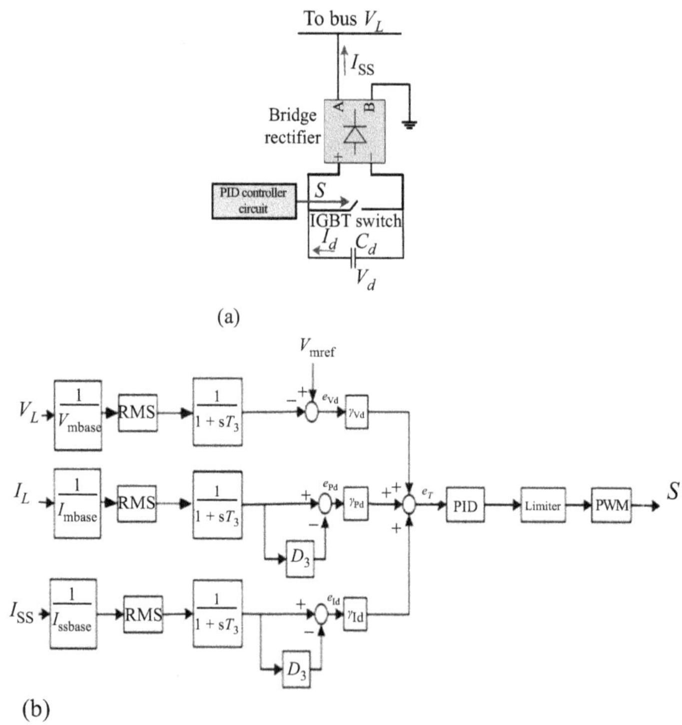

Figure 19.19 Controller scheme of D-STATCOM: (a) equivalent circuit of D-STATCOM and (b) three-level error-driven PID of D-STATCOM

system conditions [15]. The equivalent circuit of D-STATCOM and the block diagram with its three-level error-driven proportional integral derivative (PID) controller are shown in Figure 19.19(a) and (b). The parameters of the controller are given in Table 19.5. PID controller gains are determined by trial and error.

Table 19.5 Parameters of D-STATCOM controller

System part	Specifications
V_{mbase}	380 V
I_{mbase}	150 A
I_{ssbase}	52.5 A
$\gamma_{\text{Vd}} = \gamma_{\text{Id}}$	1
γ_{Pd}	0.5
$T_3 = D_3$	15 ms
$T_4 = D_4$	40 ms
PWM frequency	3,000 Hz

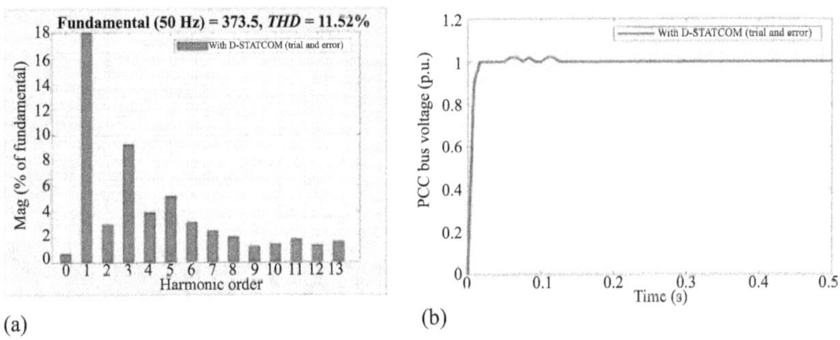

Figure 19.20 System response: (a) frequency spectrum and (b) voltage variation at PCC with D-STATCOM

Their values are $K_p = 72.244$, $K_i = 9.325$ and $K_d = 2.452$. The total error of D-STATCOM is given by

$$e_t = \gamma_{\text{Vd}}(e_{\text{Vd}}) + \gamma_{\text{Id}}(e_{\text{Id}}) + \gamma_{\text{Pd}}(e_{\text{Pd}}) \tag{19.4}$$

and the results are represented by both of frequency spectrum and voltage variation as shown in Figure 19.20(a) and (b).

19.5 Application of SFC

The switched filter compensator (SFC) as reported in [16] has been introduced at the load bus to enhance the performance of the MG-connected wind system. The SFC device is a switched/modulated filter formed by a fixed shunt capacitor bank connected to the AC side of the arm of an uncontrolled rectifier.

The configuration of the SFC is shown in Figure 19.21(a). One mode of operation is defined for the FACTS device with a controlled switch S installed on the rectifier side. The switch operation indicates the ON–OFF state of the shunt capacitor. The ON state of S means that the SFC can compensate the reactive power and mitigate harmonics because of the series-connected inductance L_f with the

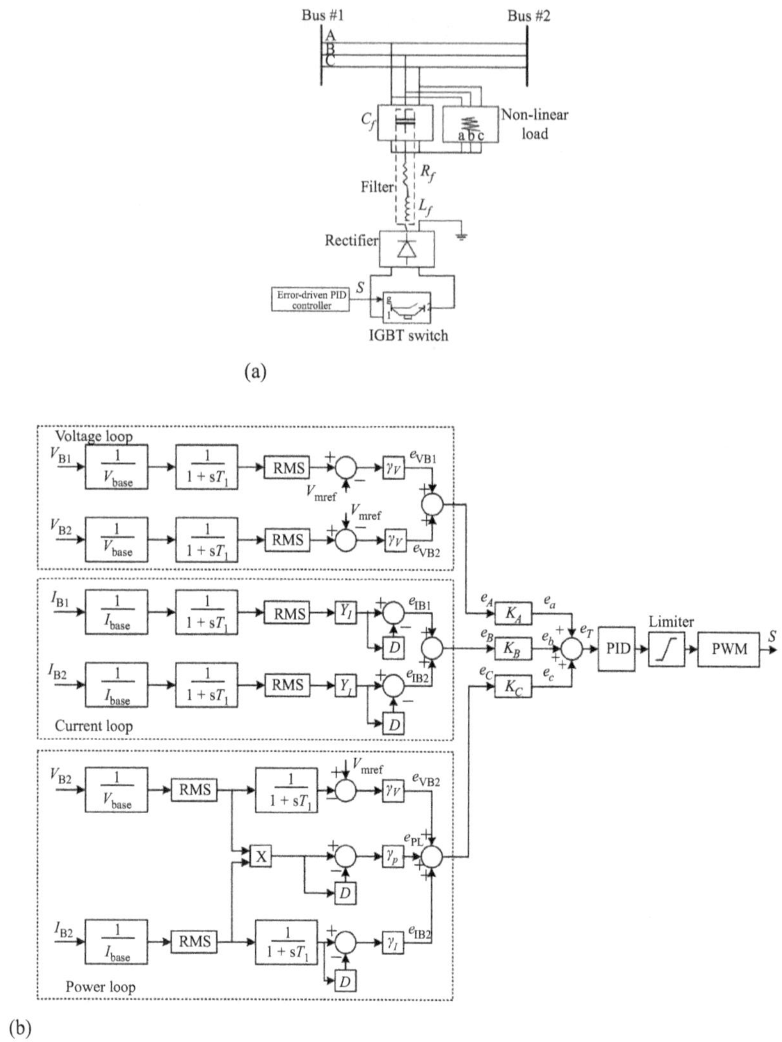

Figure 19.21 SFC/PID controller circuit: (a) SFC configuration and (b) PID controller

capacitor C_f. Filter parameters are selected based on a trade-off between design goals, and trial and error.

The pulse-width modulation (PWM) switching of the SFC scheme is based on a dynamic multi-loop error-driven controller that is designed to enhance voltage stability and PQ, improve power factor (PF) and diminish the effects of voltage transients and inrush currents.

The primary goals are improving the energy use, stabilizing voltage in a dynamic manner, enhancing PQ and improving the PF at the common interface AC buses of the MG-connected DERs. The SFC-switched filter with the modulated capacitive compensator uses PWM switching strategy triggered by a multi-regulation loop error-driven controller (Figure 19.21(b)). The controller is comprised of three loops: (i) voltage stabilization loop to compare the voltage of both sides of the SFC with the reference value giving the signal error, e_A, (ii) current limiting loop to compare the currents of both sides of the SFC with a reference value giving the signal error, e_B, and (iii) dynamic power loop to compare the measured voltage, current and calculated power at Bus #2 with their reference values giving the error e_C.

Based on Figure 19.21(b), the total error, e_T, which is the input to the PID controller can be calculated as in the following.

The error, e_A, produced by voltage stabilization loop is given by

$$e_A = \gamma_{\mathrm{VL}}(e_{\mathrm{VL1}}) + \gamma_{\mathrm{VL}}(e_{\mathrm{VL2}}) \tag{19.5}$$

where

$$e_{\mathrm{VL1}} = V_{\mathrm{mref}} - \frac{V_{L1}\left(\frac{1}{1+sT_1}\right)}{V_{\mathrm{mbase}}} \tag{19.6}$$

$$e_{\mathrm{VL2}} = V_{\mathrm{mref}} - \frac{V_{L2}\left(\frac{1}{1+sT_1}\right)}{V_{\mathrm{mbase}}} \tag{19.7}$$

Similarly, the error, e_B, produced by current limiting loop can be calculated by

$$e_B = \gamma_{\mathrm{IL1}}(e_{\mathrm{IL1}}) + \gamma_{\mathrm{IL2}}(e_{\mathrm{IL2}}) \tag{19.8}$$

$$e_{\mathrm{IL1}} = \frac{I_{L1}}{I_{\mathrm{mbase}}}\left(\frac{1}{1+sT_1}\right)\left(1 - \frac{1}{1+sT_2}\right) \tag{19.9}$$

$$e_{\mathrm{IL2}} = \frac{I_{L2}}{I_{\mathrm{mbase}}}\left(\frac{1}{1+sT_1}\right)\left(1 - \frac{1}{1+sT_2}\right) \tag{19.10}$$

Finally, the error produced by dynamic power loop can be obtained as

$$e_C = \gamma_{\mathrm{VL2}}(e_{\mathrm{VL2}}) + \gamma_{\mathrm{IL2}}(e_{\mathrm{IL2}}) + \gamma_{\mathrm{PL2}}(e_{\mathrm{PL2}}) \tag{19.11}$$

where e_{VL2} and e_{IL2} are determined by (19.7) and (19.10), respectively, and e_{PL2} is given by

$$e_{\mathrm{PL2}} = \left(\frac{V_{L2}}{V_{\mathrm{mbase}}} \times \frac{I_{L2}}{I_{\mathrm{mbase}}}\right)\left(1 - \frac{1}{1+sT_2}\right) \tag{19.12}$$

Thus, the total error e_T can be calculated as summation of (19.5), (19.8) and (19.11), that is,

$$e_T = e_a + e_b + e_c \tag{19.13}$$

where

$$e_a = e_A \times K_A \tag{19.14}$$

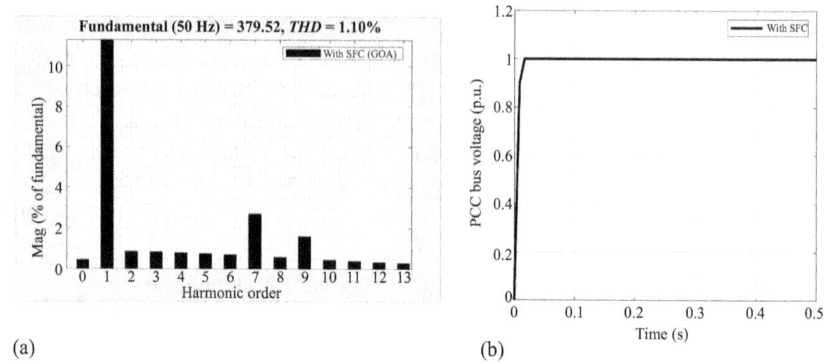

Figure 19.22 System response: (a) frequency spectrum and (b) voltage variation at PCC with SFC

$$e_b = e_B \times K_B \tag{19.15}$$

$$e_c = e_C \times K_C \tag{19.16}$$

where K_A, K_B and K_C are, respectively, selected weightings for the three regulation loops to allow adjustment of the control burdens. In addition, SFC-PID gains are determined by trial and error as $K_p = 38.6$, $K_i = 7.5$ and $K_d = 9.3$.

A PID controller processes the difference between the signals to obtain the phase angles (delta) required to drive the error to 0. Finally, the control signal of the PWM of the SFC scheme has the following form *u*(*t*) in the time domain for the PID controller:

$$u(t) = K_{pt}e_T(t) + K_{it}\int_0^t e_T(t)dt + K_{dt}\frac{d(e_T(t))}{dt} \tag{19.17}$$

where $u(t)$is the control variable; $e_T(t)$ is the selected system error; and K_p, K_i and K_d are the PID gains, respectively.

Accordingly, the SFC scheme is applied as a PQED to mitigate the harmonic distortion and stabilize the voltage at the PCC of the MG using the MATLAB/Simulink environment. The results illustrate a significant reduction of THD_V where it becomes 1.1% rather than 19.07% for the MG without PQED or 11.52% for the MG with D-STATCOM and the voltage is kept constant at the nominal value (Figure 19.22(a) and (b)). It may, however, be concluded that the D-FACTSs play an important role in improving the PQ and ensuring efficient energy utilization with RESs.

References

[1] Justo J. J., Mwasilu F., Lee J. and Jung J. W. 'AC-microgrids versus DC-microgrids with distributed energy resources: A review'. *Renewable and Sustainable Energy Reviews, Elsevier*. 2013;24:387–405.

[2] Huang W., Lu M. and Zhang L. 'Survey on microgrid control strategies'. *Energy Procedia.* 2011;12:206–12.

[3] Shuai Z., Sun Y., Shen Z. J., *et al.* 'Microgrid stability: Classification and a review'. *Renewable and Sustainable Energy Reviews*. 2016;58:167–79.

[4] Olivares D. E., Mehrizi-Sani A., Etemadi A. H., *et al.* 'Trends in microgrid control'. *IEEE Transactions on Smart Grid*. 2014;5:1905–19.

[5] KYOCERA high-efficiency photovoltaic module datasheet [online]. Available from https://www.kyocerasolar.com [Accessed 29 Jan 2020].

[6] Wolf S. and Enslin J. H. R. 'Economical, PV maximum power point tracking regulator with simplistic controller'. *Proceedings of IEEE Power Electronics Specialist Conference - PES'93*; 1993, pp. 581–7.

[7] Cardona S. and Lopez M. M. 'Performance analysis of a grid-connected photovoltaic system'. *Energy.* 1999;24:93–102.

[8] Esram T. and Chapman P. 'Comparison of photovoltaic array maximum power point tracking techniques'. *IEEE Transactions on Energy Conversion.* 2007;22(2):439–49.

[9] Cheikh S. M. A. and Larbes C. 'Maximum power point tracking using a fuzzy logic control Scheme'. *Revue des Energies Renouvelables.* 2007; 10(3):387–95.

[10] Villalva M. G. and Ruppert F. E. 'Analysis and simulation of the P&O MPPT algorithm using a linearized array model'. *Proceedings of 35th Annual Conference of IEEE Industrial Electronics*; Porto, Portugal, Nov 2009. IEEE; 2009, pp. 231–6.

[11] Safri A. and Mekhilef S. 'Incremental conductance MPPT method for PV systems'. *The 24th Canadian Conference on Electrical and Computer Engineering (CCECE)*; Niagara Falls, ON, Canada, May 2011. IEEE; 2011, pp. 1–3.

[12] Bennett T., Zilouchian A. and Messenger R. 'Comparative analysis of the perturb-and-observe and incremental conductance MPPT methods'. *The 8th International Symposium on Advanced Topics in Electrical Engineering (ATEE)*; Bucharest, Romania, May 2013, pp. 1–4.

[13] Endurance Energy MFTG Ltd. Wind turbine models [online]. Available from https://en.wind-turbine-models.com/manufacturers/247endurance [Accessed 30 Jan 2020].

[14] Rolls Battery Ltd. Battery specifications sheet [online]. 2018. Available from https://www.rollsbattery.com/battery/6-ehg-31p/?pdf=7925 [Accessed 30 Jan 2020].

[15] Li S., Li Y., Cao Y., Tan Y. and Keune B. 'Capacity optimization method of distribution static synchronous compensator considering the risk of voltage sag in high-voltage distribution networks'. *IET Generation, Transmission and Distribution* 2015;9(16):2602–10.

[16] Gandoman F. H., Sharaf A. M., Abdel Aleem S. H. E. and Jurado F. 'Distributed FACTS stabilization scheme for efficient utilization of distributed wind energy systems'. *International Transactions on Electrical Energy Systems* 2017;27(11):1–20.

Part IV

Associated systems

Chapter 20
Communications in power systems

20.1 Perspective of future power grid

As explained in Chapter 1, great attention is continuously paid by power engineers and researchers in making the power grid a smarter grid (SG). The power grid is getting more complicated in configuration as it includes not only the conventional elements, such as generation, transmission and distribution systems with uni-directional power flow, but now also the distributed generation (DG) units and energy storage systems. DG units can be directly connected to the transmission and/or distribution system or integrated together constituting microgrid(s), which can be connected to the distribution system at different locations.

This developing configuration makes the power flow bi-directional rather than uni-directional. To achieve the desired technical, environmental and economic benefits, e.g. reliability and flexibility enhancement, emission reduction and power quality improvement, the power grid must be augmented with assets and systems that will enable it to operate in an optimal manner keeping customer satisfaction. Smart metering, communication and cybersecurity systems are of utmost importance to be incorporated into the power grid with a goal of modernizing the legacy electricity network and achieving smart characteristics. Cybersecurity is discussed in Chapter 21, whereas a brief introduction to smart metering and communication systems is given in this chapter.

Advanced metering infrastructure (AMI). It is a key factor in SG architecture where advanced meters called smart meters (SMs) are used rather than the conventional meters which just read the power consumption. SMs can record the power consumption online and can communicate with the utility or other monitoring and control centres through the power grid. This facilitates (i) the utility to collect information for load monitoring, billing process, online control and managing power quality and (ii) the customers to recognize how much power they are using and accordingly could manage their consumption and, thus, carbon dioxide emission as well as they can participate in managing the peak load.

Communication systems. SG as an anticipated next generation power grid should provide monitoring, protecting and automatically optimizing the operation of the interconnected components in the power grid. It is characterized by bi-directional connection of electric power and information flows to establish an automated wide distributed delivery network. In addition, it covers different types of generation

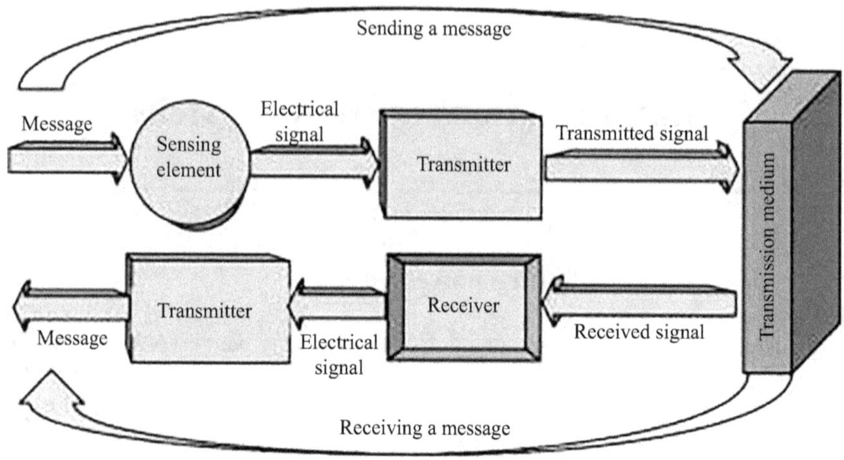

Figure 20.1 Main components of electronic communication system

(central conventional, renewable and non-renewable DGs) through transmission and distribution networks to customers (industrial, commercial and residential/home users). Each of these elements has its own controls; for instance, renewable sources are inverter-based sources equipped with power conditioning system, and home users with their thermostats, electric vehicles (EVs) and intelligent appliances. Based on all these aspects, two-way communication systems are critically required to be a complementary part of the power grid for information flow in real time from and to each node to another. Electronic communication system comprises three primary components: a source, a destination and a transmission medium, by which four basic processes are implemented as in the following [1,2]:

- conversion of data into signals in a form that is adequate for transmission;
- modulation of data signal for transmission over selected medium;
- determination of adequate transmission medium; and
- retrieving information for the modulated signals.

A simplified block diagram of an electronic communication system is shown in Figure 20.1.

SG can be envisioned as an interconnected multi-area system in which each area can incorporate a communication technology that meets specific requirements such as reliability, latency, bandwidth and security. Some of these areas are local area network (LAN), home area network (HAN), building area network (BAN), neighbourhood area network (NAN) and wide area network (WAN). This classification, from communication engineering point of view, helps in developing a reliable communication infrastructure for establishing robust real-time data transportation through WANs [3].

Communications is a discipline by itself. The intent of this chapter is to simply provide awareness and a general understanding of the communications systems from the perspective of a practising power systems engineer rather than an in-depth treatment of the communications technology.

20.1.1 Classification of communication networks

First, it is important to understand what a network is. It is simply two or more devices linked with a goal of communicating or sharing resources. This connection can either be wired through use of cables and telephone lines or be wireless through use of satellites and radio waves [4,5]. Second, communication networks can be classified based on the size of the network or the number of devices included in its area, in addition to the communication technology required for communicating the network entities. Regarding SGs, a brief description of such networks is given in the following:

LAN: It is a network connecting a cluster of devices in a local area through use of Ethernet cables with length limited by the communication speed. Wi-Fi can also be used to make LAN wireless through using a wireless router. LAN is the most common and basic type of communication network.

HAN: It is a network connecting devices within a user's home and is used to provide monitoring and control over energy usage. It connects to the utility's electric meter and monitors energy usage to enable the home owner to recognize how much energy is being used and where. It is also used to give the utility a way to control, for example, the heating and air conditioning to save energy. HAN also provides a way to conveniently turn appliances off and on and controlling all smart, digital and thermostatic devices at home such as washing machine, dish washer, refrigerator, computers and peripherals, EV and TVs with a goal of saving energy and optimizing the consumption. HAN may use wireless or communications over the AC power line, as well as a router to which some home assistants are connected. A schematic diagram of a general vision for a HAN is depicted in Figure 20.2. It can be considered as a type of LAN [6].

BAN: A BAN is a LAN that covers an entire building. It may be a collection of smaller LANs. For example, if each floor is considered a single LAN, then the combination of each floor LAN is considered a BAN. A BAN is also known as a business area network [7].

NAN: It is an offshoot of Wi-Fi hotspots and wireless LANs (WLANs) which enable users to connect to the Internet quickly and at very little expense. A NAN is generally installed by an individual to serve a family or some of neighbours. NANs cover only a small number of blocks close to a specific access point. With the help of an omnidirectional antenna, a single access point could cover a radius of more than half a mile (0.8 km). Users who wish to connect to a NAN can then make use of a directional antenna to get an improved signal from the access point.

NAN providers are usually individuals or a group that join to share an Internet broadband connection. If a user with a broadband connection, either digital

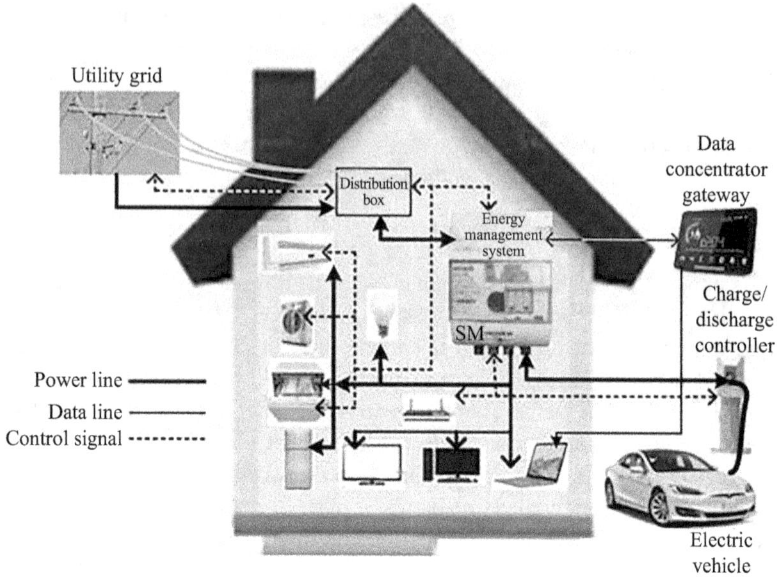

Figure 20.2 HAN vision

subscriber lines (DSLs) or cable modem, plans to share it, a NAN makes it possible to share this with anyone within range. For the receiver (usually neighbours within reach) to connect wirelessly to the shared Internet, they need to have a Wi-Fi-enabled laptop. This concept differs from the wireless deployment for hotspots. Hotspots are usually commercial Internet access points with a reach of only 100 m. They are used to attract tech-savvy customers to a coffee shop, airport or restaurant. NANs, on the other hand, offer a much wider radius of Internet connection. Therefore, commercialization of NANs is an effective way to speed up the extension of neighbourhood Wi-Fi networks. NANs allow the users to reduce their Internet expenditure by sharing a connection with neighbours. The downside is that this strategy reduces the bandwidth speed and sometimes leads to the violation of the Internet service provider's agreement. Some service providers do not allow individual broadband users to share their connection, making NANs a violation of that agreement. A typical structure of a NAN network is shown in Figure 20.3.

WAN: In contrast to the smaller LAN and HANs, a WAN is a wide area network, covering any distance necessary. The Internet could be considered a WAN that covers the entire Earth.

The SG includes many networks (domains) with various boundaries that must be connected to supply end-to-end services. The challenge is to design network architectures that can meet the interoperability requirements for inter-domain and intra-domain communications. Such networks, according to the US Energy Independence and Security Act of 2007, should supply secure and reliable end-to-end two-way communications. In this respect, one of the most important challenges is to supply a reliable last unit distance communication that covers the connectivity from the meters

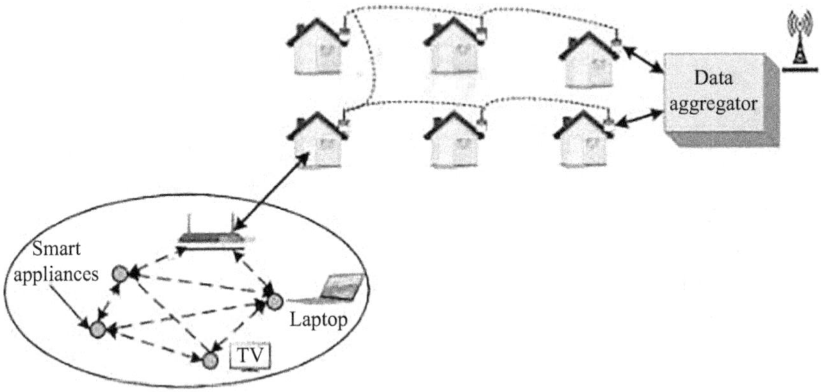

Figure 20.3 A typical NAN structure

to the AMI headend. The SG communication infrastructure may include different area communication networks. For instance, an SG that distributes electricity between generators (conventional and DG sources) and consumers (commercial, industrial and residential) may form HAN, BAN, NAN and data centres using bi-directional information flow for monitoring and controlling the devices as shown in Figure 20.4. The design of such networks depends not only on the application layer requirement, but also on the nature of its medium access control. These are some of diverse communication technologies and standards deployed for NAN [8].

Various technologies of communications in power systems can be used to transfer the information aggregated by SMs from device to others. Wired and wireless communication technologies in respect of SG needs are briefly overviewed in the next sections [9].

20.2 Wired communications

20.2.1 Power line communications

The technology of power line communication (PLC) system implies the use of existing power network, which transmits the power from generation sources to a variety of customers over a wide region, for communication purpose without the elevated cost and implementation required by the other solutions based on wireless and optical fibres. Since many years, amplitude-modulation carrier-based communication has been used by the power network operators to transfer status, and alert messages between power stations and substations as well as operator internal telephony. Nowadays, with the continuous development of communication technologies, the PLC can be used for data transmission [10].

The power network includes power lines (overhead and underground cables), transformers (step-up and step-down) and protection and control devices. Power

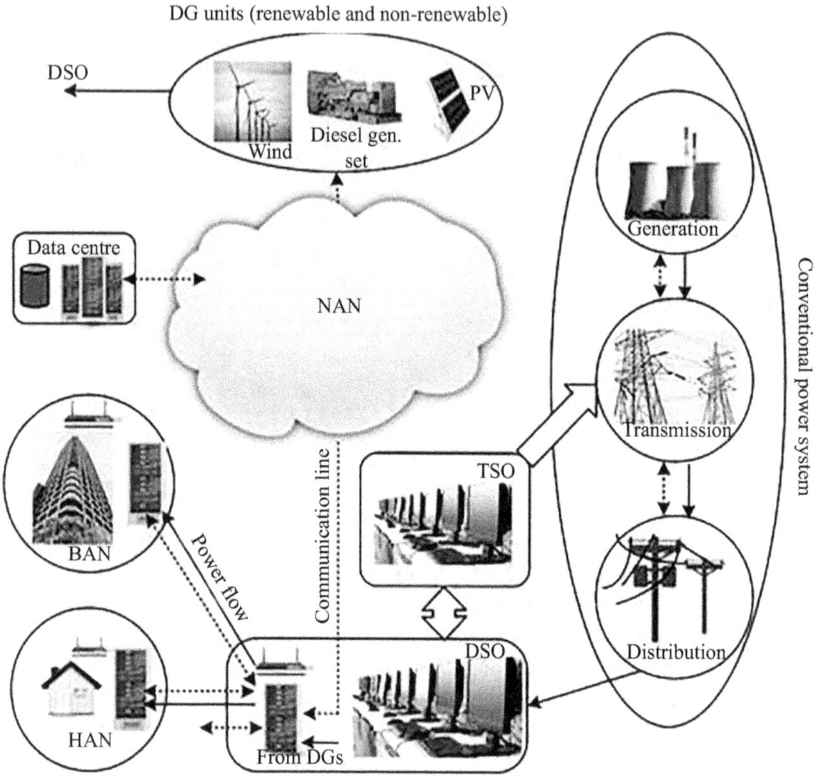

Figure 20.4 A typical SG communication infrastructure

lines work at different voltage levels: extra-high voltage (EHV), high voltage (HV), medium voltage (MV) and low voltage (LV) [11].

- EHV and HV lines: They transmit electricity from generating stations to distribution substations at voltage levels, typically, of the order of hundreds of kilovolts over long distances of the order of tens of kilometres. In the beginning, the PLC technology using long-wave frequencies (e.g. in the range of 24–500 kHz) with bandwidth of a few kilohertzs was enough for transmitting messages or internal telephony. Currently, the availability of optical fibre communication links in underground cables or overhead lines increases the bandwidth significantly and allows high data rate.

  HV lines are excellent carriers for radio frequency (RF) energy because of the presence of open wire equipment with very few crossovers. Small transmitted power (~10 W) is enough to overcome distances more than 1,000 km. In addition, because of the rapid progress of digital communications automation and through modern digital modulation and coding schemes, the bandwidth efficiency for the carrier-frequency system has been highly improved.

Typical lines of voltages in the range from 110 to 380 kV are used for nation-wide or even international power transfer and consist of long overhead lines with little or no branches. This makes them acceptable wave guides with less attenuation per line length as for their MV and LV counterparts. However, their potential for broadband SG communication services has up to the present day been limited. Time-varying HV arcing and corona noise with noise power fluctuations in the order of several tens of decibels and the practicalities and costs of coupling communication signals in and out of these lines have been an issue. Further, there is a fierce competition of fibre optical links. In some cases, these links might even be spliced together with the ground conductor of the HV system [12].

- MV lines: These connect the distribution substations to transformers' kiosks or pole-mounted transformers. The voltage levels are of the order of a few kilovolts and they run over distances of the order of a few kilometres. Numerous cross connections and several types of cables and overhead lines characterize the MV lines leading to hindering the RF signal propagation as the attenuation is high in addition to the impedance matching problems. It is to be noted that fibres in MV networks are rarely included in the power cabling and the MV substations are connected to the communication network mainly through DSL, private pilot cables or cellular radio techniques. Meanwhile, communications to substations may be realized by PLC.

  Typical MV lines, overhead lines or underground cables, working at voltages in the range from 10 to 30 kV, are connected to zone transformer substations. They are used to distribute electricity between cities and large industrial consumers. For monitoring and control purpose, MV lines are directly connected to intelligent electronic devices (IEDs) that require low data rates, and thus, NB-PLC can be relevant and economic communication solution for this task. The IEDs such as capacitor banks, reclosers, sectionalizers, ring main units (RMUs) and phasor measurement units (PMUs) are connected to the MV lines forming an MV distribution network with fewer branches than those of LV distribution network.
- LV lines: They have the same nature as the MV lines. Recently, more attention is being paid for data communication over LV lines due to the growth in telecommunication industry and the advances in digital devices that use either low or high bandwidth.

  LV lines, with voltages in the range of 110–400 V, are connected to the MV lines via MV/LV transformer substations. A communication signal on an MV line can pass through these transformers to the LV line with heavy attenuation in the order of 55–75 dB. Consequently, a special coupling device (inductive, capacitive) or a PLC repeater is frequently required to establish a high data rate communication path.

The frequency bands used by the PLC systems have a direct impact on signal reach and effective bandwidth, which define the data rate and then the applications of PLC to the grid. Therefore, based on the frequency band, PLC technology for SG can be classified into three classes: ultra-narrowband PLC (UNB-PLC), narrowband PLC (NB-PLC) and broadband PLC (BB-PLC) [13,14].

- UNB-PLC systems use very narrow bandwidth for data transmission in frequencies less than 3 kHz. These systems are usually used for data transmission when the electrical signals cross zero to be not affected by the high amplitude of the main signal and harmonics. Because of the very low frequency, these systems are less affected by transmission losses and can reach long distances without repeaters. However, the main drawback of this technology is the low data rate (a few hundred bits per second, bps).
- NB-PLC systems work with medium data rate in frequencies from 3 to 500 kHz. Thus, this class of technology can be categorized as both low-data rate and high-data rate technologies. For the first, the technology is based on single-carrier modulations conveying data rates of a few kbps, whereas the second is based on multi-carrier modulations and send data rates of hundreds of kbps.
- BB-PLC includes a large variety of systems that work with high-data rates in frequencies from 1 to 250 MHz.

Role of PLC in SGs: In the beginning, the use of PLC was for telecommunication, tele-protection and tele-monitoring between electrical substations through EHV and HV power lines at voltages such as 110, 220 and 400 kV. The carrier frequency range was used for audio signals, protection and a pilot frequency.

Nowadays, applications of PLC in SGs also pertain to power distribution systems and communication between the consumers and the utility. Current examples of these applications are AMI systems, EV charging systems, tele-control applications, smart cities and DG systems. For these applications of PLC in SGs, both NB-PLC and BB-PLC technologies are used in addition to some other applications of grid topology connectivity, cable health monitoring and fault location. NB-PLC and BB-PLC technologies provide real-time data to the devices connected to the grid, enabling an easy understanding of the network, as well as an efficient management of events and failures.

A typical example of using PLC system in SG can be envisioned as shown in Figure 20.5. The SG encompasses conventional generation, transmission through HV power lines, and MV and LV distribution through overhead lines and underground cables as well as incorporating renewable energy sources at the distribution level [15].

Regarding the HV power lines, the phase conductors are used for signal transmission by applying a coupling scheme to define how the communication equipment is connected to the power line. The PLC channel is characterized based on power line characteristics, coupling devices and line trap unit (LTU) characteristics, and coaxial linking cable characteristics. It consists of signal path between the transmitter and coupling device on the transmission side (power line conductor) and coupling device and receiver on the receiving side. LTU is used to block the signal propagation in the unwanted direction. It is to be noted that the PLC communication system requires additional equipment (not needed for the

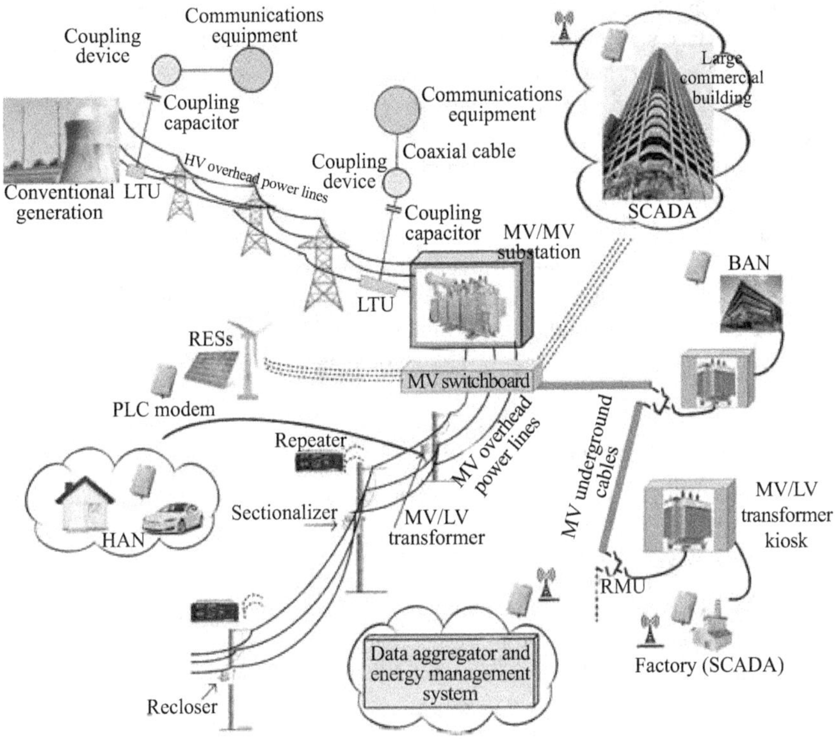

Figure 20.5 Vision of PLC in power system

other technologies) to protect communication devices and personnel from the HV and to match impedances between the coaxial cable and the power line.

NB-PLC and BB-PLC solutions can be implemented for the management of MV substations, EVs and EV charging points, RMUs and data aggregators, smart homes and distributed energy resources. In this case, PLC can be used in both MV and LV sections. The major driver for the deployment of PLC market is the increase of SG installations. Nowadays, hundreds of millions of PLC devices are deployed all over the world. The objective of achieving secure and reliable communications with utility control centres is a challenge. The main problems for that are caused by noise generated by electrical devices, which may affect the proper performance of the communications. In AMI systems, especially, the electrical noise can block the communication between the meters and the headend systems in the worst case and produce datagram losses. Other factors that may or have slowed down the deployment of the PLC technology are security issues related to the regulations in different countries, interoperability of different PLC solutions and network topology of each country; for example, a meshed electricity network, neutral secondary substations connectivity and others could challenge the deployment of PLC in SG if not properly engineered [12].

The LV lines lead directly or over street cabinets to the end customers' premises. It is to be noted that a considerable regional topology difference exits. For example, in the USA a smaller secondary transformer on a utility pole might service a single house or a small number of houses. In Europe, however, it is more common that up to 100 households get served from a single secondary transformer substation. Further, significant differences exist between building types. They may be categorized as multi-flat buildings with riser, multi-flat buildings with common meter room, single family houses and high-rise buildings. Their different electrical wiring topologies influence signal attenuation as well as interference between neighbouring PLC networks.

20.2.2 Optical communications

Technologies of optical communications are widely used for building the communication backbone that interconnects the transformer substations with control centres. They have the following advantages:

- Ability of data transmission over long distance (several kilometres) with a bandwidth of tens of Gbps.
- Adequacy of use in HV environment because of the robustness against electromagnetic and radio interference.
- Optical power ground wire as a special type of optical cables combines the functions of grounding and optical communications as well as allowing high data rate transmitted over long distance.
- Expandability for an application of providing SG services directly.
- Use in access networks, known as fibre-to-the-home (FttH) through using the passive optical network (PON) technologies, which do not require electrical power switching equipment. They use optical splitters to separate and collect optical signals.
- Ability of serving multiple premises in a point-to-multi-point fashion by using PON technology.
- Enabling the use of the standard Ethernet communication protocol over an optical network by applying the Ethernet PON technology which is one of the PON technologies. Consequently, the interoperability with existing Internet protocol (IP)-based network is facilitated.

20.2.3 Digital subscriber lines

DSL enables digital data transmission over telephone lines. As a technology, it allows the electric utilities to interconnect residential consumers to control centres without any additional cost of deploying their own communication infrastructure, just paying a communication fee to the telecommunications operator. DSL has different categories based on the range of data rate. For instance, asymmetric DSL (ADSL) supports up to 8 Mbps in the downstream and up to 640 kbps in the upstream, ADSL2+ with a maximum download speed of 24 Mbps and upload speed of 1 Mbps, and very-high-bit-rate DSL (VDSL or VHDSL) supports faster data transmission over copper wires up to 52 Mbps downstream and 16 Mbps upstream, and on coaxial cable up to 85 Mbps downstream and upstream.

20.3 Wireless communications

20.3.1 Microwave radio systems

Generally, radio frequencies, specifically the microwaves, are the transmission medium, generally described as waves with frequencies that range from approximately 500 MHz to 300 GHz or more. Microwave radios propagate signals through Earth's atmosphere between transmitters and receivers often located on top of towers spaced about 90–180 km apart. Microwave radio systems have the advantage of having the capacity to carry thousands of individual information channels between two points without the need for physical facilities such as coaxial or optical fibre cables. The main advantages of microwave radio systems are the following: (i) because of the high operating frequencies, they can carry large quantities of information; (ii) high frequencies mean short wavelengths, which require relatively small antennas; (iii) minimum delay times are introduced; and (iv) minimal crosstalk exists between voice channels.

Microwave radio system capacities range from less than 12 voice-band channels to more than 22,000 channels. Early microwave systems carried frequency-division-multiplexed voice-band circuits and used conventional, non-coherent frequency modulation techniques. Most recent developed microwave systems carry pulse-code time-division-multiplexed voice-circuits and use more modern digital modulation techniques, such as phase-shift keying or quadrature amplitude modulation [16–19]. A block diagram of a microwave radio system, transmitter and receiver is shown in Figure 20.6. A local oscillator is used with a mixer to convert the intermediate frequency (IF) or the baseband into RF as an up-converter during signal transmission and as down-converter by changing the RF signal to IF signal during receiving process.

20.3.2 Cellular communication networks

Cellular systems (e.g. the fourth generation '4G') are more public than the other wireless communication technologies because of their larger coverage area. These systems use technologies such as global system for mobile communications 'GSM', general packet radio services 'GPRS' and enhanced data rates for GSM evolution 'EDGE'. Utilities have used such technologies for data communications in supervisory control and data acquisition (SCADA) and advanced meter reading systems. New generations continuously arise aiming at supporting higher data rates, more data communication systems, reducing the

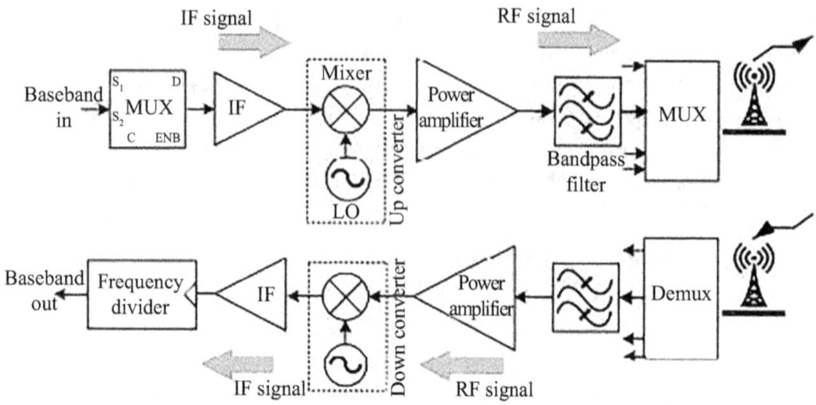

Figure 20.6 Simplified block diagram of a microwave radio system: transmitter and receiver

expenses and improving the latency performance that depends on the number of users served by the same base station.

20.3.3 Satellite communication systems

Satellite systems have the capability of supporting communications with variable bandwidth and latency performance by stationing satellites on orbits at different altitudes. Electric utilities can consider such satellite communication systems for SCADA systems as well as services supplied to rural or remote areas that may be out of the coverage area of the present communication network or the high expenses required to reach these areas with fibre optic cables.

20.4 Internet access

Internet architecture presents several benefits such as connecting large number of devices, transmitting data over multiple media, connecting multiple types of systems and maintaining reliability. That is because IP has the capability of

- offering straightforward addressing and routing for large networks;
- running over any link layer network (encompassing ethernet, wireless networks, etc.);
- identifying any type of system, when the data is sent from a point of access and delivered to multiple destinations; and
- having more tools and applications to help manage the network.

Therefore, similarities between Internet facilities and requirements of SG communication structure can be recognized as the SG communications network needs interconnection between numerous devices, integration of highly heterogeneous areas (or individual electric subsystems) and supporting different degrees of autonomous control at various times and scales. Thus, Internet technologies may be an adequate solution for the problem of SG communication interoperability. However, that is accompanied with the huge challenges of inherent threats and vulnerabilities of the Internet and must be addressed [20].

20.5 Concluding remarks

Based on the anticipated SG infrastructure, a few concluding remarks are given as follows:

- SG has the data communication network integrated into the electrical grid.
- In an SG, data about transmission, distribution and consumption are collected and analysed in near real time.
- In addition to the integration of communication network into the power grid, security of data is of utmost importance.
- Communications network in an SG includes different communication subnetworks such as HAN, NAN, BAN and WAN, or SCADA. Each is dedicated to a specific area, which has its own characteristics and requirements.
- Communications technology must be properly selected to be relevant, technical and economical, for each area or subnetwork.
- Based on the structure of the electric power system, the following steps should be considered when designing the infrastructure of the communications network:
  - Specifying the main devices of monitoring and control for generation sources (conventional, DGs), transmission and distribution systems.
  - Identifying the functions and location of control centres at each system.
  - Specifying the connections through the power system, such as connections at transformer substations, switchgear, distribution transformers, RMUs, PMUs and so on.
  - Determination of system parameters that may affect the signal propagation, especially when using wired communication systems.
  - Specifying the functions and locations of SMs required through the power system and at consumers' premises.
  - Decomposition of the power system, from communications perspective, into subsystems based on which subsystem is relevant, e.g. HAN or NAN, etc. Each subsystem covers a specific area [21].
  - Selection of communications technology that is well fitted to each subsystem based on requirements such as the bandwidth, speed of data transmission and data rate.
  - Integration of different communication technologies through, for example, specific protocols as each area in the system may have a technology that differs from others.
  - As the function of the communication system is data transmission, security is of utmost importance.

References

[1] Yousuf M. and Vajuhudeen M. 'Current communication media in power systems'. *International Journal of Research in Electrical Engineering (IJREE)*. 2017;4(2):145–7.

[2] Ancillotti E., Bruno R. and Conti M. 'The role of communication systems in smart grids: Architectures, technical solutions and research challenges'. *Journal of Computer Communications, Elsevier*. 2013;36:1665–97.

[3] Yan Y., Qian Y., Sharif H. and Tipper D. 'A survey on smart grid communication infrastructure: Motivations, requirements and challenges'. *IEEE Communication Surveys & Tutorials*. 2013;15(1):5–10.

[4] Frenzel Jr L. E. *Electronics explained: Fundamental for engineers, technicians, and makers*. 2nd edn. Oxford, UK: Elsevier Inc, Newnes; 2018, pp. 217–42.

[5] Ho Q. D. and Le-Ngoc T. 'Smart grid communications networks: Wireless technologies, protocols, issues, and standards' in Obaidat M. S., Anpalagan A. and Woungang I. (eds.). *Handbook of Green Information and Communication Systems*. Oxford, UK: Elsevier Inc, Academic Press; 2013, pp. 115–46.

[6] Beal V. *HAN-home area network* [online]. Available from https://www.webopedia.com/TERM/H/HAN.html [Accessed 4 Mar 2020].

[7] Sechilariu M., Wang B. and Locment F. 'Building-integrated microgrid: Advanced local energy management for forthcoming smart power grid communication'. *Journal of Energy and Buildings, Elsevier*. 2013;59:236–43.

[8] Wang W., Xu Y. and Khanna M. 'A survey on the communication architectures in smart grid'. *Journal of Computer Networks, Elsevier*. 2011;55:3604–29.

[9] Ruiz-Romero S., Colmenar-Santos A., Mur-Perez F. and Lopez-Rey A. 'Integration of distributed generation in the power distribution network: The need for smart grid control systems, communication and equipment for a smart city – Use cases'. *Journal of Renewable and Sustainable Energy Reviews, Elsevier*. 2014;38:223–34.

[10] Berganza I., Bumiller G., Dabak A., Lehnert R. and Sendin A. 'PLC for smart grid' in Lampe L., Tonello A. M. and Swart T. G. (eds.). *Power Line Communications: Principles, Standards and Applications from Multimedia to Smart Grid*. 2nd edn. New York: John Wiley & Sons; 2016, pp. 509–61.

[11] Oborkhale L. I. and Shoewu O. 'Power line communication technology'. *The Pacific Journal of Science and Technology*. 2007;8(2):398–405.

[12] Zajc M., Suljanović N., Mujčić A. and Tasič J. F. 'High voltage power line constraints for high-speed communications'. *Proceedings of the 12th IEEE Mediterranean Electrotechnical Conference*; Dubrovnik, Croatia, May 2004, pp. 285–8.

[13] Galli S., Scaglione A. and Wang Z. 'For the grid and through the grid: The role of power line communications in the smart grid'. *Proceedings of IEEE*. 2011;99(6):998–1027.

[14] Berger L. T., Schwager A. and Escudero-Garzás J. J. 'Power line communications for smart grid applications'. *Journal of Electrical and Computer Engineering. Special Issue; Power-line communications: Smart grid, Transmission, and Propagation*. 2013;(2013):1–16.

[15] Kabalci E., Kabalci Y. and Develi I. 'Modelling and analysis of a power line communication system with QPSK modem for renewable smart grids'. *Journal of Electrical Power and Energy Systems, Elsevier*. 2012;34:19–28.

[16] Abdalla J. H., Carvalho P., Luis F., Evangelista, C. and Bermundez L. 'Tools for microwave radio communications system design'. *Journal of Microwaves and Optoelectronics*. 2003;3(1):26–38.

[17] Office of the National Coordinator for Smart Grid Interoperability. *NIST framework and roadmap for smart grid interoperability standards*. Release 1.0. U.S. Department of Commerce, Report, 2010, pp. 1–145.

[18] U.S. Department of Energy. *Smart grid system report*. Report, 2009, pp. 1–84.

[19] NETL. West Virginia smart grid implementation plan (WV SGIP) project, Advanced Power and Electricity Research Center (APERC). *Assessment of as-is grid by non-utility stakeholders*. Report, 2009, pp. 1–28.

[20] Gao J., Xiao Y., Liu J., Liang W. and Philip Chen C. L. 'A survey of communication/networking in smart Grids'. *Future Generation Computer Systems, Elsevier*. 2012;28:391–404.

[21] Kuzlu M., Pipattanasomporn M. and Rahman S. 'Communication network requirements for major smart grid applications in HAN, NAN and WAN'. *Journal of Computer Networks*. 2014;67:74–88.

Chapter 21

Cybersecurity in power systems

21.1 Introduction

The future smarter grid as envisaged in Chapters 1 and 20 illustrates that the grid of the future will be subject to energy diversification, digitalization and advanced metering infrastructure (AMI) systems that will include numerous new devices to monitor, control and protect the grid. In addition, all grid components will be able to communicate and collaborate through a high-speed two-way communication system integrated into the power grid. This is because of the presence of conventional, and renewable and non-renewable distributed energy sources as well as their interdependency on information and communication technology (ICT) to enhance grid efficiency and performance.

Power system is a massive and complex system. Being highly dependent on ICT, with communications and networking systems, exposes the grid to potential vulnerabilities to physical attack or cyberattack [1]. Thus, with the need to mitigate the risk of compromising reliability and security of power system operation, and surmounting the probable severe consequences from customer information leakage to cascade failures that may lead to complete blackout and eradication of the infrastructure, cybersecurity is becoming an issue of utmost importance in the design of the information network. Therefore, cybersecurity issues are highlighted in this chapter with the aim of getting the power engineers and power system operators acquainted with the issues within the knowledge domain of cybersecurity.

Cybersecurity refers to the technologies, processes and practices designed to protect networks, devices, programs, and data from attack, damage or unauthorized access. The source of cyberattacks in electric power systems is mainly malicious threats in the integrated communication networks [2]. Consequently, attack countermeasures should be integrated into network protocols to achieve reliable information exchange and real-time message delivery. Relevant transmission control protocol/Internet protocol (TCP/IP) is described in the next section.

21.2 Communication network protocols in power grids

The structure of network protocol suites is mostly a series of layers, sometimes also referred to as a protocol stack. Each layer is designed for a specific purpose and exists in both the sending and the receiving systems. A specific layer on one system sends or receives the same object that another system's peer process sends or receives. Layer activities occur independently from activities in layers above or below the layer under consideration or other layers in the same system. Meanwhile, each layer acts in parallel with the same layer on the other systems [3].

The basic unit of information is formed as a packet that is transferred across a network. The basic packet consists of a header with the addresses of the sending and receiving systems, and a body, or payload, with the data to be transferred. As the packet travels through the TCP/IP protocol stack, the protocols at each layer either add or remove fields from the basic header. When a protocol on the sending system adds data to the packet header, the process is called 'data encapsulation'. Moreover, each layer has a different term for the altered packet.

The life cycle of a packet, which starts when a send message command is issued, is summarized in this section. The life cycle finishes when processing through all layers is completed and the appropriate application on the receiving system receives the packet. Protocol layers of communication network in power grids are shown in Figure 21.1 with functions described as follows:

- *Application layer:* It defines standard Internet services and network applications that anyone can use. These services work with the transport layer to send and receive data. Many application layer protocols exist such as TCP/IP, domain name system (DNS), simple network management protocol (SNMP) and so on.
- *Transport layer:* The protocols at this layer start the process of data encapsulation when the data arrive. The transport layer encapsulates the application data into transport protocol data units. It ensures that packets arrive in sequence and without error, by swapping acknowledgements of data reception and retransmitting lost packets. This type of communication is known as end to end. TCP is one of the protocols used at this level.

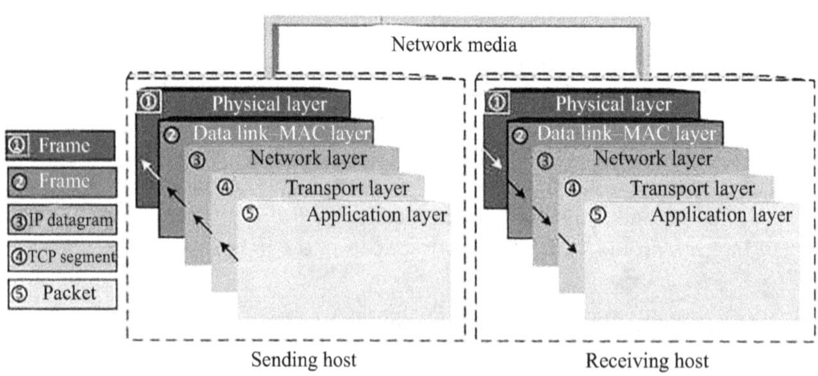

Figure 21.1 Typical TCP/IP model packet travel through protocol layers of communication network in a grid

- *Network layer (IP layer):* It accepts and delivers packets for the network. This layer includes the powerful IP, the address resolution protocol (ARP) and the Internet control message protocol (ICMP).

  IP prepares them for delivery by formatting them into units called IP datagrams. IP then determines the IP addresses for the datagrams, so that they can be delivered effectively to the receiving host.

- *Data-link layer:* It is also called 'medium access control (MAC) layer'. It identifies the network protocol type of the packet (e.g. TCP/IP) and provides error control and formatting the IP datagram into a frame in accordance with protocols such as IEEE 802.2 framing and the point-to-point protocol (PPP). These protocols attach a third header and a footer to 'frame' the datagram. The frame header includes a cyclic redundancy check field that checks for errors as the frame travels over the network media. Then, the data-link layer passes the frame to the physical layer.
- *Physical network layer:* It specifies the physical characteristics of the hardware used in the communication network media. For instance, the physical network layer on the sending host receives the frames and converts the IP addresses into the hardware addresses appropriate to the network media. The physical network layer then sends the frame out over the network media.

21.3 Cybersecurity objectives and requirements

As reported in [4], the three major objectives of cybersecurity for the grid are availability, integrity and confidentiality of data. Availability is the primary objective to ensure timely and reliable access to the information. Time latency of the availability is application dependent, for example, data availability for protective relaying is less than 4 ms, for substation supervisory control and data acquisition (SCADA) is in seconds, for market pricing in minutes and for meter reading and longer-term pricing information in hours.

Data integrity implies source and quality of data. Modification or destruction of original data leads to loss of data integration, which may also occur due to the intrusion in the cyber domain by the attackers, unauthorized employees or human error. Loss of data integration degrades the system reliability rapidly, especially for a complex system such as the modern power grid.

Data confidentiality mainly pertains to protecting personal privacy and proprietary information. It becomes especially important in systems involving interaction with customers (e.g. demand response and AMI networks).

To satisfy the objectives of cybersecurity, various issues, such as (i) attack detection through monitoring the communication network traffic status to identify and detect abnormal events due to attacks, if any, (ii) identification authentication and access control to verify the identity of a device or user to be allowed to access the grid resources as well as ensuring that the resources are accessed only by the appropriate person, (iii) secure and efficient communication protocols to satisfy time-criticality and security of message delivery in the grid, etc., should be addressed.

Referring to the National Institute of Science and Technology (NIST), the grid infrastructure model is based on different segments [5]: centralized generation, power transmission, energy distribution, distributed renewable and non-renewable energy sources, HV/MV and MV/LV substations, system operations and control, market, service providers, and the industrial, commercial and residential classes of customers. Each class has its own characteristics, needs and corresponding communication network such as HAN, BAN and IAN (Figure 21.2).

These area networks are interconnected through communication infrastructures for forwarding any information such as alarms, commands, readings of voltage values and energy consumption to system operations and control by means of IP-based protocols (e.g. IEC 61968, IEC 61850, IEEE 1815 and ZigBee) as well as controlling the message routing through use of gateways and routers. Gateways are used for telecommunication networks to allow data to flow from one discrete network to another irrespective of their protocol. They have the capability of converting one protocol to another protocol. Consequently, they are enabling the traffic flow and communication between dissimilar networks. The router is a device that provides the function of routing IP data packets between similar networks. A packet is typically forwarded from one router to another router through similar networks until it reaches its destination node [6]. Therefore, a gateway can be viewed as a router, but a router does not have to be a gateway.

21.4 Network attacks in power grids

Security attacks in communication networks come from either misbehaving users or malicious operators. Misbehaving users are those who violate communication

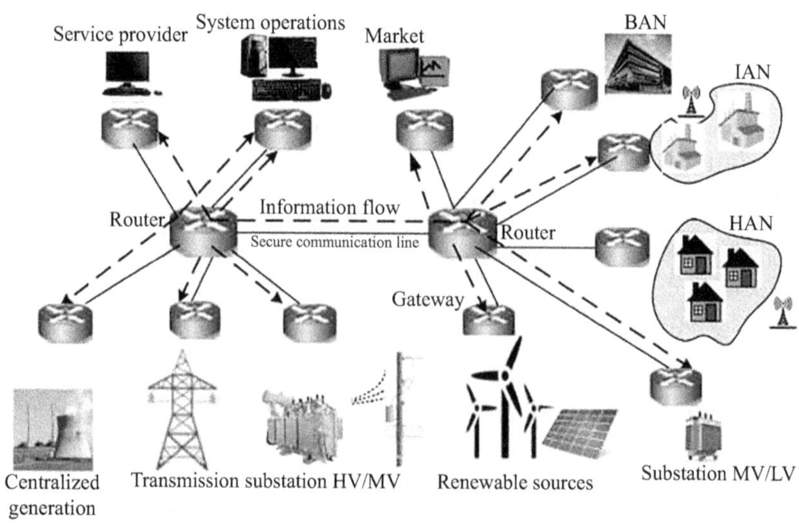

Figure 21.2 Cybersecurity of smarter grid infrastructure

protocols with the aim of obtaining more network resources than those allocated for the legitimate users. In contrast, malicious operators are dedicated to illegally acquiring, modifying or disrupting information in the network. However, irrespective of the users' type, security threats must be met. Countermeasures against attacks by misbehaving users can be incorporated in the numerous electronic computing devices used for monitoring and control purposes. Malicious operators' attacks may result in severe damage to power supplies and probably widespread power outage. Based on specific objectives, malicious attacks can be categorized and described as in the following subsections.

21.4.1 DoS attacks

Denial-of-service (DoS) attacks encompass a series of actions that prevent the functioning of any part of an information system concerning a specific device [7]. DoS can simultaneously attack an array of devices, and in this case, it is named as 'distributed denial-of-service (DDoS)'. Thus, these attacks attempt to delay, block or corrupt the communication resource in the power grids such as bandwidth; that is, they target the availability objective of the network through effects that can range from delay in service response time to complete inaccessibility.

DDoS is a distributed attack over a group of dispersed compromised devices in the grid (called 'bots' or 'slaves') controlled by an attacker (called 'bot-master') through command and control (C&C) channels and simultaneously used to attack a target victim. The impact of the individual attack of compromised bots can be aggregated to magnify the effect against the victim through communication network infrastructure. The network of compromised bots is called a 'botnet'. A server/client model exposed to a DDoS attack is shown in Figure 21.3. The botnet includes an array of clients that represents bots. The bot-master compromises the bots by infecting them, e.g. with a Trojan or a back-door program in addition to controlling them through C&C channels. The compromised bots are used to attack the server, which is the target victim [8].

In power networks, DDoS attacks against communication system can happen in a variety of communication layers as described in the following:

- *Physical layer:* One of the most efficient ways to launch physical layer attacks is channel jamming as it is a very easy method for attackers, especially for wireless communications. It becomes the primary physical layer attack in such networks and can lead to a wide range of damage to the network performance of, say, power substation systems, because of the delayed delivery of time-critical messages as a result of jamming.
- *MAC layer:* Spoofing is a harmful threat at the MAC layer. The spoofing attacker can open the address fields in the MAC frame and masquerade itself as another device to send fake information to other devices. As an application of this attack to a power substation network, a malicious node can broadcast forged ARP packets to shutdown connections of all intelligent electronic devices (IEDs) to the substation gateway node.

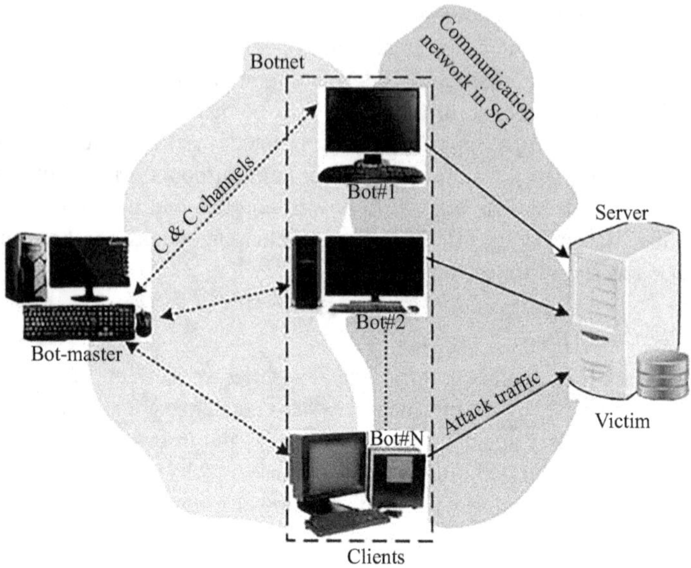

Figure 21.3 A typical server/client model with DDoS attack

- *Network and transport layers:* These two layers need to provide reliable control for information delivery over communication networks according to TCP/IP protocol model. Thus, DDoS attacks, such as traffic flooding and buffer flooding, can highly degrade the end-to-end communication performance of the power systems.
- *Application layer:* Attacks at this layer are mainly on transmission bandwidth in communication channels, computers or routers. Thus, application layer attacks can easily overflow the computers with limited computing resources by computation flooding. Accordingly, the computing and communication devices in the grid can be potential victims of this attack as they are equipped with limited abilities.

21.4.2 Attacks targeting integrity and confidentiality

These attacks occur at the application layer as they attempt to manipulate data information in the power system. Attacks targeting integrity attempt to modify data to corrupt information exchange for either consumers, for example, pricing information and account balance, or status values of power systems such as voltage readings and device running status. Therefore, integrity-check methods should be applied by utilities to protect data integrity against any false data injection from attackers.

Attacks targeting confidentiality have least impact on the grid reliability, but because of deployment of AMI and decentralized control within the grid, an increased attention is paid to protect the grid against such attacks.

21.5 Grid privacy

As reported by cybersecurity working group [9,10], privacy is not the same as confidentiality and personal information is not the same as confidential information. This is because access to confidential information is limited to only those with a business need to know and consequently may result in compromise to a system, data, application or other business function if inappropriately shared. Privacy can simply be the right to be left alone. Privacy in the power system can be described by four dimensions:

1. *Privacy of personal information:* It is any information relating to an individual, who can be identified, directly or indirectly, by that information. In addition, it involves the right to control when, where, how, to whom and to what extent an individual can share their own personal information.
2. *Privacy of the person:* It is the right to control the integrity of one's own body. It covers things such as physical requirements, health problems and required medical devices.
3. *Privacy of personal behaviour:* It is the right of the individuals to keep any knowledge of their activities from being shared with others.
4. *Privacy of personal communications:* It is the right to communicate without undue monitoring or censorship.

Smart meter (SM) reading at consumers' premises is conveyed through the communication network. Information can be exchanged between utilities and consumers. More interoperability between consumers and service providers is required for intelligent control and energy consumption management. Unprotected energy-related data will allow invasions of privacy, for example, radio waves in AMI may reveal information about where people were and when, and what they were doing. Hence, such invasions are not accepted by regulators and consumers. Privacy issues may include the following [11–13]:

- Personal information such as name, address, account number, meter reading, home electrical appliances, life style, meter IP and service provider. Personal information should be secured and monitored in a proper way.
- Privacy concerns with secure mechanisms to prevent privacy attacks on local data. Attacks may present the following:
  - fraud of energy consumption, which can be surmounted by not allowing any personnel abuse or modifying the collected data;
  - data in the SM could help deduce certain activities of home appliances, which should not be recognized and tracked by malicious intruder; and
  - near-real-time data of energy consumption can reveal some privacy of the individual power usage, which is not desired by the public. Thus, data transfer from home meters to the data centre should be secured.

21.6 Security of grid components

Power system is a large and complex arrangement, and it encompasses many different interlinked components. Security risks of some major components are highlighted in this section to briefly illustrate the impact of cyberattacks on such components [14,15]. The information flows between the different nodes in the grid, and the system operations and control centre through the communication network is shown in Figure 21.4. Each node in the network represents the location of IEDs such as SMs, sensors or remote terminal units to monitor their own system states.

21.6.1 Process control systems

Process control systems (PCSs) or operations and control centres are the components responsible for monitoring and control of physical properties of the electrical power system. In power networks, these PCSs monitor and control large geographical areas of the power grid as they are traditionally designed to work for a specific system. Consequently, many entry points exist to get into the grid. SCADA system is one of the many types of PCSs being commonly used in the electrical power system.

Security of PCSs is a paramount important factor to keep PCSs function properly, especially as they control the physical aspects of the power grid. For instance, compromising a computer loaded with data may cause some hardware damage. Similarly, when a PCS is compromised, a number of equipment can be physically damaged in addition to data being lost. The most important security objective is the availability. So, the PCS must always be available to satisfy power system availability. Lack of ability to deliver correct decisions based on false input

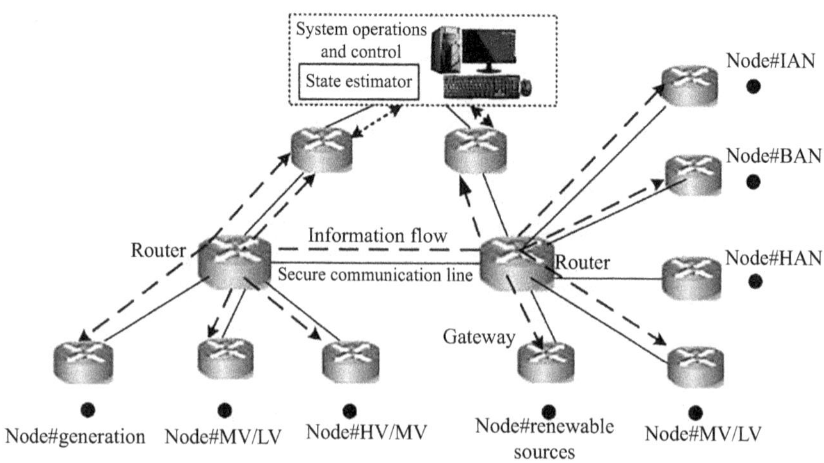

Figure 21.4 Typical information flow through communication network in power systems

data can in turn affect the integrity of decisions, the second security objective. Regarding confidentiality, the third security objective, it may be too time-consuming to meet latency as the PCS needs to run in real time. Accordingly, PCSs must address security concerns of large networks [16].

21.6.2 Power system state estimation

Power system state estimation (SE) model is a tool that the grid PCSs use to model sensor and agent data. It is important because it is used by the operators to maintain the electric power system. As the security objectives important to PCSs are also important for SE, it means that the security of SE model is also important. Availability is the most important followed by the objectives of integrity and confidentiality that adds overhead to a real-time system.

Maintaining the security of power system SE model is a challenge because of the possibility of receiving false input data into the model. System instability and financial gain are a couple of motivations for attackers. PCSs may have the false data injection security issue, and distinguishing between actual and false data is a difficult problem. There are mechanisms to distinguish bad data from normal data, but this is not enough as they must also be effective against false data attacks.

21.6.3 Smart meters

SMs are devices installed at a customer's site to measure the amount of power used and send the information back to the power utility at regular intervals. Developed electronic version of the current power meters are recommended not just for recording the amount of energy consumption but are also used to give the utility a feedback mechanism that can be used to model power usage requirements at a much more detailed level for energy management.

The security of SMs is important because altered readings from the device, especially if the device installed at a customer's site is vulnerable to attack, can lead to incorrect billing and false power usage approximations, as well as providing attackers with monetary gains. It is important that, to satisfy the security objective of integrity, the SM readings are correct and not modified. Confidentiality of the SM reading is also important. Tools that can profile a user's electrical usage readings to determine which household appliances are being used are already available. This information can be used by different companies and individuals, and it is a privacy concern. The availability of SMs is more flexible than other grid components. The maximum latency of SMs is in terms of hours, far greater than the real-time latency requirements that are in terms of milliseconds.

SM security is a challenge because it is easy to gain physical access to the SM device and there are instant monetary gains from altering these systems. The integrity of SMs and their data need to be verified before use. Confidentiality of the SM readings is also a challenge. SMs need to be networked to the power suppliers, so that they can carry out their function. This means that it may be possible for anyone else connected to that same network to observe others' power usage readings. This information can then be used to profile a user's behaviour.

21.6.4 Communication protocol

The network comprises various intercommunication components to provide proper functions. Each component has its own communication requirements and a set of security needs. Thus, the network will need several communication protocols to meet the varying connection requirements. The security of communication protocols is important as the network communication is the backbone of the modern grid. Many of the major grid functions cannot take place without communication. The security objectives that are important depend on which components are communicating, and what data they are exchanging.

The network communication protocol security is a challenge because of the presence of many different components that may communicate with different communication protocols. Another issue is that the new grid technology needs to integrate with legacy power systems, and many of these devices have constraints that must be considered. Legacy devices can typically introduce security vulnerabilities into the system because of a lack of security support.

21.6.5 False data injection attacks

During power system operation, SE is important for power system planning, analysis and studies such as optimal power flow calculations, contingency analysis and automatic generation control. SE plays a vital role for smooth operation of different energy management system applications. Basically, these studies take the SE output data as an input to make an intelligent decision. For instance, SE receives data from the SCADA network for processing purposes. Due to a false data injection attack, SCADA may send the SE wrong information and make the smart grid vulnerable.

21.7 Cloud computing

Integration of distributed generation (DG) sources, especially renewable sources such as solar and wind, plays a big role in meeting energy demand. To intensify the visibility of DG within power markets, the idea of virtual power plant (VPP) has emerged. VPP combines various small DG units to form a single virtual generating unit that can act as a conventional one and is capable of being manageable on individual bases [17]. VPP is outlined in Chapter 18, Section 18.3.1.

Based on the features that should be verified in the grid, huge amount of information must be collected, stored and processed, and information exchange between millions of devices is needed. The computing resources can be hardware such as networks, servers and storage, as well as software such as applications and services. Therefore, a model based on providing the required computation as a service, whereby storage, software and information are provided to the computers and other devices as a commodity over the Internet, called cloud computing, has been developed. A single cloud computing data centre might have storage and computing capabilities tens or hundreds of times greater than the multiple combined supercomputing facilities, which will be more expensive.

Cloud providers own a large number of networked servers with low expenses. This infrastructure includes massive pooled systems linked together and works with virtualization techniques to provide high performance of data storage units and runs alongside a local network connection that can run from a few to trillions of computations per second depending on the demand.

As reported by NIST in [18], the term 'cloud computing' is defined as 'a model for enabling ubiquitous, convenient, on-demand network access to a shared pool of configurable computing resources (e.g. networks, servers, storage, applications and services) that can be rapidly provisioned and released with minimal management effort or service provider interaction'.

This cloud model is composed of (i) five essential characteristics – on-demand self-service, broad network access, resource pooling, rapid elasticity and measured service; (ii) three service models – software as a service (SaaS), platform as a service (PaaS) and infrastructure as a service (IaaS); and (iii) four deployment models – private cloud, community cloud, public cloud and hybrid cloud as shown in Figure 21.5 [19].

Cloud-based systems are sharing of an enormous amount of information technology infrastructures, such as computational and database resources in the form of service, which focus on maximization the efficiency of operation, scalability, maintainability and reliability by decreasing cost.

NIST identifies seven domains within the power grid: generation, transmission, distribution, operations, customers, markets and service providers as shown in Figure 21.6 [5]. Accordingly, power grids should support both electrical power

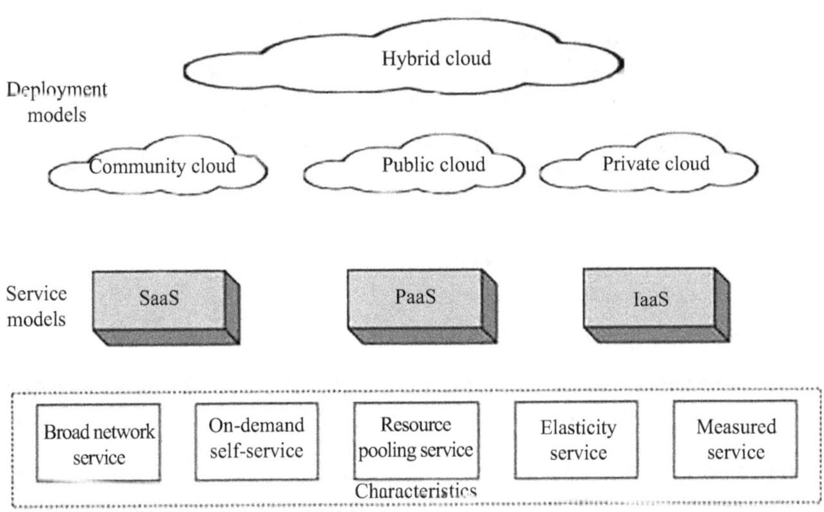

Figure 21.5 Cloud computing model

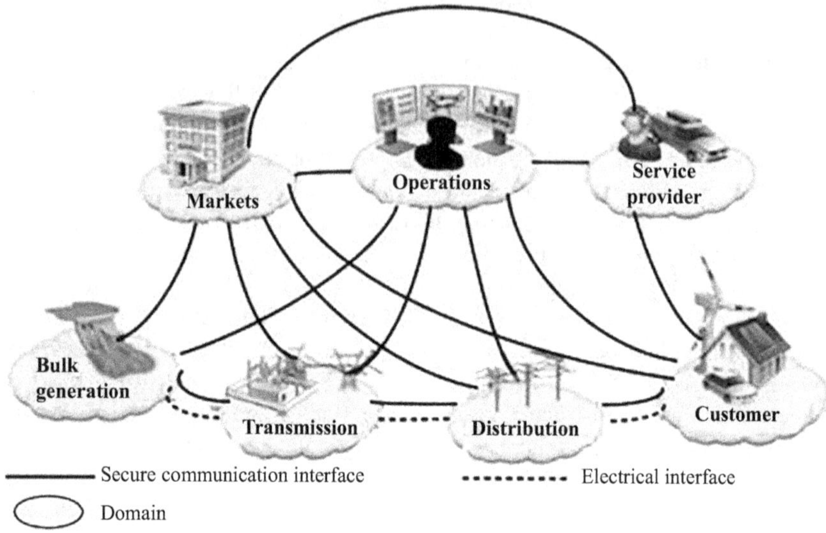

Figure 21.6 NIST – power grid model [5]

flows (generation, transmission and distribution of energy) and information processing flows (collecting, processing and distributing data).

Information flow in a power network is required to monitor and control the flow of energy (like SCADA systems) as well as monitoring and controlling future and new energy-based services.

Electric utilities are focused on (i) improving the efficiency and reliability of the delivery systems, (ii) getting solutions to empower the end consumer and (iii) to better manage the demand and supply balance along the distribution chain, besides the application of monitoring and diagnostic technologies to optimize assets use and reduce the risk of catastrophic failure of electrical infrastructure.

21.7.1 Advantages, disadvantages and threats of cloud computing

Cloud computing has strengths and weaknesses that must be understood to recognize its applicability and how to develop this approach.

Advantages of cloud computing:

- Customers' data are stored externally at the location of the cloud computing provider.
- Clouds manage themselves in case of failures or the performance degradation; that is, they introduce high reliability services.
- Customers pay for the service rather than buying software and hardware.

- Cloud services can be used instantly and can be scaled up and down easily.
- Sharing of computing resources between different customers.

Disadvantages of cloud computing:

- Dependency on Internet connectivity as constant connection is required.
- Pay per use is varied as the cost of computing differs with the time.
- Security concerns.

Threats to cloud computing:
The top seven prime threats to cloud computing are listed as follows [20]:

- Abuse and immoral use of cloud computing.
- Insecure application programming.
- Interfaces.
- Malevolent insiders.
- Shared technology liabilities.
- Data losses or leakages.
- Account, service and traffic hijacking.

References

[1] Arghandeh R., Meier A., Mehrmanesh L. and Mili L. 'On the definition of cyber-physical resilience in power systems'. *Journal of Renewable and Sustainable Energy Reviews, Elsevier*. 2016;58:1060–69.

[2] Pearson I. L. G. 'Smart grid cyber security for Europe'. *Journal of Energy Policy, Elsevier*. 2011;39:5211–18.

[3] Amin S. M. and Giacomoni A. M. 'Smart grid – safe, secure, self-healing'. *IEEE power & energy magazine*. 2012. IEEE; pp. 33–40.

[4] Kuzlu M., Pipattanasomporn M. and Rahman S. 'Communication network requirements for major smart grid applications in HAN, NAN and WAN'. *Journal of Computer Networks*. 2014;67:74–88.

[5] NIST. *NIST framework and roadmap for smart grid interoperability standards, Release 2.0*. NIST Special Publication 1108R2 report, 2012.

[6] Wang W. and Lu Z. 'Cyber security in the smart grid: Survey and challenges'. *Journal of Computer Networks, Elsevier*. 2013;57:1344–71.

[7] US Committee on National Security Systems. *National Information Assurance (IA) Glossary*. CNSS, Instruction 4009 report, 2006.

[8] Bhatia S., Behal S. and Ahmed I. 'Distributed denial of service attacks and defense mechanisms: Current landscape and future directions' in Conti M., Somani G. and Poovendran R. (eds.). *Versatile Cybersecurity. Advances in Information Security*. Springer, Cham Press; 2018, pp. 55–97.

[9] The Smart Grid Interoperability Panel, Cyber Security Working Group. *Guidelines for smart grid cyber security. Vol. 2, Privacy and the smart grid*. NIST and US Department of Commerce report, 2010.

[10] The Cyber Security Coordination Task Group. *Smart grid cyber security strategy and requirements*. NIST report, 2009.
[11] Liu J., Xiao Y., Li S., Liang W. and Philip Chen C. L. 'Cybersecurity and privacy issues in smart grids'. *IEEE Communications Surveys & Tutorials*. 2012;14(4):981–97.
[12] Siddiqui F., Zeadally S., Alcaraz C. and Galvao S (eds.). 'Smart grid privacy: Issues and solutions'. *21st International Conference on Computer Communications and Networks (ICCCN)*; Munich, Germany, Jul/Aug 2012. IEEE; pp. 1–5.
[13] Lu R., Liang X., Li X., Lin X. and Chen X. 'EPPA: An efficient and privacy-preserving aggregation scheme for secure smart grid communications'. *IEEE Transactions on Parallel and Distributed Systems*. 2012;23(9): 1621–32.
[14] Anwar A. and Mahmood A. 'Cyber security of smart grid infrastructure' in Anwar A. and Mahmood A. (eds.). *The State-of-the-art Intrusion Prevention and Detection*. USA: CRC Press, Taylor & Francis Group; 2014, pp. 449–72.
[15] Kundur D., Feng X., Kiu S., Zourntos T. and Butter-Purry K. L. (eds.). 'Towards a framework for cyber-attack impact analysis of the electric smart grid'. *First IEEE International Conference on Smart Grid Communications*; Gaithersburg, MD, USA, Oct 2010. IEEE; pp. 244–49.
[16] Mahapatra K. and Chaudhuri N. R. 'Online robust PCA for malicious attack-resilience in wide-area mode metering application'. *IEEE Transactions on Power Systems*. 2019;34(4):2598–610.
[17] Markovic D. S., Zivkovic D., Branovic I., Popovic R. and Cvetkovic D. 'Smart power grid and cloud computing'. *Renewable and Sustainable Energy Reviews, Elsevier*. 2013;24:566–77.
[18] Mell P. and Grance T. *The NIST definition of cloud computing*. NIST Special Publication 800-145 report, 2011.
[19] Bera S., Misra S. and Rodrigues J. P. C. 'Cloud computation applications for smart grid: A survey'. *IEEE Transactions on Parallel and Distributed Systems*. 2016;26(5):1477–94.
[20] Sivapragash C., Thilaga S. R. and Kumar S. S. (eds.). 'Advanced cloud computing in smart power grid'. *Third International Conference on Sustainable Energy and Intelligent Systems (SEISCON 2012), VCTW*; Tiruchengode, Tamil Nadu, India, Dec 2012, pp. 349–54.

Appendix A

Probability density and Weibull distribution functions

It is essential to obtain the probability density function (*pdf*) of the actual power output of wind or solar photovoltaic energy systems, especially for planning and analysing a power system incorporating such renewable sources of energy.

Probability density function is a statistical expression that defines a probability distribution for a continuous random variable as opposed to a discrete random variable.

A.1 Random variables

A random variable is a real-valued function that assigns a numerical value to each possible outcome of the random experiment. For instance, the random variable X assigns the value 0 to the outcome *GenOutage*, the value 1 to the outcome *GenInService* and so on. Hence, the random variable X is a mapping function from the sample space $S = \{GenOutage, GenInService, \cdots, GenRestore\}$ to the real numbers. That is,

$$X : S \rightarrow \mathrm{R}$$

Random variables usually are written in capital letters such as X, Y and Z. As a random variable is a function, it has a range. The range of a random variable X is denoted by Range(X) or RX and is the set of possible values for X.

There are two important classes of random variables: discrete random variables and continuous random variables.

A.1.1 Discrete random variables

A random variable is discrete if its range is a countable set. The set A is countable if either

- A is a finite set such as $\{1,2,3,4\}$, or
- it can be put in one-to-one correspondence with natural numbers (in this case, the set is said to be countably infinite).

A.1.2 Continuous random variables and their distributions

One big difference that is noticed as opposed to discrete random variables is that the distribution function of a random variable X is the function $F_X(x)$ of x, which is called 'cumulative distribution function (*cdf*)' and given by $F_X(x) = p(X \leq x)$ for all values x in the range of X. It is a continuous function; that is, it does not have any jumps. That jumps in the *cdf* correspond to points x for which $P(X = x) > 0$. Thus, the fact that the *cdf* does not have jumps is consistent with the fact that $P(X = x) = 0$ for all x, where probability cannot be assigned to individual values of x; instead, probability is assigned to intervals. The continuous random variable is defined as follows: 'A random variable X with *cdf*, $F_X(x)$, is said to be continuous if $F_X(x)$ is a continuous and differentiable function for all $x \in R$'.

A.2 Probability density function

To determine the distribution of a discrete random variable, probability mass function (*pmf*) or *cdf* should be provided. For continuous random variables, the *cdf* is well defined, so it can be provided. However, the *pmf* does not work for continuous random variables. Instead, the *pdf* can usually be defined where it is the density of probability rather than the probability mass. The *pdf* can be thought of as the infinite limit of a discrete distribution, i.e. a discrete distribution with an infinite number of possible outcomes. Specifically, suppose a discrete distribution with N possible outcomes is created, each corresponding to a range on the real number line. Then, suppose N is increased towards infinity, so that each outcome shrinks to a single real number; a *pdf* can be considered as the limiting case of this distribution.

The concept is very similar to mass density in physics: its unit is probability per unit length. To get a feeling for *pdf*, consider a continuous random variable X and define the function $f_X(x)$ as follows (wherever the limit exists):

$$f_X(x) = \lim_{h \to 0^+} \frac{P(x < X < x + h)}{h} \tag{A.1}$$

where

$$P(x < X \leq x + h) = F_X(x + h) - F_X(x) \tag{A.2}$$

The function $f_X(x)$ gives the probability density at point x. It is the limit of the probability of the interval $(x, x + h]$ divided by the length of the interval as the length of the interval goes to 0. Substituting (A.2) into (A.1) obtains

$$\begin{aligned} f_X(x) &= \lim_{\Delta \to 0} \frac{F_X(x + h) - F_X(x)}{h} = \frac{dF(x)}{dx} \\ &= F'_X(x) \; if F_X(x) is\ differentiable\ at\ x \end{aligned} \tag{A.3}$$

Thus, the definition of the *pdf* of continuous random variables follows as given next.

Considering a continuous random variable X with a continuous *cdf*, $F_X(x)$, the *pdf*, $f_X(x)$, is given by

$$f_X(x) = \frac{dF(x)}{dx} = F_X'(x) \text{ } if F_X(x) is \text{ } differentiable \text{ } at \text{ } x \tag{A.4}$$

As the *pdf* is the derivative of the *cdf*, the *cdf* can be obtained by integrating (A.4) as follows:

$$F_X(x) = \int_{-\infty}^{x} f_X(u)du \tag{A.5}$$

From (A.2), the following form can be written:

$$P(a < X \leq b) = F_X(b) - F_X(a) = \int_a^b f_X(u)du \tag{A.6}$$

and integration over the entire real line must get 1, that is,

$$\int_{-\infty}^{\infty} f_X(u)du = 1 \tag{A.7}$$

That is, the area under the *pdf* curve must be equal to one. However, considering a continuous random variable X with *pdf*, $f_X(x)$, the properties of the *pdf* can be summarized as

$$\left. \begin{aligned} f_X(x) \geq 0 \; for \; all \; x \epsilon R \\ \int_{-\infty}^{\infty} f_X(u)du = 1 \\ P(a \leq X \leq b) = F_X(b) - F_X(a) = \int_a^b f_X(u)du \\ F_X(x) = \int_{-\infty}^{x} f_X(u)du \end{aligned} \right\} \tag{A.8}$$

According to (A.6), $f_X(u)$ can be plotted for *RX*, and then, the probability that x falls within the range can be obtained by computing the area under the curve, $f_X(u)$ versus u. It is important to note that a probability density is not a probability per se. For one thing, there is no requirement that $F_X(x) \leq 1$. Moreover, the probability that x attains any one specific out of the infinite set of possible values is always zero. For example,

$$F_X(x=5) = \int_5^5 f_X(x)dx = 0 \text{ for any } pdf, \; f_X(x)$$

Joint distributions are defined in a natural way. For two variables X and Y, the joint *pdf*, $f_{X,Y}(x,y)$, defines the probability that (x, y) lies in a given domain D:

$$f_{X,Y}(x, y\epsilon D) = \int_{x,y\epsilon D} f_{X,Y}(x,\ y)dxdy \tag{A.9}$$

For instance, the probability that a 2D-coordinate (x, y) lies in the domain $(1 \leq x \leq 2, 1 \leq y \leq 2)$ is $\int_{x=1}^{2}\int_{y=1}^{2} f_{X,Y}(x,y)dxdy$. The *pdf* over a vector may also be written as a joint *pdf* of its variables.

Conditional distributions are defined as well; $f_X(x|\mathrm{A})$ is the *pdf* over x, if the statement A is true. This statement may be an expression on a continuous value, say $y = 5$. As a shorthand, $f_X(x|y)$ can be written to provide a *pdf* for x for every value of y. It must be the case that $\int f_X(x|y)dx = 1$, as $f_X(x|y)$ is a *pdf* over values of x.

A.3 The Weibull distribution function

The Weibull distribution is widely used for several applications such as mathematical modelling for system analysis, failure analysis, warranty analysis, reliability growth, life-cycle cost and production process control. Weibull models are used to represent *pdf* and *cdf* [1, 2]. The non-negative random variable X with distribution function

$$F_X(x) = \begin{cases} 1 - e^{-\left(\frac{x}{c}\right)^k}, & x > 0 \\ 0, & x \leq 0 \end{cases} \tag{A.10}$$

is said to have a Weibull distribution with shape parameter $k > 0$ and scale parameter $c > 0$. The *pdf* of the Weibull distribution is

$$f_X(x) = F_X'(x) = \begin{cases} \frac{k}{c^k} x^{k-1} e^{-\left(\frac{x}{c}\right)^k}, & x > 0 \\ 0, & x \leq 0 \end{cases} \tag{A.11}$$

It is to be noted that the exponential distribution is a special case of Weibull distribution with $k = 1$.

The expectation and variance of a Weibull distribution could be computed by

$$E(X) = \int_0^\infty \frac{k}{c^k} t^k e^{-(x/c)^k} dx = c\Gamma\left(1 + \frac{1}{k}\right) \tag{A.12}$$

$$E(X^2) = \int_0^\infty \frac{k}{c^k} t^{k+1} e^{-(x/c)^k} dx = c^2\Gamma\left(1 + \frac{2}{k}\right) \tag{A.13}$$

$$Var(X) = E(X^2) - (E(X))^2 = c^2\left(\Gamma\left(1 + \frac{2}{k}\right) - \left[\Gamma\left(1 + \frac{1}{k}\right)\right]^2\right) \tag{A.14}$$

A.3.1 Calculation of mean and variance

- For discrete random variables, the formula of the mean, expected, value is based on the probability function. It is calculated by

$$E(X) = \sum_X x p_X(x) \tag{A.15}$$

and the variance is defined by

$$Var(X) = E\left\{[X - E(X)]^2\right\} \tag{A.16}$$

- For continuous random variables with *pdf*, $f_X(x)$, the expectation of a function $g(x)$ is defined as

$$E\{g(X)\} = \int_{-\infty}^{\infty} g(x) f_X(x) dx \tag{A.17}$$

Hence, the mean is given by

$$E(X) = \int_{-\infty}^{\infty} x f_X(x) dx \tag{A.18}$$

The best way, in practice, to calculate the variance is by using the relations given in (A.19) and (A.20):

$$Var(X) = E\left(X^2\right) - \{E(X)\}^2 \tag{A.19}$$

and

$$E\left(X^2\right) = \int_{-\infty}^{\infty} x^2 f_X(x) dx \tag{A.20}$$

Example A.1 The lifetime, measured in years, of a brand of mobile device has a Weibull distribution with parameters $k = 2$ and $c = 13$. Compute the probability of a mobile device failing before the expiration of a 2-year warranty.

Solution

$F_X(x \leq 2) = 1 - \exp\left[-2/13^2\right] \approx 0.0234$.

In general, the computation program to provide the Weibull probability distribution function versus X at $k = 2$ and $c = 4$, 7, 10 and 13 is as in the following:

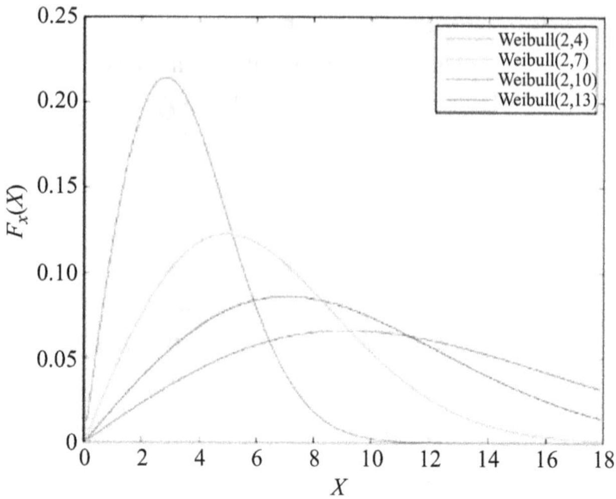

Figure A.1 Weibull probability distribution function with c = 2 and k = 4, 7, 10 and 13

$X = 0.1 : 0.2 : 18$

$k = 2; c = 4; \mathrm{Y1} = (k/c \wedge k) * \mathrm{X}. \wedge (k-1). * \exp(-(\mathrm{X}/c). \wedge k)$

$k = 2; c = 7; \mathrm{Y2} = (k/c \wedge k) * \mathrm{X}. \wedge (k-1). * \exp(-(\mathrm{X}/c). \wedge k)$

$k = 2; c = 10; \mathrm{Y3} = (k/c \wedge k) * \mathrm{X}. \wedge (k-1). * \exp(-(\mathrm{X}/c). \wedge k)$

$k = 2; c = 13; \mathrm{Y4} = (k/c \wedge k) * \mathrm{X}. \wedge (k-1). * \exp(-(\mathrm{X}/c). \wedge k)$

plot(X,Y1,'m-',X,Y2,'g-',X,Y3,'b-',X,Y4,'r-') legend('Weibull(2,4)','Weibull (2,7)','Weibull(2,10)', 'Weibull(2,13)')

$\mathrm{title}(\text{'Weibull}(k,k) : f(x) = (k/c \wedge k)\mathrm{x} \wedge \{k-1\}\exp[-(\mathrm{x}/c) \wedge k]')$

The results are plotted as shown in Figure A.1.

References

[1] Lai C. D., Murthy D. N. and Xei M. 'Weibull distributions and their applications'. *Springer Handbook of Engineering Statistics, Part A*. 2006, pp. 63–78. Available at: http://www.researchgate.net/publication/37628953.

[2] Khan M. S., Pasha G. R. and Pasha A. H. 'Theoretical analysis of inverse Weibull distribution'. *WSEAS Transactions on Mathematics*. 2008;7(2): 30–38.

Appendix B
Test system

The modified IEEE 24-bus system without hydro generation is taken as a test system. Its transmission network consists of 24-bus bars connected by 38 lines and transformers and the generation system includes 26 thermal units at locations as shown in the single-line-diagram (Figure B.1).

The characteristics of generating units are given in Tables B.1 and B.2. The hourly peak load and bus loads as a percentage of the system load are listed in Table B.3.

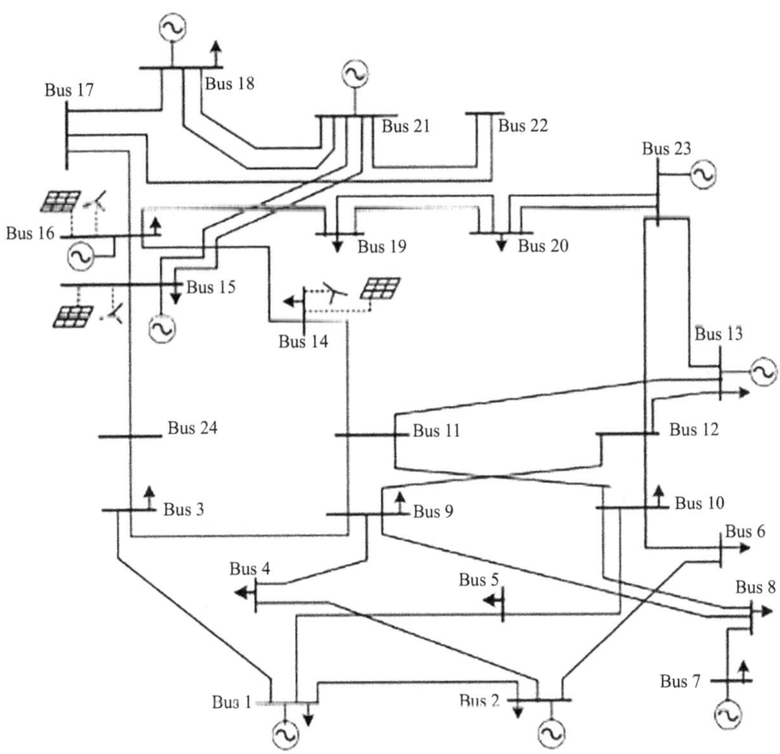

Figure B.1 Single-line diagram of the test system

Table B.1 Generating unit data

Unit	P_i (MW)	$\overline{P_i}$ (MW)	a_i (k$/MW^2)	b_i (k$/MW)	c_i (k$)	Bus #
1	2.4	12.0	0.02533	25.5472	24.3891	15
2	2.4	12.0	0.02649	25.6753	24.4110	15
3	2.4	12.0	0.02801	25.8027	24.6382	15
4	2.4	12.0	0.02842	25.9318	24.7605	15
5	2.4	12.0	0.02855	26.0611	24.8882	15
6	4.0	20.0	0.01199	37.5510	117.755	1
7	4.0	20.0	0.01261	37.6637	118.108	1
8	4.0	20.0	0.01359	37.7770	118.457	2
9	4.0	20.0	0.01433	37.8896	118.820	2
10	15.2	76.0	0.00876	13.3272	81.1364	1
11	15.2	76.0	0.00895	13.3538	81.2980	1
12	15.2	76.0	0.00910	13.3805	81.4641	2
13	15.2	76.0	0.00932	13.4073	81.6259	2
14	25	100.0	0.00623	18.0000	217.895	7
15	25	100.0	0.00612	18.1000	218.335	7
16	25	100.0	0.00598	18.2000	218.775	7
17	54.25	155.0	0.00463	10.6940	142.734	15
18	54.25	155.0	0.00473	10.7154	143.028	16
19	54.25	155.0	0.00481	10.7367	143.317	23
20	54.25	155.0	0.00487	10.7583	143.597	23
21	68.95	197.0	0.00259	23.0000	259.131	13
22	68.95	197.0	0.00260	23.1000	259.649	13
23	68.95	197.0	0.00263	23.2000	260.176	13
24	140.0	350.0	0.00153	10.8616	177.057	23
25	100.0	400.0	0.00194	7.4921	310.002	18
26	100.0	400.0	0.00195	7.5031	311.910	21

Table B.2 Generating units operating parameters

Unit	Min. Up (h)	Min. Down (h)	UH (h)	DH (h)	UR (MW/h)	DR (MW/h)	α_j (h)	β_i (h)	τ_i (h)
1–5	0	0	0	0	48.0	60.0	0	0	1
6–9	0	0	1	0	30.5	70.0	20	20	2
10–13	3	−2	2	1	38.5	81.0	50	50	3
14–16	4	−2	2	2	51.0	74.0	70	70	4
17–20	5	−3	3	2	55.0	78.0	150	150	6
21–23	5	−4	4	2	55.0	99.0	200	200	8
24	8	−5	5	3	70.0	120.0	300	200	8
25–26	8	−5	8	4	50.5	100.0	500	500	10

Table B.3 Hourly load demand and bus load as a percentage of total load

Hour	Load (MW)	Hour	Load (MW)	Hour	Load (MW)	Bus #	Bus load (%)	Bus #	Bus load (%)
1	2,000	10	2,257	19	2,539	1	3.8	10	6.8
2	1,847	11	2,308.5	20	2,488	2	3.4	13	9.3
3	1,744	12	2,334	21	2,411	3	6.3	14	6.8
4	1,693	13	2,308.5	22	2,360	4	2.6	15	11.1
5	1,642	14	2,257	23	2,231.5	5	2.5	16	3.5
6	1,643	15	2,231.5	24	2,077.5	6	4.8	18	11.7
7	1,693	16	2,231.5	–	–	7	4.4	19	6.4
8	1,795.5	17	2,334	–	–	8	6.0	20	4.5
9	2,052	18	2,565	–	–	9	6.1	–	–

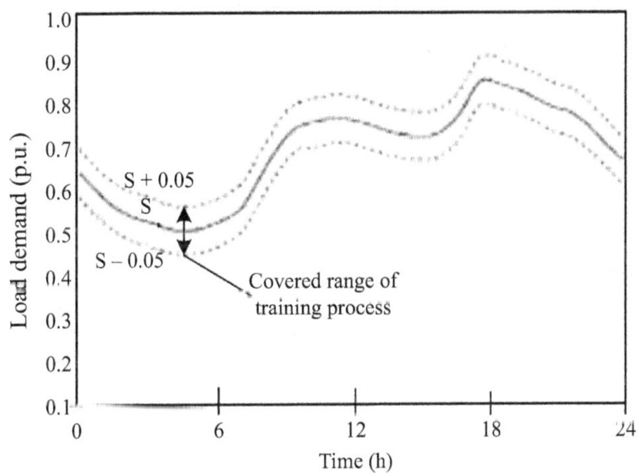

Figure B.2 Load demand over 24-hourly time periods

The patterns of load demand over 24-hourly time periods with a range of ±5% load forecast error are taken as samples for training the ANN (Figure B.2).

The data of the transmission network are reported by the reliability test system task force of the application of probability methods subcommittee and published in IEEE Transactions Power Apparatus and Systems, vol. PAS-98, no. 6, Nov./Dec. 1979.

$$FC_i(P_{ij}) = a_{ij}P_{ij}^2 + b_iP_{ij} + c_i.$$

$\underline{P_i}$, $\overline{P_i}$ = rated lower and upper generation limit, respectively.

$$SC_{i,j} = u_{ij}\left[\alpha_i + \beta_i\left[1 - exp\left(-T_{i,j}^{off}/\tau_i\right)\right]\right].$$

UH, DH = ramp-up and ramp-down time, respectively.
UR, DR = ramp-up rate limit and ramp-down rate limit, respectively.

B.1 Data of renewable energy sources

1. Each wind turbine of the wind farm of 30 MW rating has the following characteristics:
   - Cut-in speed = 5 m/s
   - Nominal speed = 15 m/s
   - Cut-off speed = 45 m/s.
2. The solar PV plant of rating 30 MW consists of 400,000 modules. Each module has the following characteristics:
   - Watt peak – 50 W
   - Open-circuit voltage = 38 V
   - Short-circuit current = 1.8 A
   - Voltage at maximum power = 38 V
   - Current at maximum power = 1.32 A
   - Voltage temperature coefficient = 194 mV/°C
   - Current temperature coefficient = 1.4 mA/°C
   - Nominal cell operating temperature = 43 °C.

Index